An Integrated Approach to Agricultural Science and Technology

Edited by **Farrell Waltz**

SYRAWOOD
PUBLISHING HOUSE
New York

Published by Syrawood Publishing House,
750 Third Avenue, 9th Floor,
New York, NY 10017, USA
www.syrawoodpublishinghouse.com

An Integrated Approach to Agricultural Science and Technology
Edited by Farrell Waltz

International Standard Book Number: 978-1-68286-268-1 (Hardback)

Printed in China.

Contents

Preface IX

Chapter 1 **Impact of Irrigation on Agriculture Productivity in Solapur District of Maharashtra State** 1
Todkari G. U.

Chapter 2 **Genetic Variability in Land Races Of Forage Sorghum {*Sorghum bicolor* (L) Moench} Collected from Different Geographical Origin of India** 4
Jain S. K. and Patel P. R.

Chapter 3 **Effect of Esterification on Moisture Absorption of Single Areca Fiber** 8
Dhanalakshmi Sampathkumar, Ramadevi Punyamurthy, Basavaraju Bennehalli and Srinivasa Chikkol Venkateshappa

Chapter 4 **Additional Gravitational Force is Acting on Some Trees Due to Presence of Underground Water Stream** 11
More B. M.

Chapter 5 **Role of Agriculture Insurance Scheme in Marathwada Region of Maharashtra, India** 14
Patwardhan P. S. and Narwade S. S.

Chapter 6 **Impacts of Drying Air Temperature, Bed Depth and Air Flow Rate on Walnut Drying Rate in an Indirect Solar Dryer** 18
Ghatrehsamani S. H. and Zomorodian A.

Chapter 7 **A Framework for Agricultural Adaptation to Climate Change in Southern Nigeria** 22
Nicholas Ozor, Madukwe M. C., Enete A. A., Amaechina E. C., Onokala P., Eboh E. C., Ujah O., and Garforth C. J.

Chapter 8 **Characterization of the Coal Humic Acids from the Candiota Coalfield, Brazil** 32
Xavier D. M., Silva A. S., Santos R. P., Mesko M. F., Costa S. N., Freire V. N., Cavada B. S. and Martins J. L.

Chapter 9 **Bush Fallow and Cowpea Crop Use as Precedent and Organic Sources of Nutrients for Rice Cultivation on Acidic Plinthosol of Central Benin in West Africa** 37
Assigbe P., Kone B., Bognonpke J. P., Toure A., Huat J. and Yao-Kouame A.

Chapter 10 **Effect of Salinity on Seed Germination and Seedling Characters of Some Forage Sorghum Cultivars** 42
Kandil A. A., Sharief A. E., Abido W. A. E. and Ibrahim M. M.

Chapter 11 **Agriculture Productivity in Solapur District of Maharashtra: A Geographical Analysis** 48
Awate S. J. and Todkari G. U.

Chapter 12 **Seed Micromorphological of Some Species *Minuartia* L. (Caryophyllaceae) in Iraq** 52
Al-Saadi S. A. A. M. and Al-Taie S. S.

Chapter 13 **Micropropagation of *Capsicum chinense* Jacq. cv. Lota Bhot Via Indirect Organogenesis** 58
Bora G., Gogoi H. K. and Handique P. J.

Chapter 14 **Pulsing with Sucrose and Silver Nitrate Enhance Water Uptake and Result in Along Vase Life in Taif Rose Cut Flowers (*Rosa damascena.* cv. Trigintipetala)** 62
Elgimabi M. N. E. E.

Chapter 15 **A Non-Destructive Technique Based on Texture Analysis for Identification of Defective Alphonso Mangoes Using Subspace Analysis Techniques** 67
Musale S. S. and Patil P. M.

Chapter 16 **Role of Co-Operative Sugar Factories in Rural Development : A Case Study of Damaji Sugar Factory, Mangalweda** 72
Todkari G. U.

Chapter 17 **Feasibility Study of a Pivoted-Plate Sensor for Silage Corn Yield Monitoring** 76
Maharlouie M. M., Loghavi M. and Kamgar S

Chapter 18 **Insecticidal Efficacy of Mintweed, Yam Bean and Celery Seed Extracts on *Aedes aegypti* L.** 82
Yongkhamcha B. and Indrapichate K.

Chapter 19 **Kinetics of Apricot Thin Layer Drying in a Mixed and Indirect Mode Solar Dryer** 88
Ghatrehsamani S. H., Dadashzadeh M. and Zomorodian A.

Chapter 20 **Characterisation of Five Siderophore Producing Actinomycetes from Soil Samples and The Use of Antibiotic Resistance to Differentiate the Isolates** 94
Ismini Nakouti and Glyn Hobbs

Chapter 21 **Effect of Seed Storage Periods, Conditions and Materials on Seed Quality of Some Soybean Cultivars** 99
Kandil A. A., Sharief A. E. and Sheteiwy M. S.

Chapter 22 **A Study on Physicochemical Properties of Soils Under Different Tea Growing Regions of West Bengal (India)** 107
Ray S. K. and Mukhopadhyay D.

Chapter 23 **Response of Some Rice (*Oryza sativa* L.) Cultivars to Germination Under Salinity Stress** 112
Kandil A. A., Sharief A. E. and Nassar E. S. E.

Chapter 24 ***Poka*-A Traditional Rice Wine of The *Galo* Tribe of Arunachal Pradesh, India** 118
Bora S. S., Lahan J. P., Madhumita Barooah and Rupjyoti Sarmah

Chapter 25 **Mapping and Assessment of Soils in Nagarjunasagar Catchment, Lingasugur Taluk of Karnataka** 122
Ram R. L., Jha P., Sharma P. K., Ahmed N., Kumar R., Sharanappa, Vasudeo Rao D. A. L. and Singh V.

Chapter 26 **Seedling Parameters of Soybean Cultivars as Influenced With Seed Storage Periods, Conditions and Materials** 129
Kandil A. A., Sharief A. E. and Sheteiwy M. S.

Chapter 27 **Evaluation of *Pongamia pinnata* Seed Extracts as an Insecticide Against American Bollworm *Helicoverpa armigera* (Hubner)** 138
Reena, Ram Singh and Sinha B. K.

Chapter 28 **Export Opportunities and Competitiveness of Vegetable Crops in Gujarat** 143
Pal V., Jadav K. S., Devi G. and Zala Y. C.

Chapter 29 **Physical and Mechanical Properties of *Citrus aurantiifolia* Swingle var. Pramalini (Kagzi-Lime)** 149
Pawar S. G., Shinde G. U. and Khodke S. U.

Chapter 30 **Evaluation Some New Hybrids for Cultivation In Conventional System in Spring Climatic Conditions of Romania** 153
Draghici E. M. and Pele M.

Chapter 31 **Decision Support Systems in A Global Agricultural Perspective - A Comprehensive Review** 160
Mir S. A., Qasim M., Arfat Y., Mubarak T., Bhat Z. A., Bhat J. A., Bangroo S. A. and Sofi T. A.

Chapter 32 **The Defense Activator from Yeast for Rapid Induction of Resistance in Susceptible Pearl Millet Hybrid Against Downy Mildew Disease** 173
Hindumathy C. K.

Chapter 33 **Application of Wireless Sensor Networks for Agriculture Parameters** 179
Awati J. S., Patil V. S. and Awati S. B.

Chapter 34 **Global Economic Crisis: Causes, Impact on Indian Economy, Agriculture and Fisheries** 182
Apu Das, Kumar N. R., Biswajit Debnath and Mandal S. C.

Chapter 35 **Evaluation of Sprouting Broccoli (*Brassicae oleraceae* var. *italica*) Genotypes for Growth, Yield and Quality** 188
Thapa U. and Rai R.

Chapter 36 **An Interactive Spreadsheet for Drip Irrigation System Uniformity Parameters Evaluation** 191
El-Nemr M. K.

Chapter 37 **Genetic Control of Seed Size In Cowpea (*Vigna unguiculata* (L.) Walp)** 196
Egbadzor K. F., Dadoza M., Danquah E. Y., Yeboah M., Offei S. K. and Ofori K.

Chapter 38 **Evaluation of Agronomic Performance of Drought - Tolerant QTL Introgression Hybrids of Millet (*Pennisetum glaucum* L R. BR.) in The Guinea Savannah Zone of Ghana** 201
Demuyakor B., Galyuon I., Kyereh S. and Ahmed M.

Chapter 39 **Maximizing Sugar Beet Yields with Decreasing Mineral Fertilization Pollution** 206
Seadh S. E.

Chapter 40 **Sugar Beet Productivity as Affected by Foliar Spraying with Methanol and Boron** 212
Abido W. A. E.

Chapter 41 **Validation of Tomato Powdery Mildew Forecasting Model Using Meteorological Data in Egypt** 218
Bakeer A. R. T., Abdel-Latef M. A. E., Afifi M. A. and Barakat M. E.

Chapter 42 **Effect of Effluent Water on Population Dynamics of Cowpea Aphid, Aphis Craccivora Koch (Hemiptera: Aphididae), on Cowpea Plants** 225
Tiroesele B., Nketso T. H. and Tshwenyane S. O.

Chapter 43 **Genotypic Differences Between Two Mungbean Varieties in Response to Salt Stress at Seedling Stage** 229
Kandil A. A., Arafa A. A., Sharief A. E. and Ramadan A. N.

Chapter 44 **Banana and Plantain (*Musa* spp.) Cultivar Preference, Local Processing Techniques and Consumption Patterns in Eastern Democratic Republic of Congo** 235
Ekesa B. N., Kimiywe J., Davey M. W, Dhuique-Mayer C., Van Den Bergh I., Karamura D. and Blomme G.

Chapter 45 **Selective Tillage Practices for Improving *In situ* Soil Moisture and Crop Yield in Cotton Based Cropping System** 243
Payal S. D. and Shinde G. U.

Chapter 46 **Land Use/Land Cover Classification and Accuracy Assessment Using Satellite Data - A Case Study of Bhind District, Madhya Pradesh** 246
Upadhyay R., Singh A., Shrivastav P. and Thakur S.

Chapter 47 **Study of Shelled Corn Shrinkage in a Microwave-Assisted Fluidized Bed Dryer Using Artificial Neural Network** 251
Momenzadeh L. and Zomorodian A.

Permissions

List of Contributors

Preface

This book has been a concerted effort by a group of academicians, researchers and scientists, who have contributed their research works for the realization of the book. This book has materialized in the wake of emerging advancements and innovations in this field. Therefore, the need of the hour was to compile all the required researches and disseminate the knowledge to a broad spectrum of people comprising of students, researchers and specialists of the field.

Agricultural research is occurring across the globe. It is constantly revolutionizing the techniques of crop production and management. Contributions of renowned experts have been compiled in this book to bring forth a global perspective on the advances in this field. Some of the significant topics covered in this book are irrigation, genetic variability, seed storage, properties of soil, agronomic performance, tillage practices, etc. This extensive text will serve as a reference guide for students, scholars, researchers, botanists, agriculture scientists and professionals engaged in the field of agriculture.

At the end of the preface, I would like to thank the authors for their brilliant chapters and the publisher for guiding us all-through the making of the book till its final stage. Also, I would like to thank my family for providing the support and encouragement throughout my academic career and research projects.

Editor

IMPACT OF IRRIGATION ON AGRICULTURE PRODUCTIVITY IN SOLAPUR DISTRICT OF MAHARASHTRA STATE

TODKARI G.U.

Dept. of Geography, Shri Sant Damji Mahavidyala Mangalweda, Dist-Solapur, MS, India.
*Corresponding Author: Email- govindtodkari@gmail.com

Abstract- Irrigation is identified as a decisive factor in Indian agriculture due to high variability and inadequacy of rainfall. Irrigation is essential for successful agriculture particularly in the area, where rainfall is inadequate uncertain, and unpredictable. These areas are prone to drought and famine condition due to partial failure and delayed arrival or early withdrawal of monsoon. Importance of irrigation has substantially increased after the adoption of High yielding varieties in developing countries. Irrigation is basic determinants of Agriculture because its inadequacies are the most powerful constraints on increase of Agricultural production. In the study region the variation of an annual rainfall from year to year is fairly large. The rainfall is irregular and uncertain in the Study region, here agriculture is gamble with monsoon. If rainfall is scare it results into crop failure. For the assure agriculture production irrigation is most important factor. There fore attempt is made here to examine the impact of irrigation on agriculture productivity in Solapur district.
Key words- Irrigation intensity, Agriculture productivity, Spatial pattern, Co-efficient index.

Introduction

Irrigation is basically an agriculture operation, supplying the need of water for plants. To agriculturalists it is a component of successful crop husbandry in dry climate, ranking high with the application of fertilizers, the control of weeds and destructive pests, cultivation and the position of drainage.

Irrigation is the most important factor in farming according to the Agriculture Department and Irrigation Officers. Irrigation has played an important role in transforming the crop cultivation and better yield .There are various any other type of irrigation such as in their well irrigation ,rivers ,tanks and canal etc. But there are additional factors such as their location , their topography , geological aspect and height , hilled area depending on various elements. In The region under study mainly two types of irrigation are practised namely well and canal irrigation .

The Solapur District is located in drought prone area , and so rainfall is low . Hence irrigation plays a vital role in agriculture. In 2004/05 the total area from all sources of irrigation is 288736 hectares The spatial pattern of irrigation is however, uneven and has been characterized by the changes over time.

Study Area

The present study deals with the geographical perspectives of the agriculture in Solapur district. The Solapur district is bounded by 17°05' North latitudes to 18° 32' North latitudes and 74° 42' East of 76° 15' East longitudes. The total geographical area of Solapur district is 14895^2 K.m. divided into eleven tahsils. The Population is 32.4 lakhs in eleven tahsils of District.(Censes 2001). It is bounded from the North by Osmanabad district and Ahmednagar district, on the North-East by Satara district and at the South & East it has common boundary of Karanataka state. Temperature is high in summer season. Rainfall varies from East to West between range of 200 to 600 millimeters. The rivers like Bhima, Sina, Man, Nira Bhogawati and many other smaller tributaries drain in the district. The soil of the district is mainly of DeccanTrap Volcanic origin. It is underlined by partially decomposed Basolatic rock material locally known as "murum".

Aims and Objective

- To understand the irrigation intensity in Solapur District.
- To illustrate the agriculture productivity in study region.
- To find out the impact of irrigation on agriculture production in study region.

Data and Methodology

Tahasilwise secondary data of irrigated area and agriculture productivity are used for present study. It is collected from District Agriculture Dept. of Solapur and District wise Statistical Information of Maharashtra state.

The statistical equation are used for irrigation intensity.

To know the agriculture productivity the Kendal's Ranking Coefficient method is used.

The statistical and cartographic techqunices are used fro representation of co-efficient index.

Explanation

Intensity of Irrigation

The intensity of irrigation expresses man's dynamic attempts to overcome the environmental limitations in the transformation of many of the barren areas in to the agriculture areas. The intensity of irrigation is the proportion of net irrigated area to net shown area of the aerial unit . It is an important indicator to determine the cropping pattern and agricultural productivity. The intensity of irrigation controlled by various factors such as source of irrigation, quantity and quality of water supply, intensity of network of water channels etc.

There are imbalances in irrigation development in solapur district. They are natural as well as created imbalance. The natural imbalances are caused due to the relative advantages and disadvantages of regions with respect to irrigation sources. These natural differences in regions can described as regional disparities.

Table 1- Irrigation intensity in Solapur District

Sr.	Irrigation Intensity	No of Tahsils	Name of tahsils
1	High	3	Pandharpur, Malshiras, Mngalwedha
2	Moderate	2	Karmala, Barshi
3	Low	6	Madha, S.Solapur, N.Solapur, Sangola, Akkalkot, Mohol

Source:- Complied by Researcher

High Intensity

The proportion of high intensity of irrigation is observed in Malshiras (75.02%) taluka , where canal and well irrigation has been developed during the last two decades .This has been followed by Mangalwedha (62.45%) and Pandharpur (60.81%). These areas captures by Bhima and Man River which has tremendously provided water for irrigation.

Moderate Intensity

Moderate irrigation intensity is observed in Karmala(49.48%) and Barshi (38.18%). These tahsils have also availability of water from either canals or wells for irrigation purpose.

Low Intensity

The Southeastern Southwestern and middle tahsils have recorded low intensity (below 40%) of irrigation giving a way to rain fed agriculture .The intensity of irrigation is very low in North Solapur (12.16%), Akkalkot (13.32%),South Solapur (14.57%)and Madha (15.24%).

Agriculture Productivity

Agricultural productivity is the interplay of a multitude of many factors, such as environmental, socio-economic and technological factors. The agricultural productivity is closely related to the per hectare yields, where as the agricultural efficiency is much more than agricultural productivity and conveys a more comprehensive meaning. Agricultural productivity is the actual performance of the land in terms of per hectare yield, where as agricultural efficiency is a ratio between the achievement in terms of agricultural production and the actual potential of the land productivity is a physical rather than a value concept and describes the changing relation between out put and one of the major inputs like land, labor, and capital.

Measurement of Productivity by Kendal's Ranking Co-efficient method

Here, the Kendal's ranking co-efficient method (1968) is used while applying this technique; seven major crops grown in all the tahsils of the district are selected. The crops have ranked in order to their yield per unit area. Then the arithmetic mean of these ranks is obtained which Kendal's called as ranking co-efficient and the same is represented cartographically. Lower the co-efficient value, higher is the productivity level of agriculture.

Three areas of agricultural productivity have been identified as shown in

Areas of high productivity.
Areas of moderate productivity
Areas of low productivity.

Table 2- Agriculture productivity in Solapur District (2005)

Sr	Agriculture Productivity	No of tahsils	Name of tahsils
1	High	4	Malshiras, Pandharpur, Sangola, Akkalkot
2	Moderate	5	Karmala, Barshi, Mohol, S.Solapur, N.Solapur
3	Low	2	Mangalwedha, Madha

Source:- Complied by Researcher

I. Areas of high productivity

This category is confined largely to northern and eastern part of Solapur district comprising the tahsils of South Solapur, North Solapur, Mohol, Barshi, and Karmala. The rivers in this area provide water for irrigation in Rabi Season. The river basins have fertile soils and many farmers in this area are adopting new inputs in agriculture by using. Fertilizers, machinery, improved seeds etc. All the favorable factors and human efforts have resulted into the high agricultural productivity in the region of Solapur district.

II. Areas of Moderate productivity

This category of productivity covers four tahsils. The co-efficient value in this area ranges from 5.5 to 6.5. Akkalkot tahsil is quite famous for pulses cultivation. Other tahsils like Akkalkot, Sangola, Pandharpur, Malshiras have fertile soil but paucity of water for irrigation and high variability of rainfall have resulted into moderate agricultural productivity.

III. Areas of low productivity

The low productivity areas cover two tahsils where the co-efficient value is above 6.5. The high variability of rainfall and low intensity of irrigation have led to low agricultural productivity in dry areas of the district.

Co-relation between irrigation intensity & agricultural productivity

Irrigation is backbone of agriculture in drought pore area. The choice of cropping pattern, variety crops, method of cultivation is also depend on irrigation. There are close relationship between irrigation intensity and agriculture productivity in study region. In Solapur District the positive co-relation largely (45.5 % to district) observed in five tahsils of district. These tahsils are Pandharpur, Malshiras, Karmala, Barshi and Madha tahsils of district. It is clearly explain that the irrigation high affects on agriculture productivity in Solapur district. If irrigation area increased, the agriculture productivity also increased.

Table 3- Co-relation among irrigation intensity and agriculture productivity in Solapur District.

Sr.	Level of irrigation and productivity	No of Tahsil	% to total	Tahsil Name
1	HI, HP	2	18.2	Pandharpur, Malshiras,
2	HI, LP	1	9.1	Mangalwedha
3	MI, MI	2	18.2	Karmala, Barshi,
4	LI, HP	2	18.2	Sangola, Akkalkot
5	LI, MP	3	27.3	S Solapur, N Solapur, Mohol
6	LI, LP	1	9.1	Madha

Source:- Complied by Researcher.

Perfect negative co-relation observed in only in Mangalweda, Sangola and Akkalkot tahsil of district. In Sangola and Akkalkot tahsil beneficial negative co-relation observed means in these tahsil irrigation intensity low but agriculture productivity is high. This tahsil are located low rainfall region and no canal or other irrigation sources developed. But the farmers of tahsils largely used micro irrigation system, mulching system, farm level water stoverages, low water crops for cultivation (fruit crops), favorable soil for fruit cultivation, etc. so the high agriculture productivity observed.

The negative co-relation is also observed in Mangalweda tahsil. In this tahsil irrigation intensity high but agriculture productivity low because in this tahsil agriculture productivity influence by irrigation as well as socio-economic and cultural factors of district. in Mangalweda irrigation is high irrigation due to Bhima and Man River flows and Nira Right Canal provide enormous water of irrigation. But low productivity due to traditional way of farming, jawar is monoculture crop, land holding size large, low fertility due dense irrigation for sugarcane, lazy farmers.

Suggestion

Agriculture productivity largely influenced by irrigation. Today, it is necessary to increase the agriculture productivity of the region for this irrigation can be increased. To achieve the growth of irrigation is the district. Few suggestions have to be suggested to individual level, institutional level and administrative level. They are:

- Give incentives to proper propaganda of irrigation management.
- Carried out research work to achieve innovative technology and methods of water management.
- Projects should be planned at micro level as poor peasant will bethe major beneficiaries.
- Watershed development program should be scientifically planned.
- Drip irrigation, sprinkler irrigation like measures should be adopted by the people. To encourage people give incentives in proportion.
- Use media for propaganda of irrigation management.
- Raise funds on local levels to complete small watershed programs.
- Rain harvesting is essential measure in drought prone areas.
- Repairing of canals to avoid seepage essential.
- Think globally act locally, to achieve sustainable water management.
- People participation should be give vital importance.
- The fund can rise through strong co-operative sector of the district.
- Rules, regulation and charges on irrigation water should be restructured.
- Administration should take care of completion of uncompleted projects in the district.
- Set local level committees to look after the progression of small project in the area. Such committees should be co-ordinate with CEO and collect orate of the district.

References

[1] Rawat D.S. (2009) *Uttar Bharat Bhoogol Patrika.*, 97-100.
[2] Dubey R.N. and Negi B.S. (1996) *Economic Geography of India*, 190.
[3] Jasbir (1974) *An Agriculture Atlas of India - A Geographical Analysis*, 67-69.
[4] Pawar C.T. (1989) *Impact of Irrigation on Agriculture.*, 13.
[5] Pawar C.T. (1989) *Impact of Irrigation: A regional perspective.*
[6] Singh Jasbir and Dhillon S.S. (2004) *Agriculture Geography.* 110,126,134,155,158,160.
[7] *Socio-Economic review and district Statically abstract of Solapur District* (2004-05).

2

GENETIC VARIABILITY IN LAND RACES OF FORAGE SORGHUM {*Sorghum bicolor* (L) MOENCH} COLLECTED FROM DIFFERENT GEOGRAPHICAL ORIGIN OF INDIA

JAIN S.K.[1*] AND PATEL P.R.[2]

Sorghum Research Station, Sardarkrushinagar Dantiwada Agricultural University, Deesa- 385535 (BK), Gujarat, India.
*Corresponding Author: Email- skjain_sdau@yahoo.co.in

Abstract- An evaluation of 102 land races of forage sorghum for days to 50 % flowering, plant height, number of leaves per plant, leaf length, leaf breadth and fodder yield per plant in augmented block design during 2009-10 and 2010-11 revealed highly significant differences among the accessions. High heritability accompanied with high GA as per cent of mean was observed for days to 50% flowering, plant height, number of leaves per plant, leaf length and fodder yield per plant suggested that these characters are under additive gene action and gives better scope for selection. Fodder yield was positively and significantly correlated with number of leaves per plant, leaf length, leaf width and panicle length. The characters leaf width, number of leaves per plant, days to 50% flowering and panicle length showed positive direct effect on fodder yield. Where as leaf length showed positive significant association with fodder yield but the direct effect of leaf length was negative with fodder yield, which may be a result of the indirect effect of this trait via other traits. Based on the results of the means of the two years considering together for the various traits, the accession E-143 and EJN-11 was found to be superior for earliness, EA-2 and GUB-50 for plant height, E-159 for leaf characters and E-203 and GGUB-39 for fodder yield. Therefore these accessions should be utilized in further breeding program for developing superior varieties..
Key words - correlation, genetic advance, heritability, path coefficient, Variability

Introduction:

Sorghum {*Sorghum bicolor* (L) Moench} is one of the most important drought tolerant crops which allow farmers to use one third less water than similar crops such as corn. In India, sorghum is cultivated in 7.53 mha, of which 2.89 mha is cultivated during rainy and kharif season with a production of 3.05 mt (Anonymous 2010). The major sorghum growing states are Andhra Pradesh, Gujarat, Karnataka, Madhya Pradesh, Maharashtra, Rajasthan and Tamil Nadu. In Gujarat the area under sorghum cultivation is 1.17 lakh ha with a production of 1.28 lakh tones and productivity being 1151 kg/ha (Anonymous, 2011). Sorghum area is fast decling for the past 10 decades yet it will continue to be an important food grain in India since its relative importance for alternate uses such as poultry and cattle feed, livestock forage, starch, sugar, alcohol and other uses will increase. It is therefore of paramount importance that technological developments are extended to increase the productivity and sustainability of sorghum production. Landraces or farmer varieties constitute the basic material for developing any variety or hybrid. An autochthonous landrace is a variety with a high capacity to tolerate biotic and abiotic stress, resulting in high yield stability and an intermediate yield level under a low input agricultural system". It is well established fact that the progress in improvement of a crop depends on the degree of variability in the desired character in the base material vis-a-vis germplasm collection. The study of relationships among quantitative traits is important for assessing the feasibility of joint selection of two or more traits and hence for evaluating the effect of selection for secondary traits on genetic gain for the primary trait under consideration. A positive genetic correlation between two desirable traits makes the job of the plant breeder easy for improving both traits simultaneously. Path coefficient analysis was performed to qualify the direct and indirect contributors of yield components and developmental traits of fodder yield. Therefore the present investigation was undertaken to study the variability in land races of sorghum during kharif 2009-10 to 2010-11.

Materials and Methods

One hundred two forage sorghum landraces collected from different geographical origin of India were used for the present study. The trial was grown in augmented block design during the two seasons namely *Kharif* 2009-10 and *Kharif* 2010-11 at Sorghum Research Station, Sardarkrushinagar Dantiwada Agricultural University, Deesa (Gujarat). Deesa is situated at latitude of 24.5° N and longitude 72° E and at an elevation of 136 M above the Mean Sea Level. The soil of the field was sandy in texture with pH value of 7.5 to 8.00 having good physical and chemical properties (Organic Carbon= 0.23, EC dsm- = 0.232, K_2O= 259.9 kg/ha and P_2O_5= 46.2 kg/ha). The experimental unit was a single-row plot of 6.75 m long, spaced at 0.45 m apart. NPK 120:40:00 fertilizers was applied as half basal dose of nitrogen and full dose of phosphorus at the time of sowing and half nitrogen applied after one month of sowing. Plots were thinned down after two weeks of crop emergence and plant-to-plant distance of 0.10 m was maintained. The experimental years showed different temperature regimes, humidity, rain fall and sunshine hours during the crop durations. The all other recommended agronomical practices were followed to raise a good crop in both the seasons. Data were taken on days to 50 % flowering, plant height (cm), number of leaves per plant, leaf length (cm), leaf breadth (cm) and fodder yield per plant (g). Statistical analysis was done according to the standard statistical procedures (Federer, 1996; Burton, 1952; Johnson et al., 1955; Al -Jibouri et al., 1958 and Dewey and Lu, 1959).

Result and Discussion

Genotypic coefficient of variation (GCV) was maximum for fodder yield per plant (36.78 %) followed by panicle length (32.77 %) and its difference with phenotypic coefficient of variation (PCV) was found less. Differences between GCV and PCV for other traits were also found to be less indicating that these traits were less affected by environmental fluctuations (Table 1). The high values of GCV and PCV for all the traits suggested that there is a possibility of improvement through direct selection for the traits. High heritability coupled with high genetic advance was observed for plant height, days to 50% flowering and leaf length indicating that these characters are controlled by additive gene action and phenotypic selection for these characters will be effective. The expected genetic advance might have been biased upward as it is based on the estimates of heritability in broad sense, secondly in the augmented design the estimation of mean square due to the error is based on the check variety only, and hence, it might have given the high estimates of genetic variances in it. High heritability and high genetic advance for these traits was also observed by earliers (Deepalakshmi and Ganesamurthy, 2007, Warked et al., 2008 and Mahajan et al., 2011).

The genotypic and phenotypic correlation coefficients worked out among different characters including fodder yield per plant revealed that in general the phenotypic correlation coefficient were similar to genotypic correlation coefficient (Table 2). In some cases the phenotypic correlation was slightly higher then the genotypic correlation coefficients, which may be a result of modifying effect of environments on the association of the characters. The fodder yield per plant showed positive and significant correlation with leaf width, number of leaves per plant, days to 50% flowering and panicle length. It is also noticed that characters that exhibited positive associations with fodder yield have also showed positive associations among them selves. Out of these traits some traits like plant height, days to 50% flowering and leaf length are having the high broad sense heritability with high expected genetic advance should be used in selection programme. Similar views have been reported by earlier workers (Sankarapandian, 2010 Jain et al., 2011).

Correlation co-efficient indicates only the general associations between any two traits without tracing any possible causes of such associations. In such situations, the path coefficient analysis at phenotypic level (Table 3 and fig. 1) is done to partition the correlation coefficients in to direct and indirect effects. Fodder yield per plant was taken as dependent variable while computing the path coefficient. The path coefficient analysis based on both the seasons revealed that the characters like leaf width, number of leaves per plant, days to 50% flowering and panicle length which had positive significant association with green fodder yield also exerted positive and high direct effects on fodder yield. Where as leaf length showed positive significant association with fodder yield but the direct effect of leaf length was negative with fodder yield, which may be a result of the indirect effect of this trait via other traits. Thus the present study revealed that these traits in determining the fodder yield and therefore, their values in constructing the selection criterion. Positive direct effect of number of leaves per plant and leaf width on fodder yield was also reported by earlier reporters (Jain et al, 2010, Prakash et al., 2010 and Jain et al., 2011).Thus the present study revealed that the plant height, leaf length and width, number of leaves per plant, days to 50% flowering and panicle length are the important traits for deciding the fodder yield but the high residual effect (0.8502) revealed that the other traits like stem girth, number of tillers per plant and leaf stem ratio also important traits. So that the consideration should be given to these traits while planning a breeding strategy for increase the fodder yield.

Based on the results of the means of the two seasons the accessions exhibited good variability in all the quantitative traits: the days to 50% flowering (38.00-105.5 days), plant height (121.00-302.25 cm), number of leaves per plant (6.00-20.50), Panicle length (4.00-36.50 cm), leaf length (37.50-103.50cm), leaf width (2.75-9.25 cm), fodder yield (0.116-0.512 kg /plant), showed wider range. Some of the accessions identified superior for the different morphological characters are EJN-11 (38.0), E-143 (42.5), ERN-13 (44.5), EG-35 (46.0) and E-186 (46.0) were for earliness, EA-2 (302.25), GUB-50 (300.75), EJN-26 (297.0), ERN-29 (292.5) and GGUB-56 (291.75) for plant height, E-186 (36.5), GGUB-47 (33.0), E-153 (31.0), ERS-25 (28.5) and EB-2 (23.0) for panicle length, ERN-11 (20.5), E-159 (20.0), GGUB-32 (20.0), GGUB-47 (19.0) and GGUB-30 (19.0) for number of leaves per plant, EB-2 (103.5), E-159 (99.0), GGUB-57 (99.0), E-195 (98.0) and EG-40 (94.0) for leaf length, GGUB-47 (9.3), E-159 (9.0), GGUB-52 (8.5), E-186 (8.3) and E-197 (7.8) for leaf width and E-203 (0.5125), GGUB-39 (0.512), EG-40 (0.438), E-1 (0.4165) and GGUB-19

(0.4125) for fodder yield (Table 4).

The conclusion that can be reached from the present study is that the leaf length and width, number of leaves per plant, days to 50% flowering and panicle length are the important component traits for fodder yield per plant. The accession viz., E-143 and EJN-11 was found to be superior for earliness, EA-2 and GUB-50 for plant height, E-159 for leaf characters and E-203 and GGUB-39 for fodder yield. Therefore these accessions should be utilized in further breeding program for developing superior varieties.

Table 1- Variability parameters in landraces of forage sorghum (Pooled over two year)

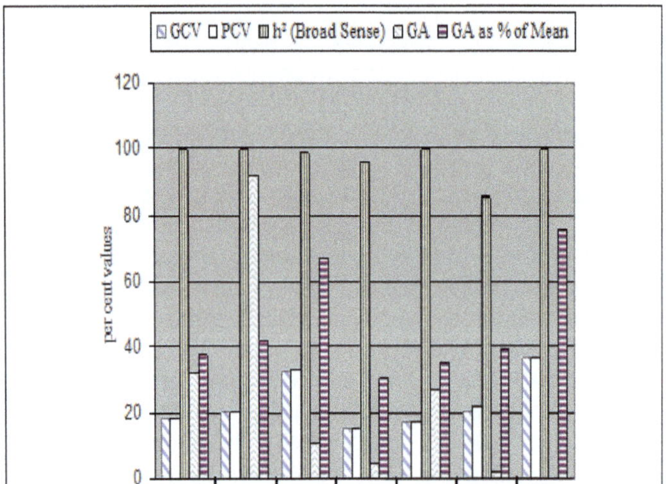

	DFF	PH	PL	NLPP	LL	LW	FYPP
GCV	18.34	20.2	32.77	15.12	17.11	20.42	36.78
PCV	18.37	20.22	33.03	15.45	17.13	22.08	36.84
h^2 (Broad Sense)	99.7	99.8	98.4	95.7	99.8	85.5	99.7
GA	32.01	91.56	10.89	4.56	27.08	2.22	0.18
GA as % of Mean	37.74	41.58	66.97	30.47	35.21	38.91	75.65

Characterstics (DFF= Days to 50% flowering; PH= plant height; PL panicle length; NLPP= number of leaves per plant; LL= leaf length; LW= leaf width; FYPP= fodder yield per plant

Table 2- Phenotypic correlation coefficient on the basis of unadjusted values between different characters in sorghum over two year

Characters	50% Flowering	Plant Height	Panicle length	Number of leaves / Plant	Leaf length	Leaf width	Fodder yield/ Plant
50% Flowering	1.00	-0.062	0.081	-0.096	0.499**	-0.029	0.106
Plant Height		1.00	-0.064	0.057	-0.005	0.13	0.041
Panicle Length			1.00	0.14	0.282**	0.215*	0.157
No. of leaves/ plant				1.00	0.206*	0.293**	0.367**
Leaf Length					1.00	0.479**	0.282**
Leaf width						1.00	0.427**
Fodder yield/Plant							1.00

Table 3. Direct (diagonal) and indirect (non diagonal) effects of different characters on green fodder yield in sorghum at genotypic level over two year

Characteristics	50% Flow-ering	Plant Height	Panicle length	Number of leaves / Plant	Leaf length	Leaf width
50% Flowering	0.1584	-0.0097	0.0129	-0.0152	0.0791	-0.0047
Plant Height	0.0058	-0.0942	0.006	-0.0053	0.0005	-0.0122
Panicle Length	0.0024	-0.0019	0.0299	0.0042	0.0084	0.0064
No. of leaves/ plant	-0.027	0.016	0.0396	0.2826	0.0581	0.0828
Leaf Length	-0.0221	0.0002	-0.0125	-0.0091	-0.0442	-0.0212
Leaf width	-0.0111	0.0485	0.0806	0.1101	0.1799	0.3757
Fodder yield/ Plant	0.1064	0.0411	0.1565	0.3673	0.2818	0.4269
Partial R^2	0.0169	0.0039	0.0047	0.1038	-0.0125	0.1604

Residual effect = 0.8502

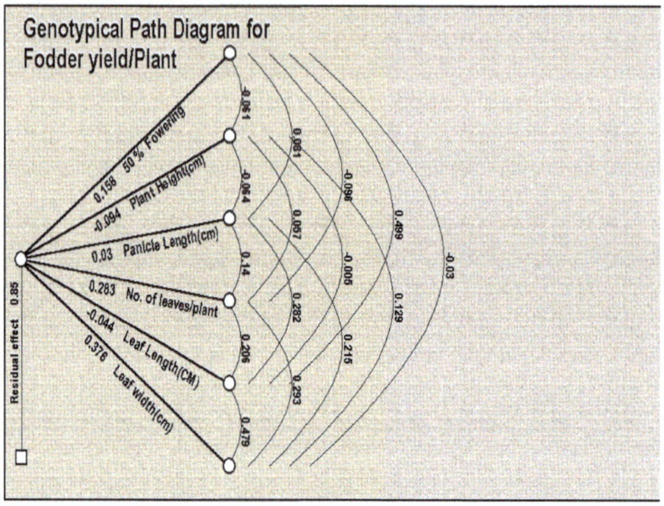

Fig. 1- Direct, indirect and residual effects of different traits on fodder yield

References

[1] Al-Jibouri H.A., Miller P.A., Robinson H.F. (1958) *Agronomy Journal*, 50: 633-637.

[2] Anonymous Ministry of Agriculture Department of Agriculture and Cooperation (2010) *Status paper on millets. Directorate of millets development.*

[3] Anonymous Gujarat.gov.in (2011) *State based on Final Forecast reports for the year.*

[4] Burton G.W. (1952) *6th International Grassland Congress* 1: 227-283.

[5] Deepalakshmi A.J., Ganesamurthy K. (2007) *Indian Journal of Agricultural Research*, 41 (3):177-182.

[6] Dewey D.R. and Lu K.A. (1959) *Agronomy Journal* 51: 515-518.

[7] Federer T. (1996) *Hawaii planters Record* 20 :191-207.

[8] Jain S.K., Patel P.R. and Elangovan M. (2011) *Indian Journal of Plant Genetic Resources*, 24(3): 327-331.

[9] Jain S.K., Elangovan M. and Patel N.V. (2010) *Indian Journal of Plant Genetic Resources*, 23:15-18.
[10]Johnson H.W., Robinson H.F. and Comstock R.E. (1955) *Agronomy Journal*, 47: 314-318.
[11]Mahajan R.C., Wadikar P.B., Pole S.P. and Dhuppe M.V. (2011) *Research Journal of Agricultural Sciences*, 2(1):101-103.
[12]Prakash R., Ganesamurthy K., Nirmalakumari A. and Nagarajan P. (2010) *Electronic Journal of Plant Breeding*, 1(3): 315-318.
[13]Sankarapandian V.R. (2010) *Andhra Agriculture Journal*, 47 (3&4):191-196.
[14]Warked Y.N., Potdukhe N.R., Dethe A.M., Kahate P.A. and Kotgire R.R. (2008) *Agricultural Science Digest*, 28 (3): 202-205.

Table 4- Range and characters mean of top five accessions over two year in the land races of forage sorghum

Characteristics	Range	Mean	Five best entries and their pooled mean over two year
50% Flowering	38.00-105.5	84.83	EG-35 (46.0), E-186 (46.0), ERN-13 (44.5), E-143 (42.5), EJN-11 (38.0)
Plant Height	121.00-302.25	220.23	EA-2 (302.25), GUB-50 (300.75), EJN-26 (297.0), ERN-29 (292.5), GGUB-56 (291.75)
Panicle Length	4.00-36.50	16.25	E-186 (36.5), GGUB-47 (33.0), E-153 (31.0), ERS-25 (28.5), EB-2 (23.0)
No. of leaves/plant	6.00-20.50	14.98	ERN-11 (20.5), E-159 (20.0), GGUB-32 (20.0), GGUB-47 (19.0), GGUB-30 (19.0)
Leaf Length	37.50-103.50	76.92	EB-2 (103.5), E-159 (99.0), GGUB-57 (99.0), E-195 (98.0), EG-40 (94.0)
Leaf width	2.75-9.25	5.72	GGUB-47 (9.3), E-159 (9.0), GGUB-52 (8.5), E-186 (8.3), E-197 (7.8)
Fodder yield/Plant	0.116-0.512	0.24	E-203 (0.5125), GGUB-39 (0.512), EG-40 (0.438), E-1 (0.4165), GGUB-19 (0.4125)

3

EFFECT OF ESTERIFICATION ON MOISTURE ABSORPTION OF SINGLE ARECA FIBER

DHANALAKSHMI SAMPATHKUMAR[1,2], RAMADEVI PUNYAMURTHY[1,2], BASAVARAJU BENNEHALLI[3*] AND SRINIVASA CHIKKOL VENKATESHAPPA[4]

[1]Research Scholar, Jawaharlal Nehru Technological University, Hyderabad-500 085, Andhra Pradesh, India.
[2]Department of Chemistry, KLE Society's BVB College of Engineering & Technology, Hubli-580 031, Karnataka, India.
[3]Department of Chemistry, Alva's Institute of Engineering & Technology, Moodbidri-574 225, Karnataka, India.
[4]Department of Mechanical Engineering, GM Institute of Technology, Davangere-577 006, Karnataka, India.
*Corresponding Author: Email- basavaraju_b@yahoo.co.in

Abstract- Environmentally beneficial composites can be made by replacing synthetic fibers with various types of cellulose fibers. The most important factor in finding good fiber reinforcement in the composite is the strength of adhesion between matrix polymer and fiber. Due to the presence of hydroxide and other polar groups in various constituents of natural fiber, the moisture absorption is high which leads to poor wettability and weak interfacial bonding between fibers and the more hydrophobic matrices. Therefore, it is necessary to impart hydrophobic nature to the natural fibers by suitable chemical treatments in order to develop composites with better mechanical properties. In the present work, the effect of esterification on the moisture absorption property of single areca fiber has been investigated.
Key words- natural fiber, moisture Absorption, chemical treatment, areca fiber, acrylic acid, sea water.

Introduction

Agro-based resources, also referred to as lignocellulosics, are resources that contain cellulose, hemicelluloses, and lignin. Lignocellulosics include wood, agricultural residues, water plants, grasses, and other plant substances. Fiber technology, high performance adhesives, and fiber surface modification utilized natural fibers in the fabrication of structural lignocellulosic composites and reported that these natural fibers-reinforced composites possess better chemical resistance, electrical resistance, good thermal and acoustic insulating properties, and high resistance to fracture. This has stimulated the use of hard natural cellulose fibers as an attractive alternative to synthetic fibers because of their low cost, less weight, and density.

High moisture absorption, poor wettability, and poor fiber-matrix adhesion are the major drawbacks of natural fiber reinforced composite materials. A better adhesion between hydrophilic fiber and hydrophobic matrix, and enhanced mechanical properties of natural fiber reinforced composite materials can be achieved by suitable chemical modification of natural fiber. In the case of lignocellulosics, cell wall polymers, extractives, and inorganics are the components that, if modified, would change the properties of the resources.

There are several ways such as acetylation, methylation, benzoylation, permanganate treatment, acrylation etc to chemically modify the lignocellulosic cell wall polymers. The most abundant single site for reactivity in these polymers is the hydroxyl group and most reaction schemes have been based on the reaction of hydroxyl groups. The hygroscopicity of the lignocellulosic material can be reduced by replacing some of the hydroxyl groups on the cell wall polymers with bonded chemical groups [1]. Acrylation is one such method for reducing the hygroscopicity of the natural fiber. The polymer hydroxyl group of the cell wall reacts with carboxylic group of acrylic acid to yield an ester so that fiber becomes more hydrophobic in nature.

Many attempts have been made to study the properties of chemically treated natural fiber-reinforced polymer composites. Vilay *et al.* investigated the effect of acrylation in bagasse fiber reinforced unsaturated polyester composites and reported that storage modulus and glass transition temperature of composites found to improve, and water absorption of composites found to decrease compared to untreated bagasse fiber composites [2].

Jannah *et al.* studied the effect of acrylation in woven banana reinforced unsaturated polyester composites and they showed, treated composite systems show improvements in flexural properties. Furthermore, they reported that the treated fiber composites reduce the water absorption compared to the untreated composite system [3].

Only a few studies on the mechanical properties of acrylic acid treated natural fiber-reinforced composite materials were reported and there is no information on the water absorption studies of acrylated single natural fiber. A better understanding of moisture absorption property of fibers will help to develop a productive use of fibers in composite materials. Therefore, this study seeks to determine the moisture absorption characteristic of acrylic acid treated single areca fiber.

Areca belongs to the species *Areca catechu Linnaeus* under the family palmecea [4]. The areca empty fruit fiber is composed of 35 -65% hemicellulose, 13-26% lignin, pectin, and protopectin [5]. The average filament length of the areca fiber is 4 cm. At present, a small quantity of areca empty fruit material is being used as a fuel in areca nut processing and plenty of unused husk which is left in the plantation causes bad odour and other decay-related problems [6].

Few researchers have studied the areca fiber-reinforced composites and found to have a good flexural strength and adhesion tensile properties. Also, reported that areca composites would be very promising material for packing and other general structural applications with a moderate duration as they shown excellent resistance moisture to absorption when compared to wood based particle boards [7-10].

Areca fiber can also be used for making value added items like thick boards, fluffy cushions, non-woven fabrics, thermal insulators, etc. [11]. Thus, the use of unmanaged areca husk as a structural material requires a detailed study of physical, chemical and thermal characteristics. One of the most important issues is the degradation behaviour of the composites exposed to environmental conditions such as humidity, sunlight or microorganisms. Also, the poor resistance of the fibers to water absorption can have undesirable effects on the mechanical properties and the dimensional stability of the composites. Therefore, it is important to study in detail the water absorption behaviour in order to estimate not only the consequences that the water absorbed may have, but also the durability of natural fibers composites aged under water.

Materials and Methods

Materials

Areca empty fruit fibers (husk) were obtained from Madhu Farm House, Nilogal, Davangere, Karnataka, India. Analytical grade reagents were purchased from Qualigens Company and used as received.

Fiber extraction

Selected areca fruit husks were used to study. Dried areca husk was soaked in deionised water for about five days. The soaking process loosens the fibers and can be extracted out easily. Finally, the fibers were washed again with deionised water and dried at room temperature for about 15 days. The dried fibers are designated as untreated fibers.

Alkali treatment

The fibers are first dewaxed by soaking in 1:2 mixtures of ethanol and benzene for 72 h, followed by washing with deionised water and then air dried. The dewaxed fibers were immersed in 20% NaOH solution, where the total volume of the solution was 15 times the weight of the fibers. The fibers were kept in alkaline solution for 72 h at room temperature; they were thoroughly washed in running water and then neutralized with 2% acetic acid solution. Lastly, the fibers were washed again in running water to remove the last traces of acid sticking to it, so that the pH of the fibers was approximately 7. Then, they were dried at room temperature for 48 h to obtain alkali-treated fibers.

Acrylic acid treatment

Alkali treated areca fiber was soaked in 5% acrylic acid (AA) solution at 50 °C for 1 h, then washing with distilled water thoroughly and dried in an oven for 24 h at the temperature of 70 °C [12].

Water absorption studies

Water absorption studies on untreated and treated fibers were carried out in various sources of water such as Pond water, River water, Bore water, and Sea water. The pH values of water samples were; Bore water: 7.40, Pond water: 8.20, River water: 7.74, and Sea water: 7.76. A bundle of fibers (2 g) was immersed in beakers containing various sources of water. After 24 h, the fiber bundle is taken out from the beakers and fibers were squeezed and dried in between the folds of filter paper. Finally the weight of the fibers was taken and the percentage of water absorption is calculated as follows:

$$\%\ of\ Water\ Absorption\ \frac{Difference\ in\ Weight}{Initial\ Weight}\ X100$$

Then again the sample was soaked in water, and the same process was repeated after every 24 h for 2400 h. A graph is plotted between % water absorption and square root of time.

Results and Discussion

Lignocellulosics are three-dimensional, polymeric composites made up primarily of cellulose, hemicellulose and lignin. In natural fibers, semicrystalline cellulose is the main reinforcement material. The cellulose is held together by amorphous hemicelluloses and fibers are cemented together in the plant by lignin which is commonly known as plant cell adhesive. Lignocellulosic fibers may differ in chemical composition, but have very similar properties such as flammability, thermoplasticity, moisture Absorption, and degradation by acids, bases and ultraviolet radiation. All of these characteristics will result in the specific end use of lignocellulosics in composite formulation [13].

Lignocellulosics change dimensions with changing moisture content because the cell wall polymers contain hydroxyl and other oxygen containing groups that attract moisture through hydrogen bonding. The hemicellulose is mainly responsible for moisture Absorption, but the accessible cellulose, noncrystalline cellulose, lignin and surface of crystalline cellulose also play major roles.

Moisture swells the cell wall, and the fiber expands until the cell wall is saturated with water (fiber saturation point). Beyond this saturation point, moisture exists as free water in the void structure and does not contribute further expansion [14].

To measure the water absorption of the areca fiber, the fiber was immersed in a beaker containing various sources of water at room temperature. After every 24 hours, the fiber was taken out and excess water on the fiber surface was removed before weighing. Three replicates were tested and the results were presented as average of three.

The Comparison of moisture absorption of untreated areca fiber in various water samples showed that areca fiber absorbs more moisture in bore water and less in sea water. Also, surface treated fiber absorbs less moisture than the untreated fiber in all water samples. Further, from Figure 1, comparison of moisture absorption property of untreated, alkali treated, and acrylic acid treated areca fiber in sea water, it can be observed that the acrylic acid treated fiber absorbs less moisture than the alkali treated fiber. This is due to the change in the surface topography of the fibers due to alkali and acrylic acid treatments.

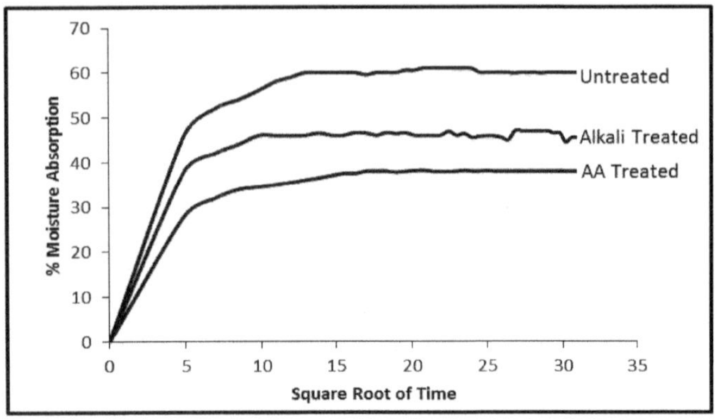

Fig. 1- Comparison of moisture absorption property of untreated, alkali treated, and acrylic acid treated areca fiber in sea water

In untreated fibers, fibrils are bound together by hemicellulose and lignin. The chemical treatments had removed most of the hemicellulose and lignin and thus areca fiber become more hydrophobic [3]. Also, in all natural cellulose fibers, a hollow cavity called lumen exists in the unit cell of the fibers [14]. The hollow cavity generates more pathways for water to start diffusing into the fiber. The velocity of the diffusion process can be reduced by reducing the void content of the fiber. The cellular structure of the fiber destroys by the chemical treatment; hence reduced the void content and this in turn reduced the moisture uptake of treated fiber.

Conclusion

The potential of using natural fibers as reinforcing agents is based on the interfacial properties between fiber and polymer matrix. Also, the poor resistance of the fibers to water absorption can have undesirable effects on the structure, dimensional and mechanical properties of the natural fiber reinforced composites. It is possible to reduce the moisture absorption property of areca fiber through its surface modification. The results presented in this work indicate that the areca fiber treated with acrylic acid absorbs less moisture than the untreated, alkali treted fiber.

Acknowledgement

The authors would like to thank Madhu Farm House, Nilogal for providing a good quality areca fruit husk. The first author would like to thank the Management of KLE Society's BVB College of Engineering and Technology, Hubli, Karnataka, India for the kind encouragement and constant support provided. She sincerely thanks Prof. Ashok Shettar, Principal, KLE Society's BVB College of Engineering and Technology, without his encouragement and motivation, this work would not have been possible. The second author would like to thank Dr. M. Mohan Alva, Chairman, Alva's Education Foundation, Moodbidri and Dr. M. Govinde Gowda, Principal, Alva's Institute of Engineering & Technology, Moodbidri, Karnataka, India, for their kind encouragement and support provided. The third author would like to thank Sri G. M. Lingaraju, Secretary, GM Institute of Technology, Davangere, India, for their encouragement and support throughout this work.

References

[1] Roger M.R. (1998) *Science and Technology of Polymers and Advanced Materials*. New York. 717-732

[2] Vilay M., Mariatti R., Mat T. and Mitsugu T. (2008) *Composites Science and Technology*, 68, 631-638.

[3] Jannah M., Mariatti M., Bakkar A. and Khalil A. (2009) *Journal of Reinforced Plastics and Composites*, 28(12), 1519-1532.

[4] Rajan A. and Kurup J.G. (2005) *Biochemical Engineering Journal*, 25, 237-242.

[5] Ramachandra T.V., Kamakshi G. and Shruthi B.V. (2004) *Renew. Sust. Energy. Rev.* 8, 1-47.

[6] Ghosh S.K., Sinha M.K. and Bandopadhya S.K. (1975) *J. Plant Crops*, 3, 29-33.

[7] Swamy R.P., Mohankumar G.C., Vrushabhendrappa Y. And Joseph V. (2004) *Journal of Reinforced Plastics and Composites*, 23(13), 1373-1382.

[8] Srinivasa C.V., Arifulla A., Goutham N., Santhosh T., Jaeethendra H.J., Ravikumar R.B. and Anil S.G. (2011) *Materials and Design*, 32(4), 2469-2475.

[9] Srinivasa C.V., Suresh Y.J. and Premakumar W.P. (2011) *Advances in Polymer Technology*, DOI 10.1002/adv.20255).

[10]Srinivasa C.V., Basavaraju B., Mownesh G.K. and Raghupatel G.R. (2010) *BioResources*, 5(3), 1845-1858.

[11]Mohankumar G.C. (2008) *World Congress on Engineering*, II, WCE 2008, London, U.K.

[12]Panigrahi S., Li X. and Tabil L. (2008) *International Conference on Flax and Other Bast Plants*, ID number: 83 (ISBN: 978-0-9809664-0-4), 399-407.

[13]Aziz S.H. and Ansell M.P. (2004) *Science and Technology*, 64, 1219-1230.

[14]Reddy N. and Yang Y. (2005) *Trend Biotechnol*, 23(1), 22-27.

4

ADDITIONAL GRAVITATIONAL FORCE IS ACTING ON SOME TREES DUE TO PRESENCE OF UNDERGROUND WATER STREAM

MORE B.M.

Brahmdevdada Mane Institute of Technology, Belati, Solapur- 413002, MS, India.
*Corresponding Author: Email- babasahebmore@gmail.com

Abstract- An unusual growth is developed with some trees. A bunch of some of their twigs is bending in the downward direction and twigs are stretched downward. Bending of twigs is due to additional gravitational force acting on these twigs. This additional gravitational force is because of presence of underground water stream. The direction of plane in which the twigs are bending indicates the ground surface line where underground water stream is present.
Key words- Unusual growth, Twigs of tree, Gravitational Force, Underground Water stream.

Introduction

All plants growth is connected with light. In trees, the trunk brings the tree up into the light filled atmosphere. But as a rule, a tree trunk grows straight up; it is not directed towards the sun as a source of light, which is in fact a move daily across the sky. This growth straight upward is known as negative geotropism. Since the trunk grows directly away from the center of earth the blossoms of numerous wild flowers, in contrast, follow the path of the sun during the day; they exemplify positive phototropism- growing toward the light source [1].

The trees genetic propensity is to capture as much light as possible, which maximize its ability to do photosynthesis, grow and produce fertile offspring, which guarantee the survival of species. Since every tree has this propensity and growth creates shade that brings about death, each tree competes with its neighbor to maximize light uptake [2].

The trees are nearly symmetrical in shapes but sometimes they are asymmetrical. You might even call them misshapen [1]. Some of the twigs of some trees are BENDING DOWNWARDS [BD] in a straight line. This bending of twigs will have some specific reason. In this publication, I propose to outline the body of research that suggests a strong evolutionary connection between the two phenomena, bending of twigs of tree and presence of underground water stream.

Growth of Trees

The buds elongate the branches and widen the crown (Branches and leaves).

The particular shape of the crown and the size of the trunk relative to the crown in an individual tree express a different asset of the tree [1,3]. The neem tree in Fig. 1 gives USUAL GROWTH [UG] of tree and the crown nea.rly symmetrical in shape. The branches are growing towards the light. The crowns of some trees are markedly asymmetrical. You might even call them misshapen or UNUSUAL GROWTH [UUG] of some branches. Sometimes the twigs of these branches are BD direction as shown in Fig. 2.

Fig. 1- Shows usual growth of Neem tree with branches nearly radial from the trunk

Fig. 2a-

Fig. 2b-

Fig. 2- Unusual growth of trees with twigs bending downwards a) Tamarind and b) Neem

New Technique of Searching of Underground water stream

Generally there are many methods, vary widely; most dowsers (also called diviners or water witches) probably still use the traditional forked stick, which may came from variety of trees. Other dowsers may use keys, wire, coat hangers, pliers, wire rods and pendulum etc [4]. Locating underground water stream is both an art and by chance [5]. For drilling of bore well, no detail maps of ground water presence and no guarantee. Therefore it is quite confusing to take a decision about whether to drill or where to drill [6].

The technique by using UUG of trees is useful for finding place for drilling bore wells. The tree which has twigs BD gives ground surface direction of UWS. This will indicate one UWS stream surely. If we want two streams, we have to observe the same tree or other nearby trees, which are having the bending of twigs with different directions. Take a cross point of these two streams direction this will give you two streams may be at different depths. The cross point for drilling of bore well is calculated as shown in Fig. 3.

Fig. 3- A circle shows point for drilling bore well

Case study

This new technique was applied to Shri Raosaheb More's land, located at Tulshi, Tal Madha, Dist Solapur. The bore well point was fixed by considering UUG of two different trees and taking a cross point of these two. The bore was drilled on 13/03/2012 and more than 200 liter per minute water source was available during the drilling of bore well. The Fig. 4 shows the photograph of out coming water source at the time of drilling.

Fig. 4- Water coming out of bore well at the time of drilling

Results and Discussion

The trees of Neem and Tamarind shows UUG, as some of twigs are BD in a straight line. It means that in addition to gravitational force, there is additional downward force acting on these twigs. This additional force is might be due to presence of UNDERGROUND WATER STREAM [UWS]. There is abundant evidence from the trees on the banks of the river or nearby the wells, to support that the bending of twigs is due to presence of underground water, where we can see that twigs are bending downwards. Because of this continuous acting additional force, the twigs are stretched in downward direction. The comparison of length of twigs from UG and that of from UUG i.e., bending twigs, shows that the length of bending twigs is larger than that of UG. But, the diameter of sub branches of twigs which are bending is smaller. This is because of continuous additional force acting in downward direction. The bending of twigs is along a straight line. If a plane is drawn parallel to the direction of bending, it gives the direction of UWS on ground surface. We have to drill a bore well at this point on the ground surface.

Conclusion

In conclusion, the bending of some of twigs of some trees indicates presence of underground water stream. The bore well drilled under study might well support the fact. As the stream will not be always in straight line, the cross point of two or three streams sometime give less guarantee. The depth of stream varies from stream to stream, many times depth is up to 100 feet but sometimes it is up to 600 feet.

Acknowledgement

Author is very thankful to Prof. Snehal Mane, the Secretary of Brahmdevdada Mane Institute of Technology, Belati, Solapur and Dr. L.P. Deshmukh, Professor in Physics, Solapur University, Solapur, for their support and encouragement. I also thanks all my colleagues, who are involved in discussion of the idea of this article.

References

[1] *http://natureinstitute.org/pub/ic/ic14/trees.html.*
[2] Darwin C. (1979) *The Origin of Species*, Middlesex, Penguin Classics, Chapter III, 116.
[3] *http:// texastreeid.tamu.edu/content/howTreeGrow/.*
[4] *http://ga.water.usgs.gov/edu/dowsing.html.*
[5] *http://dew.drop.com/purewateringroundstreams.html.*
[6] Daneil S., Rex B. and Marios S. (2006) *Kansas Geological Survey*, Public information circular no. 23.

5

ROLE OF AGRICULTURE INSURANCE SCHEME IN MARATHWADA REGION OF MAHARASHTRA, INDIA

PATWARDHAN P.S. AND NARWADE S.S.*

Department of Economics, Dr. Babasaheb Ambedkar Marathwada University, Aurangabad- 431004, MS, India.
*Corresponding Author: Email- ssnarwade22@gmail.com

Abstract- The Present paper is an attempt to study the role of National Agriculture Insurance Scheme (NAIS) in Marathwada region of Maharashtra state of India. The study showed that farmers covered under NAIS in Maharashtra in kharif season increased at the rate of 1.50% during the period under study. But the area covered and farmers benefitted declined by -6.05% and 4.14% respectively in Maharashtra during 2000-2010. The farmers covered under NAIS in Marathwada in kharif season increased at the rate of 10.67% but area declined by -1.90% during the period under study. In Marathwada claims paid (10.43%) and farmers benefitted (8.72%) in kharif season increased during the period under study.

Keywords- NAIS (National Agriculture Insurance Scheme), Area Covered under NAIS, Farmers Benefitted under NAIS

Introduction

Weather variability and uncertainty of crop yields is a basic risk faced by agriculturalists worldwide. However the magnitude and intensity of this is particularly high in India owing to extreme dependence of the farm sector on weather conditions and the poor economic condition of the majority of farmers who have extremely limited means and resources to cope with the disastrous consequences of a crop failure. The impact of crop failure is more severe in the dryland areas, which purely depend on rainfall and climatic conditions.

Floods ravage more than 40 million hectares of land masses and 68% of the total area is vulnerable to periodical droughts [1]. The enormous dependency of crop production on rainfall hardly needs reiteration. According to crop loss data for the period 1985 to 2000 more than 70% of crop loss is a result of drought and about 20% owing to excess rainfall [2]. India has about 47 million hectares of drylands out of 108 million hectares of rain fed area. The dryland farming is practiced in low rainfall region where the irrigation facilities cannot be extended beyond the static level due to existing geographical position. Such farming is mainly practiced in state of Rajasthan Tamil Nadu, Andhra Pradesh, Madhya Pradesh, Maharashtra etc. Natural calamities and adverse seasonal conditions are grossly impacting the level of agriculture productivity in dryland areas, which in turn leads to destabilization of farm incomes, thus affecting the farmer's economic condition adversely. The question arises, is how to minimize such losses.

Agriculture insurance is important mechanism to effectively mitigate the risk of natural perils in farm production and has resulted in the evolution and improvement in the crop insurance programme over the years. Credit for pioneering work on crop insurance in India goes to J.S. Chakravarti, who during early part of twentieth century proposed rain insurance scheme based mainly on the rainfall approach to insure farmers from drought in the then Mysore state [3]. A concrete step for introducing crop insurance at the National level was taken only in October 1965 with crop Insurance Bill and model scheme of crop insurance [4]. The comprehensive crop insurance scheme was started in 1985, which was replaced by NAIS in 1999-2000 [5]. The agricultural Insurance Company formed in December 2002, took over the implementation of NAIS from General Insurance Company in 2003.

Dryland contributes 42% of the total food grain production of the country. These areas produce 75% of pulses and more than 90% of sorghum, millet, groundnut and pulses from arid and semi-arid regions [4]. Coarse cereals, Pulses, Oilseeds and cotton are the principal dryland crops. Overall dryland farming plays a dominant role in agricultural production. The Present paper is an attempt to study the role of NAIS in dryland i.e. Marathwada region of Maharashtra state of India.

Maharashtra stands first in the country in drip irrigation with 60% of the total area under drip irrigation out of the total irrigated farmlands in the state i.e. 15.4% [6]. The state has one of the lowest potential for irrigation in the country, that is, even after building extensive irrigation systems based on canals and dams, only about 25% of all agricultural land can be irrigated. Marathwada region with about

98% of the agriculture is under dryland farming. The major crops grown in this area are drought resistant namely cotton, oilseeds, pulses and coarse grains such as jowar, bajra and millets etc. This region comes under rain shadow zone, with poor irrigation facilities, weather aberration, and lack of natural resources thus having high risk of crop failure and low crop yields which results into drought conditions and downfall of farmer's income.

Objectives

The following are the objective of present study

- To find out total farmers covered & Area covered under NAIS during study period in Maharashtra and Marathwada.
- To find out the total claim paid under NAIS during study period in Maharashtra & Marathwada.

Hypothesis

The following are the hypothesis frame for the study

- The area and farmers covered under NAIS has increased during study period in Maharashtra & Marathwada.
- The claim paid under NAIS has increased during study period in Maharashtra & Marathwada.

Research Methodology

There are many economic problems of dryland farmers; as such they could not afford the new technology for the development of drylands. The dryland farmers suffer from low income and heavy debts. The present study is an attempt to find out the role of NAIS in development Marathwada region of Maharashtra. The research work is based on secondary data. The required secondary data has been collected from various sources like State Government reports on Agriculture, Report of NAIS from District Agriculture office, data collected from Agriculture Insurance Company of India from Mumbai, Economic survey of Maharashtra. The secondary data required for the study is collected for the period 2000-2010. We analysed the performance of NAIS by studying the factors like number of farmers covered, Area covered, Sum insured, premium received, claims paid and benefited farmers for the period 2000-2010. To study the performance we have computed compound annual growth rates of the farmers covered, Area covered, claims paid and benefited farmers by using the semi-log function of the form $Y= a + b^t$.

Performance of the National Agricultural Insurance Scheme (NAIS) in Marathwada

Indian government has implemented various schemes for the development of agriculture and farmers through, NABARD, Indian council for agriculture research, National agricultural development Programs, Debt waiver schemes, National Horticulture Mission, Financial assistance through short term loan, Kisan credit cards, Agricultural Insurance Schemes etc. General Insurance Corporation of India catered the agriculture insurance needs through various schemes. The comprehensive crop insurance scheme started in 1985 was replaced by NAIS in 1999-2000. The agricultural Insurance Company formed in Dec 2002, took over the implementation of NAIS from General Insurance Company in 2003.

National Agricultural Insurance Scheme is implemented in Maharashtra State since 1999-2000, which covered 16 kharif and 10 rabi crops with, an objective to provide insurance coverage and financial support to farmers in event of failure of crops due to natural calamities and restore the credit eligibility in years of disaster. This scheme is very much helpful to small and marginal farmers as it provides subsidy in premium.

The data given below shows the statistics for the period 2000 to 2010 for Rabi and kharif seasons of farmers covered under NAIS in Maharashtra [Table-1] & Marathwada region [Table-2].

Table 1- Performance of NAIS (Maharashtra State) for the period 2000 to 2010 (Rabi and Kharif)

Year	Farmers covered (in lakhs)	Area (in lakhs hect.)	Sum insured (in lakhs)	Premium	Claim Paid (in Lakhs)	Benefited Farmers Actual Number
			Kharif Season			
2000	25.29	28.17	177752	4384.42	14074.6	1058137
2001	26.67	31.19	181409	8026.85	9315.16	551947
2002	17.93	22.75	101282	5205.4	2842.01	228339
2003	17.27	18.52	92090.6	3584.86	10095.4	517802
2004	20.7	19.28	112137	4604.31	12607.5	679374
2005	23.02	18.75	119748	4659.07	2878.39	233485
2006	16.38	12.68	83388.3	3809.96	14057.5	620044
2007	18.91	12.12	85398.6	3152.53	8631.08	259806
2008	34.51	22.33	163590	7415.38	46971.1	1715142
2009	31.02	21.43	243226	10135	37487.2	1328729
2010	21.24	12.54	126033	5003.26	1440.31	82502
Total	252.94	219.76	1486055	59981.1	160400	7275307
CAGR (%)	1.5	-6.05	--	--	2.18	-4.14
			Rabi Season			
2000	4.21	3.79	46398.1	513.09	2669.26	174185
2001	0.86	0.61	2026.59	50.47	212.39	19973
2002	2.48	2.74	8381.78	274.62	1633.53	121434
2003	10.35	11.6	34865	1017.62	19421.7	917105
2004	1.41	1.04	6468.16	199.05	395.23	21948
2005	2.39	2.38	9680.3	212.95	354.33	26413
2006	0.38	0.45	1534.02	29.49	0.59	504
2007	0.92	0.69	2522.86	59.25	0.25	145
2008	0.52	0.52	3090.51	72.65	544.62	10304
2009	0.82	0.96	7497.23	169.79	138.54	1361
2010	0.59	0.61	6434.73	169.95	0	0
Total	26.1	26.59	141861	2907.74	25878.8	1328303
CAGR (%)	-17.27	-15.7	--	--	-43.87	-46.85

CAGR Compound Annual Growth Rate, 2000-2010; Source: NAIS [7].

It is observed from the [Table-1] that in Maharashtra state in Kharif-2000, total farmers covered under NAIS were 25.29 lakhs with the area of 28.17 lakh hectares for the sum assured of ₹177752.25 lakhs and premium paid was of ₹4384.42 lakhs. The actual numbers of farmers benefited were 1058137, i.e. 41.84 percent of the total covered farmers with the claims paid of ₹14074.59 lakhs. From 2002 to 2007, it was observed that the total number of farmer's covered, total area covered, sum assured, premium received, claims paid and total number of actual farmers benefited is less as compared to year 2000.

Later the period 2008 to 2009 registered a rise in total farmers covered, area covered, sum assured, premium received, claims paid, and actual number of farmers benefited. The total numbers of farmers benefited in year 2008 were 49.69% and in year 2009 it was 42.83%. Again in 2010 NAIS performance declined on all the counts, with less number of farmers covered and numbers of farmers benefited were only 3.88% of the total farmers covered.

For Rabi-2000 the total farmers covered were 4.21 lakhs with the area of 3.79 lakh hectares for the sum assured of ₹46398.13 lakhs and premium paid was ₹513.09lakhs.The actual numbers of farmers benefited were 174185, ie. 41.37% of the total farmers with the claims paid of ₹2669.26 lakhs. The number of farmers covered, Area covered, Sum insured, premium received, claims paid and benefited farmers decreased registered increase during 2002 and 2003, but considerably declined during later period.

The study shows that farmers covered under NAIS in Maharashtra in kharif season increased at the rate of 1.50% during the period under study. But the area covered and farmers benefitted declined by -6.05% and 4.14% respectively in Maharashtra during 2000-2010. Farmers, area covered and benefitted farmers under NAIS all declined by -17.27%, -15.70% and -46.85% respectively in rabi season during the period under study.

In Marathwada region, for Kharif-2000, the total numbers of farmers covered were 956909, with the area covered of 965975 lakh hectares, Sum assured of ₹84299.39 lakhs and Premium paid was ₹1840.88 lakhs. The total numbers of farmers benefited were 315925, i.e. 33.01 percent, and the total amounts of claims paid were ₹4974.14 lakhs. In year 2001 the total number numbers of farmers covered were 1122548, with area covered of 1233595 lakh hectares, premium received 3104.88 lakhs, and claims amount paid 6081.30 showed a growth as compared to 2000, while the actual number of farmers benefited were 258602, i.e. 23.03% of the total farmers covered show a decline of 9.98 percentage points as compared to year 2000.

Table 2- Farmers covered under NAIS (Marathwada Region) for the period 2000 to 2010 (Rabi and Kharif)

Year	Farmers covered (in lakhs)	Area (in lakhs hect.)	Sum insured (in lakhs)	Premium	Claim Paid (in Lakhs)	Benefited Farmers Actual Number
			Kharif Season			
2000	956909	965975	84299.4	1840.88	4974.14	315925
2001	1122548	1233595	82936.5	3104.88	6081.3	258602
2002	376985	539246	19244.3	1016.78	125.49	9677
2003	890832	738173	40088.6	1503.25	6863.02	369512
2004	1264591	968273	55233.6	2177.83	7941.43	541447
2005	1571336	1018338	64927.7	2507.36	681.5	84386
2006	840673	48225	34084.7	1150.81	12477.5	527755
2007	1473455	785585	59214.2	2136.33	8554.3	256570
2008	2427768	1343912	104611	4300.2	30429.8	1209118
2009	2124501	1151057	145161	5677.06	22780	968196
2010	1553209	687575	71155.4	3099.23	476.42	44876
Total	1.5E+07	9479954	760956	28514.6	101385	4586064
CAGR (%)	10.67	-1.9	--	--	10.43	8.77
			Rabi Season			
2000	85092	94475	15027.7	152.87	317.05	16193
2001	45812	30382	3317.35	9.25	69.93	4725
2002	7532	6458	259.68	6.55	7.53	1441
2003	473569	356585	14828.1	398.85	6844.24	422135
2004	82839	49843	2855.09	67.42	85.92	13448
2005	148965	128399	5344.52	97.59	42.05	11310
2006	1549	1091	57.32	1.38	0.12	17
2007	6389	4142	220.01	6.34	0.03	10
2008	20732	13894	992.14	180.47	2.04	321
2009	7718	5074	671.54	26.06	4.94	228
2010	7492	4247	657.25	28.15	0	0
Total	887689	694590	44230.8	974.93	7373.85	469828
CAGR (%)	-23.05	-25.99	--	--	-52.66	-49.45

CAGR : Compound Annual Growth Rate, 2000-2010; Source: NAIS [7].

During 2003 to 2009 a growth is observed in number of farmers benefited as compared to year 2000 and 2001. In year 2008, total numbers of farmers benefited were 49.80% and in year 2009 it was 45.57 percent of the total farmers covered, thus year 2008 and 2009 shows the growth in numbers of farmers covered during the decade. The year 2010 shows a sharp decline in total number of farmers as compared to year 2008 and 2009, while the total number of farmers benefited were 2.88% only.

For Rabi crops during 2000, the numbers of farmers covered were 85092, with the area covered of 94475 lakh hectares for sum assured of ₹15027.74 lakhs and premium paid was ₹152.87 lakhs, the actual number of farmers benefited were 16193 i.e. 19.02%, with the claims amount of ₹317.05 lakhs. The period 2001 and 2002 showed a decline on all the counts, whereas in year 2003 the numbers of farmers benefited were 422135, i.e. 89.13% of the total farmers. The year 2004 to 2010 registered a decline on all counts as compared to the earlier periods. Number of farmers benefitted were just 1.54% in 2009 and 2.95% of the total farmers covered.

It can be seen from the [Table-2] that farmers covered under NAIS in Marathwada in kharif season increased at the rate of 10.67% during the period under study. But the area covered under NAIS declined by -1.90% in Marathwada during 2000-2010. In Marathwada claims paid (10.43%) and farmers benefitted (8.72%) under kharif season increased during the period under study. Farmers, area covered and benefitted farmers under NAIS all declined by -17.27%, -15.70% and -46.85% respectively in rabi season during the period under study.

Conclusions

Dryland farming purely depends on rainfall and Climatic conditions. The impact of financial losses due to vagaries of nature and climatic condition can be minimized by participating in the National Agriculture insurance scheme. This scheme can be very helpful for dryland farmers to restore the credit eligibility in years of disaster and for developing the farms by implementing new techniques for better yield.

The study shows that farmers covered under NAIS in Maharashtra in kharif season increased at the rate of 1.50% during the period under study. But the area covered and farmers benefitted declined by -6.05% and 4.14% respectively in Maharashtra during 2000-2010. It is observed that that farmers covered under NAIS in Marathwada in kharif season increased at the rate of 10.67% but area declined by -1.90% during the period under study. In Marathwada claims paid (10.43%) and farmers benefitted (8.72%) under kharif season increased during the period under study. All these factors are drastically reduced for the rabi season at Maharashtra as well as Marathwada level during the period under study.

References

[1] Sarkar S. and Sharma A. (2006) *Economic and Political Weekly,* XII(35), 2-28.

[2] Parchure R.V. (2002) *Bonds and Options, Capital Market Solutions for Crop Insurance Problems*, National Insurance Academy, Pune, India.

[3] Rao K.N. (2008) *Crop Insurance in India*, Agriculture Insurance Company of India Limited (AIC), India.

[4] Siddharth Sinha (2006) *Economic and Political Weekly,* XXXIX (25), 2605-2612.

[5] Bhende M.J. (2008) *Glimpses of Indian Agriculture - Macro and Micro Aspects*, Academic Foundation, New Delhi, India.

[6] Govt. of Maharashtra (2012) *Economic Survey of Maharashtra-* 2011-12, 83, Mumbai, India.

[7] National Insurance Company of India (2011) *Districtwise Coverage under NAIS from Rabi* 1999-2000 *to Rabi* 2010-11, Mumbai Regional Office, Mumbai, India.

[8] Central Research Institute for Dryland Agriculture (2011) *CRIDA Vision-* 2030, Hyderabad, India.

6

IMPACTS OF DRYING AIR TEMPERATURE, BED DEPTH AND AIR FLOW RATE ON WALNUT DRYING RATE IN AN INDIRECT SOLAR DRYER

GHATREHSAMANI S.H. AND ZOMORODIAN A.

Agricultural Engineering Department, Shiraz University, Shiraz, Iran.
*Corresponding Author: Email- Shirin63_saman@yahoo.com

Abstract- Walnut is one of the world's top 20 agricultural products and is consumed mostly as dried [3]. This product must be dried with great care to maintain its quality. Walnut products are traditionally spread on open space for a long time to reduce the kernel moisture content to a safe level. Due to direct sun radiation and some environmental impacts such as long time exposure the quality of the products may be degraded. Application of renewable energy such as solar energy may be an appropriate substitution for traditional methods.
In this research a new pilot size walnut indirect solar batch type dryer was designed and fabricated. The effect of crucial factors affecting drying time namely: drying air temperature (T1:37° C; T2:39° C and T3:41°C), drying air flow rate (F1: 0.065 m^3/s, F2: 0.075 m^3/s, F3: 0.09 m^3/s) and bed depth (D1: 2 layers, D2: 4 layers and D3: 6 layers) were evaluated. The experimental results showed that, the effect of above factors on walnut moisture loss are significant. With increasing drying air temperature, drying air flow rate and number of layers, the amount of average moisture loss decreased from 37% (d.b) to 9% (d.b) at 3 hrs. Performance of dryer was determined at temperature 41°C, air flow 0.09m^3/s, 2 layers depth.
Key words- Walnut, Solar drying, Performance, air flow rate, drying air temperature, bulk depth

Introduction

Walnut is one of the important nut fruits in the world with essential component of high nutrition values [1]. This fruit sets on the world's top 20 agricultural products and is consumed mostly as dried [3]. Walnut is mainly produced in China, USA, Turkey and Iran. Iran provides more than 10 % of world production of 300,000 ton per year [3].
Fresh walnut contains a moisture content of about 35-40 %(dry basis) which should be reduced to 4-8 % by drying in order to maintained the oil content. The traditional drying methods usually followed in Iran are sun drying in which the quality of the product is degraded [10]. Sun drying is time consuming and continues several days (about 7 days) and if the weather is not suitable the nut produced would be contaminated with fungi resulted in a grey rancid product. Furthermore, sun drying needs more space, labor intensive and dust attack can be damaged the quality of production as well as quantity. Moreover, microorganisms can increase the acid content, cause rancidity and reduce the amount of extractable nut resulting in poor-quality products [2]. On the other side it should be noted that in drying process walnut cannot tolerate drying air temperature more than maximum 43°C (110°F). For agricultural products, solar drying systems have been adopted successfully and several experimental and theoretical studies have been produced [4,5,9].The main objective of the present

work was to study the drying characteristics of walnut produced in the solar dryer.

Material and Methods

The experiments were conducted in agricultural engineering department of Shiraz University, Shiraz, Iran during August to September 2010. Walnut was purchased from local market. In order to protect walnut fruit from direct solar radiation beam, indirect solar dryer was used in this research. The main parts of the experimental rig were: "Fig. (1)".

a

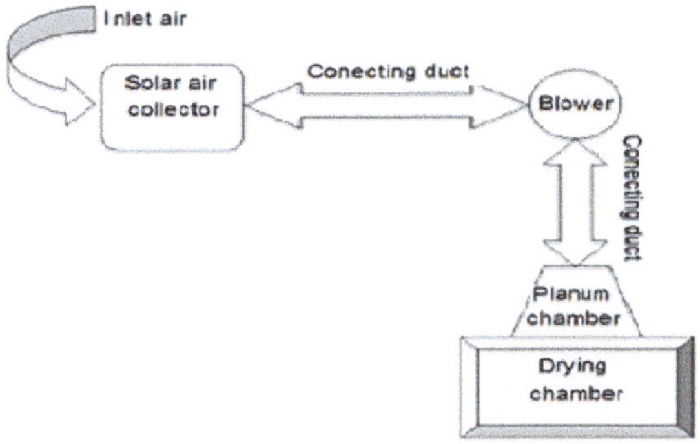

b

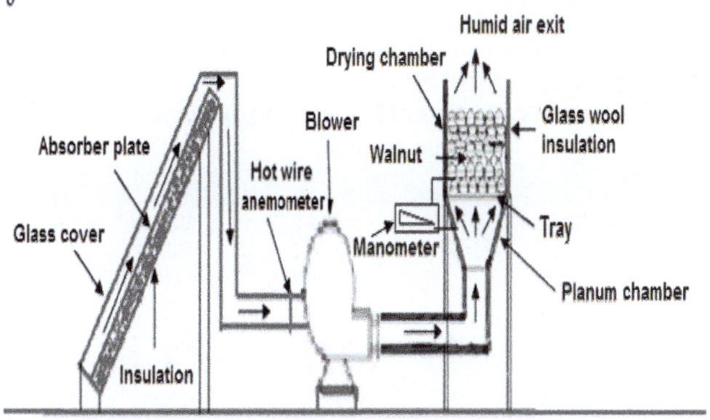

Fig. 1- Schematic layout of experimental rig (a), schematic view of the solar drier used for walnut drying (b).

Indirect Solar Dryer

The solar drier consisted of a solar air collector with $2m^2$ (2m×1m) surface area connected to an air circulating duct. In order to provide the hot drying air a single glazed solar air heater (absorber plate; 1×2m, Al plate, coated with matt black color) was employed. A centrifugal fan (1400 RPM, 50 HZ) was inserted in the rig for circulating the drying air. The dryer body was made of 5mm thickness wood with dimensions of width, depth and height of 0.3m×0.5m×0.4m, respectively.

In order to monitor the drying air temperature throughout the drying system, four calibrated thermocouples (Pt 100) with ±0.25°C accuracy were inserted at different locations of drying system. A power metering device (±0.1% accuracy) was used in the system to measure the consumption of blower electrical energy. A monometer was coupled across the drying bin to record the static pressure drop. The velocity of drying air was recorded well apart the solar air collector exit using hot wire anemometer having ±0.1 m/s accuracy. Solar radiation was measured applying a solar intensity meter (Casella) having an accuracy of about $\pm 1mv/m^2$.

Experimental Procedure

The dryer bin was loaded by fresh walnut fruit (2, 4 and 6 layers, natural filling). Three levels of drying air flow velocities (8m/s, 9.5m/s and 11m/s) and three dying air temperatures (37°C, 39°C and 41°C) were adopted for the experiments. The drying air temperature can be adjusted by covering partly the collector surface area according to ambient condition. The moisture content of walnut was determined using Eq. (1) [6]. Variations of moisture content of the samples were determined 0.5, 1.5 and 3h after commencing the experiments. The experiments were only conducted during day times (9 A.M- 4P.M) in three replications. In this research the effect of each factor (bed depth, drying air temperature and flow rate) on drying time was investigated using SPSS.

Data Analysis

In order to measure the initial moisture of the product, ten randomly selected kernels were peeled and kept in a convective electrical oven (Heraeus T5050) at 105±1°C for 24h. The initial (W_i) and final (W_f) weight of the samples was record using an electronic balance (Sartorius, 2354, ±0.01gr). The sample moisture contents (M_{wb}) were calculated using Eq. (1):

$$Mwb = \frac{(Wi - Wf)}{Wi} \times 100 \qquad (1)$$

The main goal of this work was defined to be the moisture ratio investigation of the walnut fruit in a batch type solar dryer. In each experiment (2,4 and 6 layers), fifty kernels were numbered and picked up randomly for measuring the moisture ratio investigation of the drying product at different operating conditions (3 levels of drying air temperature and flow rate).

Results and Discussions

Effect of Drying Air Temperature on Moisture Ratio

Table 1 shows the average moisture ratio changes of the walnuts during a period of drying experiments (3 hrs). These results were presented graphically in "Fig. (2)". the drying process was performed totally on falling rate period which is usually stated for agricultural materials [2,12]. The experimental data were applied to find the constants in drying exponential equations, table 2. The results illustrated a good agreement with what other researcher reported in the literature such as: peanuts [7] and hazelnuts [8].

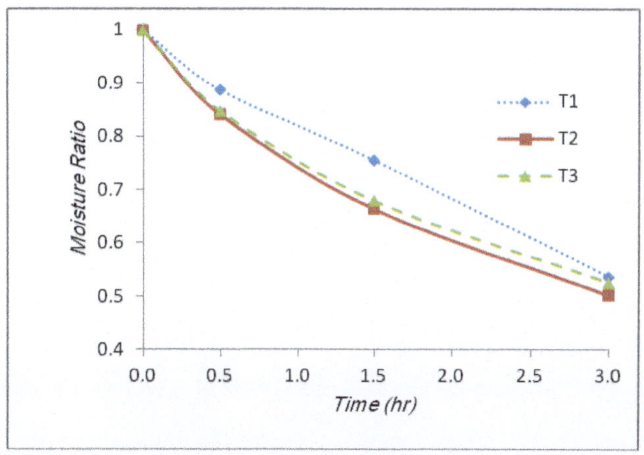

Fig. 2- Graphical representation of moisture ratio vs drying time at different drying air temperatures

Table 1- The average moisture ratio against drying time at different air temperatures

Time (hr.)	T_1	T_2	T_3
0.5	0.887	0.848	0.841
1.5	0.755	0.68	0.665
3	0.537	0.526	0.503

Table 2- Moisture ratio prediction equations at different drying air temperatures

Drying air temperature	Predicted equation
T1	$M.R = 0.9989e^{-0.203t}$
T2	$M.R = 0.9636e^{-0.209t}$
T3	$M.R = 0.9635e^{-0.223t}$

Effect of Depth on Moisture Ratio

The effect of bed depth on moisture ratio during the drying experiments was presented on table 3. As the drying process precedes the difference between the moisture ratios of the 2-kernel depth and the other depths showed a major variations. This effect may be attributed to the rate of moisture removal from the thinner layer of the walnut compared with thicker ones. The variations of moisture ratio for different drying depth were shown in "Fig. (3)". Based on experimental values, the moisture ratio prediction equations at different drying bed depth were displayed in table 4. The reports of other researchers such as Moss, 1989 on peanut and Ozdemir, 2004 on hazelnut confirm the result of present research.

Table 3- The average moisture ratio against drying time in different bed depths

Time (hr.)	D1	D2	D3
0.5	0.859	0.863	0.9
1.5	0.7	0.69	0.788
3	0.522	0.537	0.649

Table 4- Variation functions of moisture ratio of walnut against drying time at different bed depths

Bed depth	Predicted equation
D1	$M.R = 0.9818e^{-0.14t}$
D2	$M.R = 0.9704e^{-0.203t}$
D3	$M.R = 0.9704e^{-0.203t}$

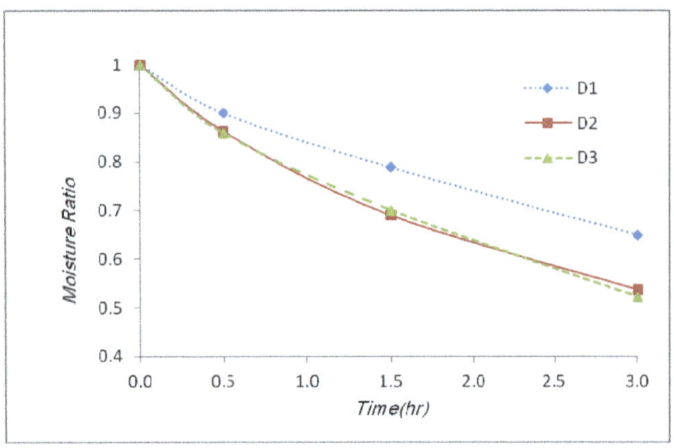

Fig. 3- Graphical representation of moisture ratio vs drying time at different bed depths

Effect of Air Flow Rate on Moisture Ratio

Although variations in air flow rates on drying rate is reported to be non-significant for many small drying kernels [11], the statistical analysis shows a significant effect of air flow rate variations on drying rate of walnut kernel, table 7. This result was illustrated schematically in "Fig. (4)". Since the sphericity of walnut kernel is high and the bulk bed porosity is low, therefore the drying air has got enough time to capture the moisture from the product. The prediction equation for moisture ratio vs time at different air flow rate depicted in table 6.

Table 5- The average moisture ratio against drying time in different air flow

Air flow rate Time (hr.)	F_1	F_2	F_3
0.5	0.88	0.84	0.812
1.5	0.76	0.656	0.609
3	0.57	0.531	0.511

Table 6- Variation functions of moisture ratio of walnut against drying time at different air flow rate

Air flow rate	Predicted equation
F1	$M.R = 0.9867e^{-0.182t}$
F2	$M.R = 0.9517e^{-0.206t}$
F3	$M.R = 0.9307e^{-0.218t}$

Table 7- The interaction effect of three factors on variation of moisture content

source	Sum of square	df	Mean square	F
Temperature (T)	11280.241	2	6540.12	402866**
Air flow rate (F)	1344.435	2	672.217	48016**
Bed depth (D)	244.85	2	122.425	8744.6**
T*F	64.425	4	16.106	1150**
T*D	33.919	4	8.48	605.7**
F*D	12.807	4	3.302	228.7**
T*D*F	64.398	8	8.05	575**
Error	0.74	54	0.014	
Total	13045.815	80		

1% significant

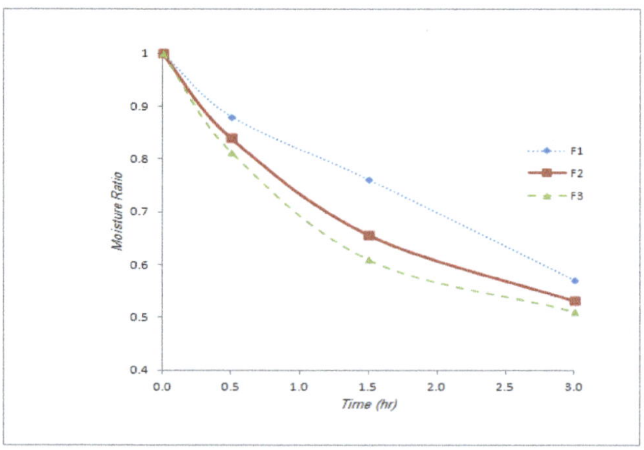

Fig. 4- Graphical representation of moisture ratio vs drying time at different air flow rates

Data analysis was performed using SPSS software (Duncan test in p<0.05). The results were displayed in table 7. The experimental results showed that, the effect of temperature on walnut moisture loss is very tangible in different layers. The amount of average moisture loss reductions were 11.03% for 2 layers, 10.67% for 4 layers and for 6 layers 10.55% of the initial drying bulk weight. In a given depths, the amount of moisture loss boosted with an increase in drying air temperature during the course of experiments (10.7% at T1, 13.42% at T2 and 15.26% at T3) and finally the different air flow rate caused reduction of moisture content of walnut reported to be 11.91% at F1, 10.84% at F2 and 10.04% at F3.

Conclusion

An indirect solar dryer was designed, fabricated and evaluated for drying walnut. It can be concluded that an indirect solar dryer is more suitable for producing high quality walnut. Maximum performance of dryer was occurred at the maximum air flow and drying air temperature and 6layer bed depth.

Nomenclature

M_{wb}	moisture content (%)
W_f	final sample mass (kg)
W_i	initial sample mass(kg)
F	air flow rate (m^3/s)
D	bed depth (layer)
T	drying air temperature (°C)

References

[1] Altunats E. and Erkolm (2009) *International Journal of Food Engineering*, 13, 1-9.
[2] Amirghasemi T. (2007) *Plant and post harvesting for Walnut*, Iran.
[3] FAO (2005) *Http://faostat.fao.org*, Accessed to web.
[4] Ivanova D. and Andonova K. (2001) *Energy Convection and Management. 42, 975-983.*
[5] Kadam D.M., Samuel D.V.K. (2006) *Biosystems Engineering*. 43, 189-198.
[6] Mohsenin N.N. (1996) *Physical properties of plant and animal materials*, Gordon, 79-127.
[7] Moss J.R., and Otten L. (1989) *Canadian Institute of Food Science and Technology Journal*, 22, 34-39.
[8] Ozdemir F., Akinci I. (2004) *Journal of Food Engineering*. 63, 341–347.
[9] Shanmugam V. and Natarajan E. (2006) *Renewable Energy*, 31, 1239-1251.
[10]Thiruchelvam T., Nimal D.A.D. and Upali S. (2007) *Journal of Food Engineering*, 78, 1446-1451.
[11]Zomorodian A. and Moradi M. (2010) *Journal of Agricultural Science and* Technology, 12, 401-408.
[12]Zomorodian A. and Allame R. (2002) *Journal of Agricultural Sciences and Natural Resources*, 6(4).

7

A FRAMEWORK FOR AGRICULTURAL ADAPTATION TO CLIMATE CHANGE IN SOUTHERN NIGERIA

NICHOLAS OZOR[1*], MADUKWE M.C.[1], ENETE A.A.[2], AMAECHINA E.C.[2], ONOKALA P.[3], EBOH E.C.[4], UJAH O.[4], AND GARFORTH C.J.[5]

[1]Department of Agricultural Extension, University of Nigeria, Nsukka.
[2]Department of Agricultural Economics, University of Nigeria, Nsukka.
[3]Department of Geography, University of Nigeria, Nsukka.
[4]African Institute for Applied Economics, Enugu.
[5]International and Rural Development Department, University of Reading, UK.
*Corresponding Author: Email- sunny_ozor@yahoo.com

Abstract- The agricultural sector which contributes between 20-50% of gross domestic product in Africa and employs about 60% of the population is greatly affected by climate change impacts. Agricultural productivity and food prices are expected to rise due to this impact thereby worsening the food insecurity and poor nutritional health conditions in the continent. Incidentally, the capacity in the continent to adapt is very low. Addressing these challenges will therefore require a holistic and integrated adaptation framework hence this study. A total of 360 respondents selected through a multi-stage random sampling technique participated in the study that took place in Southern Nigeria from 2008-2011. Results showed that majority of respondents (84%) were aware that some climate change characteristics such as uncertainties at the onset of farming season, extreme weather events including flooding and droughts, pests, diseases, weed infestation, and land degradation have all been on the increase. The most significant effects of climate change that manifested in the area were declining soil fertility and weed infestation. Some of the adaptation strategies adopted by farmers include increased weeding, changing the timing of farm operations, and processing of crops to reduce post-harvest losses. Although majority of respondents were aware of government policies aimed at protecting the environment, most of them agreed that these policies were not being effectively implemented. A mutually inclusive framework comprising of both indigenous and modern techniques, processes, practices and technologies was then developed from the study in order to guide farmers in adapting to climate change effects/impacts.
Key words- Framework, Agricultural adaptation, Mitigation, Climate change, Effects, Impacts, Southern Nigeria.

Introduction

There is wide acceptance by stakeholders that climate change will affect all nations especially the developing ones like those in Africa. The Intergovernmental Panel on Climate Change (IPCC) already predicted that Africa is the most vulnerable to climate change impacts [1]. However, it is saddening to note that Africa contributes less than 4% of the greenhouse gas emissions that contribute to climate change and yet is most vulnerable to the impacts of the change. The adverse effects on poor people and poor countries in the continent are particularly severe because these people and countries depend more directly on natural resources and are less able to adapt to climate variations and extreme weather. Poor people in developing countries especially in sub Saharan Africa are affected by environmental change because they are much more exposed to existential risks such as disease, hunger, low income and most importantly poor adaptive capacities [2].

Climate change will affect all economic sectors of Africa and will therefore present unprecedented challenges for the continent, particularly in terms of meeting its sustainable development objectives, including the millennium development goals (MDGs). Indeed, climate change is already eroding decades of hard-won national and international development gains, thus the need for concerted and coherent efforts in urgently tackling the development challenge. This challenge has further exacerbated poverty in Africa as rising temperature and sea level results in undue flooding, droughts, and salinization in low lying areas. Reconnaissance surveys and pan-African stakeholder consultations carried out by the African Technology Policy Studies Network (ATPS) and its partners, including the United Nations Environment Program (UNEP) in 2007, 2008 and 2009 show that the sectors that are most likely to be affected by climate variability in Africa are agriculture, water and biodiversity. These sectors were prioritized as most critical as they will have direct impacts on rural livelihoods.

Incidentally, agriculture is the mainstay of many countries' economies in Africa contributing between 20-50% of the gross domestic product (GDP), employing over 70% of the population, and mostly dependent on weather and climate [3, 4]. This means that any change in weather and climate will have catastrophic consequences on agricultural production, economy, employment and overall livelihoods of the most affected groups in the continent including farmers, pastoralists, foresters, fisher folks and hunters. The consultation exercise led by the ATPS also found that other factors such as increased intensity, frequencies, prevalence and uncertain patterns of land degradation, pests and diseases, droughts, flooding events and most importantly low technical and innovation capacities to adapt to these changes are the most significant indicators of farmers' inability to implement, create, alter, and implement multiple adaptive measures. In other words, the individual, community and institutional capacity to adapt to the shifting climate change impacts while continuing to function effectively i.e. their capacities, are very low. Even where this resilience exists, they are not well documented especially in Africa so as to create the opportunity for up scaling and replication in other vulnerable areas.

It is then less difficult to believe that agricultural productivity under the prevailing climate change situation in most developing countries will be very low. Consequently, the low crop yield will lead to unavoidable shocks to the already fragile economies in African countries. Food prices are expected to rise, worsening the food insecurity and poor nutritional health conditions in the continent. The impacts of climate change on food production, prices and food security depend on regional climate change, biological effects of increasing atmospheric carbon dioxide, changes in floods, droughts and other extreme events, existing agricultural systems, adaptive capacity, changes in population, economic growth and technological innovation [5]. A clear case example is the drought and food crisis situation that has been ravaging the horn of Africa (particularly Somalia, Northern Kenya, Ethiopia, Djibouti and Eritrea) since the second quarter of 2011 which has claimed many human lives, led to the death of millions of animals and livestock and has predisposed millions of people to health and nutritional challenges.

Article 2 of the United Nations Framework Convention on Climate Change (UNFCCC) is aimed towards 'ensuring that food production is not threatened'. In assessing the climate change impacts on food production, the adaptive capacity of agricultural systems has to be taken into consideration [6]. This capacity differs substantially between regions. Unfortunately, the regions affected most are the ones with the least adaptive capacity – i.e. the developing countries [7].

Addressing Africa's climate change challenges will therefore require a holistic and integrated adaptation framework [4]. While adaptation is the process of responding or adjusting to actual and potential impacts of changing climate conditions in ways that moderate harm or take advantage of any positive opportunities that it may afford, it remains the most popular option to manage the impacts of climate change on agriculture in the world today. It also includes policies and measures to reduce exposure to climate variability and extremes and the strengthening of adaptive capacity. It requires changes in knowledge, attitudes, resilience capacities, and skills of the people. However, neither adaptation nor mitigation actions alone can prevent significant climate change impacts; taken together, they can significantly reduce food security risks. While mitigation is necessary to reduce the rate and magnitude of climate change, adaptation is essential to reduce the damages from climate change that cannot be avoided [8].

The pertinent questions to ask therefore are: Are farmers aware of climate change? What are the impacts of climate change on agriculture in the study area? What are the human activities in the area that exacerbate climate change incidence? Are farmers in rural areas capable of adapting to the impacts of climate change? If yes, what are the strategies they have adopted in coping with the incidences of the change? What socioeconomic factors influence farmers' adaptation capacity? Which adaptation framework can be propagated to increase adaptation capacity of farmers in the area? This paper focuses on providing answers to these questions.

Purpose and Objectives

The overall purpose of the study was to develop a framework for agricultural adaptation to climate change in Southern Nigeria. Specifically, the study was meant to:

1. ascertain the awareness level of farmers to climate change;
2. examine the effects/impacts of climate change on agriculture in the area;
3. identify the human causal factors to climate change in the area;
4. identify the strategies for agricultural adaptation to climate change and determine the socioeconomic factors influencing farmers' adaptation to climate change;
5. identify existing government policies and programmes on climate change adaptation; and
6. develop a framework for agricultural adaptation to climate change.

Methodology

The Study Area

The study area is Southern Nigeria (Fig. 1) comprising of three geopolitical zones namely; South East, South West and South-South. Its climate is characterized by strong latitudinal zones, becoming progressively drier as one moves north from the coast. Rainfall is the key climatic variable, and there is a marked alternation of wet and dry seasons in most areas. Two air masses control rainfall-moist northward-moving maritime air coming from the Atlantic Ocean and dry continental air coming south from the African landmass.

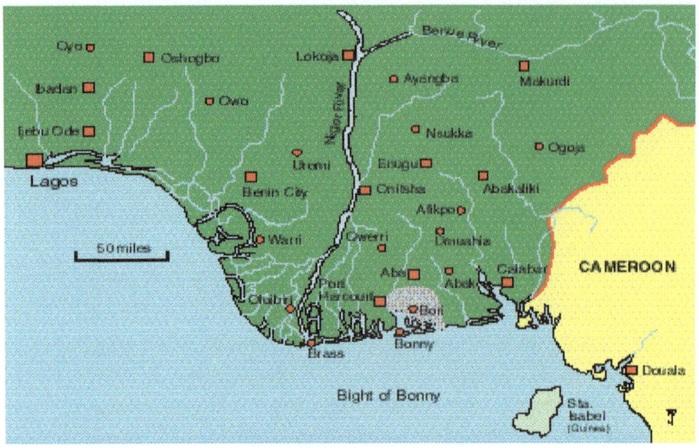

Fig. 1- Map of Southern Nigeria

The rainy season usually begins in February or March as moist Atlantic air, known as the southwest monsoon, invades the country. The beginning of the rains is usually marked by the incidence of high winds and heavy but scattered squalls. By April or early May in most years, the rainy season is under way throughout most of the area. The usual peak of the rainy season occurs through most of southern Nigeria in July with a dip in precipitation during the month of August. Although rarely completely dry, this August dip in rainfall, which is especially marked in the southwest, can be useful agriculturally, because it allows a brief dry period for grain harvesting.

From September through November, the northeast trade winds generally bring a season of clear skies, moderate temperatures, and lower humidity for most of the country. From December through February, however, the northeast trade winds blow strongly and often bring with them a load of fine dust from the Sahara. These dust-laden winds, known locally as the harmattan, often appear as a dense fog and cover everything with a layer of fine particles.

The greatest total precipitation is generally in the south-south; along the coast around Bonny (south of Port Harcourt) and east of Calabar, where the mean annual rainfall is more than 4,000 millimetres. Most of the rest of the south-south and southeast receives between 2,000 and 3,000 millimetres of rain per year, and the southwest (lying farther north) receives lower total rainfall, generally between 1,250 and 2,500 millimetres per year. Mean annual precipitation at Lagos is about 1,900 millimetres; at Ibadan, only about 140 kilometres north of Lagos, mean annual rainfall drops to around 1,250 millimetres. Moving north from Ibadan, mean annual rainfall in the west is in the range of 1,200 to 1,300 millimetres.

Temperatures throughout Nigeria are generally high; diurnal variations are more pronounced than seasonal ones. Highest temperatures occur during the dry season; rains moderate afternoon highs during the wet season.

The economy of Nigeria historically was based on agriculture, and about 70% of the workforce is still engaged in farming (largely of a subsistence type). The chief crops are cocoa, peanuts, palm oil, corn, rice, sorghum, millet, soybeans, cassava, yams and rubber. In addition, cattle, sheep, goats and pigs are raised.

The distribution of vegetation in Southern Nigeria is dependent mainly on the climate, which becomes increasingly drier further inland from the coast. Climatic zones, therefore, run roughly parallel to the coast, widening or narrowing as geographical features alter the steepness of the climatic gradient. This climatic zonation has resulted in a vegetation zonation, comprising the rain forest zone, the mixed deciduous forest zone, and the parkland zone. The first two are climax systems, but the parkland zone is probably caused by anthropogenic conversion of forest and is maintained by annual bush fires. The natural vegetation of the parkland zone would probably be mixed deciduous forest. Typical mean annual rainfall varies from 2,000 to 2,500 millimeters (mm) in the rain forest zone near the coast to 1,500 to 2,000 mm in the mixed deciduous forest zone.

Sampling Procedure and Data Collection

Multistage random sampling technique was employed in the selection of respondents for the study, which covered the three geopolitical zones of southern Nigeria and was conducted in 2009/2010. In each zone, two states were randomly selected to make a total of six states. These were Abia and Enugu in Southeast, Cross River and Delta in south-south, and Ogun and Ondo in southwest. In each state, two agricultural zones were randomly selected to make a total of twelve agricultural zones. These were Umuahia and Aba in Abia State, Enugu and Nsukka in Enugu State, Ikom and Calabar in Cross River State, Delta Central and Delta North in Delta State, Rainforest and Savanna in Ogun State, Ondo central and Ondo North in Ondo State. With the assistance of the respective state extension service departments, one farming community was randomly selected from each agricultural zone, to make a total of twelve communities. These were Oboro and Azumini in Abia State, Umulumgbe and Edem in Enugu State, Ugep and Nkpatum in Cross River State, Okpe and Okoamako in Delta State, Imala and ijebu-ode in Ogun State and Emureile and Adofure in Ondo State. In each community, also with the assistance of the local extension personnel, a list of farm households was compiled and then thirty farmers randomly selected, making a total of three hundred and sixty (360) farmers for the study.

Data were collected using structured interview schedule and questionnaire. The data collected included status of awareness of climate change, effects/impacts of climate change on agriculture, farmers' adaptation strategies and factors influencing their adaptation activities.

Data Analysis

Data from the study were analysed using both descriptive and inferential statistics. Objective one was analysed using percentage and mean scores while objectives two and three were analysed using mean scores. Objective four was analysed using mean and probit regression. Objective five was analyzed using percentage scores while objective six was described using key results from the study

Results and Discussions

Awareness of Climate Change

Results on farmers' awareness to climate change show that, 84% of them were aware of it. This suggests a high level of awareness on the subject matter in the area and can be attributed to extensive awareness creation made through the print and electronic media and through other social and religious networks. The respondents' awareness of climate change was further demonstrated by their response to the question on how they would describe climate change to a friend. Some of the descriptions from the farmers include – *prolonged bad weather, change of weather conditions, situation of volatile weather, short rainfall duration and prolonged dry season, thunderstorm and heavy rains, too much rain and too much sun, unpredictable change in the pattern of rainfall, unpredictable start and end of rains, unstable weather, and variation in rainfall pattern and sunshine intensity*. The respondents maintained that climate change is said to occur when such conditions as described above persists for a very long period usually in years. The awareness of climate problems and the potential benefits of taking action is an important determinant of adoption of agricultural technologies [9]. It was earlier found out that farmer awareness of change in climate attributes (temperature and precipitation) is important to adaptation decision making [10]. For example, it was earlier reported that farmers' awareness and perceptions of soil erosion problem as a result of changes in climate, positively and significantly affect their decisions to adopt soil conservation measures [11,12].

Similarly, on the question of whether climate change will affect agriculture, the respondents overwhelmingly (96%) said yes. The national government in Nigeria already has agencies charged with managing environmental issues such as climate change; for example, the National Emergency Management Agency (NEMA). Such agencies mostly sensitize the people through the radio, newspapers and television in addition to on the ground contacts with vulnerable people. These efforts may explain why there was a high level of awareness of climate change by the respondents.

Regarding changes in climate change variables, results showed that the uncertainties in the onset of farming season have all been increasing in the area (Table 1). These uncertainties include; early rains that are not sustained (77%), crops planted become smothered by heat waves (69%), crop planting and replanting (67%) and shifts in the start or end of rains (78%). This was also the trend (increasing incidences) for extreme weather events such as thunderstorms (51%), heavy winds (54%), floods (62%), drought (61%), heat waves (68%) and heavy rainfall (74%). Other trends that showed increasing incidences were pests (67%), diseases (59%) and weeds (73%). The trends for land degradation properties showed a higher proportion of increases in sheet erosion (42%), rill erosion (44%), gully erosion (46%), wind erosion (42%), and declining fertility (82%). It was observed that the effects of extreme weather events and uncertainties in the onset of the rainy season on agriculture are particularly more pronounced in the developing world [13]. The elements of climatic change that affect agricultural productivity includes prolonged drought, thunderstorms, flooding of crops fields, erosion of fertile soils, landslides and the falling of tender crop such as maize by wind [14].

The respondents were also asked to indicate the extent to which they think that climate change was responsible for the changes in the variables stated above. Their mean responses showed that all the changes discussed above regarding uncertainties in the onset of the farming season, extreme weather events, pests, diseases and weeds, and land degradation variables, have all to do with climate change (Table 1). The IPCC [1] had already reported that there have been noticeable impacts of climate change on plant production, insect, disease and weed dynamics. Moreover, rising atmospheric CO_2 concentrations, higher temperatures, changes in annual and seasonal precipitation patterns and the frequencies of extreme events are the usual features of climate change phenomenon [15]. It was also observed that seasonal changes in rainfall and temperature, which are features of climate change, could impact agro-climatic conditions, altering growing seasons, planting and harvesting calendars, water availability, pest, weed and disease populations [16].

Table 1- Data on direction of change of climate change phenomena and extent of contribution of phenomena to climate change (n=360)

S. No.	Phenomena you think have been getting worse over the past ten years	Increasing (IC)	Decreasing (DC)	No change (NC)	Mean extent of contribution of climate change to the phenomenon
	Uncertainties in the onset of farming season				
(i)	Early rains that are not sustained	77	18	50	4.2
(ii)	Crops planted become smothered by heat waves	69	21	9	3.9
(iii)	Crop planting and replanting	67	20	13	3.7
(iv)	Shifts in the start or end of rains	78	16	6	4
	Extreme weather events				
(i)	Thunderstorms	51	32	17	3.5
(ii)	Heavy winds	54	29	17	3.6
(iii)	Floods	62	18	20	3.5
(iv)	Drought	61	17	22	3.6
(v)	Heat waves	68	6.9	25.1	3.8
(vi)	Heavy rainfall	74	15	11	4
Pests		67	14	19	3.8
Diseases		59	18	23	3.7
Weeds		73	10	17	4
	Land degradation				
(i)	Sheet erosion	42	29	29	3.3
(ii)	Rill erosion	44	29	27	3.4
(iii)	Gully erosion	46	19	35	3.3
(iv)	Wind erosion	42	24	34	3.3
(v)	Declining fertility	82	11	7	4.1

Cut off point= 50%, Cut off= 3.0

Effects/Impacts of Climate Change on Agriculture

Table 2 shows the perceived effects/impacts of climate change on agriculture in the study area ascertained using a five point Likert-type scale. Results showed that all the effects/impacts observed have worsened in the past ten years as reported by farmers. The most significant effect/impact of climate change on agriculture was declining soil fertility ($\bar{x}$=4.15) followed by increased weed infestation ($\bar{x}$=4.10), increased heat waves ($\bar{x}$=4.08), increased drought events ($\bar{x}$=3.97), declining yield/output from crops ($\bar{x}$=3.96) and increased rainfall intensity ($\bar{x}$=3.93). Rising atmospheric CO_2 concentration, higher temperatures, changes in annual and seasonal precipitation patterns and in the frequency of extreme events are the usual features of climate change phenomenon and these features will affect the volume, quality, quantity, stability of food production and the natural environment in which agriculture takes place [15]. It was earlier found out that heat stress might affect the whole physiological development, maturation and finally reduce the yield of cultivated crops [17].

Table 2- Mean responses on the extent of effects/impacts of climate change on agriculture (n=360)

Effects/Impacts of Climate change on agriculture	Mean	Std. Deviation
Poor soil drainage	3.85	1.09
Increased soil erosion	3.83	1.16
Declining soil fertility	4.15	1.11
Increased weed infestation	4.1	0.98
Increased pest infestation	3.87	1.13
Increased diseases infestation	3.77	1.23
Premature ripening of crops	3.26	1.22
Erratic changes in production cycle	3.72	1.14
Declining yields/outputs from crops	3.96	1.17
Poor storage quality of crops	3.75	1.23
Increased vulnerability of wildlife ecosystem	3.69	1.04
Increased drought events	3.97	1.1
Increased flooding events	3.65	1.3
Increased rainfall intensity	3.93	1.16
Irregular rainfall distribution	3.89	1.17
Reduced water levels in rivers/streams	3.45	1.28
Decreasing fish population in rivers/streams	3.37	1.28
Reduced pasture availability	3.47	1.15
Increased heat waves	4.08	1.01
Reduced soil moisture	3.63	1.15
Increased wind storm	3.81	1.2
Increases in human health related problems	3.76	1.08
Sea level rise	3.3	1.23
drop in the production cycles of livestock	3.49	1.21
Declining vegetation covers	3.62	1.26
Increases in land areas reclaimed (in ha) in the past 10 years	1.75	2.45
Increases in the area of land lost (in ha) in the past 10 years	1.74	2.44

Cut off point=3.0

On the area of land gained in the past ten years, the result of the analysis show that an average of 1.75 ha of land was gained by the respondents. Similarly, about the same area of land was lost by the respondents in the past ten years. Some of the reasons given by the respondents for land gains include: area expansion, inheritance, purchase, renting, water level reduction and flood while those for loss of land include: building, land dispute, erosion, flooding, no rainfall and infertility issues.

Activities of Farmers That Contribute To Climate Change

The respondents were asked to indicate the extent to which they practice some suggested farm related activities that could contribute to climate change on a three point Likert-type scale. Results showed that the most dominant farm activity that could contribute to climate change was bush burning ($\bar{x}$=2.66) (Table 3). Other activities include deforestation ($\bar{x}$=2.42) and the use of chemical fertilizers ($\bar{x}$=2.30). Bush burning is generally the preferred traditional means of clearing farmland for seedbed preparation, which increases the concentration of greenhouse gases and particulate matter in the atmosphere. With incidences of declining soil fertility being one of the most observed effects, farmers tend to increase their rate of use of chemical fertilizers in order to obtain meaningful harvests. This however exacerbates the greenhouse gas concentration in the atmosphere. The International Federation of Organic Agriculture Movement (IFOAM) [18] reported that conventional agricultural activities of farmers contribute to climate change because excessive amounts of nitrogen fertilizer is released as nitrous oxide and mines the earth of the nutrient needed to sustain production. Additionally, rainforest clearing through slash and burn techniques reduce carbon storage and releases huge amounts of carbon dioxide from burning vegetation. Moreover, burning of fossil oils ($\bar{x}$=2.20), continuous cropping ($\bar{x}$=2.16) and urine and other droppings from farm animals ($\bar{x}$=2.02) were also practiced by the farmers, all of which are possible sources of greenhouse gases. The rest of the activities as shown in Table 3 were not practiced by farmers significantly.

Table 3- Mean responses of farmers to the extent to which farmers engaged in practices that contribute to climate change (n=360)

Farmers' practices attributable to climate change	Mean	Std. Deviation	Farmers' perceived extent to which the activities could contribute to climate change
Bush burning	2.65	0.59	3.68
Continuous cropping	2.16	0.81	2.89
Over grazing	1.62	0.8	2.82
Urine and other droppings from farm animals	2.02	0.81	2.48
Swamp rice production	1.6	0.84	2.12
Burning of fossil oils (petrol, gas, kerosene)	2.2	0.82	3.25
Use of Chemical fertilizers	2.29	0.83	2.81
Deforestation	2.41	0.7	3.69
The use of herbicides	1.95	0.83	2.48
The use insecticides/pesticides	1.97	0.84	2.55
Urbanization	1.97	0.95	3.08
Industrialization	1.81	0.93	3.06
Cement production	1.65	0.93	2.65
Gas flaring	1.76	0.97	2.91

Cut off mark=2.0, Cut off mark=3.0

On the extent to which farmers think that each of the above activities contribute to climate change, results showed that farmers ranked the two farm activities mostly practiced by them namely; deforestation ($\bar{x}$=3.69) and bush burning ($\bar{x}$=3.68) as highest contributors to climate change. This suggests some level of awareness by the farmers that some of their activities contribute to climate change. This is particularly important because, awareness of climate problems and the potential benefits of taking action is an important determinant of adoption of agricultural and climate change adaptation technologies [9, 10]. Other activities that contribute to climate change according to the farmers are burning of kerosene and other fossil fuels such as gas and petrol ($\bar{x}$=3.25), urbanization ($\bar{x}$=3.08) and industrialization ($\bar{x}$=3.06). Other activities, as their mean rankings suggest, do not contribute to climate change.

Strategies for Agricultural Adaptation to Climate Change

Table 4 presents the adaptation strategies adopted by farmers in the area in cushioning the effects/impacts of climate change. Results showed that the most intensively used adaptation practice was increased weeding ($\bar{x}$=3.97). Other significant strategies adopted by farmers include; changing the timing of land preparation activities ($\bar{x}$=3.84), multiple cropping ($\bar{x}$=3.79), processing crops to minimize post-harvest losses ($\bar{x}$=3.78), increased use of farm inputs such as manures and seeds ($\bar{x}$=3.78), crop replacement, as earlier ones wither away due to unfavourable weather conditions ($\bar{x}$=3.73), changing the planting dates ($\bar{x}$=3.67), mixed farming ($\bar{x}$=3.58), use of resistant varieties ($\bar{x}$=3.46), changing the timing of harvesting dates ($\bar{x}$=3.37), mulching/use of cover-cropping ($\bar{x}$=3.06) and relay cropping–planting and harvesting in succession ($\bar{x}$=3.04).

Table 4- Mean responses of farmers on the extent to which they apply the adaptation practices (n=360)

Adaptation strategies	Mean	Std. Deviation
Changing the timing of land preparation activities	3.84	1.23
Changing the planting dates	3.67	1.23
Changing the harvesting dates	3.37	1.44
Water storage in ponds	2.24	1.39
Groundwater harvesting	2.11	1.31
Mulching/ use of cover crops	3.06	1.49
Use of wetlands/ river valleys (e.g. Fadama)	2.84	1.45
Irrigation schemes	2.22	1.5
Multiple cropping	3.79	1.32
Mixed farming	3.58	1.39
Relay cropping- planting and harvesting in succession	3.04	1.36
Intercropping- main crops planted with subsidiaries at low densities	3.75	1.28
Contour cropping, across hill slopes	2.19	1.43
Planting deeper or shallower than the usual planting depth	2.21	1.4
Zero or minimum tillage	2.72	1.51
Construction of drainage systems	2.22	1.43
Afforestation - planting of trees	2.5	1.47
Use of resistant varieties	3.46	1.43
Decreasing animal stock	2.63	1.42
Culling of infected animals	2.98	1.38
Processing crops to minimize post-harvest losses	3.78	1.34
Destruction of infected farms	2.6	1.51
Crop replacement	3.73	1.28
Use of weather forecasts	2.25	1.34
Shortened fallow	2.92	1.52
Lengthened fallow	2.83	1.6
Expansion of cultivated land area	2.99	1.46
Increased use farm inputs e.g. manures, seeds	3.78	1.35
Cultivation on marginal lands	2.58	1.58
Change from crop production to livestock production	1.93	1.34
Change from animal production to crop production	1.95	1.37
Agro-forestry practice	2.53	1.61
Change from crop/livestock production to marketing of agricultural products	1.93	1.37
Intensive manure application	3.38	1.38
Increased weeding	3.97	1.27
Use of law enforcements	1.76	1.24
Total change from farming to other occupations	1.84	1.25
Planting Shallower than the usual planting depth	1.92	1.27
Out Migration from Climate risk zones	1.89	1.32
Construction of elevated homesteads and farmsteads	2.1	1.36
Group approach to climate risk management	2.01	1.38
Construction of dams within the farm/household	1.7	1.26

Cut off mark=3.0

One of the most significant effects of climate change on farms is increased weed infestation. To respond to this challenge, farmers noted that they have to weed their farms frequently if they are to obtain any meaningful harvest. In a similar study, farmers reported that they had to weed their farm plots more than three times in order to obtain any significant harvest from the farm [19]. This was against one or two times that they had previously weeded a typical farm plot some two decades ago. Previous studies have also shown that some proven adaptation practices of farmers were through the diversification of crops planted, use of resistant varieties, multiple cropping and increased use of inputs [20]. Moreover, processing crops generally enhances their value and storability. For instance, it was discovered that processing makes cassava roots easier to transport, gives them longer shelf-life, removes the cyanogenic compounds and improves their palatability [21].

Household Socioeconomic Determinants Influencing Farmers' Adaptation to Climate Change

Table 5 presents the results of the determinants of adaptation practices using the probit regression model. The practices (dependent variables) were grouped into: (a) agronomic; for adaptations like adjusting the timing of farm operations in response to weather variations, multiple/intercropping, mulching etc, (b) technological; for those that pertain to the use of some kind of technologies such as groundwater harvesting, irrigation, construction of drainage systems etc, (c) environmental; such as afforestation and destruction of infected farms, (d) institutional; such as use of law enforcement agencies and weather forecasts, (e) economic; such as processing crops to minimize post-harvest losses and changing from farming to other occupations and (f) social practices, for those that pertain to the use of group actions. The analysis was done using the probit regression model essentially because the dependent variable was categorized into 1, if farmer used the adaptation practices in each of the above groups and 0 otherwise. The results of the analysis show that the explanatory powers of the specified variables seem low, but this is not uncommon in cross-sectional analysis. A similar coefficient of determination was earlier reported [22]. The overall goodness of fit as reflected by prob>chi2 was however very good (<0.001) in almost all the specifications. [*See Table 5 on Page 251*]

Age was positively and significantly related with the use of agronomic adaptation practices. Older farmers generally have more experience and are able to take healthier production decisions than younger ones [23]. This is particularly relevant for agronomic practices that have to do with routine farm specific duties. Farming as a major occupation was negatively and significantly related with agronomic adaptation practices. This is surprising and counter-intuitive. The expectation is that those with farming as their major occupation will devote more of their time to the farm and hence acquire deeper knowledge of the farm than those with other major occupations, which should help them to adjust to climate change matters faster. However, the adaption practice being considered need not require all that much experience because it pertains to routine farm duties as noted earlier. Farm size was positive and significantly related with agronomic adaptation practices. This is to be expected as larger farms will generally entail greater farm investments, which should drive greater adaptation practices to ensure reasonable returns. The two variables on awareness of climate change were all positive and significantly related with agronomic adaptations. These are *whether a farmer has had about climate change before and whether he/she thinks climate change will affect agriculture*. The awareness of climate problems and the potential benefits of taking action is an important determinant of adoption of agricultural technologies [9].

On technological adaptation practices, marital status was positive and significantly related with the dependent variable. Technological adaptation will most likely be labour intensive as some of them require some kind of construction. Married farmers may therefore have the necessary family labour, which is also cheap, to engage in such practices. Farming as a major occupation was positively and significantly related with technological adaptation. In contrast with agronomic adaptation, technological adaptation may require a lot of dedication to farm duties, which only full time farmers can provide. Income was also positively and highly significantly related with the dependent variable. Technological adaptation may not only be labour intensive but also material intensive, which only farmers with enough money can provide. A similar study earlier reported that poverty is a major constraint to farmers' climate change adaptation practices [23]. In addition, it was observed that for farmers to apply purchased inputs in the right quantity and adopt innovations, they should be sufficiently empowered financially [24].

For environmental related adaptation practices, household size was positive and significantly related with the dependent variable. As in the case of technological practices, this may also have to do with cheap farm labour, which large households could provide for the farmers. For instance, afforestation, which is a component of this, will require much labour, not only for planting the trees but also for tending them to maturity. Income was also positively and highly significantly related with environmental related adaptation practices. The above explanation for income under technological adaptation also applies for environmental adaption practices. Experience was positively and significantly related with the dependent variable. Environmental adaptation may involve delicate decisions like the timing of destruction of infected farms, which can only be perfected with experience.

Farming as a major occupation was positively and highly significantly related with economic adaptation practices. It is only those whose major source of income is farming that will be highly sensitive to fluctuations in the fortunes of farming and hence will largely bother with economic adaptation practices. This will help ensure a stable source of income for them. Income was also positively and significantly related with economic adaptation. Economic adaptations could be resource intensive, for instance, processing of crops to avoid post-harvest losses. In such cases, the same amount of resources may be needed for processing, as it is for production. It was earlier reported that as much labour is required for processing cassava roots into *gari* is also used in producing the roots themselves [22]. Awareness of climate change was also positive and significantly related with the dependent variable.

Gender (1 if male and 0 otherwise) was positively and significantly related with institutional adaptations. This is to be expected as men generally listen to the radio and television, where weather forecasts are usually broadcast, more than women. Farming as a major occupation was positively and significantly related with the dependent variable. In general, only those whose major source of income is farming can go into the extra expenses of using law enforcement agencies for instance, to secure their farm against trespassers and animal invaders. Income was also positively and significantly related with institutional adaptation. This is not surprising because, only farmers who have the means can afford to buy radios and televisions in order to listen to weather forecasts. Awareness of climate change was also positive and significantly related with institutional adaptation. Radio and television are also sources of climate change awareness. Experience was however negatively and significantly related with the dependent variable. It does not require experience for a farmer to use weather forecasts, for instance. The only requirement for this is to have the information, which can be obtained from a variety of sources like radio, television, internet, friends etc.

The level of education of the farmer was positive and significantly related with the use of social adaptation practices. In group interactions, a farmer is likely to be presented with a variety of alternative ways of dealing with the issue of climate change. Education equips one with a better understanding of how to process information provided by different sources regarding new farm technologies, thereby increasing his/her allocative and technical efficiency. Farming as a major occupation was also positively and significantly related with the dependent variable. The membership of cooperatives in Enugu state is predominantly rural smallholder farmers, whose major source of income is in most cases farming [24]. Income was positive and significantly related with the adaptation practices. Being an active and functional member of a group will necessarily require money.

Government Policies and Programmes on Climate Change Adaptation

Results on government policies and programmes on climate change adaptation revealed that majority (87%) of the respondents indicated that they were aware of the government ban on indiscriminate bush burning, but about half of them (50%) said that the policy is not effectively being implemented (Table 6). This is not surprising because the farmers most often use bush burning as a means of clearing farmland for cultivation, and it ranked as the highest farm practice above. On gas flaring, majority of the farmers (65%) said they were not aware of the policy. Similarly, 88% of the farmers indicated that the policy is not effectively being implemented. Although majority of the respondents (65%) agreed that they were aware of government policy on promotion of afforestation, majority of them (59%) however said that the policy was not effectively being implemented. This could be because, not only were the farmers involved in direct deforestation for farmland area expansion and firewood harvesting, but also involved in bush burning which is a deforestation process. The respondents were also mostly (65%) aware of government ban on indiscriminate tree felling but majority (64%) also indicated that the policy is not effectively being implemented.

Table 6- Percentage distribution of respondents based on their perception of government policies/programmes on climate change adaptation (n=360)

SNo	Government Policies/Programmes		
1	Awareness of government policies/programmes on:	Aware (%)	Unaware (%)
	b) Ban on gas flaring	35	65
	c) Promotion of afforestation	65	35
	d) Ban on indiscriminate tree felling	65	35
2	Effective implementation of policies/ programmes on:	Effectively implemented	Not effectively implemented
	a) Ban on indiscriminate bush burning	50	50
	b) Ban on gas flaring	12	88
	c) Promotion of afforestation	41	59
	d) Ban on indiscriminate tree felling	36	64

Framework for Agricultural Adaptation to Climate Chang

Based on the key findings from the research study, the series of observations and interactions with stakeholders, the multiple focus group discussions held with farmers and other stakeholders, and extensive review of relevant literature on the subject matter, the following framework (Fig. 2) for agricultural adaptation to climate change in Southern Nigeria is hereby proposed. Note that the mitigation and adaptation options combine both indigenous and modern techniques, processes, practices and technologies. However, the framework does not in any way suggest a hierarchical order of either impacts or adaptation options nor is it suggesting that all the components could be applicable to a particular farmer. This is because any or some of them could manifest for a particular farmer. It is therefore necessary for individual, community and national bodies to firstly conduct situational analysis of their respective peculiarities before adopting any of the adaptation or mitigation options [*See Fig. 2 on Page 252*].

Conclusion and Recommendations

The research has shown that most farmers in Southern Nigeria are aware of climate change as manifested in increasing flooding and drought events; changes in rainfall intensity and distribution; increase in temperature; high incidences of pests, diseases and weeds; declining soil fertility and decreases in crop yields among others. However, their understanding of the concept of global climate change with regard to greenhouse gas emissions, ozone layer depletion, etc is limited. Such terminologies are poorly understood and there is no standard translation of what climate change means in local languages. Although farmers adopted numerous adaptation strategies to climate change effects/impacts, they noted that most of the strategies are increasingly tasking and might not stand the test of time especially with the increasing impacts of climate change. Grouping the adaptation strategies into agronomic, technological, environmental, economic, institutional and social, and applying the probit regression model, the major household level factors driving their adoption were farming as a major occupation and level of income of the farmer. Even though farmers were aware of some government policies and programmes aimed at environmental protection such as the ban on indiscriminate bush burning, the ban on indiscriminate felling of trees, and the tree planting programme, they confirmed that these policies and programmes were not being effectively implemented.
Based on the key findings from the research study, the following recommendations are proffered:

- There is need to increase the awareness level of farmers and the public as a whole on climate change issues. The people's ability to effectively respond to climate change challenges is determined by the quality of the information available to them and how easily they can access it. It is based on the information the people have that they can make informed judgements and decisions for adaptation. The role of awareness creation is that of everybody, but a responsive government should take the lead.
- There is need for a systems approach involving all stakeholders- science experts and researchers; policy makers and governments; private sectors; non-governmental and civil society organizations; farmers groups; youth and women- to work together in turning the challenges posed by climate change into opportunities. The systems approach will lead to the development of effective climate change resilient capacity, knowing that systems thinking espouse the principles of collaboration which breeds innovation.
- While the indigenous practices adopted by farmers over the years have helped them cope with the changing climate, it is necessary to apply the modern technologies and practices for effective adaptation to take place. This is because, with time, some of the indigenous methods and practices of adaptation will not be effective again. A framework for agricultural adaptation to climate change in Southern Nigeria (Fig. 2) developed through this research study can provide useful reference material for adaptation options in the area.
- There is need for climate change policy at national, state, and local government levels in the country. Such policies should streamline roles and responsibilities, strategies for adaptation, vulnerability scenarios, and stakeholder's involvement in a systematic manner. Regular debates, workshops, conferences, international affiliations should be used to provide updates on climate change issues.
- There is need to climate-proof all courses in universities. This means that the environmental implications of courses taught in universities should be emphasized as part of curriculum review. Besides, new courses in relevant disciplines should be offered especially at postgraduate levels to provide in-depth knowledge on climate change science; adaptation and mitigation; agrobiodiversity; pastoralism; climate extension and economics; global warming, policy issues; etc.
- Finally, there is need to utilize the power of the media- radio, television, newspapers, posters, internet, hand bills, bill boards, etc- in strengthening climate change awareness and in communicating effective response strategies to climate change. Communicating climate change involves the provision of information, facilitating policy and public dialogue, and encouraging accountability.

Acknowledgements

We wish to thank the British Department for International Development (DFID) for providing the funds used to conduct this research under the Development Partnerships in Higher Education (DelPHE) programme. We appreciate the British Council, Nigeria for the effective implementation of the DelPHE programmes in Nigeria and all the participating institutions in the DelPHE 326 Project including; University of Nigeria, Nsukka (UNN); African Institute for Applied Economics (AIAE), Enugu; and the University of Reading (UoR), United Kingdom.

References

[1] Parry M.L., Canziani O.F., Palutikof J.P., Van der Linden P.J. and HansonIntergovernmental C.E. (2007) *Contribution of Working Group II to the Fourth Assessment Report of the Intergovernmental Panel on Climate Change*, 976.
[2] German Advisory Council on Global Change (WBGU) (2005) *World in Transition, Fighting Poverty through Environmental Policy.* Earthscan, London.
[3] Ozor N. (2009) *African Institute for Applied Economics* (*AIAE*), Enugu, Nigeria, 25-42.
[4] Ozor N., Madukwe M.C., Enete A.A., Amaechina E.C., Onokala P., Eboh E.C., Ujah O. and Garforth C.J. (2010) *Journal of Agricultural Extension,* 14(1), 114-124.
[5] Pittock A.B. (2005) *Climate Change, Turning up the Heat.* London, Earthscan, 1-23.
[6] German Advisory Council on Global Change (WBGU) (2003) *Climate Protection Strategies for the 21st Century, Kyoto and beyond.* Special Report. Berlin, Germany, WBGU, 1.

[7] Intergovernmental Panel on Climate Change (IPCC) (2001) *Contribution of Working Group II of the Intergovernmental Panel on Climate Change to the Third Assessment Report of IPCC.* London.

[8] Ozor N. and Nnaji C. (2010) *Journal of Agricultural Extension*, 14(2), 106-122.

[9] Hassan R. and Nhemachena C. (2002) *AFJARE*, 2(1), 85-104.

[10]Maddison D. (2007) *Centre for Environmental Economics and Policy in Africa (CEEPA)*. 10, CEEPA, University of Pretoria.

[11]Araya B. and Adjaye J.A. (2001) *Indian Journal of Agricultural Economics*, 56, 239-252.

[12]Anim F.D.K. (1999) *Journal of Agricultural Economics*, 50, 336 -345.

[13]Science in Africa (2007) *Africa's first on-line science magazine.*

[14]Magadza C.H.D. (2000) *Environmental Monitoring and Assessment,* 61, 163-205.

[15]Brussel S.E.C. (2009) *Commission of the European communities. Commission working staff working document accompanying the white paper,* 147.

[16]Mark W.R., Mandy E., Gary Y., Lan B., Saleemul H. and Rowena V.S. (2008) *Federal Ministry for Economic Cooperation and Development*, Germany.

[17]Khanal R.C. (2009) *The Journal of agriculture and environment*, 10, 100-110.

[18]International Federation of Organic Agriculture Movement (IFOAM) (2007) *Organic agriculture's role in countering climate change.* IFOAM, Germany.

[19]Ozor N. and Nnaji C. (2011) *Journal of Agricultural Extension and Rural Development,* 3(3), 42-50.

[20]Benhin J.K.A. (2006) *CEEPA Discussion,* 21. CEEPA, University of Pretoria, South Africa.

[21]Enete A.A., Nweke F.I. and Tollens E. (2004) *Journal of International Agriculture,* 43(1), 57-69.

[22]Nweke F.I. (1996) *COSCA,* 14, COSCA, IITA, Ibadan, Nigeria.

[23]Enete A.A., Madu I.I., Mojekwu J.C., Onyekuru A.N., Onwubuya E.A., and Eze F. (2011) *ATPS,* 53.

[24]Enete A.A. and Achike I.A. (2008) *Outlook on agriculture,* 37 (2), 131-134.

Table 5- Parameter Estimate of the Probit Regression Model

Explanatory Variable	Agronomic	Technological	Environmental	Economic	Institutional	social
Household size	0.32(1.04)	0.02(0.69)	0.05(1.75)*	0.03(0.90)	-0.02(-0.58)	0.02(0.57)
Education	-0.01(-0.42)	0.03(1.49)	-0.003(-0.14)	0.01(0.37)	-0.002(-0.09)	0.05(1.77)*
Marital status (1 if married, else 0)	0.35(1.46)	0. 50(1.86)*	0.30(1.24)	0.27(1.02)	0.30(1.14)	0.29(0.92)
Gender (1 if male, else 0)	-0.23(-1.06)	-0. 33(-1.49)	-0.08(-0.39)	-0.26(-1.20)	0.38(1.78)**	0.08(0.32)
Occupation (1 if farming was major, else 0)	-0.41(-1.65)*	0. 97(3.54)***	0.18(0.81)	0.72(2.75)***	0.63(2.52)***	0.74(2.40)**
Income	-2.34e-08(-0.06)	1.53e-06(3.95)***	1.22e-06(3.11)***	8.67e-07(2.71)***	9.59e-07(2.84)***	7.96e-07(2.38)**
Farm size	0.20(2.71)***	-0.04(-0.89)	0.05(1.12)	0.03(0.77)	-0.01(-0.15)	-0.05(-0.94)
Aware of climate change (Yes/No)	0.59(2.31)**	0. 06(0.23)	0.19(0.79)	1.24(3.32)***	0.48(1.71)*	0.21(0.61)
Climate change will affect agric (Yes/No)	2.02(3.55)***	-0.37(-0.71)	0.21(0.43)	0.12(0.20)	-0.22(-0.44)	_
Experience	-0.01(-0.81)	-0.02(-1.49)	0.02 (2.02)**	-0.02(-1.61)	-0.02(-1.70)*	-0.002(-0.19)
Intercept	-3.00(-3.71)***	-2.29(-3.05)***	-1.73(-2.54)***	-3.28(-3.93)***	-1.11(-1.63)*	-3.18(-4.41)***
Statistics: No. of obs.	243	243	243	243	243	234
LR chi^2	43.59	40.01	37.18	43.13	34.02	20.62
Prob>chi^2	<0.001	<0.001	<0.001	<0.001	<0.001	0.02
Pseudo R^2	0.162	0.141	0.113	0.149	0.118	0.101

*Figures in parenthesis are Z-ratios. *** = significant at 1%, ** = significant at 5% and * = significant at 10% probability.*

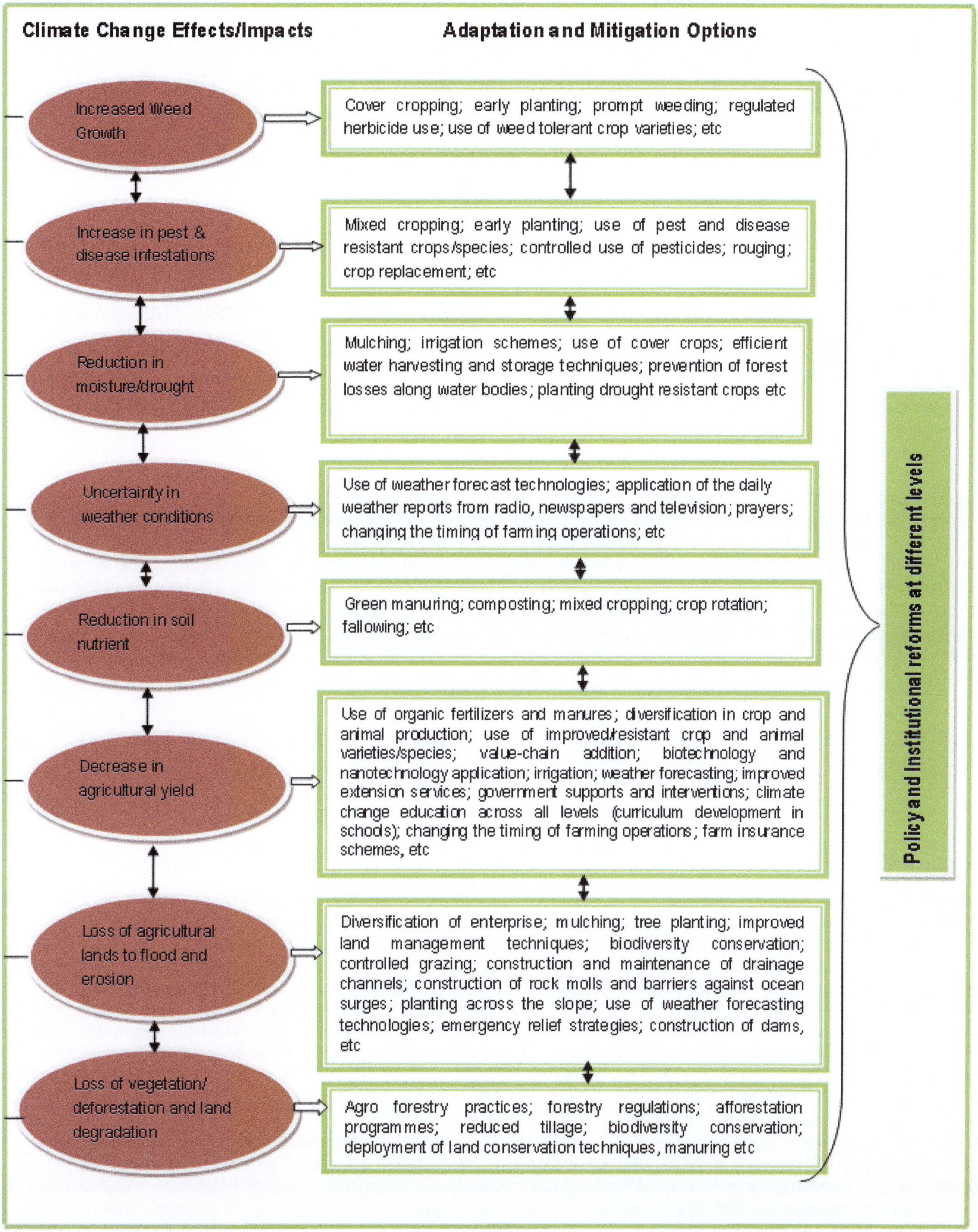

Fig. 2- A Framework for mitigating and adapting to climate change impacts on agriculture in Southern Nigeria

8

CHARACTERIZATION OF THE COAL HUMIC ACIDS FROM THE CANDIOTA COALFIELD, BRAZIL

XAVIER D.M.[1], SILVA A.S.[1], SANTOS R.P.[2], MESKO M.F.[1], COSTA S.N.[3], FREIRE V.N.[3], CAVADA B.S.[4] AND MARTINS J.L.[1]*

[1]Departamento de Química Analítica e Inorgânica, Universidade Federal de Pelotas, Pelotas, Rio Grande do Sul, Brazil.
[2]Laboratório de Engenharia de Materiais e da Computação de Sobral (LEMCS), Universidade Federal do Ceará, Sobral, Ceará, Brazil.
[3]Departamento de Física, Universidade Federal do Ceará, Fortaleza, Ceará, Brazil.
[4]Laboratório de Moléculas Biologicamente Ativas, Departamento de Bioquímica, Universidade Federal do Ceará, Fortaleza, Ceará, Brazil.
*Corresponding Author: Email- jmartins.martins@gmail.com

Abstract- Humic acid (HA) is a principal component of humic substances, which are the major organic constituents of soil, peat, coal, many upland streams, dystrophic lakes and ocean water. HA has an important impact on a variety of environment processes. This work aimed to characterize the HA extracted and purified from the Candiota coalfield (Brazil) by visible (VIS), Fourier transform infrared (FTIR), energy dispersive X-ray (EDX), inductively coupled plasma absorption emission (ICP-AES), and X-ray diffraction (XRD) spectroscopies, thermogravimetry (TG), differential thermal analysis (DTA), and scanning electron microscopy (SEM). The results of these analyses showed that Candiota HA presents a high degree of humification, which is considered to be type A with a turbostratic structure. The elemental analysis (EDX/ICP-AES) indicated the presence of low concentrations of heavy metals.
Key words - Humic acid, coal, characterization, spectroscopy, thermal analyses.

Introduction

The organic matter of soils can be divided into non-humic and humic substances. Carbohydrates, amino acids, protein, lipids, nucleic acids, and lignins are non-humic substances that originate from plants and other organisms. However, humic substances (HSs) are materials synthesized during the decomposition of plants and animal residues with or without the assistance of micro-organisms (humification). HSs are defined as hydrophilic, acidic, high molecular weight, amorphous, and colloidal polydispersed substances with a yellow to brown-black color [1]. HSs are not only found in soils, but are also distributed among several environments, such as streams, rivers, lakes, oceans, sediments, and geologic deposits [2,3,4]. HSs are major importers, exporters and transporters of solutes in soils and natural waters, and they play a much greater role than clays and minerals in this respect [5].

Based on their solubility properties, HSs are generally classified into the three following categories: humic acids (HAs), fulvic acids, and humin [6]. These three humic fractions are structurally similar and their properties differ, especially with respect to their molecular weight, their ultimate analysis, and the number of functional groups present [7]. The predominant fractions of humic substances are HAs [8]. HAs are defined as organic macromolecules with high molecular mass, larger carbon contents, and smaller oxygen contents due to their phenol and carboxylic groups and are soluble at alkaline pH values [4]. HAs have an important role in soil fertility and to bind many heavy metallic ions through complexation [4,9,10]. HAs are present in larger amounts in terrestrial and geologic humic matter and in smaller amounts in aquatic humic matter [1].

The mineral coal, which results from the accumulation of the remains of organic material, contains HSs in its composition. The coal organic matter contains significant amounts of oxygen, nitrogen and sulfur. HSs also contain those elements in their composition in the form of functional groups such as carboxylic acid, ketones, and amines [11]. The most important mineral coalfield in Brazil is the Candiota coalfield, which is located in the Rio Grande do Sul state. Candiota contributes 37.9% of the national mineral coal because the Candiota field is the largest deposit of mineral coal in Latin America [12].

The characteristics of HAs are associated with their origin [13]. The physiochemical characterization of the HAs is indispensable [4,9,10]. Several methods (destructive and non-destructive) are used for the characterization of HAs. Different techniques supply important information for understanding the structure, composition and properties of these materials [14]. Fourier transform infrared (FTIR), ultraviolet-visible (UV-VIS), inductively coupled plasma absorption emission (ICP-AES) and solid-state ^{13}C nuclear magnetic resonance (CP-MAS 13C NMR) spectroscopies; X-ray diffraction (XRD); scanning electron microscopy (SEM); thermogravimetry (TG); and differential thermal analysis (DTA) are some of the commonly techniques used for the characterization of HAs [15,16].

The purpose of this study is to characterize humic acids extracted from mineral coal from the Candiota coalfield (Rio Grande do Sul state, Brazil) using visible (VIS) absorbance, Fourier transform infrared (FTIR), energy dispersive X-ray (EDX), inductively coupled plasma absorption emission (ICP-AES), and X-ray diffraction (XRD) spectroscopies, thermogravimetry (TG); differential thermal analysis (DTA); and scanning electron microscopy (SEM).

Materials and Methods

In this study, samples of mineral coal from the Candiota coalfield (Rio Grande do Sul State, Brazil) were collected directly from the mine in August of 2009. The samples were air-dried, ground, and sieved through a 1-mm sieve. Humic acid (HA) was isolated from the samples using conventional methods [4]. In brief, humic substances were extracted using 1.0 M NaOH. The extracted humic substances were then separated into HA and fulvic acid (FA) fractions by acidifying the extract to pH 1-1.5 using 6.0 M HCl. The extracted HA was purified according to the procedure reported by Stevenson [4]. The HA fraction was suspended in a solution of 0.1 M HCl and 0.3 M HF to remove mineral impurities and then dialyzed until Cl^- was eliminated. The purified HA from the mineral coal from the Candiota coalfield (CHA) was freeze-dried for chemical analysis.

For absorbance measurements, the CHA samples (1 mg) were dissolved in 100 mL of 0.1 M NaOH, and the absorbance (Abs) was measured at 400, 465, 600 and 665 nm with a J&M TIDASDAQ spectrometer. The results (mean value of ten repetitions) were used to calculate the E_4/E_6 ratio (Abs_{465}/Abs_{665}) and Δ *Log K* coefficient ($Log\ Abs_{400} - Log\ Abs_{600}$) according to Giovanela [16]. These parameters are widely used for the characterization of humic substances.

The Fourier transform infrared (FTIR) spectra were recorded in the transmission mode by a Shimadzu IR Prestige-21 spectrophotometer using a KBr pellet containing 1% CHA. The infrared spectra were recorded at 4 cm^{-1} resolution from 5000 cm^{-1} to 450 cm^{-1}. The spectra were automatically corrected for the KBr background to minimize the CO_2 and H_2O absorptions.

The thermal analyses (TGA and DTA) were performed in a nitrogen atmosphere, with a 10 °C/min heating ratio, from 23.8 °C to 1000 °C, using Shimadzu DTG 60 equipment.

The elemental composition of the CHA was determined by an energy dispersive X-ray analysis (EDX) and inductively coupled plasma atomic emission spectrometry (ICP-AES). Qualitative measurements were performed using a Shimadzu Ray Ny-EDX 720. The concentrations of Al, Ca, Cd, Co, Cu, Fe, K, Mg, Mn, Mo, Na, Ni, Pb, V, Zn, and Sr were determined (mean value of three repetitions) by ICP-AES with a Perkin-Elmer ICP-AES instrument, model Optima 4300DV, after digestion (humid) in the microwave with a closed system using concentrated HNO_3.

The X-ray diffraction (XRD) patterns for the mineralogical analyses were obtained at room temperature (300 K) using a Shimadzu XRD-6000 powder diffractometer. A Bragg-Bretano geometry was used with Cu-ka radiation. The tube was operated at 40 kV and 40 mA. The diffraction data were collected over the range of $10° \leq 2\theta \leq 80°$ with 0.02° steps and an integration time of 2 s per point.

The SEM micrographs of the CHA powder were obtained with a Shimadzu SSX-550 scanning electron microscope. The freeze-dried samples were previously coated with a thin gold layer in a Quick Coater SC-701 sputtering system for 10 min before analysis.

Results and Discussion

The values of the E_4/E_6 ratio and the *D log K* coefficient for the CHA were 3.20 ± 0.10 and 0.42 ± 0.02, respectively, which are considered to be type A (*D log K* < 0.6 - high degree of humification), according to Kumada [14]. These results suggest that the HA-C presents a high structural condensation (larger amount of aromatic structures than aliphatic structures) and a large molecular size [17]. The measured E_4/E_6 and *D log K* values are among the smallest values reported [4,16,18].

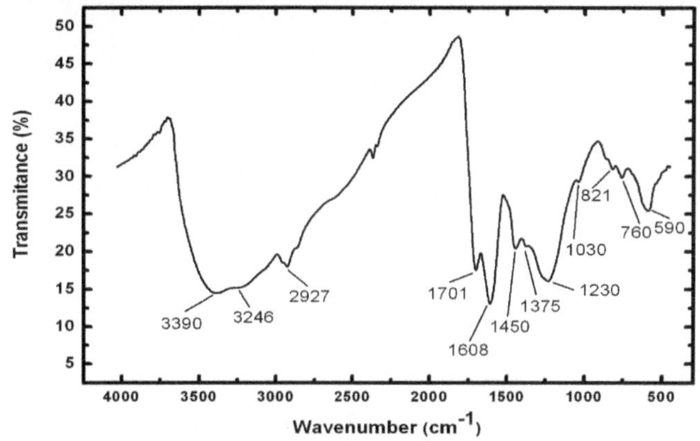

Fig. 1- FTIR spectra of the CHA showing the principal absorption bands.

Fig. 1 shows the FTIR spectrum for the CHA. The broad bands detected in the 3680-3000 cm^{-1} region were attributed to the O-H stretching of alcohols and/or phenols, as well as N-H stretching of amines and/or amides [19]. The bands between 3000 and 2855 cm^{-1} correspond to the asymmetric and symmetric C-H stretching of methyl and methylene groups of aliphatic and nonstrained cyclic hydrocarbons [16]. This region shows the degree of saturation of the sample, revealing a low percentage of aliphatic carbons in the CHA. The sharp band at 1701 cm^{-1} was attributed to the C=O stretching vibrations due to carboxylic groups [20]. The strong band detected at 1608 cm^{-1} indicated the presence of aromatic ring C=C stretching [21]. Absorption at 1450 cm^{-1} was due to aliphatic C-H deformation [8]. The absorption band attributed to the C-H bending of CH_3 and C-H deformation of CH_2 and CH_3 groups was measured at 1375 cm^{-1} [16]. The bands at 1230 and 1030 cm^{-1} corresponded, respectively, to the aromatic ether C-O-C and C-O stretching of polysaccharide [19,22]. Absorption bands attributed to the aromatic C-H out-of-plane deformation of aromatic structures were observed in the 860-720 cm^{-1} region [23]. The Si-O stretching absorption occurred at 590 cm^{-1} [24]. This band may represent silicate impurities [25]. The bands observed in the FTIR showed that the CHA is rich in aromatic structures, according to results from VIS spectroscopy.

The thermal analyses (Fig. 2) indicated the occurrence of three events: the first, between 28 and 140 °C, corresponded to a 7.7% mass loss; the second, between 140 and 360 °C, corresponded to an 8.8% mass loss; and the third, between 360 and 535 °C (exothermic DTA peak at 450.7 °C), corresponded to a 75% mass loss. The first two events were identified only in TG. The last was identified in TG as well as in DTA. The first event was attributed to the loss of free water (dehydration) from the CHA sample [26]. The second event was associated with the degradation of polysaccharides, the decarboxylation of acidic groups and the dehydration of aliphatic alcohols [27]. The third event (which showed a larger mass loss) was attributed to the decomposition of aromatic structures and cleavage of C-C bonds [28]. Consequently, the strong exothermic peak at approximately 450 °C showed a larger number of aromatic structures relative to the aliphatic structures in the CHA (high degree of humification). This characteristic has been observed in HAs extracted from coals [29].

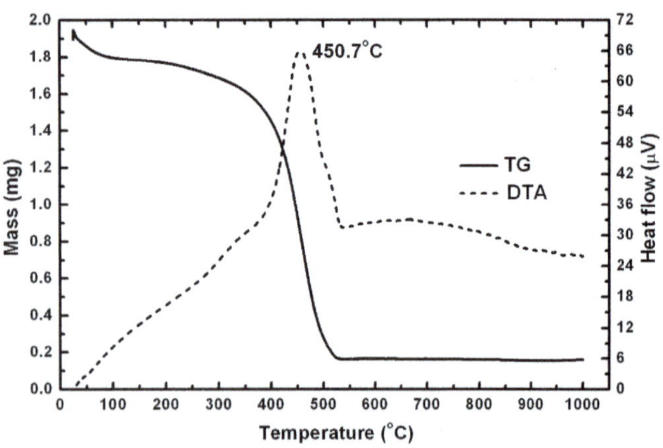

Fig. 2- TG and DTA curves for the CHA. The value indicated represents an exothermic peak

With respect to the elemental composition of the CHA, EDX qualitative measurements indicated the presence of Br, S, Rh, Ba, Ti, Fe, Ni, Mo, Na, Al, Si, Ca, Te, and V. Other elements may be present but not detectable by EDX because of low concentrations. Table 1 shows the results of the ICP-AES analysis. ICP-AES allows for the determination of small concentrations (ppb) of metallic elements in a sample. All the elements investigated showed low concentrations. Na was the most prevalent element (approximately 1.05%), followed by Al (approximately 0.12%). Other elements showed concentrations below 0.03% or concentrations below the detection limit (Cd, Co, and Pb). Unpurified Has generally contain a large quantity of Na and Ca, which originate from the alkaline treatment during the extraction and high rates of occurrence of Si, Fe and Al in the form of impurities [30]. The low concentrations observed by ICP-AES revealed the effect of purification of the CHA. The purification process performed after the extraction causes a significant loss of cations, thereby decreasing the elemental concentration. Very few of the metallic elements detected can be incorporated into the CHA structure because of the capacity of the CHA to bind many heavy metallic ions through complexation [31,32]. Other elements such as Si can originate from impurities. S should be incorporated into the CHA structure [33].

Table 1- Result of ICP-AES measurements of HA-C. Concentrations below the detection limit are marked as ""*

Metal	Concentration ($\mu g g^{-1}$)	Metal	Concentration ($\mu g g^{-1}$)
Al	1259 ± 55	Mo	60.6 ± 2,5
Ca	201 ± 28	Na	10552 ± 62
Cd	*	Ni	11 ± 2
Co	*	Pb	*
Cu	15.5 ± 0,9	V	10.1 ± 0,4
Fe	137 ± 15	Na	4.72 ± 0,22
K	44.8 ± 2,0	Sr	3.46 ± 0,25
Mg	45.4 ± 9,0	Mn	5.74 ± 1,30

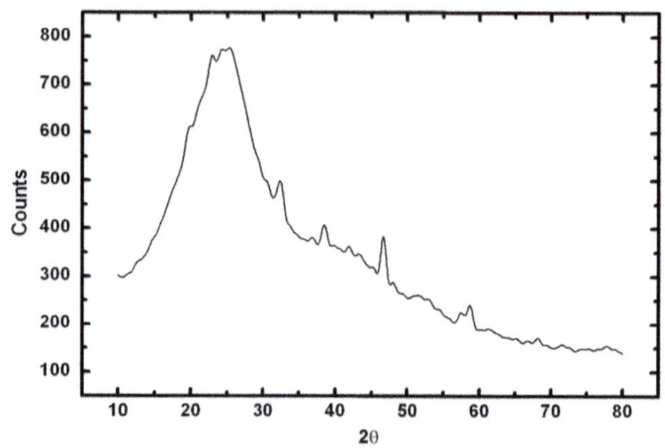

Fig. 3- XRD pattern for the CHA. Some peaks associated with crystalline phases can be observed

The X-ray diffraction pattern is shown in Fig. 3. A broad and symmetric peak was observed around at 2q = 24.80°, corresponding to an interplanar spacing of approximately 3.59 Å. A broad and feeble peak was also observed at approximately 2q = 40.40°, corresponding to an interplanar spacing of approximately 2.25 Å. These peaks were attributed to the (002)- and (01)-bands of graphite, respectively [34]. HAs with these bands contain stacked fractions of several layers with interplanar spacing between 3.4 and 3.6 Å of condensed rings composed of several benzene rings (turbostratic structure) [14,15]. The type A HAs show this structure, which is associated with the high degree of humification this type of HAs [14]. Consequently, the CHA presumably has a turbostratic structure. The X-ray diffraction pattern of the CHA was similar to the pattern of Sochiken-1 HA [14]. Sharp peaks (unidentified) associated with crystalline phases were also were observed on the diffractogram. These phases (impurities) presumably are composed of the elements detected in the EDX/ICP-AES analyses. Further experiments will be performed to identify these crystalline phases.

Fig. 4 shows SEM images of the CHA particles at different magnifications. Particles with different shapes and sizes can be observed (Figs. 4a, 4b). At high magnification (Figs. 4c, 4d), the surfaces of these particles revealed a rough structure consisting of aggregates of particles smaller than 90 nm. The presence of some porous particles with a mean size of 110 nm was also observed. These values were measured directly on the image at high magnification. Evidently, the values obtained are approximate because we are at the SEM resolution limit. The mechanisms involved in the molecular aggregation process can be weak, such as London and van der Waals forces, or strong, such as charge transfer and hydrogen bonding [16]. High structural condensation and large-size molecules (suggested by the characterization techniques) can favor the strong interactions between the CHA molecules. These interactions would promote the cohesion of the molecules in small particles to form larger aggregates (particles observed in low magnifications) during the drying process after purification. This hypothesis still needs to be tested.

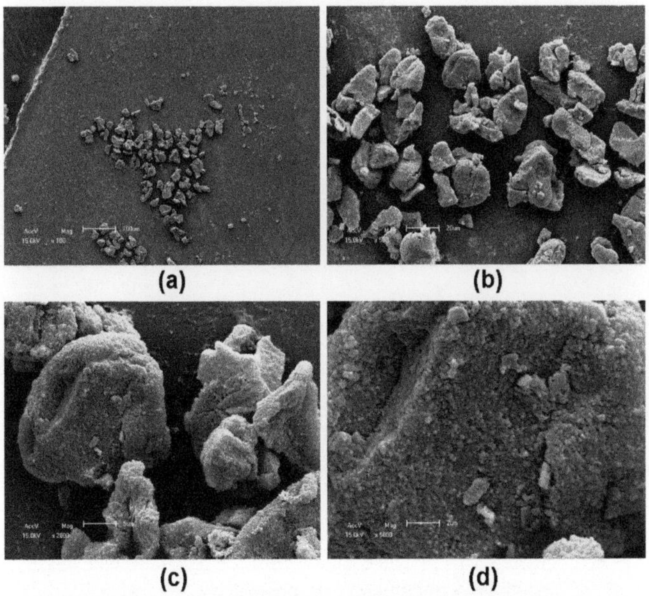

Fig. 4- SEM images of the CHA. Magnifications: (a) 100x, (b) 500x, (c) 2000x, and (d) 5000x. Small pores can be observed on the rough surface at high magnification

Conclusions

Characterization by absorbance (*E_4/E_6* ratio and Δ *Log K* coefficient), FTIR, TG/DTA, XRD, and SEM showed that the purified humic acid from Candiota (CHA) presents a high degree of humification, which is considered to be type A with a turbostratic structure. The E_4/E_6 and *D log K* values are among the smallest values reported for humic acids, presumably indicating that the CHA is relatively rich in carboxylic and phenolic groups. The functional groups that contribute most to surface charge and reactivity of humic substances are phenolic and carboxylic groups. The presence of carboxylic and phenolic groups gives the humic acids the ability to form complexes [4]. Consequently, the CHA can exhibit a high capacity to complex with heavy metallic ions. The elemental analysis (EDX/ICP-AES) indicated the presence of low concentrations of heavy metals

Acknowledgments

We would like to thank the professors, technicians and students of the Biotechnology Center of Sobral (NUBIS) and Material Science and Technology (LCTM) of the Federal University of Ceará, CE, Brazil and the Environmental Chemistry Laboratory of the Federal University of Pelotas, RS, Brazil, for the research facilities, technical support and important suggestions for the elaboration of this work. Benildo Sousa Cavada (B. S. Cavada), Ricardo Pires dos Santos (R. P. Santos), and Valder Nogueira Freire (V. N. Freire) are investigators of CNPq (Brazil). This work received partial financial support from National Council of Research (CNPq) - projects nos. 501221/2009-3 and 473911/2010-8.

References

[1] Tan H.K. (1998) *Principles of soil chemistry, 3rd ed.,* Marcel Dekker, New York.
[2] Schnitzer M. (1978) *Soil Sci.*, 8, 1-64.
[3] Hayes M.H.B. and Clapp C.E. (2001) *Soil Sci.*, 166, 723-737.
[4] Stevenson F.J. (1994) *Humus Chemistry, Genesis, composition, reactions*, *2rd ed.*, John Wiley & Sons, New York.
[5] Ghabbour E.A. and Davies G. (2005) *Humic substances, Nature's Most Versatile Materials,* Taylor & Francis, New York.
[6] Steinberg C.E.W. (2003) *Ecology of Humic Substances in Freshwaters,* New York.
[7] Pansu M. and Gautheyrou J. (2006) *Handbook of soil analysis,* New York.
[8] Mengchang H., Yehong S. and Chunye L. (2008) *J. Environ. Sci.*, 20, 1294-1299.
[9] Vaughan D. and Malcolm R.E. (1985) *Influence of humic substances on growth and physiological processes. In: Soil Organic Matter and Biological Activity.*
[10]Da Silva R.M., Jablonski A., Siewerdt L. and Silveira P. (2000) *R. Bras. Zootec.*, 29, 1623-1631.
[11]Lawson G.J. and Stewart D. (1989) *Coal Humic Acids. In: Humic Substances II. In Search of Structure.*
[12]Departamento Nacional de Produção Mineral (1996) *Informativo Anual da Indústria Carbonífera. Ministério de Minas e Energia, Brasília.*
[13]Olivella M.A., del Rio J.C., Palácios J., Vairavamurthy M.A. and las Heras F.X.C. (2002) *J. Anal. Appl. Pyrolysis,* 63, 59-68.
[14]Kumada K. (1987) *Chemistry of Soil Organic Matter,* Tokyo.
[15]Agarwal S.P., Awer M.D.K., Khanna R., Ali A., Sultana Y. and Serb J. (2010) *Chem. Soc.*, 75, 413-422.
[16]Giovanela M., Crespo J.S., Antunes M., Adamatti D.S., Fernandes A.N., Barison A., da Silva C.W.P., Guégan R., Motelica-Heino M. and Sierra M.M.D. (2010) *J. Mol. Struct.*, 981, 111-119.
[17]Sanches S.M., Campos S.X. and Vieira E.M. (2007) *Eclet. Quím.*, 32, 49-56.
[18]Shirshova L.T., Ghabbour E.A. and Davies G. (2006) *Geoderma,* 133, 204-216.
[19]Amira S., Jouraiphyb A., Meddichb A., Gharousb M., Wintertonc P. and Hafidid M. (2010) *J. Hazard. Mater.*, 177, 524-529.

[20]Erdogan S., Baysal A., Akba O. and Hamamci C. (2007) *Polish. J. Environ. Stud.*, 16, 671-675.
[21]Siddiqui Y., Meon S., Ismail R., Rahmani M. and Ali A. (2009) *Int. J. Agric. Biol.*, 11, 448-452.
[22]Çimen F., Sozudogru S., Kayran C., Demirci S., Ozenc D.B. and Ozenc N. (2007) *Biodegradation*, 18, 295-301.
[23]Senesi N., Meano Y.M. and Matin J.P. (1987) *Biol. Fertil. Soils*, 5, 120-125.
[24]Jayaganesh S. and Senthurpandian V.K. (2010) *Asian J. Earth Sci.*, 3, 130-135.
[25]Xu D., Zhu S., Chen H. and Li F. (2006) *Colloids Surf. A*, 276, 1-7.
[26]Francioso O., Ciavatta C., Montecchio D., Tugnoli V., Sanchez-Cortes S. and Gessa C. (2003) *Bioresour. Technol.*, 88, 189-195.
[27]Fernández J.M., Hockaday W.C., Plaza C., Pólo A. and Hatcher P.G. (2008) *Chemosphere*, 73, 1838-1844.
[28]Francioso O., Montecchio D., Gioacchini P., Cavani L., Ciavatta C. and Trubestskoj O. (2009) *Geoderma*, 152, 264-268.
[29]Fernandes A.N. (2007) Ph.D. *Thesis, Universidade Federal de Santa Catarina, Florianópolis*, p 149.
[30]Terkhi M.C., Taleb F., Gossart P., Semmoud A. and Addou A. (2008) *J. Photochem. Photobiol. A*, 198, 205-214.
[31]Courtijn E., Vandescasteele C. and Dams R. (1990) *Sci. Total Environ.*, 90, 191-202.
[32]Pourret O., Davranche M., Gruau G. and Dia A. (2007) *Chem. Geol.*, 243, 128-141.
[33]Simpson A.J., Kingery W.L., Hayes M.H.B., Spraul M., Humpfer E., Dvortsak P., Kerssebaum R., Godejohann M. and Hofmann M. (2002) *Naturwissenschaften*, 89, 84-88.
[34]Schnitzer M., Kodama H. and Ripmeester J.A. (1991) *Soil Sci. Soc. Am. J.*, 55, 745-750.

BUSH FALLOW AND COWPEA CROP USE AS PRECEDENT AND ORGANIC SOURCES OF NUTRIENTS FOR RICE CULTIVATION ON ACIDIC PLINTHOSOL OF CENTRAL BENIN IN WEST AFRICA

ASSIGBE P.[1], KONE B.[2]*, BOGNONPKE J.P.[3], TOURE A.[4], HUAT J.[4,5] AND YAO-KOUAME A.[2]

[1]Institiut National des Recherches Agricoles du Benin, B.P. 226 Bohicon, Benin.
[2]Felix Houphouet Boigny University, Cocody, soil sciences department,Abidjan, Côte d'Ivoire.
[3]Nagui Abrogba University, Department of plant biology and improvement, Abidjan, Côte d'Ivoire.
[4]Africa Rice Center (WARDA), Cotonou. 01 BP 2031 Cotonou, Benin.
[5]CIRAD, UPR Hortsys, F-34398, Montpellier Cedex 05, France.
*Corresponding Author: Email- kbrahima@hotmail.com

Abstract- Tropical soils have low fertility and chemical fertilizers are not affordable for most of the farmers, especially in West Africa. The development of low cost technologies for amending agricultural soils in this environment is needed. Biological and organic fertilizers were tested in 20 farmers' fields in central Benin for rainfed rice production on acidic plinthosol under monomodal rainfall pattern conditions. Short time matured cowpea and native grass fallow were preceding NERICA1 and Ina Okpè rice cultivars that are interspecific (*Oryza glaberrima* × *Oryza sativa*) and local varieties respectively. Cowpea and bush residues were incorporated into soil before rice seeding. Soil nutrients (C,N,K, Ca and Zn) contents were significantly depleted whereas null balance was observed for soil available-P content. An increased of soil content in Fe (>130 mg kg^{-1}) was significantly observed in all the treatments. Highest yield of 1.6 t ha^{-1} was recorded in the treatment composed of cowpea and NERICA1. The study technology was deemed suitable in deficient-P acid soil whereas the land foot slope position was excluded to avoid potential increasing of Fe content (iron toxicity of rice) in soil of the subsequent lowland. Minimum chemical fertilizer was required for improving rainfed rice cultivation on plinthosol in monomodal rainfall pattern ecology.

Keywords- Biology Nitrogen Fixation, organic matter, fertilizer, rice, grain legume crop

Introduction

Soil nutrients depletion is among the major threats of crop production in Africa. In West Africa, because of land-use intensification and limited use of nutrient inputs, the fragile upland rice production systems contributing to a third of the total rice production, are experiencing intense soil degradation, nutrient depletion and low rice productivity [6]. Thus, rice yields in the upland production systems are seldom above 1 t ha^{-1} despite the reported potential yield of up to 4 t ha^{-1} [20]. A number of solutions have been proposed to address this declining of productivity.

Becker and Johnson [4,5] reported that the use of N2-fixing legumes including *Mucuna spp.*, *Stylosanthes guianensis*, *Canavalia ensiformis* grown as preceding fallow cover crops increased upland rice productivity and suppressed weed growth under intensified land use in Côte d'Ivoire (West Africa). But this technology was not widely adopted by farmers. Farmers in West Africa are often reluctant to adopt legume cover crops that are not for human consumption or without a direct economic gain, in spite of the benefits for restoring soil fertility [31]. However, the grain legumes are infinitely more adoptable when integrated into cereal-based systems [25] for household food security and cash income. Somado, et al. [27] proposed the combined use of legume cover crops and sparingly-soluble indigenous phosphate rock as a low-cost technology to increase the supply of both N and P and enhance upland rice productivity on acid soils in Côte d'Ivoire (West Africa). However, phosphate rock has a limited availability in West african countries meanwhile, grain legume crops can fix atmospheric-N_2 and their residues can also enrich soil in nutrient by 38.9 g N kg^{-1}, 4 g P kg^{-1}, and 21 g K kg^{-1} only from the leaf dry matter as observed for cowpea [1]. Therefore, legume as preceding crop to upland rice and organic source of nutrients after incorporation can improve soil fertility and upland rice yield. However, unlike the maize-based systems, the contribution of cowpea to upland rice-based systems has not been extensively studied in West Africa [21].

Yet, the occurrence of NERICAs (New Rice for Africa), the interspecific rice varieties released by Africa Rice Center can be more adapted to such practice of fertility strategy. In fact, interspecifics were known to be low-management rice type, adequate for resource-limited small holder [10,16].

The actual study was initiated in Benin for the cropping of NERICA1 and a traditional cultivar after cowpea cropping and the growth of native grass fallow. The objective was to identify the effects of cowpea growth and its residues on soil fertility and rice yield in legume-rice based system without chemical fertilizer compared to native bush effects.

Material and Methods

Experiment Site

On-farm experiments were conducted in farmers' field and by farmers in the area of Glazoué, a mountainous area in the Centre Benin (West Africa). It is a Guinea savanna zone characterized by a degraded wood savanna. The rainfall pattern was monomodal with an annual average of 918.9 mm in 2001 and 1089.9 mm in 2002. Ten farmers were concerned in each selected site [Dassa- (7°45'00"N, 2°11'00"E) and Glazoué- (7°58'25"N, 2°14'24"E)]. The bed rock was granite gneiss with a plinthosol as dominant soil of the plateau. It was a sandy loamly soil, acidic and poor in organic carbone, total nitrogen and available P in the 0-20 cm depth. But, it has a suitable content of Ca, Mg and K [Table-1].

Table 1- Soil physico-chemical characteristics in the 0-20 cm depth before the experiment

	Mean	SD	Minimum	Maximum
Clay (g kg^{-1})	118	23.2	80	170
Sand (g kg^{-1})	758	67.2	610	880
Silt (g kg^{-1})	125	51.3	40	240
pHwater	5.8	0.4	4.9	6.7
C (g kg^{-1})	5.2	1.5	3.2	8.6
N (g kg^{-1})	0.4	0.08	0.33	0.66
P (mg kg^{-1})	7.57	7.8	1.3	94.2
Ca (cmol kg^{-1})	1.86	0.7	0.75	16.6
Mg (cmol kg^{-1})	0.36	0.25	0.1	1.45
K (cmol kg^{-1})	0.19	0.06	0.1	0.3
ECC (cmol kg^{-1})	3.3	1.9	2	6.7
Zn (mg kg^{-1})	14.6	6.06	5.1	31.1
Fe (mg kg^{-1})	28.5	27.7	8.1	125.4

Previous Land use Before Rice Cropping

Herbaceous cultivar of cowpea (Vigna unguiculata (L) Wap. unguiculata) locally named "tontouin" were used by all the farmers. It is a short duration matured cultivar of 65-70 days cycle. Every year, early in April, 2 plots of 5 m × 10 m were choose randomly in farmers' field for cowpea cropping. After manual cleaning and tillage, the cowpea was sown per hill of 2 grains in dimensions of 30 cm × 60 cm. The hand weeding was applied at 21 and 45 days after emergency. At maturity, the cowpea was harvested and plant residues were left in the field, sundried during a week and incorporated to the soil during the tillage operation for rice seeding. In the meantime, two other plots of 5 m × 10 m occupied exclusively by bush (dominated by *Rottboellia cochinchinensis*, *Digitaria horizontalis*, *Ageratum conyzoides* and *Tridax procumbens*, *Imperata cylindica*, *cyperus* spp) were cleaned and the native grass residues were also incorporated into the soil.

Rice Plots

Early in June of every year of experiment, cowpea and grass residues were incorporated in the soil of their respective growing plots of 5 m × 10 m dimensions. A completely randomized blocks design was applied using a local rice variety named Ina Okpè (V1) and NERICA 1 (V2) rice variety. Each rice variety was sown in a plot previously occupied by legume and bush fallow. Rice was sown at 3 grains per hill spaced by 20 cm × 20 cm ten days after residues incorporation. Hand weeding was applied at 21 and 45 days after emergency. At maturity, the rice was harvested, sundried before threshing. The grain yield was calculated for standard moisture content of 14%.

Studied Treatments

Two cropping systems were considered including rotation of cowpea-rice (Syst-1) and bush-rice (Syst-2) within the same cropping season. Two cultivars named Ina Okpè and NERICA1 were respectively noted V1 and V2 for the experiment. Therefore, four treatments were defined as T1 (Syst-1 + V1), T2 (Syst-1 + V2), T3 (Syst-2 + V1) and T4 (Syst-2 +V2).

Soil Sampling and Laboratory Analysis

Soil samples were taken in 0-20 cm depth of soil in each corner and the center of treatment plots (T1, T2, T3 and T4). The five samples were mixed before taking for a composite sample of soil in a given treatment before the experiment in 2001. A sample was taken from each treatment plot in all the location of experiment. Similar operations were done in 2002 before the second experiment. The composite samples were dried, ground and sieved (2mm) before the laboratory analysis was carried out. Soil pHwater (1: 2.5) and soil contents in Carbone-C (Wakley and Black), total nitrogen-N (Kjeldahl), P (Bray I), K, Ca, Mg and Zn (1 N NH4OAc (pH 7.0)) were determined as described by American Society of Agronomy-ASA and Soil Sciences Society of America-SSSA [22].

Statistical Analysis

By descriptive statistic, mean value of soil nutrient contents, the minimum and maximum values as well as the standard deviation were determined in 2001 before the experiment. A mixed model analysis was used to calculate the difference (2001-2002) of nutrient content in the soil of a given treatment from 2001 to 2002. By the same method, the balances of cation contents (2001-2002) were determined between treatments. The balances of C, N, P and K contents were also determined by mixed model within the same plot from 2001 to 2002.

Analysis of variances was use to determine mean values of rice grain yield for each cultivar (V1 and V2) in a given cropping system (Syst-1 and Syst-2) and for each treatment (T1, T2, T3 and T4). Pearson correlation was performed to evaluate the relationship between nutrients and rice grain yield. These analysis were done for $\alpha = 0.05$ with SAS software.

Results

Soil Status

[Table-2] is showing the changes in soil C, N, P and K contents according to the difference between treatments from 2001 to 2002. No significant ($P > 0.05$) effect of treatments was observed on soil nutrient content at α = 0.05. Nevertheless, treatment effect could be observed on organic C content in soil at α = 0.10 (P = 0.08). More organic C content was noticed for T4 than T1 and T3 at this significant level. But, the difference observed between T2 and T4 was not significant at all.

Table 2- Differences of soil contents in C,N,P and K between different treatments (T1,T2,T3 and T4) before the experiment of 2002

Treatments	C (g kg^{-1})	N (g kg^{-1})	P (mg kg^{-1})	K (cmol kg^{-1})
T1 - T2	-0.065	0.025	-0.677	0.004
T1 - T3	0.053	0.006	-1.599	-0.01
T1 - T4	-0.916***	0.002	-0.617	-0.005
T2 - T1	0.061	0.025	0.677	-0.004
T2 - T3	0.119	0.032	-0.883	-0.014
T2 - T4	-0.851	0.023	0.059	-0.008
T3 - T1	-0.053	-0.006	1.599	0.01
T3 - T2	-0.119	-0.032	0.882	0.014
T3 - T4	-0.970***	-0.009	0.942	0.005
T4 - T1	0.916***	0.002	0.617	0.005
T4 - T2	0.851	-0.023	-0.059	0.008
T4 - T3	0.970***	0.009	-0.942	-0.005
Pr>F	0.08	0.518	0.574	0.523

****:highly significant*

[Table-3] is showing the difference of soil nutrient contents from 2001 to 2002 according to a given treatments. Significant depressed N contents were observed in T1 (0.01 g kg^{-1}) and T2 (0.009 g kg^{-1}). Similar effects were observed respectively for T3 and T4. It is also observed significant decreased of soil content in Ca (1.39 cmol kg^{-1}-1.67 cmol kg^{-1}) for all the studied treatments. Only the treatment T3 has induced significant decreased of soil content in Mg from 2001 to 2002. Potassium content in soil was also significantly decreased in treatments T2 and T3 indifferently to cultivars and previous land occupation. Zinc contents in soil on 2002 were decreased in all the treatments in opposite to Fe contents which were increased from 2001 to 2002 according to the negative values observed for the differences.

Table 3- Chemical changes in soil (2001-2002) of each treatment from 2001 to 2002

	Difference (2001-2002)			
	T1	T2	T3	T4
Ph	-0.046*ns*	0.100*ns*	0.012*ns*	-0.082*ns*
C (g kg^{-1})	0.580*ns*	1.960*ns*	2.191*ns*	0.438*ns*
N (g kg^{-1})	0.010***	0.009***	0.014***	0.009***
P (mg kg^{-1})	0*ns*	0*ns*	0*ns*	0*ns*
Ca (cmol kg^{-1})	1.42***	1.55***	1.676***	1.396***
Mg (cmol kg^{-1})	0.05*ns*	0.119*ns*	0.229**	0.135*ns*
K (cmol kg^{-1})	0.028*ns*	0.068**	0.041*	-0.0002*ns*
Zn (mg kg^{-1})	9.772***	8.459***	9.845***	9.899***
Fe (mg kg^{-1})	-132.18***	-145.3***	-125.53***	-132.22***

*ns: not significant; *: significant; **: highly significant; ***: very high significant*

Rice Yield

[Table-4] shows the coefficient (R) values of Pearson correlation between rice grain yield obtained in 2001, and soil nutrients content, particularly the cations (Ca, Mg, K, Zn and Fe) in 2002, before the second trial. Significant (*P* <0.05) R-value of 0.38 was observed between K content in soil and rice grain yield obtained in the treatment T2. The treatment T3 has also induced a significant positive R-value (0.41) between soil content in Ca and rice grain yield recorded. In opposite, in the same treatment (T3), significant negative R value (-0.57) was observed between soil content in Fe and rice grain yield. No significant relationship was observed in treatments T1 and T4 respectively related to legume crop and bush fallow as precedent of rice.

Table 4- R values of Person correlation between rice grain yield and soil content in cations (Ca, Mg, K, Zn and Fe) for each treatment in 2002

	Coefficient (R)			
Nutrients	T1	T2	T3	T4
Ca	-0.141	0.139	0.410**	0.185
Mg	0.079	-0.055	0.132	0.213
K	-0.109	0.386*	0.321	0.039
Zn	-0.235	0.181	0.579	0.065
Fe	0.285	0.019	-0.576***	0.164

: significant; **: highly significant; *: very high significant*

[Fig-1] shows the grain yield harvested for both rice cultivars in the two cropping systems including cowpea (Syst-1) and native grass (Syst-2) as precedent of rice respectively. In Syst-1 and Syst-2, rice grain yields were significantly higher for V2 (NERICA 1) than V1 (Ina Okpè). Moreover, the yields observed in Syst-2 were lower than those in Syst-1. This result was confirmed by [Fig-2] showing significantly (*P*= 0.002) higher average rice grain yield for Syst-1 (1.41 t ha^{-1}).

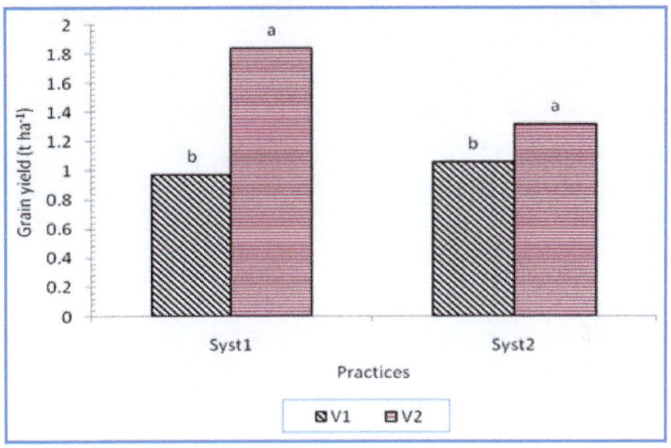

Fig. 1- Average grain yield significantly (P<0.05) obtained for each cultivar in the two different studied systems
(Syst-1= legume incorporation in soil; Syst-2 = grass incorporation in soil; V1 = Ina Opke, V2= NERICA1).

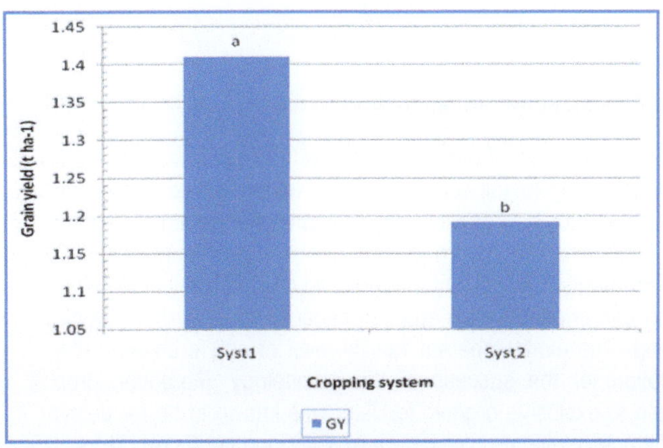

Fig. 2- Average rice grain yield observed (P= 0.002) by cropping system (lsd was 0.13 t ha-1)

Lower correlation coefficient (R) values were observed between cropping system and grain yield than the one observed with cultivar [Table-5]. The coefficient values were also significantly negative with the cropping system while they were positive with cultivar. Increasing values of R were observed from 2001 to 2002 for Syst (-0.10; -0.34) and cultivar (0.50; 0.66). Across year mean values indicated significant ($P<0.0001$) higher value of correlation coefficient (0.61) for cultivar while, -0.24 was noticed studied systems (Syst).

Table 5- Coefficient (R) of Pearson correlation between rice grain yield and the cropping system as well as the cultivar in 2001 and 2002 likewise across year.

		System	Cultivar
2001	R value	-0.1	0.57
	P>I r I	0.03	<0.0001
2002	R value	-0.34	0.001
	P>I r I	0.66	<0.0001
Mean	R value	-0.24	0.002
	P>I r I	0.61	<0.0001

Significant difference was observed between mean values of grain yields obtained according to treatments as illustrated in [Fig-3]. In 2001 and 2002, the lowest yields were respectively observed in T1. The treatment T3 has also induced similar lower yield in 2002. The highest yields are observed invariably in T2 every year of the experiment. Moderate rice yield was observed in T4 in 2001 and 2002.

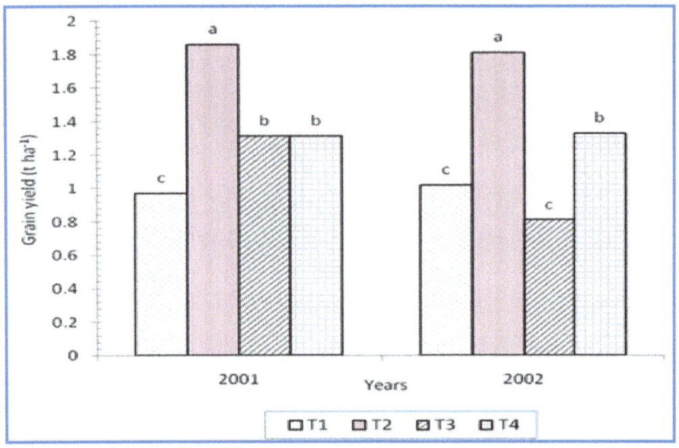

Fig. 3- Rice grain yield induced by each treatment in 2001 and 2002

Discussions

Cropping System Effects in the Soil

Exclusive application of organic matter as fertilizer should be considered according to soil chemical status. Infact, nutrients released by organic matter can be slow depending on temperature, oxygen supply, soil moisture, hydrogen ion concentration and inorganic nutrient contents, C/N ratio of the mulched organic matter [19]. This situation can impaired crop nutrition because of delay in nutrients supplying. Therefore, inherent fertility level of soil is an essential contribution for the success of this technology. Moreover, crop response to exclusive organic fertilizer was known to be far below the satisfaction level required to meet word food need [14]. Thus, it is required a minimum chemical fertilizer in association with organic matter for well amendment of agricultural soil. Minimum N and P were also required for cowpea growing and biological N2 fixation in soil [2,24] in order to improve yield and biological amendment of soil. Therefore, the use of cowpea as preceding crop to rice without fertilizer on a poor soil in N and P could not affect significantly nutrients balance in a positive way as observed in the soil of actual study. Consequently, soil nutrient contents were depleted after rice cropping in 2001, especially, for soil C content [Table-2]. In addition to the low organic matter production by cowpea [8], there is a loss of about 78% of initial cowpea C from the incorporation period to the end of the rainy season [12]. This can explain the restriction of soil organic C builds up in our study. The treatment including native grass incorporation and NERICA1 has the highest C content, probably, because of a low mineralization rate of this organic matter.

From 2001 to 2002, there was a possibility of P release up to 7.57 mg kg^{-1} according to the null value of P balance in the soil of all the treatments. Therefore, the incorporation of vegetal residue as cowpea or native grass can supply sufficient P to soil for rainfed rice production but cannot restore the fertility of soil. In fact, it has been reported that legumes incorporation can improve bioavailability of sparingly soluble soil P [30] and organic matter did so for native P in soil [28]. Thus we conclude that the incorporation of cowpea or native grass residues can prevent from P content depletion in acid soil. The effect native grass of P as observed in actual study is an essential finding that can help in management of agricultural land in tropical zone where P-deficient in acid soil in among the major limiting factors [18].

In all the studied treatments, decreasing of Ca and Zn contents were observed after rice growing in 200, but Fe contents were increased. This results as two implications: incorporation of plant residues i) can help to correct Fe-deficient occurring in upland rice as chlorosis (*rain-induced scorching*), ii) increasing soil Fe content in upland can provoke contamination of lowland throughout runoff creating iron toxicity that is also a serious threat of rice [11]. From this analysis, we assert that the studied technology should be recommended for foot slope ecology because to limit the possible enrichment of subsequent lowland soil in Fe.

Adoption of Biological and Organic Fertilization in Rice-based Systems

Cowpea as preceding crop of rice has induced highest grain yield compared to the grass fallow in a proportion of 16%. Although this result is highly significant, the output of such study should depend on the nature of spontaneous, because of the variability of plant residue content in nutrients as describe by Tian [29]. Infact, some species like *Chromolaena Odorata* was known to have excellent effects on most of rice growth parameters and yield when their residues were incorporated into soil [17]. However, most of the grass growing cycle occupied all the cropping season, hampering a second cropping within the same land. The growth and incorporation of legume crops are therefore recommended as green manure in ranfed area for subsequent crops [3,23] as they can grow within a short time. Variance of actual result can also be observed depending on cowpea cultivar abilities for well nodulation [13] and their maturity cycle (biomass production). Indeed, Schulz, et al. [26] showed that aboveground residue dry matter increases with the maturity class of the cowpea. In the studied ecology where the raining season occurred from April to October, the cowpea should be early matured cultivar (70 days) to ensure the subsequent rice cropping during 70-90 days. In such conditions, Dakora, et al. [9] suggested a rapid incorporation of cowpea residues before rice sowing in order to improve the suitability of this technology. However, the rice grain yield of 1.6 t ha^{-1} induced by cowpea as previous crop is closed to the yield observed under the use of chemical fertilizers in South Benin on Acrisol while, about 3 tha^{-1} were harvested in the Centre Benin on Ferralsol Ferric [15]. Therefore, the use of cowpea as preceding crop in rice-base systems is more relevant to the ecological zone as South Benin where the technology will not induce a yield gap. The actual study revealed that interspecific upland rice as NERICA1 can be use successfully for this.

To enhance rice production in the Centre Benin by this technology, it should be recommended the application of minimum chemical fertilizer in order to improve N fixation ability et dry matter production of cowpea. In fact, the low fertility of the studied soil could affect these parameters [7] reducing the benefit to the subsequent rice.

Conclusion

The study has revealed the interspecific cultivar-NERICA1 as more relevant for bio- and organic fertilizer on acidic plinthosol. The use of grain legume as preceding crop and organic source of nutrients was better than grass fallow practice. However, the thechnology has low benefit to soil and induced rice yield was limited as much as on Acrisol in bimodal rainfall ecology.

Because of the increased of soil content in Fe and the yield observed, this technology was recommended for bimodal rainfall excluding the soil on foot slope to avoid potential iron toxicity of rice in the subsequent lowland.

Applying minimum chemical fertilizer to cowpea was recommended for improving the impact of the technology for soil and rice yield in plinthosol under monomodal rainfall ecology.

Reference

[1] Ali M. (2004) *Tropical Pulse Crops, In World Fertilizer Use Manual, Ed. International Fertilizer Industry Association*, 5.

[2] Amadji L.G., Koné B., Aholoukpé H.N.S. (2007) *J. Rech. Sci. Univ. Lomé (Togo), A*9(2), 109-120.

[3] Becker M., Ladha M.K., Ali M. (1995) *Plant Soil,* 174, 181-194.

[4] Becker M. and Johnson D.E. (1998) *Biology and Fertility of Soils,* 27, 358-367.

[5] Becker M. and Johnson D.E. (1999) *Nutrient Cycling in Agroecosystems,* 53, 71-81.

[6] Buresh R.J., Smithson P.C. and Hellums D.T. (1997) *SSSA Special Publication,* 51, SSSA, Madison, WI, USA.

[7] Carsky R.J., Singh B.B., Oyewole B. (2001) *Biological Agriculture and Horticulture,* 18, 303-315.

[8] Carsky R.J., Vanlauwe B., Lyasse O. (2002) *World Cowpea Conference III held at the International Institute of Tropical Agriculture (IITA), Ibadan, Nigeria,* IITA, Ibadan, Nigeria.

[9] Dakora F.D., Aboyinga R.A., Mahama Y., Apaseku J. (1987) *MIRCEN Journal,* 3, 389-399.

[10]Dingkuhn M., Jones M.P., Johnson D.E., Sow A. (1998) *Field Crops Research,* 57, 57-69.

[11]Dufey I., Hiel M.P., Hakizimana P., Draye X., Lutts S., Koné B., Dramé K.N., Konaté K.A., Sié M., Bertin P. (2012) *Crop Sci.,* 52, 539-550.

[12]Franzluebbers K., Juo A.S.R., Manu A. (1994) *Plant and Soil* 166, 255-265.

[13]Giller K.E. (2001) *Nirogen Fixation in Tropical Cropping Systems, 2nd ed.*, CAB International, Wallingford, UK.

[14]Hiyami Y. and Ruttan V. (1985) *Agricultural Development: An International Perspective.*, John Hopkins Univ. Press, Baltimore, MD.

[15]Koné B., Amadji G.L., Igué M., Ayoni O. (2009) *Jounal of Animal and Plant Science,* 2(4), 156-162.

[16]Koné B., Yao-Kouamé A., Sorho F., Diatta S., Sié M., Ogunbayo A. (2010) *Int. J. Biol. Chem. Sci.,* 4(3), 563-570.

[17]Koné F., Kouadio J.Y., Yapo E.S., Visser M. (2010) *Journal of Applied Sciences Research,* 6(11), 1581-1588.

[18]Koné B., Oikeh S., Diatta S., Somado A., Kotchi V., Sahrawat K.S. (2011) *Arc. of Agronomy and Soil Science,* 57(4), 421-32.

[19]Kumar R., Reddy R.M., Sinha P.S., Tirkey J., Singh M.K., Prasad B.C. (2010) *Journal of Entomology,* 7(4), 219-226.

[20]Oikeh S.O., Nwilene F., Diatta S., Osiname O., Touré A., Okeleye K.A. (2008) *Agronomy Journal* 100 (3), 735-741.

[21]Oikeh S., Houngnandan P., Robert A., Niang A., Toure A., Koné B. (2010) *Journal of Agricultural Science and Technology* 4(1-26), 54-61.

[22]Page A.L., Miller R.H. and Keeney D.R. (1996) *Methods of Soil Analysis, Chemical and Microbiological Properties*, Part-2, *ASA Monograph,* 9, *2nd ed., American Society of Agronomy,* USA.

[23]Rafael M., Navarro C., Ariza D., Gonzalez L., Campo A., Arjona M., Ceacero C. (2009) *Soil Tillage Res.,* 102, 38-44.

[24]Sanginga N., Lyasse O., Singh B.B. (2000) *Plant and Soil,* 220, 119-128.

[25]Sanginga N., Dashiell K.E., Diels J., Vanlauwe B., et al. (2003) *Agriculture Ecosystems Environment,* 100, 305-314.

[26]Schulz S., Carsky R.J. and Tarawali S. (2001) *Soil Fertility Maintenance in West Africa,* ASA, Madison WI.

[27]Somado E.A., Becker M., Kuehne R.F., Sahrawat K.L., Vlek P.L.G. (2003) *Agronomy Journal,* 95, 1172-1178.

[28]Tandon H.L.S. (1987) *Fertilizer Development and Consultation Organization,* New Delhi, India.

[29]Tian G. (1992) *Biological Effects of Plant Residues with Constrasting Chemical Compositions on Plant and Soil under Humid Tropical Conditions,* PHD Thesis, Wageningen Agricultural University, Pays-Bas.

[30]Vanlauwe B., Diels J., Sanginga N., Carsky R.J., Deckers J., Merckx R. (2000) *Soil Biology and Biochemistry.,* 32, 2079-2090.

[31]Vanlauwe B., Aihou K., Aman S., Iwuafor E.N.O., Tossah B.K., Diels J., et al. (2001) *Agronomy Journal,* 93, 1191-1199.

EFFECT OF SALINITY ON SEED GERMINATION AND SEEDLING CHARACTERS OF SOME FORAGE SORGHUM CULTIVARS

KANDIL A.A.[1], SHARIEF A.E.[1]*, ABIDO W.A.E.[1] AND IBRAHIM M.M.[2]
[1]Department of Agronomy, Faculty of Agriculture, Mansoura University, Egypt.
[2]Ministry of Agriculture, Egypt.
*Corresponding Author: Email- shariefali42@gmail.com

Abstract- In order to investigate the effects of osmotic stress on forage sorghum varieties at critical stage of germination to salinity stress. A study was conducted at the Seed Technology Laboratory of Mansoura University, Egypt during April to June 2012. Three forage sorghum cultivars (Mabrouk, Horas, Pionner-8581) were evaluated in the salinity concentrations of 0, 2, 4, 6, 8, 10, 12, 14 and 16 dSm^{-1} NaCl. Four parameters namely final germination percentage, germination index, energy of germination and seedling vigour index were estimated. Eight seedling characters namely shoot and root length, shoot and root foliage weight, shoot and root dry weight, seedling height reduction and relative seedling dry weight were measured. Forage sorghum cultivars significantly varied in means of final germination percentage, germination index, seedling vigor index, energy of germination, shoot and root lengths, shoot and root fresh weights, shoot and root dry weights, seedling height reduction and relative dry weight. Horas cultivar surpassed other cultivars in all of the studied characters followed by Mabrouk cultivar which came in the second rank, while Pioneer 858 cultivar came in the last rank. Pioneer 858 cultivar recorded highest values of seedling height reduction and relative dry weight followed by Mabrouk cultivar, while, Horas cultivar came in the last rank for both characters. Increasing salinity levels from 0 to 15 dSm^{-1}significantly decreased germination percentage, germination index, seedling vigor index, energy of germination, shoot and root lengths, shoot and root fresh weights, shoot and root dry weights and relative dry weight, the control treatment recorded the highest averages of these characters. Thus, the lowest averages of these characters were recorded with highest salinity levels *i.e.* 15 dSm^{-1}. Seedling height reduction percentage was increased significantly with increasing salinity levels from 0 to 15 dSm^{-1}. Results clearly showed that germination index, seedling vigor index, shoot and root lengths, shoot and root fresh weights, shoot and root dry weights, seedling height reduction and relative dry weight significantly affected by the interaction between forage sorghum cultivars and salinity concentrations. It could be concluded that Hoars cultivar was more tolerant to salinity stress followed by Mabrouk cultivars which must be put in breeding program of forage sorghum under salinity conditions.

Keywords- Forage sorghum cultivars, Germination parameters, Seedling characters, Salinity levels, seedling height reduction and relative dry weight.

Introduction

Seed germination is one of the significant factor that process during the stages of forage sorghum life cycle in arid and semi-arid region [11]. Seed germination and early seedling growth are dangerous stage for the establishment of plant populations under saline conditions [18]. The salinity of water and soil is one of the stress factors, especially in the regions where irrigation is unavoidable [28]. Sorghum is appropriate for providing forage because of having high producing amount in the unit of level and the more efficiency of using water, the special morphology features of this plant leads to know that as index of crop resistance to drought [9]. In terms of salinity tolerance degree it also be known as a moderately tolerant crop with the almost salinity tolerance of 6.8 dSm^{-1} [10]. But among varieties of one agricultural species, bearing of salinity factor can be very dissimilar. These differences in agricultural species can be important as an index for assessor salinity tolerance [12]. Seed with rapid germination under salt stress and normal conditions may be predictable to achieve a rapid seedling establishment and more salt tolerance [4]. Xiu-Ling W., et al [30] found that "Rio" and "Theis" cultivars are most salinity tolerant and "BJK236", "Mer-72-2" cultivars are most salinity sensitive among thirty-nine sweet sorghum varieties. Germination rate, germination index, germination energy, vigor index, root length and root fresh weight of sweet sorghum varieties under salt stress are significantly different from those under control. This cultivar was found to be the most tolerant cultivar. Whereas, Jitian2 cultivar was the least tolerant cultivar. Asfaw K.G. [2] indicated that accessions such as 235461, 69239, 223550, 69029 and 23403 were found to be salt tolerant, whereas, accessions 22885, 233247, 237264, 237265 and 237267 were found to be salt sensitive. El Naim A.M., et al [5] pointed out that the cultivar Wad Ahmed (V2) was found salt tolerant with respect to seedling biomass production. Kausar A., et al [16] showed that sorghum lines JS-2002 and Sandalbar were categorized as tolerant, Hegari- sorghum and JS-263 as medium tolerant while Noor as medium sensitive and FJ-115 and PSV-4 as sensitive ones. Rani C.R., et al [23] indicated that CSV-15 and HC-171 cultivars showed the utmost shoot and root lengths, and total dry weight. They also showed that there is a substantial intra-specific variation in salinity tolerance.

Salt stress adversely affects plants at all stages of their life cycle. Salinity affects seed germination by creating an external osmotic potential that prevents water uptake or due to the toxic effects of sodium and chloride ions on the germinating seed [17]. Plants differ in their ability to develop under saline conditions [10]. Salinity compact root length and plant height as the level of salt increased, however, the decrease in the length of the roots was more prominent as compared to the shoots [3,25]. This decrease may be due to the toxic effect of NaCl as well as to an imbalance of nutrients in the plants. Low level of salinity (2 dSm^{-1}) augmented seed germination percentage, while the high levels (4, 8 and 16 dSm^{-1}) significantly inhibited seed germination and seedling growth i.e. shoot and root lengths and shoot and root fresh and dry weights and they added that the solidest inhibition of germination occurred at the higher salt concentrations [3,5,23,26,29]. Sadeghi H., et al [26] showed that aggregate salt concentrations reduced germination and seedling growth in all the cultivars and the extent of decrease varied with cultivars and salt concentrations. The presence of high salt concentration (10 and 12 EC), CSV-15 and HC-171 cultivars showed the utmost shoot and root lengths and total dry weight. The shoot growth of sorghum cultivar was more adversely affected compared to the root growth by salt stress. The specific objectives of this investigation were to compare tolerance levels of three forage sorghum to salinity as well as to determine the relative salinity, germination and seedling parameters responses of the forage sorghum varieties to different levels of salinity.

Materials and Methods

A laboratory experiment was carried out in Seed Science and Technology Laboratory, Agronomy Department, Faculty of Agriculture at Mansoura University during the period from April to Mai 2012. The aim of this investigation was study the response of some forage sorghum cultivars to germinate under salinity levels and to confirm the seedling growth performance for salinity tolerance among forage sorghum cultivars.

Treatments and Experimental Design

The experiment was arranged in factorial experiment in randomized complete block design (RCBD) with four replications. The experiment included two factors, the first factor included three different forage sorghum cultivars, which were obtained from department of Forage Crops, Agricultural Research Center, Ministry of Agriculture and Land Reclamations, Egypt *i.e.* Horas and Mabrouk The third cultivar Pioneer 858 cultivar was obtained from private seed company. All cultivars were stored under normal conditions in paper bags. The second factor included eight salt concentrations of NaCl dSm^{-1} includes 0, 3, 5, 7, 9, 11, 13 and 15 dSm^{-1} NaCl. Seeds of cultivars were surface sterilized by immersion for 5 minutes in sodium hypochlorite solution, then repeatedly washed with deionized water. Fifty seeds of uniform size in each treatment for each cultivar were allowed to germinate on a filter paper in 9 cm diameter Petri dishes. Each filter paper was moistened with a water solution at eight different NaCl concentrations. Thus, the whole experiment comprised 96 Petri dishes and Petri dishes were placed in a growth chamber for 12 days at 25 ±1 °c for germination.

Studied Characters

After 12 days ten seedlings were selected from each replicates and then seedlings were evaluated as follows:

A. Germination Parameters

I. Final Germination Percentage (FGP): The number of seeds germinated at the end of experiment was considered and expressed as percentage as described in the following equation [6,24]:

$$(FGP) = \frac{\text{Number of germinated seeds after 12 days}}{\text{Number of seeds tested}} X100$$

II. Germination Index (GI): It was calculated according to the following equation [15]:

$$(GI) = \frac{\text{\%Germination in each treatment}}{\text{\%Germination in the control}}$$

III. Seedling Vigor index (SVI): it was calculated using following equation [1]:

$$(SVI) = (\text{Average shoot length} + \text{Average root length}) \times \text{Germination percentage}$$

IV. Energy of germination (EG): Energy of germination was recorded at the fourth day as the percentage of germinating seeds 4 days after sowing relative to the number of seeds tested [24].

B. Seedling Characters

I. Shoot length (cm).

II. Root length (cm).

III. Shoot fresh weight (g).

IV. Root fresh weight (g).

V. Shoot dry weight (g).

VI. Root dry weight (g).

VII. Seedling height reduction (SHR): It was calculated using the following equation [14].

$$SHR(\%) = \frac{\text{Plant height at control - Plant height at saline condition}}{\text{Plant height at saline condition}} X100$$

VIII. Relative dry weight (RDW): It was calculated according to the following equation [14]:

$$RDW (\%) = \frac{\text{Total dry weight under saline condition}}{\text{Total dry weight under control condition}} X100$$

Statistical Analysis

All data of this study were statistically analyzed according to the technique of variance (ANOVA) for factorial experiment in Randomized Complete Block Design (RCBD) as published by Gomez K.A., et al [8] by using means of "MSTAT-C" computer software package. Least Significant Differences (LSD) method was used to test the differences between treatment means at 5% and 1% levels of probability [27].

Results and Discussion

Cultivars Performance

Results presented in [Table-1], [Table-2], [Table-3] showed that forage sorghum cultivars a significantly affected final germination percentage (FGP), germination index (GI), seedling vigor index (SVI), energy of germination (EG), shoot and root lengths (cm), shoot and root fresh weights (g), shoot and root dry weights (g), seedling height reduction (SHR %) and relative dry weight (RDW%). Horas cultivar surpassed other cultivars in final germination percentage, germination index and seedling vigor index, energy of germination, shoot and root length (cm), shoot and root fresh weight (g), shoot and root dry weight (g) followed by Mabrouk cultivar, which came in the second rank. While, Pioneer 858 cultivar recorded the lowest values of these characters. It can be noticed that Horas cultivar surpassed other cultivars in averages of FGP, GI, SVI, and EG by 41.67, 22.40, 56.17 and 54.71%, respectively than Pioneer 858 cultivar and by 12.8, 8.23, 18.23 and 13.22%, respectively compared with Mabrouk cultivar. Horas cultivar surpassed shoot and root length, shoot and root fresh weight, shoot and root dry weight, by 31.54, 30.99, 46.66, 44.44, 44.44 and 33.33%, respectively compared with Pioneer 858 cultivar and by 8.48, 13.37, 10, 11.11, 11.11 and 0%, respectively compared with Mabrouk cultivar.

Table 1- Means of final germination percentage (%), germination index (%), seedling vigor index and energy of germination as affected by forage sorghum cultivars and salinity concentrations

Treatments	Final germination percentage (%)	Germination Index (%)	Seedling vigor index	Energy of germination
	Sorghum cultivars performance			
Mabrouk	71.31	77.46	901	60.12
Pioneer 858	47.75	65.5	483	31.37
Horas	81.87	84.41	1102	69.28
F-Test	**	**	**	**
L.S.D. 5 %	2.54	3.35	37	5.16
L.S.D. 1 %	3.38	4.45	50	4.75
	Salinity concentrations effect			
0 dSm-1 NaCl	87.33	100	1862	71.33
3dSm-1 NaCl	79.33	90.37	1320	63.33
5dSm-1 NaCl	74.33	84.17	1051	60
7dSm-1 NaCl	69	77.97	821	57.66
9dSm-1 NaCl	64	72.33	615	51
11dSm-1 NaCl	59.33	66.76	487	45.75
13dSm-1NaCl	53.83	60.36	324	41.66
15dSm-1 NaCl	48.66	54.37	151	38
F-Test	**	**	**	**
L.S.D. 5 %	4.15	5.47	61	5.84
L.S.D. 1 %	5.52	7.27	81	7.77
C-Interaction F-Test	NS	**	**	NS

**, ** and NS mean significant and insignificant respectively at levels of probability 0.05 and 0.01.*

Table 2- Means of shoot length, root length, shoot fresh weight and root fresh weight as affected by forage sorghum cultivars and salinity concentrations

Treatments	Shoot length (Cm)	Root length (Cm)	Shoot fresh weight (g)	Root fresh weight (g)
	Sorghum cultivars performance			
Mabrouk	6.15	5.31	0.027	0.024
Pioneer 858	4.6	4.23	0.016	0.015
Horas	6.72	6.13	0.03	0.027
F-Test	**	**	**	**
L.S.D. 5 %	0.21	0.21	0.001	0.001
L.S.D. 1 %	0.28	0.29	0.002	0.002
	Salinity concentrations effect			
0 dSm-1NaCl	10.96	0.07	0.064	0.06
3dSm-1 NaCl	8.49	7.99	0.042	0.03
5dSm-1 NaCl	7.19	6.7	0.03	0.02
7dSm-1 NaCl	6.01	5.52	0.02	0.01
9dSm-1 NaCl	5.09	4.44	0.015	0.01
11dSm-1NaCl	4.15	3.51	0.011	0.01
13dSm-1NaCl	3.02	2.44	0.008	0.007
15dSm-1NaCl	1.69	1.1	0.003	0.003
F-Test	**	**	**	**
L.S.D. 5 %	0.34	0.35	0.002	0.002
L.S.D. 1 %	0.46	0.47	0.003	0.003
C-Interaction F-Test	**	**	**	**

**, ** and NS mean significant and insignificant respectively at levels of probability 0.05 and 0.01.*

Table 3- Means of shoot dry weight, root dry weight, seedling height reduction (SHR) and relative dry weight (RDW) as affected by forage sorghum cultivars and salinity concentrations.

Treatments	Shoot dry weight (g)	Root dry weight (g)	Seedling height reduction	Relative dry weight
	Sorghum cultivars performance			
Mabrouk	0.008	0.012	47.84	31.78
Pioneer 858	0.005	0.008	48.87	35.71
Horas	0.009	0.012	44.12	31.84
F-Test	**	**	**	**
L.S.D. 5 %	0	0.002	1.86	2.61
L.S.D. 1 %	0.001	0.003	2.48	3.48
	Salinity concentrations effect			
0 dSm-1 NaCl	0.02	0.03	0	100
3dSm-1 NaCl	0.01	0.01	21.83	57.41
5dSm-1 NaCl	0.007	0.01	34.08	37.33
7dSm-1 NaCl	0.005	0.007	45.25	23.5
9dSm-1 NaCl	0.005	0.005	54.25	18.33
11dSm-1 NaCl	0.003	0.004	62.5	14.41
13dSm-1NaCl	0.002	0.003	72.91	9.25
15dSm-1 NaCl	0.002	0.006	84.75	4.66
F-Test	**	**	**	**
L.S.D. 5 %	0.001	0.004	3.05	4.27
L.S.D. 1 %	0.001	0.006	4.05	5.69
C-Interaction F-Test	**	*	**	**

**, ** and NS mean significant and insignificant respectively at levels of probability 0.05 and 0.01.*

Results in [Table-3] clearly showed that highest averages of seedling height reduction and relative dry weight were produced from Pioneer 858 cultivar followed by Mabrouk cultivar which came in the second rank. While, the lowest values of these characters were produced from Horas cultivar. It could be stated that Pioneer 858 cultivar surpassed in seedling height reduction by 9.71% than Horas cultivar and by 2.10% compared with Mabrouk cultivar and surpassed in relative dry weight by 11% compared with Mabrouk cultivar and by 10.83% compared with Horas cultivar. The presence of broad intraspecific genetic variation among sorghum accessions for salt tolerance [7]. Salinity reduced germination parameters as the salt level was increased. The toxic effects of Na^+ and Cl^- ions on germinating seeds affect germination percentage, germination rate, energy of germination and seedling vigour index. Salinity exerts its undesirable effects through osmotic inhibition and ionic toxicity [21]. There are reports of genetic diversity in response to osmotic stress in a wide variety of plant species, which seems at variance with a central tenet of the two-phase hypothesis of salinity resistance [20]. The differences in germination parameters due to forage sorghum cultivars may be due the genetical factors and heredity variation among the three forage sorghum cultivars. These results are in good accordance with those obtained by [2,5,16,30].

Salinity Concentrations Effect

Results in [Table-1], [Table-2], [Table-3] reported that salinity concentrations had a significant effect on averages of FGP, GI, SVI, EG, shoot and root length, shoot and root fresh weight, shoot and root dry weight, seeding height reduction and relative dry weight. Increasing salinity levels from 0 to 15 dSm^{-1} significantly decreased all these character. Vice versa, seedling height reduction was increased with increasing of salinity concentrations. Results in Table 3 through 5 clearly showed that increasing salinity levels from 0 to 15 dSm^{-1} NaCl significantly reduced FGP, GI by (9.16 and 9.63%), (14.88 and 15.83%), (20.98 and 22.03%), (26.71 and 27.67%), (32.06 and 33.24%), (38.36 and 39.64%) and (44.28 and 45.63%), respectively as the salinity levels order compared with the control treatment. While increasing salinity levels from 0 to 15 dSm^{-1} NaCl reduced SVI and EG by (29.10 and 11.21%), (43.55 and 15.88%), (55.90 and 19.16%), (66.97 and 28.50%), (73.84 and 35.86%), (82.55 and 41.59%) and (91.89 and 46.72%), respectively as the salinity levels order compared with the control treatment for both characters, respectively. Moreover, increasing salinity levels from 0 to 15 dSm^{-1} NaCl reduced shoot and root length by (22.53 and 20.65%), (34.39 and 33.46%), (45.16 and 45.18%), (53.55 and 55.90%), (62.13 and 65.14%), (72.44 and 75.76%) and (84.58 and 89.07%), respectively as the salinity levels order compared with the control treatment. In addition, increasing salinity levels from 0 to 15 dSm^{-1} NaCl reduced shoot and root fresh weight, shoot and root dry weight by (34.37, 50.0, 50.0 and 66.66%), (53.12, 66.66, 65.0 and 66.66%), (68.75, 83.33, 75.0 and 67.66%), (76.56, 83.33, 75.0 and 83.33%), (82.81, 83.33, 85.0 and 86.66%), (87.50, 88.33, 90.0 and 90.0%) and (95.31, 95.0, 90.0 and 80.0%), respectively as the salinity levels order compared with the control treatment in each of the mentioned characters. Results revealed that increasing salinity levels from 0 to 15 dSm^{-1} NaCl reduced relative dry weights by 42.59, 62.67, 76.50, 81.67, 85.59, 90.75 and 95.34%, respectively as the salinity levels order compared with the control treatment. Increasing salinity levels from 0 to 15 dSm^{-1} NaCl increased seedling height reductions by 56.11, 107.88, 148.51, 186.30, 233.98 and 288.22, respectively as the order of salinity levels compared with the control treatment. Plant for resistance against salinity increasing of proline, sugar and chlorophyll content in leaves are useful for retaining and improving of photosynthesis quality in plants [22]. Increasing salinity levels decreased germination parameter and seedling growth which is directly related to the amount of absorbed water by the seeds and the toxic effects of Na^+at high salt concentrations might have caused physical damage to roots thereby decreasing their ability to absorb water and nutrient, which may resulted in poor growth [13]. Increment in salinity level caused reduction in final germination percentage FGP in most sorghum cultivars. Likewise, highest salinity levels had delayed germination (increased days of emergence). The delay was pronounced in salt-sensitive accessions than in the intermediate and salt-tolerant ones which is in agreement with previous findings in sorghum [19]. These results are in agreement with those obtained by [3,5,23,25,26].

Interaction Effects

Regarding to the interaction effects data illustrated in [Fig-1], [Fig-2], [Fig-3], [Fig-4], [Fig-5], [Fig-6], [Fig-7], [Fig-8], [Fig-9], [Fig-10] clearly showed that GI, SVI, shoot and root length (cm), shoot and root fresh weight (g), shoot and root dry weight (g), SHR % and RDW % were significantly affected by the interaction between forage sorghum cultivars and salinity concentrations. Vice versa, averages of FGP and EG did not reach the levels of significant. Results clearly indicated that highest percentages of GI was produced from the control treatment in all cultivars. While, the lowest percentages of germination index was produced from the salinity level of 15 dSm^{-1} NaCl with sown Pioneer 858 cultivar [Fig-1]. Highest seedling vigour index was produced from the control treatment, with sown Horas cultivar followed by the control treatment with Mabrouk cultivar without significant differences between them. While, the lowest values of seedling vigour index was produced from the salinity level of 13 or 15 dSm^{-1} NaCl with sown Pioneer 858 without significant differences between them [Fig-2]. In addition, sown Mabrouk cultivar and treated with highest salinity level of 15 dSm^{-1} NaCl also recorded the lowest values of seedling vigour index [Fig-3]. Horas cultivar surpassed in germination index, seedling vigour index, shoot and root length and shoot and root fresh weight under all salinity levels than the other two cultivars. Moreover, Mabrouk and Horas cultivars surpassed in root length under all salinity levels Pioneer 858 cultivar without significant differences between them [Fig-4], [Fig-5], [Fig-6], [Fig-7], [Fig-8], [Fig-9]. Increasing salinity concentrations to 15 dSm^{-1} NaCl and sown Pioneer 858 or Mabrouk cultivars significantly produced highest percentages of seedling height reduction as illustrated in [Fig-10] and lowest percentages of seedling height reduction without significant differences between them. These results are in good accordance with those obtained by [3,5,23,26,29].

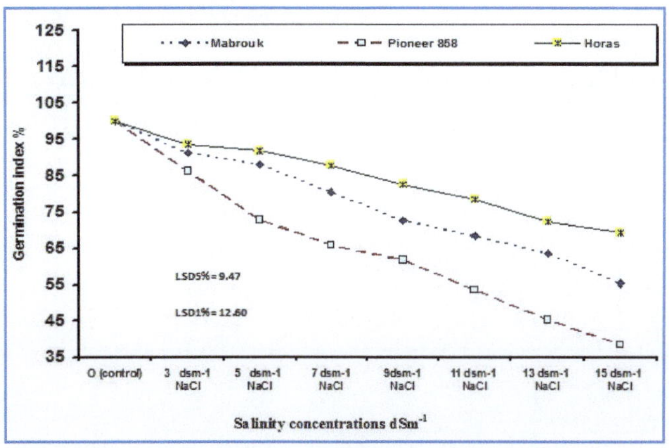

Fig. 1- Means of germination index as affected by the interaction between cultivars and salinity concentrations

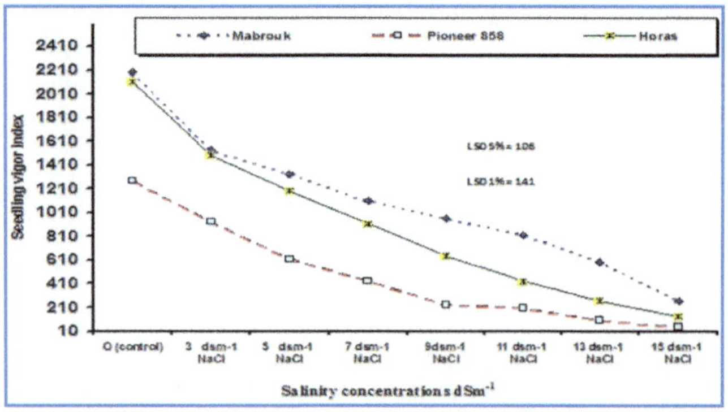

Fig. 2- Means of seedling vigor as affected by the interaction between cultivars and salinity concentrations

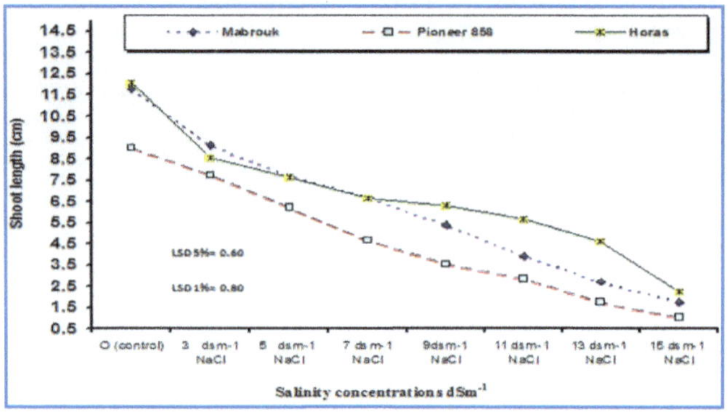

Fig. 3- Means of shoot length as affected by the interaction between cultivars and salinity concentrations

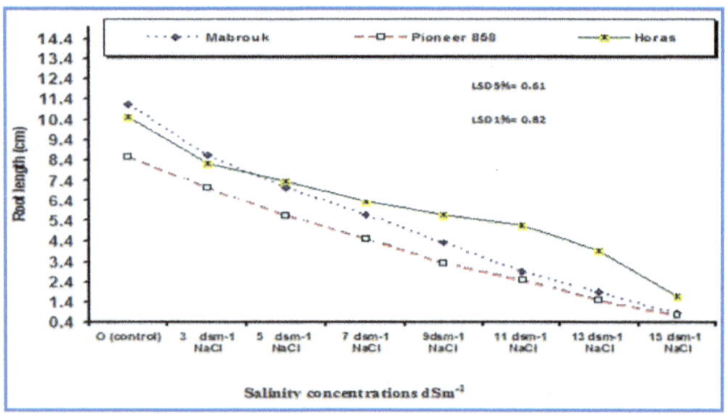

Fig. 4- Means of germination index as affected by the interaction between cultivars and salinity concentrations

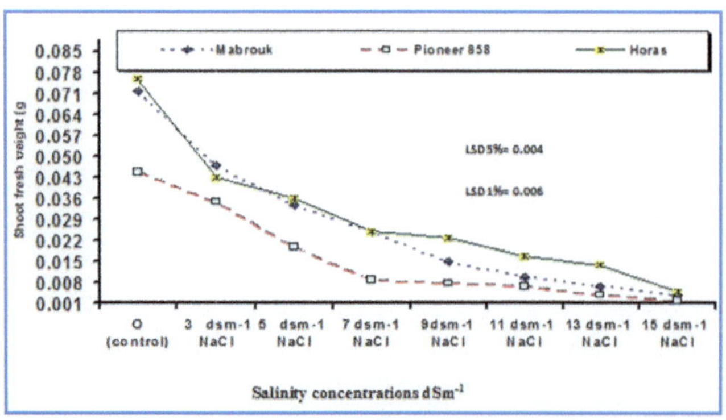

Fig. 5- Means of shoot fresh weight as affected by the interaction between cultivars and salinity concentrations

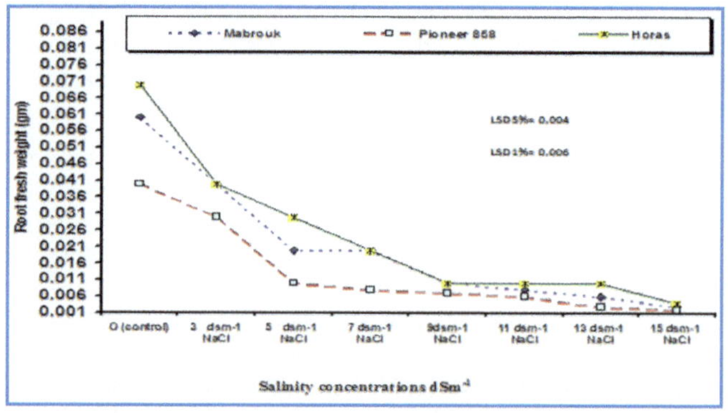

Fig. 6- Means of root fresh weight as affected by the interaction between cultivars and salinity concentrations

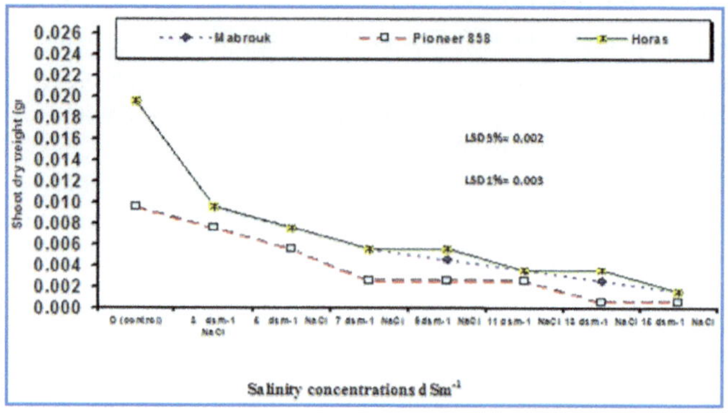

Fig. 7- Means of shoot dry weight as affected by the interaction between cultivars and salinity concentrations

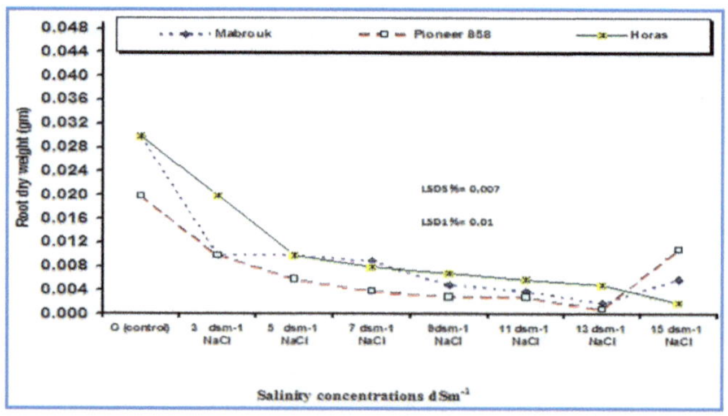

Fig. 8- Means of root dry weight as affected by the interaction between cultivars and salinity concentrations

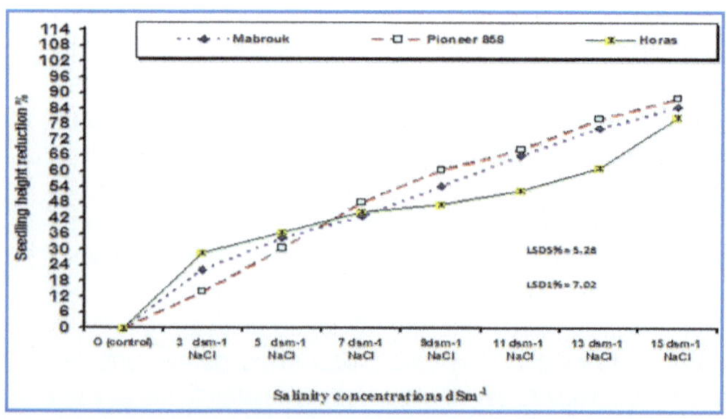

Fig. 9- Means of seedling height reduction as affected by the interaction between cultivars and salinity concentrations

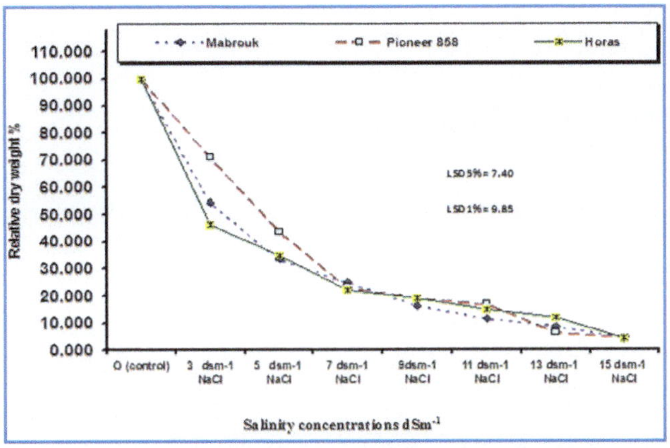

Fig. 10- Means of relative dry weight % as affected by the interaction between cultivars and salinity concentrations

Conclusion

It could be concluded that increasing salinity is associated with decreases in all the studied characters, germination or seedling parameters. To maximize forage sorghum germination percentage and seedling parameters under salinity stress, confirm achieving by using Horas cultivar due to its tolerance to high salinity concentrations and surpassing other cultivars in giving highest values of all the studied characters. This cultivar was more tolerant to salinity up to 15 $dSm^{-1}NaCl$ and is proved to limit or reduce salinity effects on germination or growth of forage sorghums. Thus, it is recommended to use Horas cultivar in breeding programs for enhancing forage sorghum production in Egypt under saline conditions.

References

[1] Abdul-Baki A.A. and Anderson J.D. (1973) *Crop Sci.*, 13, 630-633.

[2] Asfaw K.G. (2011) *Asian Journal of Agri. Sci.*, 3(3), 242-249.

[3] Bashir F., Ali M., Hussain K., Abdul Majeed and Nawaz K. (2011) *Bot. Res. Intl.*, 4(1), 1-3.

[4] Bybordi A. and Tabatabaei J. (2009) *Notulae Botanicae Horti Agrobotanici Cluj-Napoca*, 37(2), 71-76.

[5] El Naim A.M., Mohammed K.E., Ibrahim E.A. and Suleiman N.N. (2012) *Sci. and Techno.*, 2(2), 16-20.

[6] Ellis R.A. and Roberts E.H. (1981) *Seed Sci. Technol.*, 9, 373-409.

[7] Geressu K. and Gezaghegne M. (2008) *African Journal of Agricultural Research*, 3(1), 044-048.

[8] Gomez K.A. and Gomez A.A. (1984) *Statistical Procedures in Agricultural Research, Wiley, 2nd ed.*, New York, 680.

[9] Gossett D.R., Millhollon E.P., Lucas M.C. (1994) *Crop Sci.*, 34, 706-714.

[10]Greenway H. and Munns R. (1980) *Ann. Rev. Plant. Physiol.*, 31, 149-190.

[11]Hadas A. (2004) *Handbook of Seed Physiology: Application to Agriculture, Food Product Press*, New York, 3-49.

[12]Hollington P.A. (1998) *Salinity Management in Agriculture*, 273-289.

[13]Iqbal A., Bakht J. and Shafi M. (2000) *Pakistan J. Biol. Sci.*, 3 (9), 1406-1408.

[14]Islam M.M. and Karim M.A. (2010) *The Agriculturists*, 8(2), 57-65.

[15]Karim M.A., Utsunomiya N. and Shigenaga S. (1992) *Japanese J. of Crop Sci.*, 61, 279-284.

[16]Kausar A., Ashraf M.Y., Ali I., Niaz M. and Abbass Q. (2012) *Pak. J. Bot.*, 44(1), 47-52.

[17]Khajeh-Hosseni M., Powell A.A. and Bingham I.J. (2003) *Seed Sci. Tech.*, 31(3), 715-725.

[18]Khan M.A. and Gulzar S. (2003) *J. Arid Environment*, 53, 387-394.

[19]Marambe B. and Ando T. (1995) *J. Agronomy and Crop. Sci.*, 174, 291-296.

[20]Munns R., Schachtman D.P. and Condon A.G. (1995) *Australian Journal of Plant Physiology*, 22, 561-569.

[21]Munns R., James R.A. and Lauchli A. (2006) *J. Exp. Bot.*, 57, 1025.

[22]Rahdari P., Tavakoli S. and Hosseini S.M. (2012) *Journal of Stress Physiology & Biochemistry*, 8(1), 182-193.

[23]Rani C.R., Reema C., Alka S. and Singh P.K. (2012) *Res. J. Recent Sci.*, 1(3), 1-10.

[24]Ruan S., Xue Q. and Tylkowska K. (2002) *Seed Science & Techno.*, 30, 61-67.

[25]Saberi A.R., Siti Aishah H., Halim R.A. and Zaharah A.R. (2011) *African J. of Biotechnology*, 10(47), 9647-9656.

[26]Sadeghi H. and Shourijeh F.A. (2012) *Asian J. of Plant Sci.*, 11 (1), 19-27.

[27]Snedecor G.W. and Cochran W.G. (1980) *Statistical Methods, 7th Ed.*, 507.

[28]Tabatabaei S.A., Ranjbar G.H. and Anagholi A. (2012) *International Research Journal of Applied and Basic Sciences*, 3(2), 305-308.

[29]Wambua J.M., Makobe M.N., Njue E.M. and Nyende A.B. (2010) *JAGST.*, 12(2), 82-91.

[30]Xiu-Ling W., Xu C. and Gui-Ying L. (2010) *Chinese J. of Eco-Agric.*, 18(6), 1239-1244.

AGRICULTURE PRODUCTIVITY IN SOLAPUR DISTRICT OF MAHARASHTRA: A GEOGRAPHICAL ANALYSIS

AWATE S.J.[1] AND TODKARI G.U.[2]

[1]Dept. of Geography, Walchand College, Solapur, MS, India.
[2]Geography Dept. Shri Sant Damaji College, Mangalwedha, Solapur, MS, India.
*Corresponding Author: Email- govindtodkari@gmail.com

Abstract- As the largest private enterprise in India, agriculture contributes one forth of the national GDP. Agriculture has been and will continue to be the life line of India economy. However, agriculture productivity and development differ very much from region to region, which needs a detail investigation. The temperature and pedagogical conditions are favorable for growing valuable crops like jawar, sugarcane, oil seeds, bajara, wheat, etc. By contrast, very low level of agriculture productivity is confined to the tahsils belonging to drought prone areas having irregular rainfall, rugged topography and poor irrigation facilities. Inadequacy of water is main hurdle in agriculture productivity. For the present investigation tahsilwise secondary data has been collected from socio-economic review and District Statistical Abstract. The data collected has been processed and method of yield co-efficient method has been employed to find out the levels agriculture productivity. The result are shown in tabular and from and are also depicted by choropleth method on map. Hence, in present paper an attempt has been made to assess the regional disparities in levels of agriculture productivity in Solapur District of Maharashtra.
Key words- agriculture productivity, yield co-efficient, Technological factors, wide variation, sustainability.

Introduction -

Agriculture has been and will continue to be the lifeline of Indian economy. The progress made by agriculture in the last four decades one of the biggest success stories of free India. Agriculture and allied activities constitute the single largest contributor to the gross domestic product. Agriculture is the means of livelihood of about two-third of the workforce in the country. This increase in agriculture product has been brought about by bringing additional area under cultivation, extension of irrigation facilities, the use of improved high variety of seeds, water management and better techniques evolved through agriculture research, pesticides and cropping practices. However, agriculture productivity differ very much from region to region in Solapur District which needs a detail investigation.

Objectives-

The present investigation aims to find out meaning of agriculture productivity, different methods of measurement agriculture productivity and evaluation the major crop productivity and spatial variation in agriculture productivity in study region.

Database and methodology

Present study mostly relies on the secondary data collected through Agriculture Department, District statistical Department of Solapur and socio-economic abstract of Solapur district in 2009/10. For the present investigation, District is selected as in general and tahsils in particular. Firstly, major crops productivity is measured and the productivity indices have been calculated on the basis of a statistical technique formulated by Jasbir Singh. The Figers and district maps are used for comparative study of spatial distribution. The yield index of each crop is calculated by the equation given below:

$$\text{Yield of Crop} = \frac{\text{Yield of crop 'a' in the areal unit} \times 100}{\text{Yield of crop 'a' in the region.}}$$

Study Area

The present study deals with the geographical perspectives of the agriculture in Solapur district. The Solapur district is bounded by 17°05' North latitudes to 18° 32' North latitudes and 74°42' east to 76°15' East longitudes. The total geographical area of Solapur district is 14895 K.m². divided into eleven tahsils.

Climate of the district is dry. The daily mean maximum temperature range between 30° C to 35° C and minimum temperature range between 18°C to 21°C. The highest temperature is 47° C recorded in the month of May. The average annual rainfall is registered 510 mm. The soil of the district essentially derived from the Deccan trap. The soil of the district can broadly classify into three groups shallow, medium and deep soil.

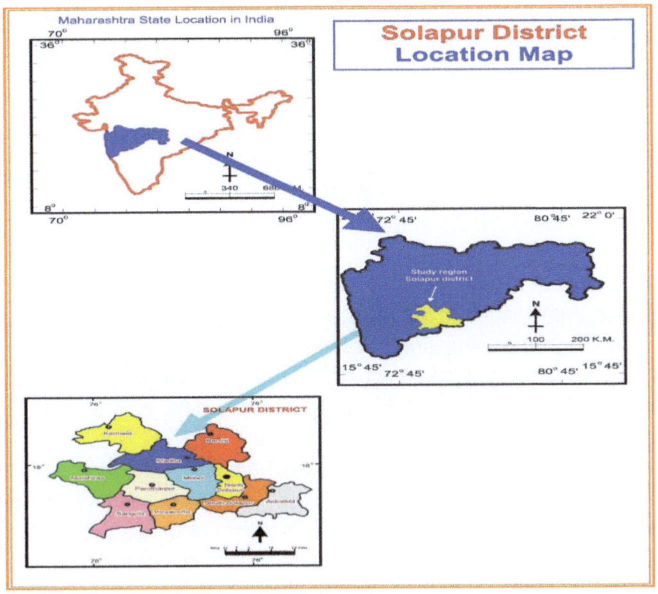

Fig. 1-

Concept of Agriculture productivity-

Agriculture productivity may be defined as the ratio of the index of total agriculture output to the index of total input used in farm production. In the conference of the Indian Economics (23rd) suggested by some economist that the yield per acre should be considered to indicate agriculture productivity. Pandit has stated that productivity is defined in economics as the output per unit of input or the getting some output from smaller input. He further suggested that increase in productivity is generally use of a more efficient use of all or some factors of production i.e. land, labour and capital.

Agriculture productivity on macro and micro level in the region closely influenced by various physical (physiography, soil, water, climate), socio-economic, political, technological and organizational factors. The traditional approaches of measurement of agriculture productivity does not take into consideration the social and environmental cost which are also helpful to increase the productivity of agriculture.

In recent decades geographers and economists have developed sophisticated tools and techniques to determine the agricultural productivity. Some of the techniques developed and used for measurement of agriculture productivity are given below

- Out per unit area
- Production per unit of farm level
- To asses agriculture production as grain equivalent (Buch, 1956)
- Input output ratio (Khusro 1964)
- Ranking co-efficient method (Kendal 1939, Stamp 1960, Shafi 1990)
- Carrying capacity of land in term of production (Stamp 1958)
- A system of four co-efficient a) productivity of co-efficient, b) ranking co-efficient, c) money value co-efficient, d) energy co-efficient (Kendal 1939)
- Giving weight age to the ranking order of the output per unit area with the percentage share under each crop(Deshpade 1964, Bhatia 1967)
- Determing an index of productivity (Shafi 1972)
- Computing the crop yield and concentration indices ranking co-efficient (Jasbir Singh 1976)
- To asses a agriculture productivity in term of money
- Assessing the net income in Rs per hectors of cropped (Jasbir Singh 1985)

Productivity of principal crops

Wheat, Jawar, Bajra, Maize, Tur, Gram, Sugarcane, Cotton and Groundnuts are the important crops of the Solapur District. Sugarcane and Jawar is accounting for more than 50 percent of the total cultivated area of the region. Jawar can be grown both a Rabi and Kharif crops in Solapur district. Wheat is more common in dry areas. Sugarcane is the principal crop of the region. It is a late arrival in the field of cash crops. Its cultivation is confined to the areas receiving adequate and timely irrigation.

The crop productivity changes have occurred in response to many technological developments during the last few decades. The adoption of seeds, fertilizers and irrigation has resulted into increase of farm production and diversifying the production pattern. The soil conservation has been an addition, for increasing the productivity. Thus, all these factors interact the changes in agricultural production.

Jawar Productivity- There are wide variations in the yield of Jawar in the region under study. The average productivity of the region as a whole is 449 Kg per hectare. The tahsil wise yield ranges from 225.4 Kg. per hectare to 812.8 Kg per hectare. In the region, the three tahsils have recorded a yield level of greater than 725 Kg. per hectare. The very high productivity is observed in Mangalvedha tahsil with an average productivity of 812.8 Kg per hectare. The improved dry farming techniques assured supply of water, use of new strains of seeds, proper also of fertilizers, and modern methods in Jawar husbandry, all of them are responsible for higher production. The moderate level of productivity is noted in Karmala, Pandharpur, Mohol and South Solapur. At the other extreme, remaining all tahsils has low productivity with output per hectare of less than 425 Kg per hectare.

Bajra Productivity-There are wide variations in the average level of productivity of bajra. The average productivity of bajra, in the region is 415 Kg per hectare. In the region understudy the four tahsils have productivity above district average and seven tahsils have below average. The major parts of Solapur district i.e. Mohol and North Solapur have high level of bajra productivity. The moderate productivity is recorded in Barshi, Karmala, Malshiras, Sangola and Mangavedha tahsils. In Madha, Pandharpur, South Solapur and Akkalkot, the productivity is low. Recently new strains and the improvement in dry farming techniques have improved the productivity of bajra.

Table 1- Tahsilwise major crops productivity (2009/10)

Sr. No.	Crops	Yield/ hectare in Kg.	Productivity tahsils		
			High	Moderate	Low
1.	Jawar	449	Mangalvedha	Karmala, Pandharpur, Mohol, South Solapur	Barshi, North Solapur, Malshiras, Akkalkot, Sangola, Madha
2.	Bajra	415	Mohol, North Solapur	Barshi, Karmala, Malshiras, Sangola, Mangavedha	Madha, Pandharpur, South Solapur, Akkalkot
3.	Wheat	1100	Karmala, Malshiras, Pandharpur, North Solapur	Madha, Mohol, south Solapur	Barshi, Akkalkot, Mangalvedha, Sangola
4.	Tur	387	North Solapur South Solapur Akkalkot	Sangola, Pandharpur, Madha, Mohol	Barshi, Karmala, Malshiras, Mangalvedha
5.	Gram	564	Barshi, Karmala, Malshiras, South Solapur	Sangola, Pandharpur, Madha, Akkalkot	North Solapur, Mangalvedha , Mohol,
6.	Groun dnut	932	South Solapur	Barshi, Karmala, Malshiras North Solapur, Mangalvedha, Mohol, Sangola, Pandharpur, Madha	Akkalkot

Source- Complied by Author

Wheat productivity-The average productivity of wheat in the region is 1100 Kg. per hectare. There are only four tahsils which have a yield above 1500 Kg per hectare. These four tahsils are namely Karmala, Malshiras, Pandharpur and North Solapur; all together make a major contribution to wheat output in the region. The moderate production of wheat is observed in Madha, Mohol, and south Solapur tahsils and low production in Barshi, Akkalkot, Mangalwedha and Sangola tahsils with an average productivity of 1000 Kg per hectare.

Tur Productivity- Tur is a more important food grain of the region. The average productivity of the region is 387 Kg per hectare. There are three tahsils namely, North Solapur, South Solapur, and Akkalkot, which have recorded productivity level higher than 500 Kg per hectare. At the extreme, the four tahsils namely Barshi, Karmala, Malshiras and Mangalwedha which have productivity levels less than 250 Kg per
hectare. Sangola, Pandharpur, Madha, Mohol tahsils have productivity ranging between 250 to 500 Kg. per hectare. The levels of tur productivity are very high in those parts of the district where tur concentration is significant. Tur is used by the people in daily diet as a pulse and hence, it is grown by the people in sizable area of the Solapur District.

Gram Productivity- Gram is known as (HARBARA) in rural areas. The average productivity of region is 564 Kg per hectare. There are four tahsils namely Barshi, Karmala, Malshiras and South Solapur which have productivity, levels higher than 500 Kg per hectare. There are three tahsils which have productivity levels, less than 250 Kg per hectare. Low productivity occurs in North Solapur, Mangalvedha and Mohol, where local factors such as poor soils, relief and low intensity of irrigation directly affect the level of gram production. Elsewhere, the level of gram production ranges from 250 to 500 Kg per hectare.

Groundnut productivity- Groundnut is the important oilseed crop of the region and its area of concentration is in southern part of South Solapur tahsil. The average productivity of region is 932 Kg per hectare. It recorded high productivity with an average productivity of 653 Kg per hectare while low productivity is in Akkalkot tahsil with an average productivity of 269 kilogram per hectare. Elsewhere, the levels of production are moderate due to agronomic limitation and the competition with grain crop.

Measurement of Productivity by Yield Index Method

The regional imbalance in agricultural productivity of Solapur district is measured by using index method. While applying this method, the seven crops grown in most of the tahsils are selected.

Areas of high productivity- It is mainly confined to the central and eastern part of district excluding south Solapur tahsil. Thus the tahsils of Mohol, Karmala and North Solapur recorded high productivity.

Areas of moderate productivity- This category of productivity is mainly observed in western part of Solapur district except Karmala. The rainfall is low in sangola tahsil. Irrigation facilities are not more developed. The average yield of crop is low. Kharif crops are dependent on monsoon and Rabi crops are grown only where water is available for irrigation.

Areas of low productivity- In the central part of Solapur district low production is recorded. This part consists of Madha, Pandharpur, Malgalvedha and Akkalkot tahsils. Hence, the yield of crops is very low. The irrigation facilities are not developed. Agricultural implements are old and traditional. Rainfall is uncertain, thus, the combined effect of all these factors leads to the low agricultural productivity.

Table 2- level of Agriculture Productivity

Sr. no.	Agriculture productivity	Co-efficient value	Share of agri. Productivity region (%)	No of tahsils	Tahsils name
1	High	Above 115	25.20	3	Karmala, N. Solapur, Mohol
2	Moderate	90-115	38.60	4	Malshirash, Sangola, Barshi, S. Solapur
3	Low	Below 90	36.19	4	Mangalwedha, Madha, Pandharpur, Akkalkot

Source- complied by Authors.

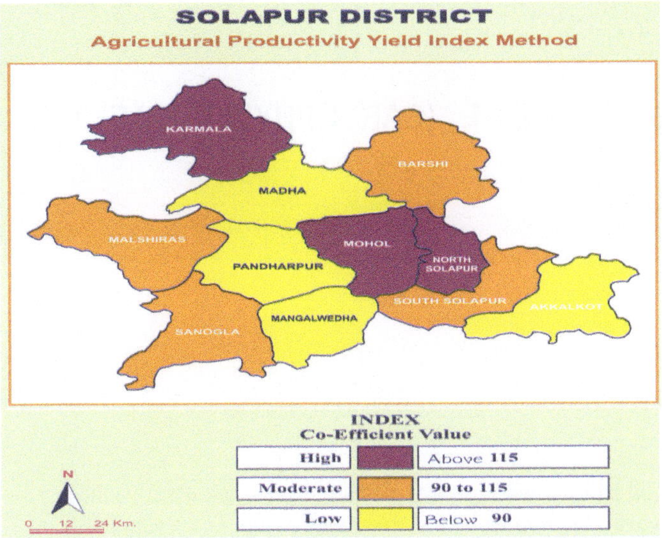

Fig. 2-

Concluding Remarks

The analysis reveals that there are great variation in level of agriculture productivity throughout the study region. Only 25.20 percent area of the study region is high productivity region, whereas 36 percent area of the study region is belong to low level of agriculture productivity. Relatively high level of productivity is observed in Karmala, N. Solapur and Mohol tahsils. The low level of productivity is confined to 4 tahsils which falls mostly in drought prone areas having inadequate, irregular and insufficient rainfall. And another group of tahsils belongs to high use of irrigation water where one crop (sugercane) is continuous cultivated.

The foresaid analysis clearly indicates that there are rare variation in level of agriculture productivity in the region. It needs to be taken care of while formulating development of productivity policies. The tahsils with low level of agriculture productivity should be given top priority so that they may come up at par with high productivity areas and the concept of social justice may be fulfilled.

References -

[1] Bharadwaj O.P. (1964) *The national Geographical Journal of India.*, 10(2).

[2] Majid Hussain (2004) *Agriculture Geography.*

[3] Mohammad Shafi (2006) *Agriculture Geography.*

[4] Noor Mohammad (1992) *New dimensions in Agricultural Geography.* 8(4).

[5] Patil N.D. (1981) *Improved crop production technology for drought prone areas of Maharashtra.*

[6] Siddiqui A.R. and Jat B.C. (2009) *The Deccan geographer*, 48 (2).

[7] Varsha Vaidya and Date V.S. (1992) *Transaction of the institute of Indian geographers,*14(2).

[8] Vishwakarma D.D. (2009) *The Deccan geographer*, 48(2).

12

SEED MICROMORPHOLOGICAL OF SOME SPECIES *Minuartia* L. (CARYOPHYLLACEAE) IN IRAQ

AL-SAADI S.A.A.M.[1]* AND AL-TAIE S.S.[2]

[1]Department of Biology, College of Science, University of Basra, Basrah, Iraq.
[2]Department of Biology, College of Science, University of Misan, Amarah, Iraq.
*Corresponding Author: Email- saharmalik2010@gmail.com

Abstract- Seed micromorphology of seven species of *Minuartia* L. were examined by light and scanning electron microscope. High variation was found in seed coat micromorphology were observed. Most of the species investigated had verrucate ornaments except *M. meyeri* and *M. aucheriana* (with verrucate-reticulate ornaments), and *M. hamate* (depressed verrucate ornaments). The shape of cell margin in seeds shows a dentate except *M.hybrida* was sinuate *M.meyeri* was serrate and *M.picta* was irregular dentate. While cell shape also various between species, it was rounded polygonal in *M.montana* and *M.intermedia*, but they are oblong in *M.meyeri* and *M.hamata*, and elongated polygonal in other species.

The testa cell contours of most studied taxa are V-shaped, but they are irregular in *M. picta*. While position of hilum appear central in all species, exception *M.aucheriana, M.picta* and *M.montana* were showed subcentral.

Keywords- Confusion matrix, Data Mining, Decision tree, Neural Network, stacking ensemble, voted perceptron

Introduction

Minuartia L. belongs to the family Caryophyllaceae (the subfamily Minuartioideae (DC.) Beilschm., tribe Alsineae DC.), according to [1], *Minuartia* is a comprises an estimated 175 species distributed in temperate and arctic areas of Asia minor, Europe, northern Africa, North and South America [2]. *Minuartia* includes seven sections and 74 taxa in the flora of Turkey [3]. The genus is represented by 57 species in Europe [4], while in Palestine with 7 species [5]. Rechinger [6] mentioned in Flora Iranica 21 species belonging to six sections in Iran. Only nine species were distribution in north and west region of Iraq.

The studies of seed morphology have taxonomic value, and providing information that facilitates identification of species and genera. Bittrch [7] pointed that the seeds of Caryophyllaceae family different forms reniform, pyriform and orbicular, the size of the seeds ranges from 0.4-3 mm length as well as in colors black, brown and white.

Seed morphology of *Minuartia* has been described in several studies The seed morphology of the genera *Arenaria* and *Minuartia* examined by Wofford [8]. Yildiz [9] studied macro and micromorphology of seed of 17 species (12 genera) of Caryophyllaceae from Turkey(North Anatolia) by scanning electron microscope(SEM) techniques including two species from *Minuartia* (*M. anatolica* var. *anatolica* and *M. erythrosephala* var. *cappadocica*, he showed that seed type of *Minuartia* (*M. anatolica* var. *anatolica* ovoid while *M. erythrosephala* var. *cappadocica* reniform. Koc, et al [10] study one species of *Minuartia* (*M. turcica* M. Koc.) and reported that the seeds oval or reniform and it surface tuberculate.

Mostafavi, et al [11] indicate that seed micromorphology of *Minuartia sabalanica* Assadi & Mostafavi, provides more effective characters for distinguishing *M. sabalanica*from *M. umbellulifera*. In addition, *Minuartia sabalanica*has an almost reniform seed shape covered with irregular polygonal cells. Two species *Minuartia hamzaoglui* M. Koc. & Aksoy and *Minuartia erythrosepala* (Boiss.) Hand. Mazz were described, the seed micromorphology as reniform and anticlinal cell wall V-undulate [12], As well as four species of *Minuartia* (*Minuartia biflora* (L.) Schinz & Thell., *M. hybrida* (Vill.) Schischkin, *M. kashmirica* (Edgew.) Mattf., *M. meyeri* (Boiss.) Bornm. were studied by Kanwal, et al [13].

Some qualitative and quantitative characters such as seed length and width, seed shape, cell shape, and cell ornamentations were studied by Mostafavi, et al [14] studied 20 species of *Minuartia* L. that represent 2 different subgenera (*Minuartia* subgen. *Spergella* (Fenzl) McNeill and *Minuartia* subgen. *Minuartia*) using scanning electron microscopy, while eight taxa of Caryophyllaceae, belonging to the genus *Minuartia* L., *Moehringia* L., *Petrorhagia* (Ser.) Link and *Silene* L., have been discussed in taxonomic, biogeographically and endemism points of view, three species belonging to the *Minuartia garckeana*, *M. bulgarica* and *M. setacea* characters the seed of *Minuartia* was small tuberculate [15].

Seed micro-morphology by scanning electron microscope (SEM)

has been applied in systematic studies about different genera of Caryophyllaceae for example *Arenaria* L. [16], *Gypsophila* L. [17] *Silene* L. [18-21] *Stellaria* [8,9,16,22-26].

No detailed reports are available on seed morphology for genus *Minuartia* from Iraq so that this study describes the amount of variation within taxa of *Minuartia* and assesses the value of seed morphology in determining interrelationships between species.

Materials and Methods

In the present study, seven taxa of the *Minuartia* have been investigated. Seeds of these species were collected from herbarium specimens deposited in BAG (National Herbarium of Iraq, Baghdad, Abo-Ghraib) during 2013 -2014. The list of specimens is shown in [Table -1]. At least three seed were studied for each taxon in order to calculate the mean and to ensure the constancy of characters. Seed were mounted on aluminum stabs and were sputter coated with a thin layer of gold-palladium. The general exomorphological features of the seeds were examined using Olympus microscope model SMZ 600 using image analysis software Dc- 2, and scanning electron microscope Model Hitachi s-3000H at an accelerating voltage of 10-15 KV.

The data of seed criteria were numerically analyzed using SPSS 17 program for creating the data matrix. The similarity of pattern was measured according to the Jaccard's coefficient as defined [27].

The terminology to describe seed coat surface sculpturing follows previous studies on Caryophyllaceae seeds [9,14,17,28, 29].

Results

Morphological features of the seeds such as size, color, shape and testa cell are very various within *Minurtia* species. A set of relevant characters that were found distinctive for the species is reported in [Table-2]. The seed morphology and sculpture patterns are shown in [Fig-1], [Fig-2], [Fig-3] & [Fig-4].

Seed Shape

The seeds shape within *Minurta* is reniform, reniform - pyriform, and elliptic. Most of the species, including *M.montana*, *M. picta* and *M. aucheriana*, have reniform- pyriform seeds. While found reniform in *M.meyeri* and *M. hybrida*. Seed shape in only two species *M. hamata* and *M. intermedia* are elliptic [Table-2], [Fig-1], [Fig-2], [Fig-3] & [Fig-4].

Size

According to the [Table-2], the largest seeds occur in *M. montana* (sect. *Minuartia*) was 0.83 ×0. 61mm. The smallest seeds are found in *M. picta* (sect. *Spergella*) was 0.51-0.43 mm, which was morphologically different from other species of *Minuartia*.

Color

Seed color in most cases was brown to dark- brown in *M. intermedia* and *M. picta*. Two species brown color represented in *M. hybrida* and *M. monata,* while other two species were different from other species, *M.aucheriana* with radish-dark brown color and *M. hamata* was yellow -brown color [Fig-1].

Ornamentations

The exomorphology of seed coat varies among studied species. Most of the species investigated had verrucate ornaments except *M. hybrid* and *M. aucheriana* with verrucate-reticulate ornaments, and *M. hamata* was depressed verrucate ornaments and appear some appendages on the side of seed [Table-2], [Fig-1], [Fig-2], [Fig-3].

Table 1- The list of the *Minuartia* species used for seed micromorphological study and their localities

Taxa	Localities	Herbarium Number
M. aucheriana	NE of Qandil - Erbil	2441
M. hamata	Jabel Singar - Mosul	52172
M. hybrida	12 km. from Sinjar to Karsi - Mosul	37731
M. intermedia	Karol N.W. Slope - Mosul	52459
M. meyeri	Avroman (Horaman) Mountains around Tawila, in Sulaimaniya Liwa	6796
M. montana	10 km. W. of Tawela (on road between Halabja &Tawela) - Sulaimaniya	22132
M. picta	10-15km. E of Erbil (to Darband)	25288

Table 2- Comparison of seed micromorphological data in *Minuartia* species.

Species	Seed length (mm)	Seed width (mm)	L/W	Ornamentation	Seed shape	Number of seeds	Seed color	Cell shape	Cell margin shape	Position of helium
M. aucheriana Bomm.	(0.5-1) 0.75	(0.4-0.7) 0.58	1.29	Verrucate reticulate	reniform -pyriform	20-25	Radish - dark brown	Elongated polygonal	Dentate	Sub-centric
M. hamata (Hausskn.) Mattf.	(1.5- 1.75) 1.58	(0.7-1) 0.85	1.85	Depressed verrucate	elliptic	1	Brown - yellow	Oblong	Dentate	Central
M. hybrida (Vill.) Schischk.	(0.4-0.6) 0.52	(0.4-0.45) 0.41	1.21	Verrucate-reticulate	reniform	6	Brown	Elongated polygonal	Sinuate	Central
M. intermedia (Boiss.) Hand.-Mazz.	(0.66-0.90) 0.79	(0.5-0.7) 0.53	1.41	Verrucate	elliptic	02-12	Brown - Dark brown	Rounded polygonal	Dentate	Central
M. meyeri (Boiss.) Bornm.	(0.6-0.8) 0.68	(0.5-0.6) 0.56	1.21	Verrucate	reniform	04-06	Brown - Dark brown	Oblong	Serrate	Central
M. montana L.	(0.7-1) 0.83	(0.5-0.7) 0.61	1.48	Verrucate	reniform-pyriform	05-08	Brown	Rounded polygonal	Dentate	Sub-centric
M. picta (Sibth. & Sm.) Bornm.	(0.50-0.55) 0.51	(0.4-0.5) 0.43	1.02	Verrucate	reniform-pyriform	03-14	Brown - Dark brown	Elongated polygonal	Irregular dentate	Sun-centric

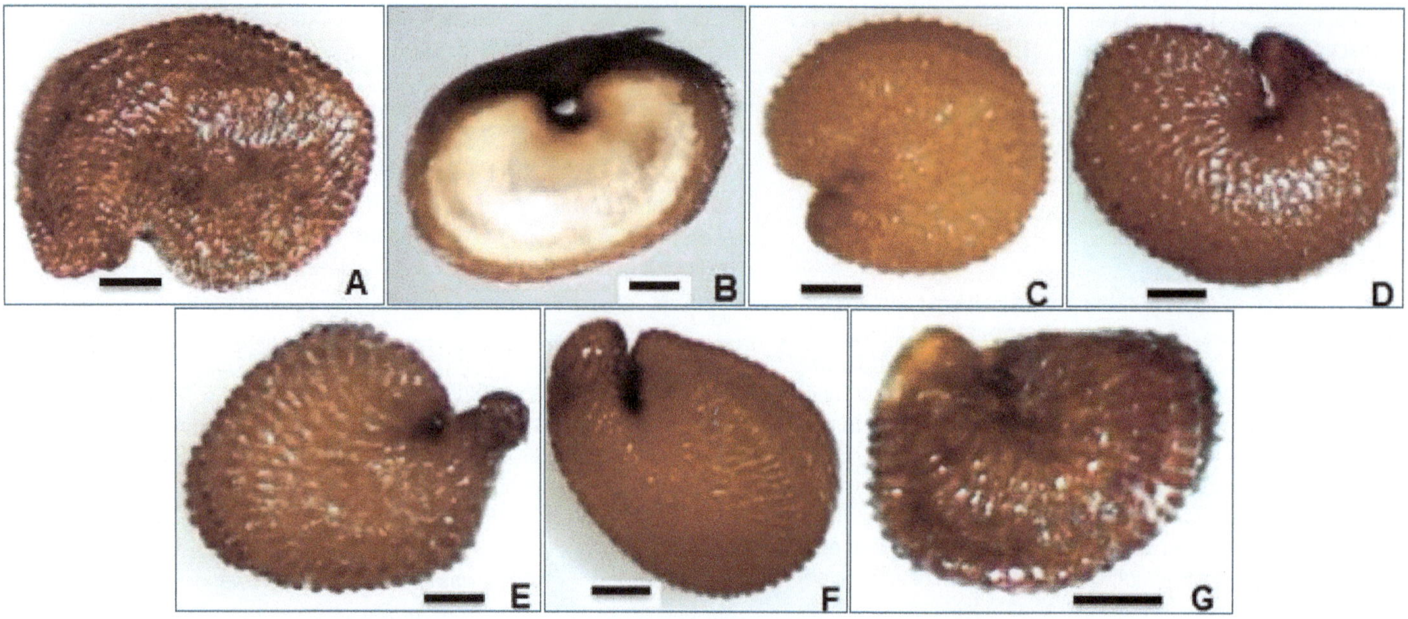

Fig. 1- Variation in seed shape of *Minuartia* (0.1 mm): **(A)** *M. aucherana*; **(B)** *M. hamate*; **(C)** *M. hybrid*; **(D)** *M. intermedia*; **(E)** *M. meyeri*; **(F)** *M. montana*; **(G)** *M. picta*.

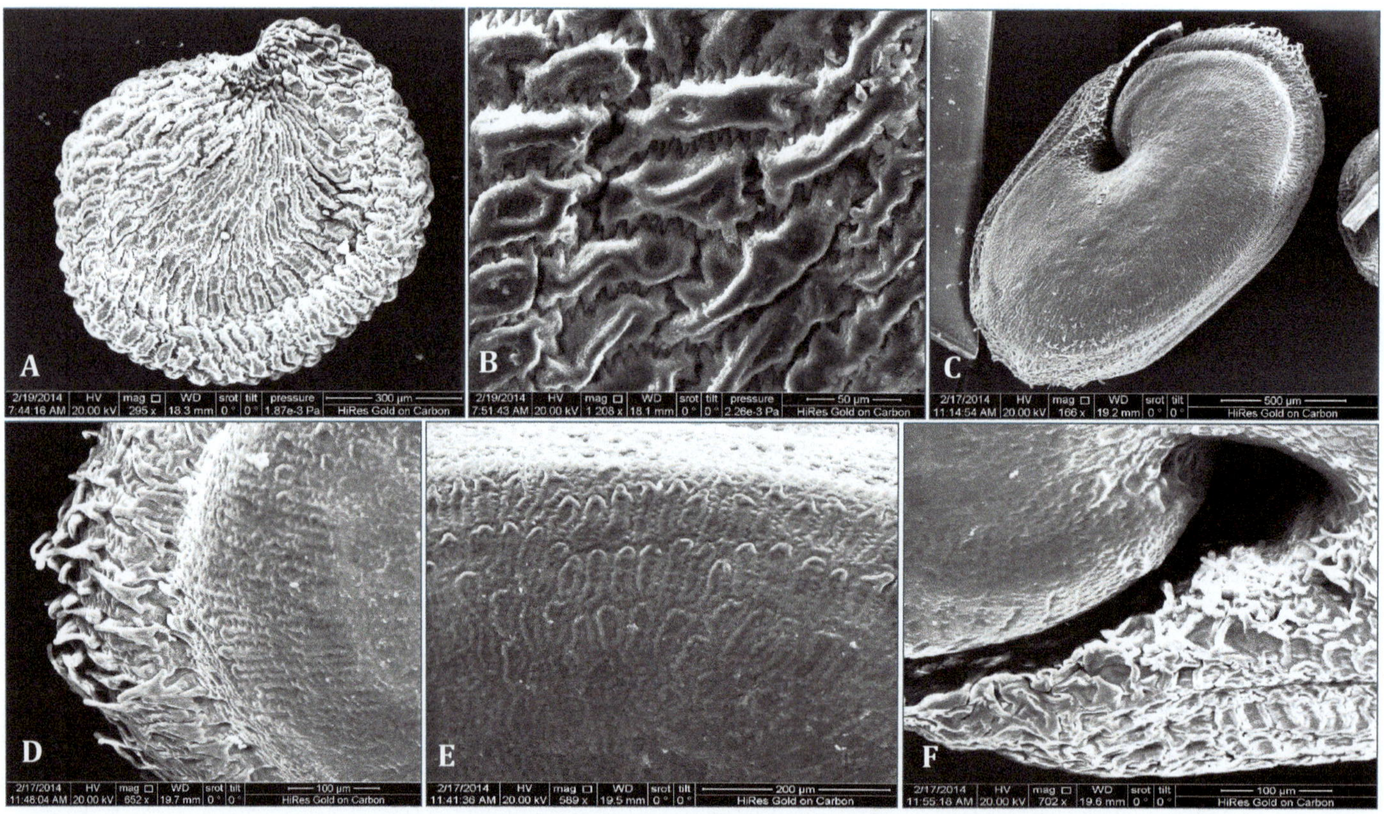

Fig. 2- SEM micrographs of seed morphology of *Minuartia* species: **A,B**: *M. aucheriana*; **C,D,E,F**: *M. hamate*

Seed Coat Cell

Most of the species investigate showed that shape of cell margin in seeds a dentate, while it was irregular dentate in *M. picta*, sinuate in *M. hybrid and serrate in M.meyeri* [Table-2], [Fig-3], [Fig-4]. While cell shape also various between species, it was rounded polygonal in *M. montana*, But they are oblong in *M. meyeri*, and elongated polygonal in other species.

The testa cell contours of most studied taxa are V-shaped, but they are irregular in *M. picta*. Most examined species is formed by convex epidermal periclinal wall and having papillae, exception was observed in seed feature of *M.aucheriana* that was corrugate. In addition *M. intermedia* and *M. meyeri* cells show a more evident secondary sculpture forming rough dots in cell margin [Fig-3]. Position of hilum appear central in all species, exception *M.aucheriana*, *M. picta* and *M. montana* were showed subcentral.

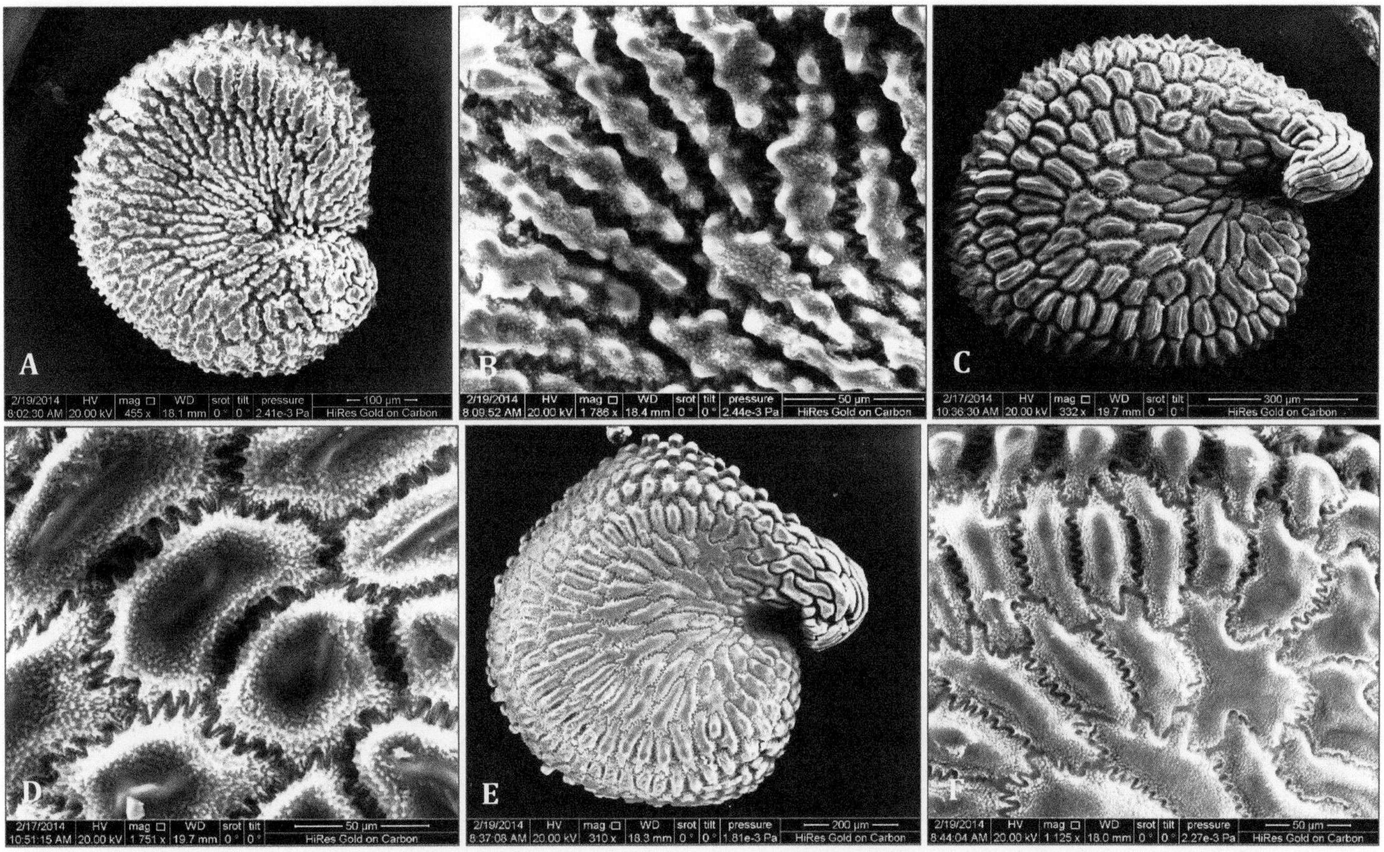

Fig. 3- SEM micrographs of seed morphology of *Minuartia* species: **A,B:** *M. hybrid*; **C,D:** *M. intermedia*; **E,F**: *M. meyeri*

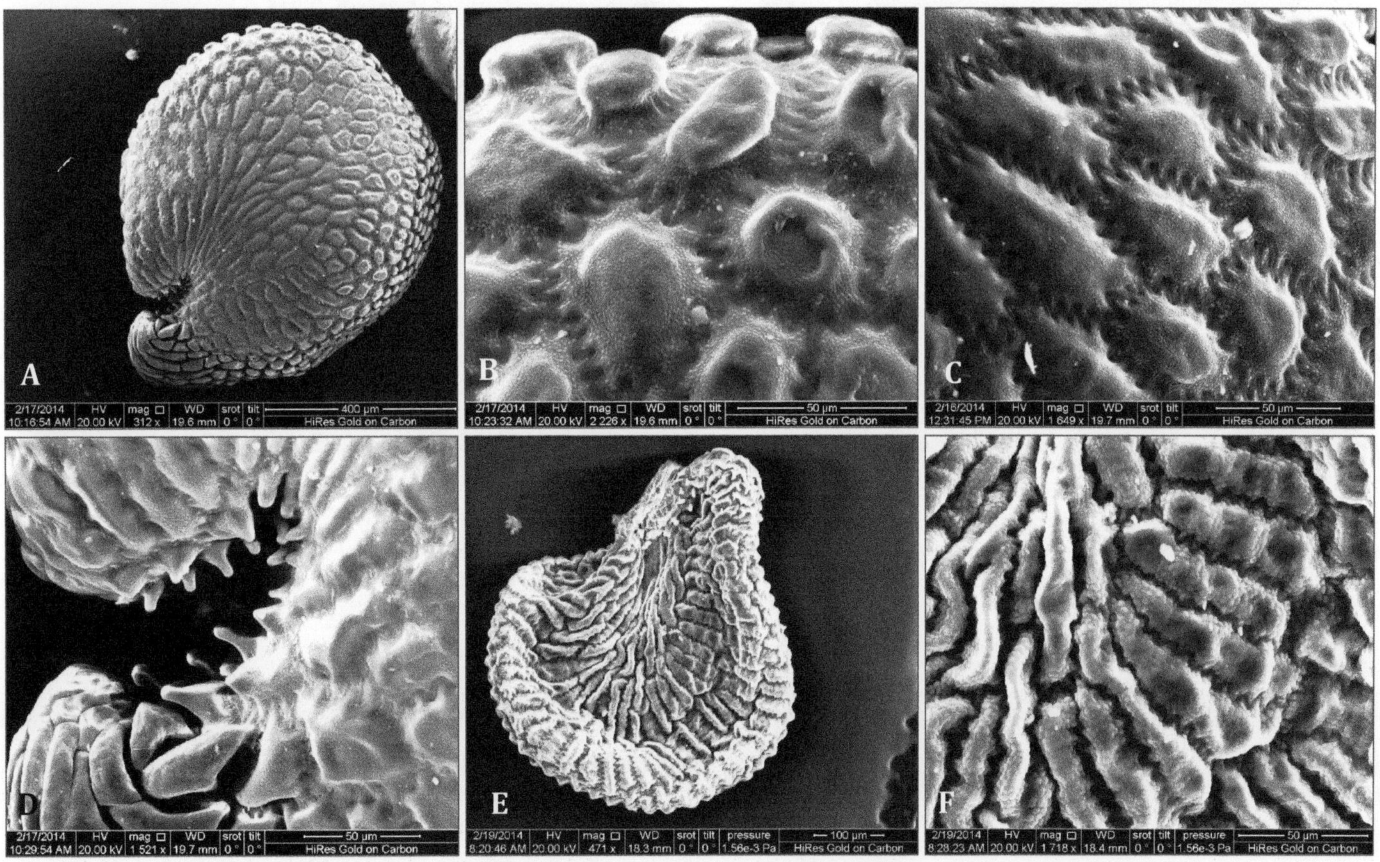

Fig. 4- SEM micrographs of seed morphology of *Minuartia* species: **A,B,C,D**: *M. Montana*; **E,F:** *M. Picta*

Number of Seeds

The number of seeds have been given a good taxonomic value between species, it differ between one seed in *M. hamata* to 20-25 seeds in *M.aucheriana* [Table-2].

Discussion

The seed micromorphology in *Minuartia* species bear a great variation in seed coat micromorphology which can be used for taxonomic purposes and provided valuable and distinctive characters for separating taxa at the species level. Results suggest that qualitative characters including seed shape, color parameters, cell ornamentation, cell margin shape and also quantitative including length and width of seed have much importance in evaluating relationships within *Minurtia* species. Seed morphology of the most examined species are reniform-pyriform in *M.montana*, *M. picta* and *M. aucheriana*, but reniform in *M. meyeri* and *M. hybrida*, and elliptic in *M. hamata* and *M. intermedia*. Moreover the color in most studied taxa brown or dark brown only *M.hamata* seeds are brown-yellow and *M.aucheriana* was radish- dark brown. Mostafavi, et al [14] showed that seed type of *Minuartia* species in Iran are three different seed shapes including reniform, reniform pyriform, and elliptic.

The *Minuartia* species studied has been divided into four sections where *M. hybrid* is placed in section Sabulina, we show some characters separating it from other species such as having characteristic sinuate of the shape of cell margin seed, and this pattern is not observed in other species of the genus [Fig-2].

The placement of *M. hamata*, *M. montana*, *M. intermedia* and *M. meyeri* are closely related species of same section *Minuartia* [6,30,31], Seed character in the above species is a useful taxonomic character for separating them, these species differ in seed shape, ornamentation and the number of seed in the capsule [Table-2]. Moreover, seed micromorphology of *M. meyrie* and *M. intermedia* is completely different from that other species (sect. *Minuartia*) presence white dotted margin cell of ornamentation, in contrast with their high similarities in morphological characters [Fig-3] [11], compared with *M. hamata* and *M. montana*. There are some differences observed in *M. hamata* such as number of seed (one), cell shaped and ornamentation [Table-2].

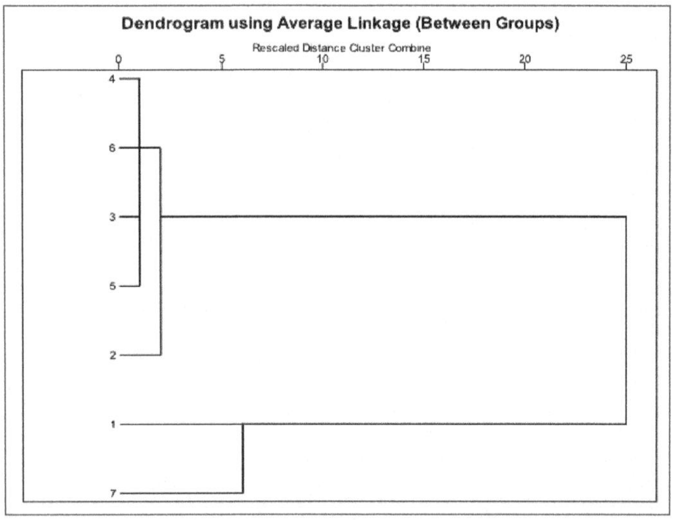

Fig. 5- Dendrogram based on the variation in data of seed criteria illustrating average taxonomic distance (dissimilarity) between the taxa studied. **1-** *M.aucheriana*; **2-** *M. hamate*; **3-** M. *hybrid*; **4-** *M. intermedia*; **5-** *M. meyeri*; **6-** *M. montana*; **7-** *M. picta*.

The dendrogram [Fig-5] resulted from the hierarchical cluster analysis of the applied seed characters, divided the studied species into two groups. The similarity between the species of the first group and the species of the second group ranged between 32.5 and 58.5%. The first group formed of two species (*M. aucheriana* and *M. picta*). These two species can easily distinguish from the species of the second group by its flat periclinal cell wall and the helium sub-centric. The second group includes the remaining five taxa divided the species in two subgroups. The species *M. hamata* separated in one subgroup, the species in this group characterized by its one seed, oblong and contain granulate on the side of seed. Within the second group *M. hybrid, M. intermedia*, *M. meyeri* and *M. montana* it have some overlapping prescriptions despite the fact *M. meyeri* oblong cell shape and *M. hybrid* serrate cell margin, deposited that they occurred in same subgroup. Taia [24] indicated that the ultramorphological characters of seeds were important in the identification of some species of Caryophyllaceae.

Conclusion

Microornamentation of the seeds was important in the identification and classification of some species of *Minuartia* was provided valuable and distinctive characters for separating taxa at the species level.

Conflicts of Interests: None declared.

References

[1] Takhtajan A. (1997) *Diversity and Classification of Flowering Plants*, Colombia University Press, New York.

[2] Rabeler R.K. & Hartman R.L. (2005) *Caryophyllaceae*, *Flora of North America North of Mexico*, 5, 3-215.

[3] Davis P.H., Mill R.R. & Tan K. (1988) *Flora of Turkey and the East Aegean Islands*, Edinburgh Univ. Press, Edinburgh.

[4] Brenan J.P.M. (1964) *Flora Europaea*, Cambridge University Press, 125-131

[5] Zohary M. (1966) *Flora or Palestine*, 1(1), 80.

[6] Rechinger K.H. (1988) *Flora Iranica*, 163, 28-53.

[7] Kubitzki K., Rohwer J.G. & Bittrich V. (1993) *Caryophyllaceae, Magnoliid, Hamamelid and Caryophyllid Families*, Springer, Germany, 206-236.

[8] Wofford B.E. (1981) *Systematic Botany*, 6, 126-135.

[9] Yildiz K. (2002) *Pak. J. Bot.*, 34, 161-171.

[10]Koc M., Aksoy A. & Hamzaoğlu E. (2011) *Turkish Journal of Botany*, 35, 337-341.

[11]Mostafavi G., Assadi M., Nejadsattari T., Sharifnia F. & Mehregan I. (2011) *Iranian Journal of Botany*,17, 220-226.

[12]Koc M. & Aksoy A. (2013) *Turk. J. Bot.*, 37, 428-433.

[13]Kanwal D., Abid A.R. & Qaiser M. (2012) *Pak. J. Bot.*, 44(1), 407-424.

[14]Mostafavi G., Assadi M., Nejadsattari T., Sharifnia F. & Mehregan I. (2013) *Turk. J. Bot.*, 37, 446-454.

[15]Dönmez A.A., Ağar O.T., Uğurlu Z., Mutlu B. & Horasan O. (2013) *Hacettepe J. Biol. & Chem.*, 41(2), 103-113.

[16]Wyatt R. (1984) *Syst. Bot.*, 9, 423-431.

[17]Amini E., Zarre S. & Assadi M. (2011) *Nord. J. Bot.*, 29, 660-669.

[18]Hong S.P., Han M.J. & Kim K.J. (1999) *J. Plant Biol.*, 42, 146-150.

[19]Yildiz K. (2006) *Pak. J. Bot.*, 38, 67-83.

[20]Perveen A. (2009) *Int. J. Biotech.*, 6(4), 219-227.

[21]Fawzi N.M., Fawzy A.M. & Mohamed A.A. (2010) *International Journal of Botany*, 6, 287-292.

[22]Crow G.E. (1979) *Brittonia*, 31(1), 52-63.

[23]Chung Y.H. & Lee E.J. (1988) *Korean J. Bot.*, 31, 51-68.

[24]Taia W.K. (1994) *Alex.J.Agric.Res.*, 39, 631-642.

[25]Zareh M.M. (2005) *Pak. J. Biol. Sci.*, 8, 714-720.

[26]Mahdavi M., Assadi M., Fallahian F. & Nejadsattari T. (2012) *Iran. J. Bot.*, 18(2), 302-310.

[27]Dunn G. & Everitt B.S. (1982) *An Introduction to Mathematical Taxonomy*, Cambridge University Press, Cambridge.

[28]Minuto L., Fior S., Roccotiello E. & Casazza G. (2006) *Plant Systematics and Evolution*, 262, 189-208.

[29]Minuto L., Roccotiello E. & Casazza G. (2011) *Plant Biosystems*, 145, 60-67.

[30]Davis P.H. (1967) *Flora of Turkey and the East Aegean Islands*, 2, Edinburgh University Press.

[31]McNeill J. (1962) *Notes from the Royal Botanic Garden,* Edinburgh, 24, 79-155.

13

MICROPROPAGATION OF *Capsicum chinense* Jacq. cv. LOTA BHOT VIA INDIRECT ORGANOGENESIS

BORA G.[1,2]*, GOGOI H.K.[1] AND HANDIQUE P.J.[2]

[1]Division of Biotechnology, Defence Research Laboratory (DRL), Tezpur- 784 001, Assam, India.
[2]Department of Biotechnology, Gauhati University, Guwahati- 781 014, Assam, India.
*Corresponding Author: Email- geetashreejht@gmail.com

Abstract- *Capsicum chinense* Jacq. is well known for its pungency factor, "capsaicin". An attempt has been made to maintain the desired pungency of *Capsicum chinense* Jacq. cv. Lota Bhot through *in vitro* propagation via callusogenesis from leaf explants under the influence of different plant growth hormones and an additional component, Silver nitrate ($AgNO_3$). The combination of Kinetin (45µML^{-1}) along with 2,4-Dichlorophenoxy Acetic Acid (2,4-D) and $AgNO_3$ at 3.5µML^{-1} and 35µML^{-1} respectively was found optimum for callus induction as well as multiple shoot induction whereas maximum shoot length was recorded at Kin (40µML^{-1}), 2,4-D (3.0 µML^{-1}) and $AgNO_3$ (30µML^{-1}). *In vitro* flower induction and fruit development was most observed under the influence of Gibberellic Acid (GA_3) at 30µML^{-1}. Murashige and Skoog (MS) medium fortified with Benzyl Amino Purine (BAP) [1]and 2,4-D at 4.0µML^{-1} and 5.5 µML^{-1} correspondingly was standardized for maximum induction of roots *in vitro*. Healthy *in vitro* regenerated plantlets were acclimatized in the potting substrate containing half strength MS medium enriched with soil, sand, vermicompost and vermiculite in all equal proportion and 78.56% of total acclimatized plants were successfully transferred to the main field.

Keywords- Lota Bhot, callusogenesis, *In vitro* flower induction, *In vitro* fruit development, Hardening, GA_3, $AgNO_3$

Introduction

Indian spices are one of the major reasons of breathing the country which helps in Indian economy as well. Among the various spices used in day to day basis, the most prevalent is chilli, or Capsicums. Chilli pepper or *Capsicum* species has been known since the beginning of civilization in the Western Hemisphere. The genus *Capsicum* is a member of the Solanaceae family that includes tomato, potato, tobacco, and petunia. So far, there exist approximately 22 wild species and five domesticated species of chilli: *C. annuum L., C. baccatum L., C. chinense Jacq., C. frutescens L.,* and *C. pubescens* R. & P. [1,2].

The one attribute most typical of chillies is pungency and must be considered one of its most important traits. The pungent factor lies within the seed and to some extent in the placenta nearby. Many innovative uses of pungency are being studied so far. Apart from having many valuable medicinal properties, it is extensively used as a barnacle repellent, to repel mice from all unwanted and hidden places and is effectively good to get rid of squirrels from eating its seeds. Moreover, to replace mace and tear gas, the pungent factor, capsaicin has been used widely as anti-mugger in more than a thousand police departments. Besides, the very effective pepper spray is to be considered as the first hand defensive tool for the protection of mankind especially women from most of the unfriendly situations.

Lota Bhot, a typical of *C. chinense Jacq.,* is basically a green, round shaped, small sized, extremely pungent fruit which is heavily infested by birds in the field during cultivation. Though the crop is endemic to Assam, it is occasionally found and rarely cultivated under natural condition and which is why it could be considered as a rare type. This highly supports the fact that its extinguishing germplasm should be conserved urgently Commensurate with this, intervention of biotechnology or to be more precise, plant tissue culture for accelerating clonal multiplication of desired clones and strains of medicinal plants through micropropagation and their conservation through establishing Tissue Banks or Gene Banks are warranted in the right earnest. Even though there are several reports of *In vitro* regeneration on different domesticated species of *Capsicum* [1-15], there is no such report on indirect organogenesis of *C. chinense cv. Lota Bhot* however.

Some of the rarely observed phenomena in the genus *Capsicum* are *in vitro* flower induction and fruit formation [16]. Growth hormone GA_3 has several reports of *in vitro* flower development in different crops [17, 18]. GA_3 is known to regulate many physiological processes in plants like seed germination, shoot development and elongation, flower development etc. However for the development of *in vitro* flowers, there are other features too like genetic, hormonal and trophic factors [17]. Response of plants towards GA_3 is also dependent on genotype, type of explants used, media composition and culture conditions including temperature and photoperiod [17]. Another potent flower inducer is silver nitrate. Silver nitrate enhances flower induction by suppressing the action of ethylene [11, 19]. It also has a profound effect on multiple shoot induction. So far there have been many reports of micropropagation in other chilli cultivars. However, this communicatio, brings into focus for the first time, the occurrence of *in vitro* organogenesis in the said chilli cultivar, Lota Bhot.

The present investigation has been designed with the idea of highlighting the regeneration potential of Lota Bhot utilizing various plant growth hormones and a vital component, $AgNO_3$ in order to obtain indirect organogenesis via leaf explants.

Materials and Methods

Young healthy leaves of *C. chinense* cultivar, Lota Bhot (cultivated in the experimental area within Defence Research Laboratory), were taken for the purpose. Initially, the leaves were treated with 0.1% Bevistin solution (w/v) and kept for 15-20 minutes for indirect organogenesis via callus phase. Green leaves were washed thoroughly with doubled distilled water for about 3-4 times, subjected to surface sterilization with 70% alcohol (5-10 sec), followed by mixing with an aqueous solution of 0.1 (w/v) $HgCl_2$ for 3 min, finally rinsing with properly with sterile distilled water for 4-6 times to remove the traces of $HgCl_2$. Leaf segments were cut aseptically from all sides in square shape [Fig-1a] and inoculated onto Murashige and Skoog (MS medium) [20] supplemented with growth hormones, KIN, 2,4-D and $AgNO_3$ in different concentrations for induction of callus [Fig-1b] [Table-1]. MS medium was fortified with 3% [m/v] sucrose (Himedia, Mumbai, India) and 0.8% (m/v) Agar was used as a gelling agent. The pH of the medium was set at 5.8 and autoclaved at 1.06 kg cm^{-2} and 121°C for 15 min prior to inoculation. The cultures were incubated in dark for 7-10 days for induction of callus [Fig-1b] and subsequently brought to light of 16 h photoperiod using cool white fluorescent tubes at 25±2°C for growth of the callus. Green and friable calli were transferred to shoot induction medium supplemented with KIN, 2,4-D and $AgNO_3$ for induction of multiple shoots [Fig-1c],[Fig-1d], [Table-1]. Healthy shoots (1.5-2cm) were aseptically trimmed and transferred onto *in vitro* flower and fruit induction medium fortified with GA_3 and $AgNO_3$ [Fig-1f], [Fig-1g], [Table-2] and subsequently inoculated onto rooting medium fortified with 2,4-D and BAP [Fig-1e], [Table-3]. For acclimatization and hardening of *in vitro* raised seedlings, *in vitro* rooted plantlets were carefully taken out from the rooting medium and washed thoroughly with tap-water to remove of the traces of medium. These plantlets were transferred to perforated plastic cups containing half strength MS mixed with soil, sand, vermicompost and vermiculite (1:1:1:1) and kept in the poly house for 30-25 days [Fig-1h]. The top of the cup was covered with small perforated polythene bags for initial 10-15 days and plantlets were watered regularly to maintain high humid condition. However in the later period humidity was reduced by removing the polythene bags. Six-week-old hardened plants were finally transferred to the main field. Data were collected and the percentage of plant survival was recorded.

All the experiments were conducted with CRD and repeated thrice with each treatment having 10 replicates. Significance between treatments was calculated through One-Way ANOVA and differences among treatment means were based on Turkey's Honesty Significant difference (HSD at 0.05).

Fig. 1- Micropropagation of Lota Bhot (*C. chinense*) via indirect organogenesis: **a)** Leaf explants; **b)** Callus induction; **c)** *In vitro* shoot induction; **d)** Multiple shoot induction; **e)** *In vitro* root induction; **f)** *In vitro* flower induction; **g)** *In vitro* fruit induction; **h)** Hardening.

Results

For maximum induction of callus and multiple shoots regeneration *in vitro*, the noble combination of Kin, 2,4-D and $AgNO_3$ at 45µML^{-1}, 3.5µML^{-1} and 35µML^{-1} respectively was recorded optimum wherein a mean of 43.1 numbers of leaf explants transformed into callus [Table-1], [Fig-1b]. A very high number of multiple shoot (8.1± 0.74) were induced from each callus at that very combination [Fig-1c], [Fig-1d]. But interestingly, maximum shoot length (6.0 ± 0.37) was achieved when the concentrations of Kin, 2,4-D and $AgNO_3$ were slightly reduced to 40, 3.0 and 30µML^{-1} correspondingly.

Flower induction and fruit formation *in vitro* from leaf explants responded effectively to the application of GA_3 at 30µML^{-1} [Table-2], [Fig-1f],[Fig-1g]. Maximum number of flower bud (14.7±1.31) per explant, highest number of flower induction (6.6±1.04) and fruit formation (5.8±1.51) *in vitro* were observed when GA_3 (30µML^{-1}) single-handedly was supplemented to MS medium. On the other hand application of $AgNO_3$ at 32µML^{-1} without GA_3 was quite significant in inducing a mean of 5.9 flowers; however only a mean of 3.3 flowers turned out into fruits finally. Furthermore it is worthwhile to mention that the combined effect of GA_3 (28µML^{-1}) and $AgNO_3$ (34µML^{-1}) reduced the number of flower induction and fruit formation to a mean of 5.2 and 4.8 respectively.

Table 1- Effect of various media composition on in vitro callus induction, multiple shoot induction and shoot length in *C. chinense* cv. Lota Bhot

S. No.	Treatments MS+ Kin+2,4-D+ $AgNO_3$ (µM/L)	No. of Leaves Forming Callus (Mean±SE)	No. of Multiple Shoots/Callus (Mean±SE)	Shoot Length (cm) (Mean±SE)
1	20.0+1.0+10.0	12.0±2.32a	2.6± 0.63a	1.4± 0.52a
2	25.0+1.5+15.0	18.2±2.67b	3.5 ± 0.55a	3.4 ± 0.26b
3	30.0+2.0+20.0	21.3±2.2c	4.2 ± 0.63b	4.4 ± 0.54c
4	35.0+2.5+25.0	30.7±4.13d	5.6 ± 0.43c	5.5 ± 0.39d
5	40.0+3.0+30.0	37.4 ± 1.59e	6.9 ± 0.67d	6.0 ± 0.37d
6	45.0+3.5+35.0	43.1 ± 1.21f	8.1 ± 0.74d	5.0 ± 0.42e
7	50.0+4.0+40.0	37.9 ± 2.76e	7.4 ± 1.40d	4.8± 0.37c
8	55.0+4.5+45.0	35.6 ± 3.23g	5.3 ± 1.60c	4.0 ± 0.83f
9	60.0+5.0+50.0	30.8± 4.02d	4.0 ± 1.35b	3.1 ± 0.49b
10	65.0+5.5+55.0	22.5 ± 4.33c	4.0 ± 0.86b	2.3 ± 0.60g
		SE(Means) = 1.2	SE(Means) = 1.33	SE(Means) = 0.59

Means followed by the same letters are not significantly different at p=0.05

Table 2- Effect of different concentrations of GA_3 and $AgNO_3$ on in vitro flower and fruit induction in *C. chinense* cv. Lota Bhot

S. No.	Treatments MS+GA_3 + $AgNO_3$ (µM/L)	No. of Flower Buds/Explant (Mean±SE)	No. of Flowers per explant (Mean±SE)	No. of Fruits per explant (Mean±SE)
1	26 + 0	7.4±1.27a	4.1 ± 0.96a	2.7 ± 0.95a
2	28 + 34	11.6±1.11b	5.2 ± 0.97b	4.8 ± 0.73b
3	30 + 0	14.7±1.31c	6.6 ± 1.04c	5.8 ± 1.51c
4	0 + 32	12.5±1.73d	5.9 ± 1.24d	3.3 ± 1.78d
5	0 + 30	8.8 ± 1.03a	2.5 ± 1.01e	2.5 ± 0.61a
		SE(Means) = 0.44	SE(Means) = 0.25	SE(Means) = 0.43

Means followed by the same letters are not significantly different at p=0.05

Regarding induction of roots *in vitro*, the extent of root initiation was almost found equivalent at two combinations: BAP (4.0µML^{-1}) and 2,4-D (5.5µML^{-1}); BAP (3.5µML^{-1}) and 2,4-D (4.5µML^{-1}) respectively. The first combination yielded maximum number of roots (54.3±7.57) and the second one generated next best possible result (54.0±7.43) [Table-3], [Fig-1e]. The first combination could also successfully generate maximum root length of 13.63 cm. Healthy rooted plantlets were best hardened in the substrates consisting of half strength MS with soil, sand, vermicompost and vermiculite (1:1:1:1), and 78.56% of total hardened plants could be successfully raised in the mail field.

Table 3- Effect of different concentrations of BAP and 2,4-D on in vitro root induction and root length in *C. chinense* cv. Lota Bhot

S. No.	Treatments MS+BAP+2,4-D (µM/L)	No. of Roots per Regenerated Shoot (Mean±SE)	Root Length (cm) (Mean±SE)
1	2.0+1.5	25.3±4.43a	3.1 ± 1.77a
2	2.5+2.5	30.4±6.60b	5.6 ± 0.86b
3	3.0+3.5	46.3±4.18c	7.5 ± 1.66c
4	3.5+4.5	54.0±7.43d	12.6 ± 3.38d
5	4.0+5.5	54.3 ± 7.57d	13.63 ± 2.48e
6	4.5+6.5	43.3 ±3.72e	11.54± 2.80f
		SE(Means) = 3.05	SE(Means) = 0.58

Means followed by the same letters are not significantly different at p=0.05

Discussion

Almost all the species of the genus *Capsicum* are reluctant to micropropagation. The appropriate amount of media and source of nutrients are the pre-requisites for the establishment of a stable regeneration protocol. Plants of Lota Bhot obtained from DRL, Tezpur had a good potential for callusogenesis when cultured onto full strength MS basal medium. Callus induction medium was standardized with different concentrations of Kin and 2,4-D along with an additional component, $AgNO_3$. Presence of $AgNO_3$ enhances the chances of multiple shoots induction (8.1± 0.74). However, a drastic negative change in the growth curve was observed both in terms of shoot bud formation and shoot length upon altering the concentrations of these components. The present investigation reports for the first time the combined effect of Kin, 2,4-D and $AgNO_3$ in the *in vitro* multiple shoots induction in *C. chinense* cv. Lota Bhot although so far there have been several regeneration protocols standardized in different species of *Capsicum*.

In the genus *Capsicum, in vitro* flower and fruit inductions are rarely observed. Earlier reports suggest the occurrence of *in vitro* flowering and fruiting in *C. frutescens* when shoot tips were cultured in a polycarbonate container coupled to an automated plant tissue culture system [16]. Effects of $AgNO_3$ and $CoCl_2$ on *in vitro* flowering in *C.frutescens* were broadly elucidated [11]. Reproductive processes like flower induction and fruit development were primarily controlled by GA_3 and to some extent $AGNO_3$ as evident from the said investigation. Addition of GA_3 to flower and fruit induction media significantly increases the number of flower bud formation and fruit development [Table-2]. GA_3 is a potent regulator of flower and fruit induction both *in vitro* and *in vivo*. Induction of floral bud in response to the application of GA_3 has been recently reported in *C. annuum* [21]. Effect of GA_3 was investigated in Lota Bhot. It was able to induce a mean of 14.7 flower bud/explants, 6.6 flowers per explant and 5. Fruits when applied alone at 30µML^{-1}. $AgNO_3$ (32µML^{-1}) when applied without GA_3 could effectively initiate flower formation (5.9 ± 1.24) however was not effective enough to generate a good amount of fruits *in vitro*. $AgNO_3$ is a potent flower inducer in *C. frutescens* [11, 22]. Off white flowers (with 5 sepals and 5 petals) remained healthy and in open condition for almost 15 days. A very high dose (above 34 µM) not only reduced the number of flower but also the size of the leaf. Concentration that was lower than 30µM/L, did not produce any flower at all. As far the fruit is concerned, the fully matured fruit attained a diameter of approximately 4 cm and girth of about 4.5 cm and remain healthy for almost 8-10 d before it started deteriorating. $AgNO_3$ is considered as a powerful ethylene inhibitor [11]. Ethylene is known to inhibit the action of an enzyme called S-adenosyl methionine decarboxylase which is responsible for production of polyamines that promotes flower induction. Thus the application of silver nitrate promotes *in vitro* flowering by inhibiting ethylene production.

The combination of BAP (4.0μML^{-1}) and 2,4-D (5.5μML^{-1}) generated highest number of roots (54.3 ± 7.57) from a single regenerated shoot. Healthy, elongated and branched type of roots were observed in the said investigation.. However, higher doses of BAP generated short and thin roots whereas doses lower than 2μML^{-1} did produce very fewer roots. Several reports confirms the effectiveness of IBA in promoting *in vitro* rooting in Chilli has been reported earlier [4,9,21] however there are few reports suggesting the significance of IAA on *in vitro* rooting in *Capsicum* [7]. These two rooting hormones were also tried in the experiment but the result obtained with BAP was found to be far more conducive. This may be due to the difference in genotype of the cultivar used in the investigation.

The *in vitro* rooted Lota Bhot plantlets were transferred to poly house for hardening and acclimatization. Inclusion of bio fertilizer (vermicompost), vermiculite along with sand not only improved the status of the soil but also the health of the plant and increased the percentage of survival (78.56%). Technique of micropropagation is quite a new concept for Lota Bhot, a very rare cultivar of *C. chinense,* endemic to Assam. Hence, this communication could be effective enough in opening a way out to explore more about it in the near future.

Conclusion

The present investigation provides a firsthand report of an efficient *in vitro* regeneration protocol via leaf explants in *C. chinense* cv.Lota Bhot. Significance of plant growth hormones and $AgNO_3$ in several aspects of *in vitro* organogenesis including *in vitro* shoot regeneration, flower and fruit induction has been elaborately described. Lota Bhot being one of the pungent chillies has remarkable potential in the field of food, medicines and pharmaceuticals. Concept of tissue culture in Lota Bhot is quite a new topic so far. The present investigation therefore, brings into light, an innovative approach for exploring more about the crop through micropropagation technique.

Acknowledgement

Authors thank the Director, Defence Research Laboratory (DRL), Tezpur, Assam for showing keen interest and providing DRDO fellowship. The DRL and Department of Biotechnology (Gauhati University) are duly acknowledged for providing necessary facilities and permission for the work.

Conflicts of Interest: None declared.

References

[1] Agrawal S., Chandra N. and Kothari S.L. (1989) *Curr. Sci.*, 57, 1347-1349.

[2] Harvell K.P. & Bosland P.W. (1997) *HortScience.*, 32, 1992

[3] Sanatombi K. & Sharma G.J. (2007) *Scientia Horticulturae*, 113, 96-99.

[4] Sanatombi K. & Sharma G.J. (2008) *Biologia Plantarum*, 52(1), 141-145.

[5] Christopher T. & Rajam M.V. (1994) *Plant Cell Tiss. Org. Cult.*, 38, 25-29.

[6] Ramirez-Malagon R. & Ochoa-Alejo N. (1996) *Plant Cell Rep.*, 16, 226-231.

[7] Bodhipadma K. & Leung D.W.M. (2003) *In Vitro Cellular and Developmental Biology - Plant*, 39(5), 536-539.

[8] Kumar A.M. & Nair A.S. (2004) *Plant Cell Biotech. Mol. Biol.*, 5 (3&4), 95-100.

[9] Peddaboina V., Christopher T. & Subhash K. (2006) *Scientia Hort.*, 107(2), 117-122.

[10]Sanatombi K. & Sharma G.J. (2006) *Journal of Food Agricultural Environment*, 4, 205-208.

[11]Sharma A., Kumar V., Giridhar P. & Ravishankar G.A. (2008) *Electronic Journal of Biotechnology*, 11(2), 1-6.

[12]Ashrafuzzaman M., Hossain M.M., Ismail M.R., Haque M.S., Shahidullah S.M. and Uz-zaman S. (2009) *African Journal of Biotechnology*, 8(4), 591-596.

[13]Song J.Y., Sivanesan I, An C.G. & Jeong B.R. (2010) *African Journal of Biotechnology*, 9(19), 2768-2773.

[14]Ostroshy M., Moradi K., Nekouei M.K. & Struik P.C. (2011) *Asian Journal of Biotechnology*, 3(1), 38-45.

[15]Kehie M., Kumaria S. & Tandon P. (2012) *3 Biotech*, 2(1), 31-35.

[16]Tisserat B. & Galleta D.P. (1995) *HortScience*, 30, 130-132.

[17]Rkhis A.C., Maalej M., Messauod S.O. and Drira N. (2006) *African Journal of Biotechnology*, 5(22), 2092-2302.

[18]Ziv M. & Naor V. (2006) *Propagation of Ornamental Plants*, 6 (1), 3-16.

[19]Beyer E.M. (1976) *HortScience*, 11, 175-196.

[20]Murashige T. & Skoog F. (1962) *Physiol. Plant*, 15, 473-497.

[21]Kumari M., Patade V.Y. & Ahmed Z. (2012) *World Journal of Science and Technology*, 2(7), 26-35.

[22]Giridhar P., Indu E.P., Vijaya R.D. & Ravishankar G.A. (2003) *Tropical Science*, 43(3), 144-146.

PULSING WITH SUCROSE AND SILVER NITRATE ENHANCE WATER UPTAKE AND RESULT IN ALONG VASE LIFE IN TAIF ROSE CUT FLOWERS (*Rosa damascena*. CV. TRIGINTIPETALA)

ELGIMABI M.N.E.E.*

Department of Biology, Taif University, Saudi Arabia.
*Corresponding Author: Email- nourde1@yahoo.com

Abstract- An experiment was carried out on Taif rose cut flowers cv. (Trigintipetala) to study the effect of pulsing with sucrose and silver nitrate ($AgNO_3$) on vase life and other parameters. In the trial, sucrose and $AgNO_3$ at 4 levels of concentrations (0, 3, 5 or 7% w/v sucrose), (0, 20, 30 or 50 ppm $AgNO_3$) were tested alone as well as combination. The cut flowers were treated in three stages of maturity [bud (B), half opening (H.O) or complete opening (C.O)]. The pulsing times were 5, 10 or 24 hours, then transferred to distilled water as control. The experiment was laid out in a completely randomized design replicated thrice. The sucrose, 7% and $AgNO_3$ 30% individually recorded higher vase life of 17.2 and 15 days respectively. Whereas, their combinations (sucrose 7% × $AgNO_3$ 30%) was significantly superior to the rest of combination in keeping higher water uptake and retarded the chlorophyll as well as the carbohydrate degradation during the postharvest life and resulting in a highest vase life of 19.1 days.

Keywords- Longevity, Taif rose cut flowers, $AgNO_3$, Sucrose, water uptake, Chlorophyll, carbohydrate

Introduction

Rose is one of the most popular cut flowers in different countries for highest economic importance in the floricultural industry [1], and a very long vase life compared to other cut flowers [2], which is one of the characteristics demanding in a commercial value of the ornamental flowers. Taif rose have an important role in the national economy for its great value in exportation of its oil to the Arab countries, ornamentation, medicinal use as well as manufacture of perfumes, [3]. Vase life termination for many cut flowers is characterized by wilting [4]. Water balance is a major factor determining quality and longevity of cut flowers. It is influenced by water uptake and transpiration and balance between two mentioned processes [5]. When the amount of transpiration exceeds the volume of water uptake, water deficit and wilting develops [6]. Low water uptake is often due to occlusions located mainly in the basal stem end [4]. Microorganisms are communally cause of stem blockage of xylem vessels which accumulated in the vase solution or in the vessels [7] and result a negative effect on the continuing water uptake and transportation by the leaves of cut flowers and stem tissue [8,9], and finally shortening there vase life [10,11], and reduces water uptake. This blockage of xylem vessels led to water stress, which was expressed in the form of early wilting of leaves or flowers [7]. Pulsing with high concentration of silver nitrate is beneficial in extending the life of many cut flowers like gerberas, gladioli, chrysanthemum carnation and roses [12,13]. Tight carnation buds were opened in 10% sucrose after impregnating the stems for 1 hour in a 1000 ppm $AgNO_3$ solution [12]. Pulsing of cut roses for 10 and 20 min with $AgNO_3$ improved the vase life up to 6.0 and 5.3 days, respectively [14]. Application of $AgNO_3$ significantly increased the vase life as well as the gain fresh weight of rose as compared to 5.3 days in control [15]. Also, sucrose acts as a preservative materials, in addition to extending the vase life of cut flowers [16]. Different concentrations of sucrose ranging from 0 to 7.5% had been investigated by Pun, et al. [17] on cut spray carnation, they found that 5.0 sucrose recorded the best vase life and delayed the climacteric ethylene in petals. Furthermore. Butt, 2005 [18] study the effect of sucrose in different concentrations on two cultivars of *Rosa hybirida* and results showed that sucrose at 25 gL^{-1} extended the vase life by 8.2 days.

On the other hands, sugars with biocides have become an important preservatives for floral several cut flowers [19], and prevented bent-neck of flower stems of {*Cara Mia*} rose cultivar [20]. Butt, [21] reported that when cut roses were treated with sucrose for 16 hours, the treatment increased the soluble sugar content, and reduced both photosynthesis rate and chlorophyll content. However, it stimulated the respiration rate of leaves, as compared with the control [12]. Also Nair, et al [22] study the effect of (Ag NO_3) plus sucrose in different concentration on vase life of gerbera cut flower, the treatment extended the vase life and delaying the head dropping and discoloration.

Material and Methods

Experiment 1

Plant Materials

Cut flowers of Taif roses were used for different trials. The flowers were obtained directly from the commercial grower. The flowering stems were trimmed to a uniform length of 25 cm to flowers at different stages and for different pulsing time. Then a smooth slanting cut were made to flower stem to facilitate the optimum up take of given solutions. Cutting had been done underwater to avoid air embolisms. All leaves on the lower section of the stem were removed. The sample had been divided into 30 groups with five replications, each replication contain one flower (to avoid contamination). They were kept at room temperature 23 $\pm$ 1°C at normal day light and natural ventilation. Flowers had been harvested in three stages" bud stage (B.S), one or two basal flowers open [half opening] (H.O.S), or when they are fully opened (C.O.S)".

Treatment Setting

Treatments were set following a completely randomized design. Each treatment represented by 5 replication.

Chemical Preparation

Chemicals were obtained from a trading company in Saudi Arabia.

Analytically pure Silver nitrate ($AgNO_3$) had been pulsed as follow:

1st Trial: Five hours only for $AgNO_3$ at concentrations of 30, 50, 70 ppm. Then flowers were transferred into 250 ml glass bottle filled with distilled water along the duration of experiment (for the three stages).

2nd Trial: Ten hours only for. $AgNO_3$ at concentrations of 30, 50, 70 ppm. Then flowers were transferred into 250 ml glass bottle filled with distilled water along the duration of experiment (for the three stages).

3rd Trial: Twenty four hours for $AgNO_3$ at concentrations of 30, 50, 70 ppm. Then flowers were transferred into 250 ml glass bottle filled with distilled water along the duration of experiment (for the three stages).

Also analytically pure sucrose had been pulsed as fallow:

1st Trial: Five hours (5h) at concentration of 3%, 5% and 7%. w/v. Then flowers were placed in 250 ml glass bottle containing distilled water, along the duration of experiment (for the three stages).

2nd Trial: Ten hours (10h) at concentration of 3%, 5% and 7%. w/v. Then flowers were placed in 250 ml glass bottle containing distilled water, along the duration of experiment (for the three stages).

3rd Trial: Twenty four hours (24h) at concentration of 3%, 5% and 7%. w/v. Then flowers were placed in 250 ml glass bottle containing distilled water, along the duration of experiment (for the three stages).

Control Treatment

Flowers was trimmed in different stages of maturity and kept in 250ml distilled water along the duration of the experiment, as a control treatment for the different trials.

Vase Life Determination

Visual rating of flowers was carried out on the basis of a scale from 1 to 4 when: 1 = entirely fresh flowers, 2 = initiation of wilting in 20% of petals and beginning of bent neck, 3 = wilting in 20-50% of petals and increasing the bent neck, 4 = wilting in 50-100% of petals.

Experiment 2

Plant Material

The same preparation of plant material mentioned in experiment (1)

Chemical Treatments

($AgNO_3$) + Sucrose Treatments

Silver nitrate ($AgNO_3$ at 30 ppm) was applied for 24 hour depending on the previous results (which the best one), whether with or without sucrose at 7% (w/v) the concentration, which attained the best results in the previous experiment. The two compounds are dissolved in sterilized distilled water in glass bottle containing 250 ml to study the effect of their interaction.

Control Treatment

Flowers are trimmed and kept in 250ml distilled sterilized water along the duration of the experiment, as a control treatment for different experiments.

Vase Life Determination

Vase life of rose cut flowers was measured as mentioned in experiment (1)

Fresh Weight Measurements

Fresh weight determination of the flowers were mad just before the immersion of flowers into the solutions and repeated on the day when the vase life of the control flowers was terminated. The flowers are taken out of solutions after 24 hours pulsing time.

Chlorophyll Determination

Chlorophyll content was measured for the best treatment of each chemical for all cut flowers under study. Chlorophyll content of sepals segments was extracted by methanol and the absorbance was determine by spectrophotometer. This was done on day 0, 5, and on the day when the vase life of the control flowers are terminate. The samples were collected separately from each replicate and the average of the three replicates was calculate. The chlorophyll concentration was calculated as mg l^{-1} fresh weight. The equations for the determination of concentrations of chl. *A* and chl *b* were Chl *a* (mg l^{-1}) = 12.21 A_{663} - 2.81 A_{646} Chl *b* (mg l^{-1}) = 20.13 A_{646} -5.03 A_{663}.

Carbohydrate Determination

Soluble carbohydrate determined on stem of the best treatment of each chemical and the best stage of maturity and best pulsing time, tested in the study. Sample were taken on the same day as mentioned in chlorophyll determination. One flowering stem from each replicate was used. Dried sample were ground together into homogenized powder, using a household crusher (10 mg of oven-dried plant material were extracted by 10 ml of borate buffer (28.63g boric acid + 29.8g KCL + 3.5g NaOH in IL of hot de-ionized distilled water), left overnight, and filtered. Using spectrophotometer to detect the different types of sucrose.

Results and Discussion

Effect of $AgNO_3$ and Sucrose on Vase Life of Taif Rose Cut Flowers

Vase Life- The statistical analysis of results showed that the $AgNO_3$ & sucrose, significantly extended the vase life of Taif rose cut flowers compared to control in different concentrations and different stages of maturity, [Table-1]. The longest vase life attained at 7% (w/v) sucrose which gave 17.2 days, followed by 15 days for $AgNO_3$ at 30ppm and 7 days for control compared to other concentrations of sucrose and $AgNO_3$ [Table-1]. Carbohydrates are most important compounds in reserve solutions, manly sucrose which sometimes, lonely present in phloem sap. Cut flowers senescence is initiated when energy required for synthesis reactions is considerably reduced. Therefore exogenous application of sucrose would be helpful in delaying senescence by maintaining the structure and activity of mitochondria [23]. Also sucrose act as food source or respiratory substrate and delays the degradation of proteins and improves the water balance of cut flowers [24]. While, Steinitz [25] opined that addition of sucrose to the solution increased the mechanical rigidity of the stem by inducing cell wall thickening and lignifications of vascular tissues. Also $AgNO_3$ is an effective inhibitor of ethylene biosynthesis [26] and act as bactericides [27]. Our results in confirm the results obtained by other researchers [28] and [29] they opined that $AgNO_3$ prevented microbial occlusion of xylem vessels in Dendrobium, thereby enhancing water uptake and increasing longevity of flowers. Awad, et al [30] also attributed the beneficial effect of $AgNO_3$ in the vase water to the production of Ag^+ions, which might inhibit the rise of ethylene precursor thereby enhancing longevity of cut flowers.

Table 1- Effect of $AgNO_3$ and sucrose on vase life of Taif rose cut flowers

Treatments	Vase life after different Pulsing time / hours								
	B			H.O			C.O		
	5h	10h	24h	5h	10h	24h	5h	10h	24h
$AgNO_3$ 30 ppm	11.8	14.5	13.8	11	13	15	8.8	12.4	12.8
$AgNO_3$ 50 ppm	11.8	14.5	12.8	10.2	13.4	13.8	9	12	13. I
$AgNO_3$ 70 ppm	10.8	11.6	12.2	10.8	13.2	13.2	7.6	9.6	11.8
Sucrose 3%	9.4	11.6	12.6	13.2	14	13.8	11.6	13.4	13
Sucrose 5%	10.2	11p	11p	13.8	13	14.2	12.2	14.6	15.8
Sucrose 7%	12.4	12	13.2	13.2	14.3	17.2	14.5	13.8	13
Control		4*			7*			6*	

Stages of Maturity

Depending on results of [Table-1], the (H.O.S) stage result in the highest longevity compared to different stages, these data are consistent with the hypothesis that, smaller buds of cut flowers may abort for lack of carbohydrate [31], similar results obtained by van Meetern, et al [32] in lilies cut flowers, thereby the lack of carbohydrate has been shown to lead to increased ethylene biosynthesis and bud abortion [33], which are consistent with this suggestion. Curiously, supplying added carbohydrate did not consistently reduce ethylene production in very yang buds. They may have insufficient sink strength to compete with expending buds for carbohydrate provide from the vase solution [Table-1], thereby short vase life in flowers had been cut in bud stages compared to other stages [Table-1], may be due to leaching of carbohydrate as reported by Vander Muisers, et al [34] in Asiatic lily, while the flowers cut in complete opening stage import more carbohydrates from the leaves and stem than that taken in H.O.S [19], which in agreement of the results of [Table-1].

Pulsing Time

The results of [Table-1] showed that the 24h pulsing time in preservative solution attained good results compared to other pulsing times, which in agreement of the recommendation of Halevy and Kofranek [35] when he pulsed Lisianthus flower in 6% sucrose for 24h the postharvest life had been improved and flower opening considerably increased [36].

The Combination Effects of $AgNO_3$ and Sucrose on Vase Life and Postharvest Quality of Taif Rose Cut Flowers

Vase Life- Sucrose or $AgNO_3$ alone was less effective as compared to their combinations with regard to vase life. When 7% sucrose was added to 30ppm $AgNO_3$ the vase life was extended to 17.8, 19.1 and 13.9 days for (B), (H.O) and (C.O) stages respectively compared to 9.9, 10.5 and 8.9 days without sucrose for the same mentioned stages, [Table-2], through 24h depending on results of [Table-1]. This may be due to the nutrient effect of sucrose [37] and antibacterial effect of $AgNO_3$ [38]. Further more when added $AgNO_3$ to sucrose may suppress autocatalytic ethylene production by inhibiting ethylene action [39], so that the two compound prevent the xylem blockage witch lead to promote water uptake by stem and finally result in a good vase life, similar results by Ohkawa, et al [38].

Table 2- Effect of the best treatment of $AgNO_3$ and sucrose on vase life and postharvest quality of Taif rose cut flowers

Treatments	Vase life after different Pulsing time / hours		
	B	H.O	C.O
$AgNO_3$ 30 ppm	13gf	15d	12h
Sucrose 7%	13gf	17b	14e
$AgNO_3$ 30 ppm 7% sucrose	17.8bc	19.1a	13.9f
Control	4k	7j	6ji

**Different letters explain the significant differences between means, according to Duncan multiple range p = 5.05.*

Chlorophyll Content

The previous treatments lead to a considerable delay in degradation of Chl *a* and Chl *b* compared to control [Table-3], [Table-4] and [Table-5]. The concentration of chlorophyll *a* was higher than chlorophyll *b* at any point of time throughout the vase life. [Table-3], [Table-4] and [Table-5].

Table 3- Effect of $AgNO_3$ with or without sucrose and sucrose compared to control on chlorophyll content for Taif rose cut flowers in bud stage. (unit was mgl^{-1} fresh weight).

Treatments	Days of determinations of chl. *a* and chl. *b*					
	1st day		6th day		10th day	
	Chl. *a*	Chl. *b*	Chl. *a*	Chl. *b*	Chl. *a*	Chl. *b*
$AgNO_3$ 30 ppm	0.66	0.19	0.86	0.20	0.17	0.05
$AgNO_3$ 30 ppm + sucrose 7%	0.93	0.21	0.79	0.17	0.20	0.04
Sucrose 7%	0.97	0.29	0.64	0.18	0.15	0.04
Control	0.13	0.06	0.54	0.14	0.48	0.28

Table 4- Effect of $AgNO_3$ with or without sucrose and sucrose compared to control on chlorophyll content for Taif rose cut flowers in half opening stage. (unit was mgl^{-1} fresh weight).

Treatments	Days of determinations of chl. *a* and chl. *b*					
	1st day		6th day		10th day	
	Chl. *a*	Chl. *b*	Chl. *a*	Chl. *b*	Chl. *a*	Chl. *b*
$AgNO_3$ 30 ppm	0.79	0.29	0.25	0.03	0.24	0.03
$AgNO_3$ 30 ppm + sucrose 7%	1.34	0.46	0.75	0.14	0.19	0.07
Sucrose 7%	0.50	0.13	1.10	0.32	0.09	0.05
Control	0.65	0.19	0.15	0.03	0.09	0.02

The data show that there was a positive role for sucrose with or without $AgNO_3$ on preserving the flowers in good condition by inhibiting the chlorophyll retention. Similar results were obtained by Tjosolved, et al [40] and Serek, et al [41], Chatterjee, et al [42], and Singh and Tiwari [26].

Table 5- Effect of $AgNO_3$ with or without sucrose and sucrose compared to control on chlorophyll content for Taif rose cut flowers in complete opening stage. (unit was mgl^{-1} fresh weight).

Treatments	Days of determinations of chl. *a* and chl. *b*					
	1st day		6th day		10th day	
	Chl. *a*	Chl. *b*	Chl. *a*	Chl. *b*	Chl. *a*	Chl. *b*
$AgNO_3$ 30 ppm	0.51	0.14	1.05	0.58	0.08	0.02
$AgNO_3$ 30 ppm + sucrose 7%	0.40	0.08	1.08	0.35	0.31	0.11
Sucrose 7%	0.54	0.15	0.44	0.13	0.19	0.05
Control	0.42	0.13	0.37	0.12	0.07	0.03

[1]mg 1^{-1} fresh weight for different stage.

Carbohydrate Content

Data of [Table-6], [Table-7], [Table-8] and [Table-9] showed that fructose, glucose and sucrose were the main soluble carbohydrates in petals and stems of Taif cut roses. Fructose was the major component in the petals as well as in stems but, generally, its value was higher than in stems. Sucrose contents in petals and stems were lower than those of glucose. The carbohydrate content significantly increased as a result of using 30ppm $AgNO_3$ + 7% sucrose till the 5th day then sharply decreased on the 12th day at which the vase life of control was terminated. While stem contents of the previous sugars increased at the beginning of the experiment, then decreased towards the end of the experiment compared to control [Table-6]. Pulsing for 24h may enhanced total starch contents of petals as well as compared to untreated cut rose flowers [43]. These results were coincided with those obtained by Zagory and Reid [44] on many cut flowers, on bird of paradise [45], cut rose flowers, [46], chrysanthemum, [47], freesia, [48] on gladiolus cut flower spikes, [49] and tuberose cut flowers spikes [50].

Table 6- Effect of $AgNO_3$ with or without sucrose and sucrose on carbohydrates content for petals of rose cut flowers in bud stage.

Treatments	Days of determination of carbohydrate contents								
	1st day			5th day			12th day		
	G	F	S	G	F	S	G	F	S
$AgNO_3$ 30 ppm	8.08	9.63	1.55	7.03	8.65	1.92	4.89	5.57	0.68
$AgNO_3$ 30 ppm + sucrose 7%	7.39	8.04	0.65	5.01	6.78	1.77	6.84	7.39	0.55
Sucrose 7%	5.08	6.08	1.81	9.57	10.38	0.81	8.88	9.25	0.37
Control	9.78	10.64	0.86	9.76	11.14	1.39	6.38	9.71	3.33

F: Fructose G: Glucose S: Sucrose (in different tables)

Table 7- Effect of $AgNO_3$ with or without sucrose and sucrose on carbohydrate content for stem of rose cut flowers in bud stage.

Treatments	Days of determination of carbohydrate contents								
	1st day			5th day			12th day		
	G	F	S	G	F	S	G	F	S
$AgNO_3$ 30 ppm	4.55	6.55	2.01	3.86	3.99	0.13	5.08	6.88	1.81
$AgNO_3$ 30 ppm + sucrose 7%	6.58	7.23	0.66	7.27	7.46	0.19	5.52	6.11	0.59
Sucrose 7%	5.18	6.89	1.71	3.07	3.26	0.19	3.48	4.09	0.61
Control	5.05	5.65	0.61	6.28	7.92	1.64	5.68	6.05	0.38

Table 8- Effect of $AgNO_3$ with or without sucrose and sucrose on carbohydrates content for petals of rose cut flowers in half opening stage.

Treatments	Days of determination of carbohydrate contents								
	1st day			5th day			12th day		
	G	F	S	G	F	S	G	F	S
$AgNO_3$ 30 ppm	3.55	5.55	2.01	3.86	2.99	0.13	4.08	6.88	1.81
$AgNO_3$ 30 ppm + sucrose 7%	5.84	7.23	0.66	7.83	8.17	0.36	8.34	10.54	2.21
Sucrose 7%	4.88	5.89	1.71	10.95	12.07	1.12	2.68	4.09	0.61
Control	5.05	5.65	0.61	7.79	8.62	0.83	9.81	12.21	2.41

Table 9- Effect of $AgNO_3$ with or without sucrose and sucrose on carbohydrate content for stem of rose cut flowers in half opening stage.

Treatments	Days of determination of carbohydrate contents								
	1st day			5th day			12th day		
	G	F	S	G	F	S	G	F	S
$AgNO_3$ 30 ppm	4.11	5.32	1.21	5.67	6.52	0.85	3.81	4.84	1.03
$AgNO_3$ 30 ppm + sucrose 7%	4.99	6.23	1.29	4.69	5.92	1.23	6.07	7.49	1.42
Sucrose 7%	7.43	9.78	2.35	4.69	4.92	1.23	10.46	11.07	0.61
Control	5.95	6.45	0.51	7.79	8.62	0.83	9.81	12.21	2.41

Table 10- Effect of $AgNO_3$ with or without sucrose and sucrose on carbohydrates content for petals of rose cut flowers in complete opening stage.

Treatments	Days of determination of carbohydrate contents								
	1st day			5th day			12th day		
	G	F	S	G	F	S	G	F	S
$AgNO_3$ 30 ppm	5.43	7.68	2.25	4.69	5.92	1.23	7.09	7.81	0.72
$AgNO_3$ 30 ppm + sucrose 7%	5.98	6.28	031	8.86	9.11	0.25	7.39	10.51	3.12
Sucrose 7%	5.95	6.87	0.92	8.83	9.45	0.62	9.38	9.99	0.61
Control	3.86	4.17	0.31	9.45	10.06	0.61	9.91	11.55	1.64

Table 11- Effect of $AgNO_3$ with or without sucrose and sucrose on carbohydrate content for stem of rose cut flowers in complete opening stage.

Treatments	Days of determination of carbohydrate contents								
	1st day			5th day			12th day		
	G	F	S	G	F	S	G	F	S
$AgNO_3$ 30 ppm	6.46	7.73	1.27	4.86	5.83	0.97	4.93	5.82	0.89
$AgNO_3$ 30 ppm + sucrose 7%	7.14	8.63	1.49	5.95	6.68	0.75	6.39	9.5	2.12
Sucrose 7%	4.65	5.97	1.32	6.04	7.67	1.63	4.67	5.92	1.25
Control	6.05	6.96	0.91	9.45	10.06	0.61	9.91	11.55	1.64

Conclusion

A significant improvement in vase life of Taif rose cut flowers was occurred when treated with 30ppm $AgNO_3$ and the effect was further improved when $AgNO_3$ at 30 ppm combined with 7% (w/v) sucrose which attained the best result compared to other concentrations of sucrose. Also chlorophyll and carbohydrate had been retarded during the postharvest life as the result of using this combination treatment which they represent one of the more important parameters of good performance of postharvest physiology of cut flowers.

Conflicts of Interest: None Declared.

References

[1] AIPH (2003) *International Statistics Flowers and Plants Union Fleurs*, Den Haag.

[2] Nukui H., Kudo S., Yamashita A. & Satoh S. (2004) *Journal of Experimental Botany*, 55(397), 641-650.

[3] Bazaid S.A. (2004) *Journal of the Egyptian Academy for Development*, 5(1), 77-90.

[4] He S., Joyce D.C., Irving D.E., Faragher J.D. (2006) *Postharvest Biol. Technol.*, 41, 78-84.

[5] Da Silva J.A.T. (2003) *Online J. Biol. Sci.*, 3, 406-442.

[6] Halevy A.H., Mayak S. (1981) *Acta Horticulture*, 3, 59-143.

[7] Marandi R.A., Hassani A., Abollahi A. and Hanafi S. (2011) J. Med. Plant. Res., 5(20), 5039-5043.

[8] Hassan F. (2005) *International Journal of Horticultural Science*, 10(4), 101-107.

[9] Kazemi M., Hadvi E. and Hekmati J. (2010) *World Applied Sci. J.*, 10, 737-740.

[10] An J., Zhang M., Wang S. & Tang J. (2008) *LWT-Food Science and Technology*, 41(6), 1100-1107.

[11] Kazemi M., Hadvi E. and Hekmati J. (2011) Am. J. Plant Physiol., 6, 106-112.

[12] Kofranek A.M. and Paul J.L. (1975) Acta. Hort., 41, 199-206.

[13] Elgimabi M.N.E. (2011) *Am. J. Agric. Biol. Sci.*, 6(1), 128-133.

[14] Reddy T., Nagarajaiah C. and Raju B. (1988) *Hort. Absts.*, 59, 2360.

[15] Asrar A.W.A. (2012) *J. Saudi Soc. Agric. Sci.*, 11, 29-35.

[16] Delaporte K., Klieber A. and Sedgley M. (2005) *Journal of Horticultural Science and Biotechnology*, 80(4), 471-475.

[17] Pun U.K., Shimizu H., Tanase K. and Ichimura K. (2005) *Acta Hort.*, 669, 171-174.

[18] Butt S.J. (2005) *International Journal of Agricultural and Biology*, 7(1), 91-99.

[19] Pun U.K., Ichimura K. (2003) *Japan Agricultural Research Quarterly*, 37(4), 219-224.

[20] Abou El-Ghait E.M., Gomaa A.O., Youssef A.S.M. & Mohamed Y.F. (2012) *Research Journal of Agriculture & Biological Sciences*, 8(2), 261-271.

[21] Butt S.J. (2003) *A Review on prolonging the vase life of rose cut flowers*, Pakistan Rose Annual, Pakistan National Rose Society, 49-53.

[22] Nair S.A., Singh V. and Sharma T.V. (2003) *Jour. Tropic. Agric.*, 41, 56-58.

[23] Coort G.D. (1973) Hort. Science, 8, 195-198.

[24] Sujatha A., Singh V. and Sharma T.V.R.S. (2003) *J. Tropical Agri.*, 41, 56-58.

[25] Steinitz B. (1982) *Gartenbouwissenschaft*, 47, 77-81.

[26] Singh A.K. and Tiwari A.K. (2002) *South Indian Horticulture*, 50 (1/6), 140-144.

[27] Han S.S. (1998) *HortScience*, 33, 731-733.

[28] Ketsa S., Piyasaengthong Y., Prathuangwong S. (1995) *Postharvest Biology and Technology*, 5, 109-117.

[29] Loubaud M. & van Doorn W.G. (2004) *Postharvest Biology and Technology*, 32(3), 281-288.

[30] Awad A.R.E., Meawad A., Dawh A.K. and El-saka M. (1986) *J. Ornamental Hort.*, 181, 177-193.

[31] Spikman G. (1989) *Scientia Horticulturae*, 39(1), 73-81.

[32] Meeteren U.V. & Proft M. (1982) *Physiologia Plantarum*, 56(3), 236-240.

[33] Woltering E.J. and Van Doom W.G. (1988) *Journal of Experimental Botany*, 39(11), 1605-1616.

[34] van der Meulen- Muisers J.J.M., van Oeveren J.C., van der Plas L.H.W., van Tuyl J.M. (2001) *Postharvest Biology and Technology*, 21, 201-211.

[35] Halevy A.H. and Kofranek A.M. (1984) *HortScience*, 19, 845-847.

[36] Deplarote K.L. and Sedgley M. (2004) *Acta Horticulturae*, 630, 77-84.

[37] Delaporte K.L., Klieber A. & Sedgley M. (2000) *Postharvest Biology and Technology*, 19(2), 181-186.

[38] Ohkawa K., Kusuhara Y. & Suh J.N. (1999) *HortScience*, 34, 112-113.

[39] Ichimura K. and Hiraga T. (1998) *J. Jpn. Soc. Hort. Sci.*, 68.

[40] Tjosvold S.A., Wu M.S. (1994) *HortScience*, 29(4), 293-294.

[41] Serek M., Sisler E., Reid M. (1994) *Journal of American for Horticultural Science*, 119(6), 1230-1233.

[42] Chatterjee S.R., Bhattacharjee S.K. and Wangi M.M. (2003) Ind. J. Hort., 60, 394-398.

[43] Mayak S., Bravdo B., Gvilli A. and Halevy A.H. (1973) *Scientia Horticulturae*, 1, 357-365.

[44] Zagory D. and Reid M.S. (1986) *J. Am. Soc. Hort. Sci.*, 111, 154-158.

[45] El-Saka M.M. (1992) *Physiological studies for increasing the longevity of some cut flowers*, Ph. D. Thesis, Faculty of Agric., Zagazig University, Egypt.

[46] Anju B., Tripathi S.N., Sehgal O.P., Bhat A. (1999) *Advances in Horticulture and Forestry*, 6, 125-131.

[47] Kwon H., Kim K. (2000) *Journal of the Korean Society for Horticultural Sciences*, 41(2), 135-138.

[48] Gendy A.S. (2000) *Physiological study on the effects of some postharvest treatments on gladiolus cut flowers*, M. Sc. Thesis, Faculty of Agric., Zagazig University, Egypt.

[49] El-Bouhy N.F. (2002) *Effect of some postharvest treatments on polyanthus tuberose cut flowers*, M.Sc. Thesis, Faculty of Agric., Zagazig University, Egypt.

[50] Mohammad Y.F. (2009) *The effect of certain agricultural postharvest treatments on some ornamental plants*, M.Sc. Thesis, Faculty of Agriculture., Moshtohor, Benha University, Egypt.

[51] El-Bouhy N.F. (2010) *Physiological studies on the effect of some postharvest treatments on some cut flowers*, Ph.D. The sis, Faculty of Agric., Zagazig University, Egypt.

[52] Gendy A.S. and Hamad E.H. (2011) *J. Product. and Dev.*, 16 (3), 397-414.

[53] van Doorn W.G. (1997) *Water relations of cut flowers*, Horticultural Reviews, 18.

15

A NON-DESTRUCTIVE TECHNIQUE BASED ON TEXTURE ANALYSIS FOR IDENTIFICATION OF DEFECTIVE ALPHONSO MANGOES USING SUBSPACE ANALYSIS TECHNIQUES

MUSALE S.S.[1]* AND PATIL P.M.[2]

[1]Sinhgad College of Engineering, Pune- 411 004, MS, India.
[2]RMD Sinhgad Technical Institutes Campus, Pune- 411 058, MS, India.
*Corresponding Author: Email- sandeepmusale@yahoo.co.in

Abstract- In many circumstances, texture is the only information that can be used in natural image analysis because it is an important property of the surface that characterizes it's nature. Texture is defined as a spatial arrangement of local (gray level) intensity attributes which are correlated within areas of visual scene corresponding to surface regions. An image region has a constant texture if sets of its local properties in that region are constant. Thus texture analysis has received considerable attention in the field of image analysis and pattern recognition. Texture exhibits some sort of periodicity of the basic pattern of Spongy Tissue in alphonso mango. This leads to use textural property to identify different patterns of Spongy Tissue in alphonso for detection of defects in alphonso mango. Visual assessment of texture made by human is time consuming and inspection made by human does not achieve a high degree of accuracy and preciseness. Automated visual inspection of the textural pattern improves the accuracy and preciseness during detection of defects in alphonso mango. In the literature, the researchers worldwide have been working in various texture analysis algorithms for different applications like detection, recognition, classification, segmentation, clustering etc. Many algorithms suffer from low sensitive detection, difficult back ground adaption and high memory requirement. Problems and limitations associated with the available techniques have been reported by many studies. Each has some drawback under all lighting conditions and no one has used a robust, reliable algorithm for detection of spongy tissue in alphonso mango under real life test environment. To develop an optimized algorithm using a non contact mechanism which will detect the defective alphonso mangoes happen to be a challenging task. The objective of the proposed research work is to obtain computationally cost effective and noncontact solution that achieve better recognition rate under various conditions in consultation with the agriculture scientist. In this paper we have proposed use of subspace analysis techniques for extraction of textural features that identifies Spongy Tissue in alphonso mango successfully. Performance of the proposed algorithm is carried out on the alphonso database [1]. This paper presents a methodology that combines the principal component analysis (PCA) and fisher linear discriminant analysis (FLD) applied for defect detection in alphonso mangoes. Performance of the proposed algorithm has been compared with that of PCA and FLD applied individually for both healthy and defective alphonso mangoes. Experimental results computed using the proposed integrated algorithm, individual PCA and individual FLD have been validated manually with the cut sections of the alphonso mangoes available in the database. It is observed that the proposed method shows significant improvement for both defective as well as healthy alphonso mangoes over PCA and FLD applied individually. The feature vectors extracted by using integrated PCA-FLD method have 99% of the highest discriminant power.

Keywords- Alphonso mangoes, Spongy Tissue, digital X-Ray imaging, Non-Destructive testing, subspace analysis, principal component analysis, fisher linear discriminant analysis, textural features

Introduction

Horticultural crops play an important role in the economy of the nation, as there are crops suitable for almost all the agro-climatic zones of the country. India has accelerated in total annual production of horticultural crops touching over 149 million tone during 1999 -2000. Today, India has emerged as the second largest producer of the fruits (46 million tone) and vegetables (91 million tone), contributing nearly 10% and 14% respectively, of the world production. The horticulture crops cover about 8% of the total area contributing about 20% of the gross agricultural output in the country. India also enjoys the crown of being number one producer of mango and banana in the world. The overall productivity of the fruit and vegetable is 12 and 15.2 tones / hec, respectively. Though the country is the second largest producer of the fruits and vegetables in the world, per capita consumption of vegetables and fruits of our country for over one billion populations is very low. Fresh fruits and vegetables have the lion's share of exports followed by processed one.

Quality assurance in agricultural products is being more crucial with the liberalized international trade system and globalization for capturing and retaining the market. In order to strengthen the competitiveness in terms of export, advanced post-harvest technology adoption is essential. The export of fruits should lead to increase in farming income. Substandard and inadequate post harvest technology and management system is a major bottleneck to the expansion of trade for India. India is the second largest producer of fruits in the world. However, India's fruit production only contributes 1% to the export earnings from agricultural products. Also, India is the largest producer of mango fruit in the world. Amongst mangoes, Alphonso mango is the leading commercial variety produced in Maharashtra State recognized nationally as well as internationally. But in recent years this variety is facing problems due to the pests, diseases and disorders like Spongy Tissue. The peculiarity of this disorder is that the external symptoms of the affected fruits are not apparent from outside either at the ripe stage or at the time of picking. The tissue affected is visible only when the ripe fruit is cut into two halves. The overall loss due this disorder is about 30%, which amounts to a loss of nearly Rs 135 million. Traditionally, chemical and biochemical methods are available to investigate compositional and metabolic differences between the healthy and the damaged tissue. The occurrence and intensity of the disorders depend upon factors like those related to location, climate and cultivars. The symptoms are manifested at the final stage of fruit growth and maturation. Padda, et al [2] by using Canonical Discriminant Analysis found that the best tools to assess changes in mango fruit during ripening were firmness, followed by flesh value and total soluble solids content. Subedi, et al [3] showed that flesh value well correlated with maturity of fruit. Color of flesh was found to be consistent in various cultivars and although it is a destructive measurement, it is used as a maturity index in several producing regions [4]. Thus, online identification of defects in fruits is highly complex and most challenging because they require real time solution. So, there is an immediate necessity to develop a system which can detect the presence of Spongy Tissue in the mangoes non-destructively prior to export of the mangoes. This paper is an effort towards the direction. In this paper we have developed an algorithm using subspace analysis techniques, which can detect the presence of spongy tissue from stored X-ray images. The textural property of the acquired X-ray images shows the presence of the defects or disorders. Principal Component Analysis is a popular technique used to derive a starting set of features. Application of PCA for defect detection that uses textural properties, does not consider the classification aspect since it is based on the optimal representation criterion in terms of mean-square error. Talukder, et al [5] pointed out that PCA alone (as an unsupervised method) is not necessarily good at drawing distinctions between patterns. To improve the PCA standalone classification performance, one needs to combine further this optimal representation criterion with some discrimination criterion. One widely used discrimination criterion is the Fisher linear discriminant (FLD), or linear discriminant analysis (LDA) [6], which defines a projection that makes the within-class scatter small and the between-class scatter large. As a result, FLD derives compact and well-separated clusters. As the original image space is high dimensional, most of these methods apply PCA first for dimensionality reduction. Subsequent FLD transformation is then used to build the most discriminating features (MDF) space for classification. For good generalization FLD requires large training sample size. One possible remedy for this drawback is to improve FLD's generalization performance by balancing the need for adequate signal representation and subsequent classification performance using sensitivity analysis on the spectral range of the within-class eigen values. The second well-known approach coming under the name of fisher face is insensitive to large variation in the conditions already enumerated above. It uses both PCA and FLD. It is worth stressing that by maximizing the ratio of between-scatter matrix and within-scatter matrix, FLD produces well separated classes even in the subspace having low dimensions under large variation in lighting conditions.

In this paper a methodology that combines the principal component analysis (PCA) and fisher linear discriminant analysis (FLD) to aid in the criteria development for selecting spectral bands for multispectral imaging applications has been proposed. This method and the individual PCA and FLD methods have been applied for defect detection in alphonso mangoes. The defect detection performance of the proposed method has been compared with PCA and FLD. Performance of the proposed algorithm is verified with the alphonso mango database [1] which is easily available for researchers in this area after sending requisition to the authors through mail. The database consists of X-ray images of healthy alphonso mangoes, artificially created defects in the mangoes as well as naturally affected spongy tissue mangoes collected using digital X-ray machine. In the captured images, the spongy tissue affected portion or artificially defect created portion of the mangoes shows dark gray levels, whereas unaffected part or healthy portion shows a uniform light grey levels. The database also consists of the digital photographs of all the alphonso mangoes along with the cut sections of the alphonso mangoes may be required for manual validation of the automated results obtained with the help of proposed algorithms. Defects in alphonso mangoes have been detected successfully using the proposed method.

Feature Extraction with Proposed PCA-FLD Method

PCA is a transform that uses orthogonal axes for dimensionality reduction by performing an eigen decomposition of the spectral covariance matrix of the data. Let C_t represent the covariance matrix of $N \times N$ dimension and computed by,

$$C_t = \frac{1}{N}\sum_{i=1}^{N} (X_i - \mu)(X_i - \mu)^T \quad (1)$$

where,

C_t : represents covariance matrix,

x_i : the ith random sample with dimension N,

μ : represents mean vector of the set of pixels,

N : the sample size.

In order to maximize the energy of the projected samples, the eigen -decomposition transform is used and computed as,

$$E_{pca} = \phi_m^T C_t \phi_m \quad (2)$$

where,

E_{pca} : represents the energy of the samples projected,

Φ_m : is one of the transform vectors used to project the data samples,

m = 1,.....,N.

C_t : the covariance matrix,

ϕ_m^T : the transpose of the transform vectors used to project the data samples

A set of eigenvectors is orthogonal and each eigenvector corresponds to a unique combination of bands. The residual mean square error between transformed data and original data is minimized by performing eigen-decomposition of the covariance matrix. The first eigenvector corresponding to the largest eigen value and provides largest variance of the data. Similarly, the first two eigenvectors explain more variance of the data than any other pair of original bands. Magnitude of eigen value indicates the energy residing in the data along the direction parallel to the corresponding eigenvector. The larger the eigen value, higher is the energy it represents. Therefore, to reduce the original *N*-dimensional data to a lesser dimension *M*, one can project the original data to the *M* eigenvectors corresponding to the largest of the *M* eigen values. Considering the first few significant compositions, this transform results in a lower-dimensional multivariate feature vector. This feature vector still preserves most of the energy in the original higher-dimension system.

FLD is an effective class-specific method which projects the scatter of data to make them more reliable for classification. Two matrices are introduced in the FLD namely, the between-class scatter matrix, denoted as S_b, and the within-class scatter matrix S_w. The between-class scatter matrix S_b is defined as [7],

$$S_b = \sum_{i=1}^{c} |x_i| (\mu_i - \mu)(\mu_i - \mu)^T \tag{3}$$

The within-class scatter matrix is defined as [7],

$$S_w = \sum_{i=1}^{c} \sum_{x_k \in x_i} (x_k - \mu)(\mu_k - \mu)^T \tag{4}$$

where,

μ_i : represents the mean vector of ith class,

μ : the mean of total samples,

C : the number of classes,

x_i : represents the ith class and

$|x_i|$: represents number of samples in i^{th} class.

The optimal data projection, denoted as E_{fld}, can be obtained by selecting the transform-vector Φ_m that maximizes the ratio of the projected between-class samples to the projected within-class samples and is defined as,

$$E_{fld} = \frac{\sum_{m=1}^{M} \phi_m^T S_b \phi_m}{\sum_{m=1}^{M} \phi_m^T S_b \phi_m} \tag{5}$$

PCA does not guarantee the feature class separation ability of the selected band, whereas, the FLD method is effective in class segmentation, sensitive to noise and may not cover enough energy from the original data.

To design a set of transformation vector-basis providing supervised classification information in a better way that simultaneously preserves enough energy from the original data cube, a method is proposed that combines (1) to (4) in order to construct an evaluation equation called integrated PCA-FLD method. A weight factor *K* introduced to adjust the degree of classification that preserves the desired energy. The constructed evaluation equation is given as,

$$E_{ev1} = \sum_{m=1}^{M} \frac{\phi_m^T [KC_t + (1-k)S_b]\phi_m}{\phi_m^T [KI_t + (1-k)S_w]\phi_m} \tag{6}$$

where,

E_{ev1} : represents the constructed evaluation,

$K\hat{I}$ [0,1],

C_t : the covariance matrix,

S_b : the between-class scatter matrix,

S_w : the within-class scatter matrix

I : the identity matrix.

The eigen-decomposition becomes inaccurate, if the within-class scatter matrix S_w in [Eq-5] becomes very small. [Eq-6] eliminates this problem. If S_w becomes very small, the effects of S_w can be ignored by adjusting the weight factor *K* towards 1, which means the principal components are more heavily weighted. The ratio between S_b and S_w dominates, if the selected value of *K* is small meaning more differential information between classes is taken into account. The integrated method enhances the advantages of PCA and FLD and reduces the disadvantages of both PCA and FLD simultaneously. In fact, FLD and PCA methods represent the extreme situations of [Eq-6], as K=0, it becomes FLD [Eq-5] and when K=1, the evaluation equation is equivalent to PCA method [Eq-2].

To find the transform that works equally well on representation and discrimination, one can find the set of Φ_m that maximizes [Eq-6] with K=0.5. The solution of Φ_m, called the generalized eigen vector, can be obtained by equating the derivative of [Eq-6] with respect to Φ_m to zero as in [Eq-7].

$$\left(\frac{KC_t + (1-K)S_b}{KI + (1-K)S_w} \right) \phi_m = \lambda_m \phi_m \tag{7}$$

where,

λ_m : represents eigen values,

Φ_m : the corresponding generalized eigen vector.

For hyper spectral dimensionality reduction, if the original feature space is having *N* dimensions, then the reduced *M* ($M < N$) dimensional features can be obtained by selecting the value of *M* generalized eigenvectors corresponding to the *M* largest eigen values obtained from [Eq-7]. The value of *K* varies between 0 to 1 that shifts the weight between PCA and FLD. The projected samples become the best in representation and classification at the proportion represented by the value of *K*. Based on application in hand, an optimal proportion between PCA and FLD can be selected by searching the optimal value of *K* that achieves best classification rate of samples. The transformation matrix *T* (*N* X *M*) is formed by selecting *M* generalized eigen vectors as the column vectors for the matrix. The linear transform is defined as in [Eq-8].

$$Y = T^T X \tag{8}$$

where,

X : represents the original feature space with given *N* x *J* dimension,

N : the features, *J* : the sample size,

Y : the transformed feature space with *M* x *J* dimension.

The transformed feature space consists of the new features (or bands) that can be obtained from the linear combination of all original features/bands for all pixels. Since the transformation matrix consists of *M* eigen vectors corresponding to the *M* largest eigen values, transformed feature space can provide a better classification and representation ability than any subset of *M* bands of the original data set. For the experimentation purpose, we have selected the value of *M* equal to 1. In order to extract the features, the eigen vector corresponding to the largest eigen value of [Eq-7] is used. In this case, the transformed feature space represents a single feature vector i.e. the first principal feature vector, generated by linear combination of original *N* features by using the integrated PCA-FLD method. Similarly, by applying PCA or FLD method individually, one can obtain the first principal feature vector of PCA or FLD, respectively. The physical meaning of the first principal feature vector in the integrated PCA-FLD method provides the best discriminative feature compared with any individual feature in the original *N* feature space. A classifier is then designed to determine the category of input image samples. The mean and standard deviation of the feature vectors are calculated for spectral feature vectors that belong to one image sample. The ratio of the standard deviation to the mean is used as numerical input variable to a *k*-Nearest Neighbor classifier algorithm. The *kNN* algorithm is used to determine the *k* samples in the training set that are close to the Xth unknown sample in the testing set. As the patterns with similar attributes should be assigned to the same class, samples with closer ratios are classified as the same category. Hence, if a majority of the *k* Nearest Neighbors belongs to the class of 'healthy alphonso mango', then the sample is classified as a healthy alphonso mango, otherwise it is classified as the defective one. For experimentation purpose, we have used k =15 in the classifier.

Results and Discussions

In the proposed method of defect detection in alphonso mangoes, the value of *K* was used to adjust the balance between the representative and classification abilities of the projection matrix, and therefore increased the flexibility of the method for Spongy Tissue detection in alphonso mangoes. For different image samples, the optimal values of *K* differs. An effective method for *K* value selection was needed.

In this work, we have searched the values of *K* from 0 to 1 with a step size of 0.05. When the *K* value was changed, the recognition rates varied. In general, by choosing the optimal *K* values, both the representation and classification effects were properly weighted for the testing samples. The small set sample tests can be considered a validating process before the training and the testing process. Those image samples were called validating image samples, which were used to determine the variable parameters in the method before final application to the training and testing processes.

Table 1- Samples used for various experiments

Experiment	Sample Type	Total number of samples	No of samples for Training Set	No of samples for Testing Set
1	Healthy and defective alphonso mangoes	200	50	150
2	Healthy and defective alphonso mangoes	200	100	100

Performance of the proposed algorithm is carried out on the alphonso database [1]. The database consists of 1000 X-ray images of both healthy and defected alphonso mangoes. Out of these 1000 images we have randomly selected 100 images of healthy and 100 images of Spongy Tissue affected alphonso mangoes (with moderate and severe degrees of damage) for our experimentation purpose. The samples of healthy and Spongy Tissue affected alphonso mangoes used for these experiments are listed in [Table-1]. In the first experiment, out of 100 X-ray images of healthy alphonso mangoes we have selected 25 samples and out of 100 X-ray images of Spongy Tissue affected alphonso mangoes we have selected 25 samples (total 50 samples) were used for training and then tested together with the remaining 150 samples. The total 200 sample testing results with *K* set to 0.60 have been shown in [Fig-1]. The results indicate that the PCA method and the integrated PCA-FLD method achieved a similar better performance compared with the FLD method. The sample classes used in the first experiment were well separated hence the solution was dominated by the PCA method at classifying data since they are well distinguishable. On the other hand, since FLD maximizes the E_{fld} ratio shown in [Eq-5], which is small in this case, it becomes unreliable and noise sensitive, and leading to the poor performance in this situation.

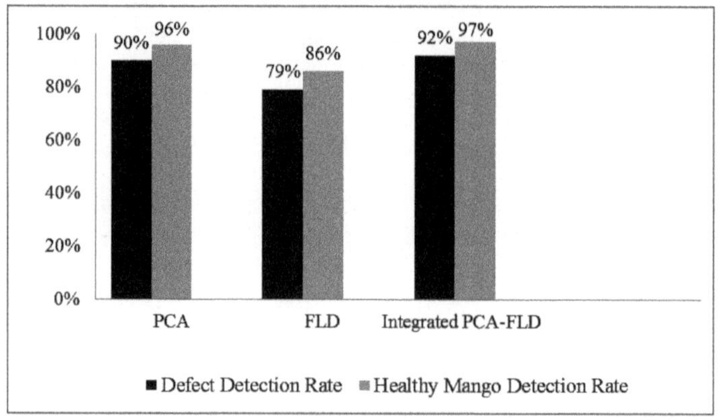

Fig. 1- Defect detection rates of Spongy Tissue affected alphonso mango and healthy alphonso mango detection rates using FLD, PCA and integrated FLD-PCA methods.

In order to test the reliability of these algorithms, number of samples during training was increased. In this experiment, 100 samples were used randomly for training, and the remaining samples were used for testing; therefore, the recognition rates for both good and faulty textile fabric images were high. The recognition results with *K* value (set empirically) set to 0.20 have been illustrated in [Fig-2]. Since these two classes had very similar spectral characteristics, the differentiating information played a more important role in classification. Therefore, the FLD solution showed better performance than PCA. By using the FLD method, the defect detection rate achieved 91%, and the healthy sample detection was 98%, which were, respectively, 6% and 1% higher than those achieved with the PCA method. It is observed that in both the experiments the results of the proposed integrated PCA-FLD method achieved a similar better performance.

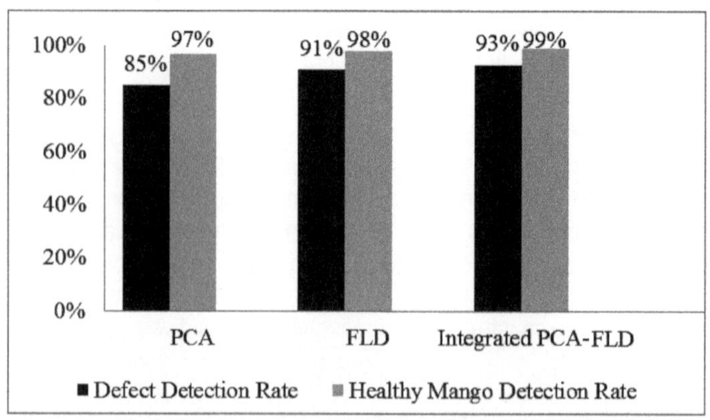

Fig. 2- Defect detection rates of Spongy Tissue affected alphonso mango and healthy alphonso mango detection rates using FLD, PCA and integrated FLD-PCA methods.

Conclusions

The number of regions in PCA and FLD depends on the size of the defect in alphonso mango. Classification rate increases as the number of segmented region increases. Also, the efficiency of the FLD method remains constant. This method does not require training collection.

The integrated PCA-FLD method for feature extraction when applied to defect detection and classification of alphonso mangoes achieves better recognition performance. The integration process was neither a simple combination of the two methods nor using them sequentially during the entire procedure. The method was derived based on a constructed evaluation equation that combined the representation and classification effects together. The principal components preserved the most energy of the original data, and provided good performance in recognizing obviously separated classes. The discriminant analysis contributed to classifying similar patterned classes. By properly adjusting the weight factor, the integrated PCA-FLD method was more flexible in processing with different sample patterns and the result became robust to noise. The results show that the integrated PCA-FLD method outperformed the PCA or FLD method. The reason is that both the representation information and class specific information are included in the integrated PCA-FLD feature. FLD provides the highest discriminant power of any feature vector. The feature vectors extracted by using integrated PCA-FLD method have 99% of the highest discriminant power. The first principal component of PCA presents the highest discriminant power among other principal components. Compared with FLD and integrated PCA-FLD, the first principal component has the lowest discriminant power.

Acknowledgment

The authors are thankful to the "Board of College and University Development (BCUD), Pune University, Pune" for supporting this non-commercial research entitled "Detection and Analysis of Defects in Agro Products," Online Proposal Number: 13ENG001578 under the "Research Proposals under University Research Grant Scheme for the year 2013-2015." The authors are also thankful to the anonymous reviewers for their valuable suggestions and comments to elevate quality of the paper.

Conflicts of Interest: None declared.

References

[1] Musale S.S. & Patil P.M. (2014) *Database development of defective and healthy alphonso mangoes*, International Conference on Food, Agriculture and Biology, Kuala Lumpur, Malaysia, 60-65.

[2] Padda S.M., do Amarante C.V.T., Garcia R.M., Slaughter D.C., & Mitcham E.M. (2011) *Postharvest Biology and Technology*, 62, 267-274.

[3] Subedi P.P., Walsh K.B. & Owens G. (2007) *Postharvest Biology and Technology*, 43, 326-334.

[4] Sivakumar D., Jiang Y. & Yahia E.M. (2011) *Food Research International*, 44, 1254-1263.

[5] Talukder A. & Casasent D. (1998) *Journal of Optical Engineering*, 37(3), 904-913.

[6] Fisher R. (1936) *Ann. Eugenics*, 7, 179-188.

[7] Belhumeur P., Hespanha J. & Kriejman D. (1997) *IEEE Transactions on Pattern Analysis and Machine Intelligence*, 19(7), 711-721.

16

ROLE OF CO-OPERATIVE SUGAR FACTORIES IN RURAL DEVELOPMENT : A CASE STUDY OF DAMAJI SUGAR FACTORY, MANGALWEDA

TODKARI G.U.

Dept. of Geography, Shri Sant Damji Mahavidyala Mangalweda, Dist-Solapur, MS, India.
*Corresponding Author: Email- govindtodkari@gmail.com

Abstract- Sugar industry occupies an important position on the industrial map of the state of Maharashtra. These factories played a major role in the socio-economic development of rural areas in Maharashtra. The growth of various urban centers is helping to solve the problem of unemployment by providing employment in the growing industries and business. For the present investigation Damaji co-operative Sugar Factory is selected. Mangalwedha is located in drought prone area but Bhima river flows in these region. The Bhīma river catchment are favorable for sugar cultivation. Remaining drought area is large than irrigation area, drought area sugar factory become key factor of development and employment. In this investigation primary and secondary data are used. Such type of study represents real the role of co-operative sugar factory in rural of Mangalwedha and helps to planners, agricultural scientists and research scholars.
Key words - health camps, economic activities, beautification, social responsibility, Agricultural family

Introduction

The sugar factories being agro-based processing units, are located near the supply of raw material namely sugarcane. These industrial units get located mainly in the rural areas and at tahsil or district centers. With the location of these processing units in the rural areas it is visualized that these factories can be the most appropriate agents for bringing about rural industrialization and change. The agro-based sugar factories with active Government aid and encouragement led to industrial dispersal in the state like Maharashtra. Along with dispersal of industrialization these factories due to their linkage effects became the growth centers for rural development.

Sugar co-operative for their expansion gave importance to better sugarcane production, supply of appropriate agricultural inputs to farmers and increased irrigation facilities in their area of operation. In addition growth of educational facilities, medical facilities, etc. were also undertaken as a part of area development by these co-operatives sled to growth of ancillary units, like paper plants, distillery units etc. which in turn increased employment and led to further industrial development in the rural areas. The success of sugar co -operatives in their commitment to area development led to growth of other agro-based processing units like dairy co-operatives, spinning mills, fruit processing units and such agro based industrialization helped in the development of agriculture and brought about the necessary transformation of the rural areas and in the process the benefits of these activities reached the farmers and they too benefited and progressed. Such process of rural transformation and change is very noticeable in western Maharashtra and more so in Solapur district where the co-operative sugar factories have indeed become the 'Growth Centers' for rural development.

Objectives

- To asses the employment opportunity in sugar factories.
- To illustrate the role of co—operative sugar factories in rural development.
- To suggest the recommendation for the development of sugar factory.

Study area

'Shri Sant Damaji Co-operative Sugar Factory, Mangalwedha' is situated in northern side of Mangalwedha city. Shri Sant Damaji Co -operative Sugar Factory Ltd Mangalwedha has started its first crushing season on 10th June, 1991 with low producing and crushing capacity. It is bordered by the Pandharpur tahsil on the whole of the northern side, by the Bathan on the east by the Bharmapuri on the south -east and Mangalwedha city on the south by the Sangola tahsil on the west. The study region i.e. Mangalwedha tahsil is one of the tahsil of Solapur district. It lies between 18 ° North latitude and 76 ° East longitude . The total area of the study region is about 1141.13 sq.km. Which is 7.62 percent of the Solapur district, consist of eighty one villages. According to geographical area, Mangalwedha tahsil is tenth number tahsil in Solapur District.

Database & Methodology

For the present investigation primary data and secondary data are used. The primary data is collected through interviews, schedule and questioners. The secondary data are collected from annual sugar factory report and Damaji Sugar Factory Office. The collected data are represented by using various statistical and cartographic techniques.

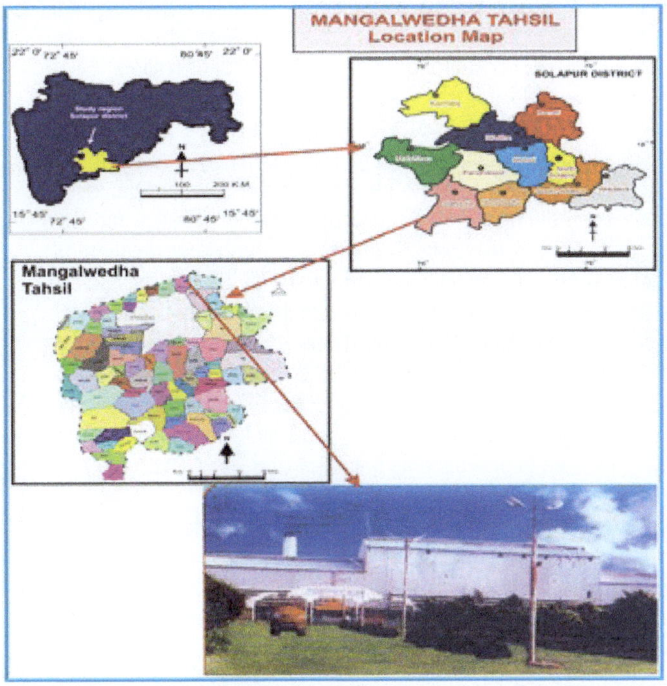

Finding & Discussion

The co-operative sugar factory is established in a rural setting it becomes possible for the organization to help generate various ancillary activities for the benefit of local farmers and other member of rural society. Sugar co-operative provides the basis for organization other economic activities such as modern poultries, dairies, irrigation schemes, banks, goober gas plants, better breeding of cows and many other such activities which contribute largely to the betterment of the economic conditions not only of the farmers but also landless labourers and other people in the area. The factory also establishes school, heath centers for medical facilities with modern hospitals and dispensaries and organizes various cultural and sports activities. A considerable amount is set apart by many co-operative factories to make the life of the local inhabitants more purposeful and meaningful. Thus diversification of production activities leads to a process of rural industrialization which also is employment generating.

Lift Irrigation Scheme-The Shree Sant Damaji Co-operative Sugar Factory Ltd Mangalwedha has sponsored the lift irrigation schemes with a view to increase the land under sugarcane and to fulfill the needs of basic raw material, i.e. sugarcane within the command area. Lift irrigation schemes is in the area are formed mainly due to the keen interest shown by co-operative sugar factory gives guarantee to the banks for extending loans for irrigation schemes.

Development of Sugarcane- Crushing capacity of the factory is 2500 metric tones. As factory crushing season is of 223 days there is necessity of 600000 metric tones of sugarcane for good quality and high production of sugar. In working field and the variety of seedling used are Kosi 671, Co 94012, Co 86032 and the VSI 434, VSI 435 etc produced by tissue culture technology by company VSI for per acre increment in sugar production. These different programmes are organised by factory for the farmers.

Supply of Good Quality Sugarcane Seeds-The factory has been supplying the standard seed of sugarcane on its own cost to cane producer. The sugarcane seed is brought from famers. Similarly, the supply of seeds is also made available on credit from the seed plots of the other producers. Factory has supplied various types of sugarcane seeds i-e CO 671, Co 86032, to the member farmers. The Shri Sant Damaji Co-operative Sugar Factory has supplied sugarcane seed to farmer members. After the year 2005 more and more farmers have taken the benefits of cane seeds. From the year 2005-06 sugar factory expend 2 lakhs on cane seed programme. And in the year 2006-07 factory expand 1.54 lakhs on sugarcane seeds. In recent year more farmers have taken benefit of sugarcane seeds.

The beautification of factory site- From last one and half years nearly 1700 trees have been planted not only to control pollution but also to give shadow of love to members. It includes different types of Mango ,Nilgiri, Alphonso, Ashoka along with it coconut, and Chikku etc. cash trees also planted. For the beautification of the campus various trees are planted e.g. Tamarind, Bottle palm, Chapa, Silver oak, Gulmohor etc. The Golden Duranda is grown on 3500 feet on the factory site to decorate the campus.

Accident Insurance for factory members- The factory has made resolution in its meeting accident insurance for the members nearly 14240 members insurance would be the beneficiary of it. Nearly fourteen members have got benefit of this accident insurance. Life Insurance Corporation of India is providing these facilities to factory.

Helping hand to heirs of martyr- Out of social accountability, the factory used to help to the heir of the martyr, who dies on the border of the nation. The help is in the form of money. The amount is up to 1 lakh. This help is not only given to factory members but given to them also, who are not member .The factory has donated to *Mukhyamantri Sahayyata Nidhi* for earth quake affected people. In the year 2000-01factory funded 100000 to Government.

Agriculture exhibition To give advanced knowledge and information of agricultural to the farmers, factory used to organize agriculture exhibition each and every year on the factory site. In this exhibition different stalls are installed e.g. agricultural technology, advanced equipments, fertilizers, seeds, animal husbandry, various government schemes etc. Many farmers visit and implement the things in their farms.

Computer Training to employees- In the 21 century, every thing is computerized. Hence the factory has arranged of the facility of computer training to the officers and other workers. Today more than 300 officers and workers are computer literate.

Cattle Camp- The factory has started the camp for the cattle's belonging to drought stricken area of the factory. Out of social responsibility the factory has started cattle camp on 2 Sept. 2003 to save live stock. It was run for 10 months and almost more than 2000 cattle's were kept in the camp. Factory expands 25 lakh on these cattle camp.

Shopping centre- In the age of IT competitor the question of employment became very severe in front of the rural youth also. The factory has decided to build shopping centre and make it available to the unemployed rural youth. It will help the rural youth for get the work to their hands.

The safe of fruit-bearing trees to farmers- To improve the financial condition of the farmers, the factory has started the scheme of providing the fruit - bearing trees to the farmers. This scheme was started in the year 2002-03. In the year 2002-03 factory had supplied 1120. Fruit plants on no profit and no loss basis and the cost of plants was Rs15590. In the year 2004-05 farmers took 515 fruit plants from the factory the cost of which was Rs7770. After the year 2004-05 the number of fruit plants supplied has increased. Thus the Shri Sant Damaji Co-operative Sugar Factory is making its attempts to develop fruit plantation and thus increase the greenery in its area of its area of operation.

Sugar distribution to members- The factory Board of Director decided to distribute sugar to the members 5 kilograms per year on concessional rate.

Sugar Factory School- The sugar school is started near factory site with the help of sugar commission for the children of sugarcane cutter labours. Factory gives the training to teachers in collaboration with 'Aastitva prakashan' Pune. Factory provides all the educational aids to all the children which are admitted in the sugar factory school. In the year 2006-2007, 150 children get benefits of it.

Felicitation of Meritorious students- The students who embark their names in Merit list of H.S.C. and S.S.C. exams belongs to Agricultural family. These students are felicitated by factory every year. The factory used to give inspiration to the meritorious students by felicitating them.

Response to emergencies- During emergencies and natural calamities the resources of the factory prove extremely useful for organizing relief and other activities. On all such occasions the leaders and shareholders have contributed generously. The Government department have now realized the importance of the factories in the life of the local people and have there for been relying increasingly on their help and co-operation. It is such easier and more effective on mobilize people for relief work when the leaders and shareholders voluntarily decide to do so instead of on the diktat of a Government official.

Supply of Molasses- Molasses has become a very valuable raw material for the chemical industry. It is also used as fertilizer into field as well as for the burning coke. Molasses is a good for the fields growing sugarcane. .In the year 2000-01 farmers had taken 22500 tones of molasses. The rate of molasses was Rs. 1333.33 per tone. In the year 2006-07 farmers took 20444.760 tones of molasses and the rate was Rs 2006 per tone. The molasses is used for fuel purpose and also as manure in the farms.

Road Construction- Efforts have been made by the sugar factory for the development of roads. This process has helped to build up infrastructural facility in the rural areas. For accelerating the transport of harvested cane to factory from the fields, the sugar co-operative has built up approach road in its area of operation. Sugar factory has been constructing cross ways across the river and streams by using home pipe to reduce the distance between the factories and the area of harvesting. Government grants are made available to the factory to make use of the bulldozers for road leveling purpose. As a result of these efforts made by the sugar factory, the sugar factory in turn gets the benefit of the higher recovery of harvested cane duke to quick transportation of cane to the factory for crushing. The above table gives details of the expenditure of road construction. In 2006-07, there was Rs18000 expenditure on road construction and in the year 2006-07, Rs 27000 expenditure was incurred on road construction. Shri Sant Damaji Co-operative Sugar Factory Ltd., Mangalwedha has built up some approach roads in the area of operation as below.

Labour Welfare Programme- There is good relation between workers and the management from the inception of the co-operative sugar factory. Factory is always ready to solve all labour problems. Sugar factory sends the workers for training and seminars. The facilities of indoor and outdoor games have been provided by the Shri Sant Damaji Co-operative Sugar Factory. These activities include volleyball, carom, chess, table tennis, cricket etc. The factory canteen is run on private basis. Tea, Snakes, meals etc., are served in the canteen. Factory has established a good modern colony on factory site for workers and their family members. One health centre is in the premises of the factory with a qualified Medical Officer and requisite staff. All workers of the factory and their family and harvesting workers are giving medical treatment of concessional rates. Library facility with book of well known writers is available for the workers and families residing at the factory site. The dailies, and Magazines which are widely popular in Maharashtra are made available for reading purpose and the workers take the benefit of it. It is the factory's pride and honor that the marginal amount of bonus (20 percent) are paid to the factory workers.

Medical Facility- The Shri Sant Damaji Co-operative Sugar Factory organized camps. The factory organizes health camps for various diseases of the eye, teeth, skin, child health, cancer etc. In these camps expert doctors in the respective fields are called and the treatment is given free of charges to the patients. If needed operations are done and some cast is borne by the co-operative sugar factory. In the year 2001 to 2007 these camps were organized and 4030 persons took benefit of the facility

Conclusion

Shri Sant Damaji Co-operative Sugar Factory is situated in Uchethan and established in 1991. This sugar factory is affected on catchment area with various ways. Various schemes have adopted by sugar factory for the surrounding rural area development. In which Agriculture exhibition, Computer Training to employee, Cattle Camp, Felicitation of Meritorious students, Response to emergencies, Supply of Molasses, supply of good quality sugarcane seeds, Accident insurance, Sugar factory school, Labour Welfare programs etc. are included. But cogeneration and distillery project are not started in this factory which can give large profit and create more employment opporchnity.

References

[1] *Annual Report of Shri Sant Damaji Co-operative Sugar Factory Ltd., Mangalwedha.*2009/10

[2] Mohite.Y.J. *Sugar Industry in Maharashtra.*

[3] Bhanje B.M. (1995) *Sugar Co-operative And Rural Transformation. A Geographical Perspective of the Command Area of Warana Sugar Factory', Unpublished Ph.D. thesis, (Geography) Shivaji University, Kolhapur.*

[4] Inamdar N.R. (1973) *Government and Co-operative Sugar Factory.*

[5] Singh G.B. (1979) *Transformation of Agriculture.*

[6] Das B.N. and Sarkar A.K. (1972) *Rural area Development Karnal Area.*

[7] More K.S. (1980) *Changing Pattern of Agriculture Landaus In Kolhapur District (Maharashtra)' unpublished Ph.D thesis, Shivaji University, Kolhapur.*

FEASIBILITY STUDY OF A PIVOTED-PLATE SENSOR FOR SILAGE CORN YIELD MONITORING

MAHARLOUIE M.M., LOGHAVI M.* AND KAMGAR S.

Shiraz University, Shiraz, Iran.
*Corresponding Author: Email- loghavi@shirazu.ac.ir

Abstract- Crop yield map is a useful means to quantify yield variations within the fields and to provide a scientific basis for implementing site -specific crop management strategies. This research explores the feasibility of developing a new silage yield monitoring system by employing a new mass flow sensor based on using the momentum of chopped material impacting a pivoted plate loaded by a retractable spring installed at the end of a chopper discharge spout. A precise rotary potentiometer was used to measure the degree of pivoted plate rotation. To save and process the output voltage of the potentiometer, a programmable circuit comprising of AVR microcontroller ATMEGA series and MAX 232 microchip was used and the output data was sent to a portable computer for further processing. To calibrate the output signal of the sensor versus instantaneous chopped material flow rate, the chopper was fed artificially with predetermined mass flow rates of 6, 8 and 10 kg s^{-1} of silage corn at five replications for each feeding rate. The results of statistical analysis showed high correlation between degrees of the pivoted-plate rotation and instantaneous feed rates with adjusted R^2 of 0.98. To evaluate the accuracy of mass flow rate estimation, a platform scale weighing system was used during the field tests. The statistical analysis did not show any significant difference between sensor mass flow rate estimation and platform scale weighing system output. By using the resulted calibration equation, yield map of a silage cornfield was generated.
Key words- precision farming, yield monitoring, chopper, silage corn, pivoted-plate sensor

Introduction

One of the most important topics in precision farming is preparing yield maps for grain or non-grain crops. Crop producers use these maps to find the existence of problems hindering their access to the full potential yield and profit in their fields. The most complicated part of yield map technology is determining the mass flow rate (mass/time) of the harvested crop during harvesting operations. Yield sensors as reported by Loghavi and Almaee (2009) are based on a wide variety of measurement methods including paddle wheel volume flow sensor (Schueller et al., 1987; Searcy et al., 1989; Schnug et al., 1993), pivoted auger (Wagner and Schrock, 1989), capacitance sensor (Stafford et al., 1991), ultrasonic sensor (Klemme et al., 1992), strain gage-based impact sensor (Borgelt, 1993), elevator-based flow sensor (Howard et al., 1993), gamma ray sensor (Stafford et al., 1991; Massey Ferguson, 1993), infrared sensor (Hummel et al.,1995), and x-ray techniques (Arslan et al., 2000)[16].

Cereal grain yield monitors are well developed and experimental methods used by researchers have been commercialized due to the availability of proven combine grain yield monitors. However, up to now limited commercial methods have been developed for non-grain crops mainly because of diverse methods of harvesting and substantial differences between these crops regarding their physical and mechanical properties. Several research projects have been conducted and many investigations are still in progress to develop practical techniques in this area.

Godwin et al (1999) developed a yield mapping system for use in non-grain crops (i.e. roots and forage) using the measurement of mass accumulation rate. The concept was realized by instrumenting a high-sided trailer to record the rate of accumulation of crop mass as the trailer followed the harvester. The location of the trailer was recorded using a differential GPS with a linked data set to log the position and the processed mass information. The mean yield values of sugar beet and potato crops taken from the yield map and from samples measured by batch harvesting were not significantly different [10]. Pelletier and Upadhyaya (1999) developed a continuous mass flow-type yield-sensing device equipped with load cells along with a differential global positioning system (DGPS) for mounting on a processing tomato harvester to collect spatially variable yield data in real-time. Their results showed high correlation of 0.997 between load cells output and clean tomato mass flow rate [22]. Cotton yield monitor systems have been developed and tested [7, 23, 24, 28, 29, 30, 32, 34, 40, 41, 46, 47], and several versions of a cotton yield monitor have come to market. All commercial cotton yield monitors use the principle of optical attenuation using a light-emitter array and a light-detector array mounted opposite to each other on a cotton picker's pneumatic duct. The light attenuation signal is converted by the data acquisition system to an estimate of the cotton mass flow rate [33]. For peanut yield monitoring, development of a load cell-based yield monitor [7, 39] and an optical-based yield monitor [35] have been reported.

For yield monitoring and mapping of root crop (i.e. potato, sugar beet, etc.) yields, the following measurement systems have been reported: radiometric [2], load cells in belt conveyors [4, 42, 43], image analysis [14, 25], pivoted table with a load cell [27], bounce plate [8, 9] and weighing frame with load cells and speed sensors [37], curved impact plate [11].

For sugar cane yield monitoring the following attempts have been reported: development of hydraulic pressure and flow measurement system [5], construction of load cell-based mass measurement trailer [26], weight scale yield sensor [3], load cell transducers and speed detectors mounted on side conveyor [21] and load cells as a billet weighing instrument set up on the harvester side conveyor just before the sugar cane billets are dropped into an in-field trailer [17].

In silage and forage crops, investigations were initiated by instrumenting a silage corn harvester with strain gauge-based torque sensors fitted on main blower shaft and main drive shaft of cutter head, feed rolls and front attachment [38]. Auernhammer *et al.* (1995) used a radioactive transmitter and receiver to find a relationship between feed rate changes of dry silage crop and changes in the intensity of radioactive wave received by the detector. Although they found good correlation between these parameters, this technique was banned due to US food industry regulations that prevented food producers from using radioactive waves during food processing [1].

Missoten *et al.* (1997) installed an impact plate sensor in spout of a drawn chopper. A deflector was used to guide chopped material toward the impact sensor. To compensate the effect of speed changes of passing material, a radar-type speed sensor was also installed. Results showed a linear relation (R^2=0.985) between mass flow rate and output data through the sensor [20].

Wild and Auernhammer (1999) equipped a round baler with strain-gauge load cells installed on the axle and drawbar to prepare a yield map. They conducted tests in stationary and on-the-go modes. The results showed that in stationary mode, the difference between actual mass flow rate and sensor output was only 1% but in on-the-go mode, the difference was 10% due to vibrations of the moving vehicle [44].

Martel and Savoie (2000) equipped a pull-type forage harvester with four different types of sensors to estimate mass flow rate and moisture content. Sensors measured feed roll displacement, crop impact force against a hinged plate, the frequency drop of a capacitance controlled oscillator and number of light beam interruptions by forage particles in the spout. When testing in a cornfield, good correlations were obtained between estimated mass-flow-rate and either the feed roll displacement (R^2 = 94%) or the crop impact force (R^2= 0.95) [18]. Savoie *et al.* (2002) in a similar investigation to Martel and Savoie (2000) equipped a forage harvester and a high–dump wagon with five types of sensors to measure mass-flow rate. Final results showed that impact force in the spout produced a very good linear correlation with mass–flow rate (R^2 = 0.951). Feed roll displacement produced the second best linear correlation with mass-flow rate (R^2 = 0.863). Torque at the PTO shaft and at the cutter-head required a moisture correction to improve correlation with mass–flow rate. The capacitance-controlled oscillator was poorly correlated with mass-flow rate but yielded the best correlation with moisture content (R^2 = 0.662) [30]. Lee *et al.* (2002) constructed a trailer equipped with load cells that could continuously measure the weight of chopped material. They used a blue-tooth data transmitter to transmit moisture data from a capacitance type moisture sensor mounted in an opening made in the middle of a spout of the silage chopper to a host computer. The masses of the empty and full silage wagon were measured with a platform scale before and after harvesting, and were compared with load cell measurements in the silage yield mapping system. By using GPS data and sensors data, they could prepare a yield map for silage crop [15].

Kumhala and Prosek (2003) developed and evaluated a torque meter and an impact force sensor simultaneously on a three-point linkage type mower-conditioner. They concluded that the impact force sensor was prefered to the torque sensor due to the lower cost and less sensitivity to the kind of crop being harvested [12].

Kumhala et al., 2007 developed a parallel plate capacitance sensor (capacitor) operating at 27 MHz frequency was designed for forage mass determination. Laboratory tests showed that moisture content as well as contact pressure of material placed between the plates influenced the sensor's output signal. The results showed a strong relationship between the weight of material and the sensor's signal. They claimed that it is possible to recommend this experimental method for further research and development [13]. Kumhala *et al.* (2010) tested a parallel plate capacitive throughput sensor for the purpose of chopped maize throughput measurement. They recognized a non-linear dependence between the sensor output signals and material throughput. They recommended that the material moisture content have to be respected during capacitive throughput sensor calibration.

As mentioned above, many studies have been focused on measuring the impact force of chopped materials impacting a deflector plate in the spout of chopper harvester, but scientist claimed that load cells were highly sensitive to machine vibrations in the field causing errors in measurements. The objective of this research was to investigate the feasibility of using a pivoted plate sensor in the spout of the silage corn chopper for real-time measurement of chopped material throughput in preparation of a site-specific yield map. The main advantage of this new method is installing the mass flow sensor without any major modification in conventional system of the harvester. The other advantage of the suggested method is using a precise potentiometer to measure angular rotation of a pivoted plate that is indirectly related to the momentum of chopped materials impacting the plate instead of measuring the impact force directly by a load cell as mentioned in reviewed literatures.

Materials and Methods

Theoretical criteria

A throughput flow rate sensor was developed and installed to meter the rate of chopped material leaving the spout of the chopper. The designed mechanism consisted of a pivoted curved plate with a re-coil spring mechanism and a precise rotary potentiometer to measure angular rotation of the curved plate due to the momentum of chopped materials leaving the spout.

Design and construction of mechanical parts

The curved plate was shaped and its main pivot and recoil spring and lever mechanism were machined and fabricated according to the designed drawing (Fig. 1.).

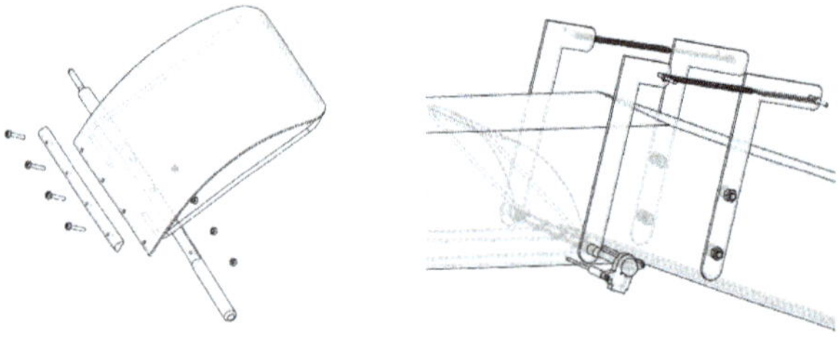

Fig. 1- Drawing map of pivoted plate and accessories (lever mechanism, spring and potentiometer)

To design the proper recoil spring, calculations were performed with the following facts and assumptions:

By assuming ground speed of 3.2 km h^{-1} (0.9 m s^{-1}), the maximum feed rate of chopper in this research was estimated to be about 10 kg s^{-1} (depends on crop yield, moisture content, swath width and harvester travel speed). By knowing some geometrical and technical specifications of the chopper harvester (CLAAS, JAGUAR62) such as cutter head diameter and rotational speed, coefficient of friction between chopped material and spout inside surface (μ= 0.49), the height and total distance between cutter head and impact plate, the velocity of material before impacting the curved plate sensor (V_{bI} as shown in Fig. 2) was calculated (Eq. 2). To calculate the material velocity after impacting the pivoted plate sensor, the minimum required velocity of materials to reach the platform scale weighing system was considered.

To facilitate the calculations, the following assumptions was considered: a) Chopped materials having a projectile motion and b) Neglecting the effect of drag force (Eq. 4). By knowing the required parameters such as before and after impact velocities and the desired linear displacement of the designed spring at the maximum assumed material feed rate (X= 0.17 m), the proper spring was selected with the specifications given in table1. The curved impact plate and its recoil spring and linkages after installation on the chopper spout are shown in figure 3.

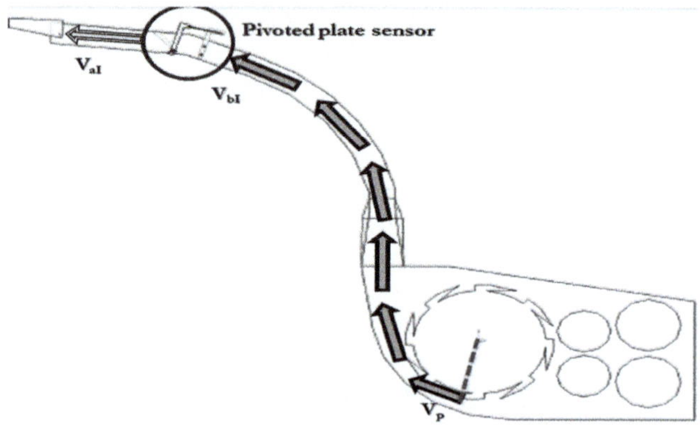

Fig. 2- Schematic Diagram of velocity changes in chopper chute (1)

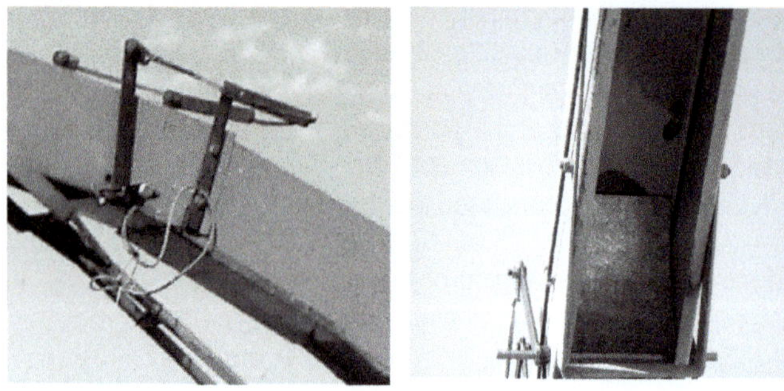

$$F_I = ma \Rightarrow F_I = m\frac{dv}{dt} \Rightarrow \int_{\Delta} F_I dt = m(v_{aI} - v_{bI})$$

$$F_I \approx \left[\frac{m(v_{aI} - v_{bI})}{\Delta}\right] \qquad (2)$$

$$\frac{1}{2}mv_p^2 = \frac{1}{2}mv_{bI}^2 + mgh + \mu.mgd \qquad (3)$$

$$v_p = \frac{2\pi n_c r_c}{60} = \frac{2\pi \times 950 \times 0.35}{60} = 34.82(m/s)$$

$$(2) \Rightarrow v_{bI} = \left(v_p^2 - 2gh - 2\mu.gd\right)^{0.5} = 33.75(m/s)$$

$$Y = \frac{1}{2}g\left(\frac{R}{v_{aI}\cos\theta}\right)^2 + R\times\tan\theta \qquad (4)$$

$$v_{aI} = \left\{\frac{g}{2}\left(\frac{R^2}{(Y - R\times\tan\theta)\cos^2\theta}\right)\right\}^{0.5} = \left\{\frac{9.81}{2}\left(\frac{36}{(2-6\times\tan 15)\cos^2 15}\right)\right\}^{0.5} \cong 22(m/s) \qquad (5)$$

$$F_I = 10\times(22-33.75)/1 = 117.5 \text{ N} \qquad (6)$$

$$F_I = k\times x \Rightarrow k = 117.5/0.17 \cong 690 \text{ N/m}$$

Design and construction of electronic circuit

To measure the degree of curved-plate rotation due to impact of chopped material, a precise rotary potentiometer (AB,800FP-POT5, 50kΩ, Mexico) was used. For real-time data recording on a portable computer, an AVR micro controller model ATMEL 16L was used. The output signal of the rotary sensor was connected to ADC port of the micro controller and sent to portable computer via max232 microchip and serial to USB convertor cable, every 0.1 second. A Schematic diagram of the electronic circuit designed for this study is shown in figure 4.

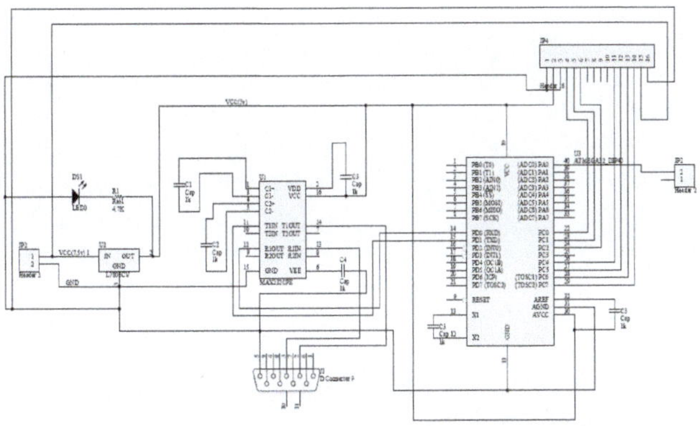

Fig. 4- Schematic diagram of the electronic circuit

Platform scale weighing system

To evaluate accuracy of the mass flow sensor in field conditions, a drawn type wagon was equipped with a sub frame and a precise weighing load cell to continuously weigh the chopped material poured into its storage bag (Fig. 5). This wagon was used as a control treatment (accurate reference) in field trials.

Fig. 5- A field test of the sensor using platform scale weighing system

System calibration and testing

In order to calibrate and test the performance of the yield monitoring system, field tests were conducted in the experimental fields of the College of Agriculture, Shiraz University, Shiraz, Iran.

To calibrate the sensor, a part of the test field was prepared with artificially selecting crop densities giving predetermined feed rates of 6, 8 and10 kg s^{-1} each in 5 replications. These feed rates were controlled by varying the speed of the tractor based on anticipated yield determined by previous sampling in the field. The instant feed rate at time n determined by subtracting accumulated mass of chopped material at time n-1 from the accumulated mass at time n. Finally, a yield map was created for a small plot with three crop rows, each 20 meters long using geographical software Surfer 8.0.

Results and Discussion

System calibration

In order to find a calibration curve between the crop feed rate and degree of the curved plate rotation, data must be normally distributed and for a more precise calibration curve, a proper statistical tool should be employed to eliminate the outliers in each treatments. To fulfill these requirements, statistical analyses were conducted using statistical software SPSS 13.0. The normal distribution of the data in each treatment was verified by conducting Kolmogorov-Simirnov normality test. A confidence interval ((1-α) ×100 = 95%) around the mean was used as a criterion to eliminate outliers in each feed rate treatment. (Eq. 7):

Confidence interval around the mean = Mean of population ± Standard Error of Mean$\times t_{\alpha/2}$ (7)

In order to use the rotation of the pivoted-plate as an indicator of the instantaneous mass flow rate in the field, a calibration curve was drawn on the experimental data after eliminating outliers as shown in figure 6.

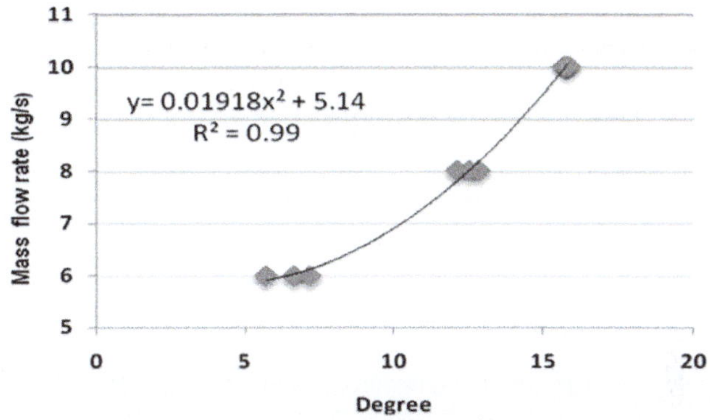

Fig. 6- Relation of pivoted plate angular rotation and instantaneous mass flow

Evaluation of the sensor output accuracy

To compare the sensor output and instant feed rates of platform scale weighing system statistically, a t-test was used. The results revealed no significant differences between sensor output and platform scale feed rates at 5% confidence level. These two data sets were correlated by correlation coefficient of 0.78 and as an another statistical tool to measure how well the estimated yield by pivoted plate sensor fits actual instantaneous yield measured by the platform scale, root mean squared error (RMSE) was calculated and it was equal to 1.505.

Result of yield map creation

The resulting yield map (Fig. 7) shows the mean yield of 5 kg m^{-2} with standard deviation of 2.0 Kg m^{-2}. The minimum yield was 2.5 kg m^{-2} and the maximum yield was 10.8 kg m^{-2}.

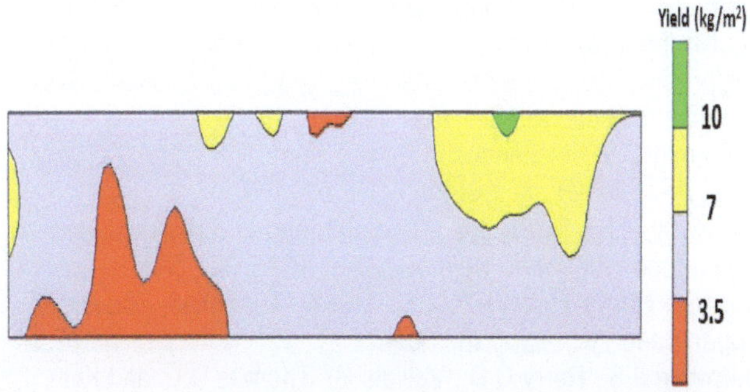

Fig. 7- A typical silage corn yield map of the test field

Conclusion

In this research, the feasibility of using a pivoted plate mass flow sensor for silage corn yield monitoring was investigated. Results of statistical analysis for calibration and field tests showed a highly correlated relation between angular rotation of pivoted plate and instantaneous mass flow rate. More investigations are needed to evaluate this suggested approach at different moisture levels and maturity stages of plant to find a more general calibration equation for mass flow prediction.

Nomenclature

F_I	Impact force of material due to pivoted plate sensor, N
V_{aI}	Speed of material after impacting pivoted plate sensor, m s^{-1}
V_{bI}	Speed of material before impacting pivoted plate sensor, m s^{-1}
V_p	Peripheral speed of cutting cylinder = Speed of chopped material leaving cutting head, m s^{-1}
n_c	Rotary speed of cutter head, rpm
r_c	Radius of cutter head, m
m	Maximum mass of material impacting the pivoted plate in impact duration (1 second period), kg
Δ	Impact duration (1 second)
g	Gravitational acceleration, m s^{-2}
h	Height of discharge spout, m
d	Total distance between cutter head and pivoted plate sensor, m
μ	Friction coefficient between chopped material and spout inside surface
k	Recoil spring stiffness coefficient, N m^{-1}
x	Desired linear displacement of recoil spring, m
Y	Vertical Displacement between chopper spout and platform scale weighing system, m
R	Horizontal Displacement between chopper spout and platform scale weighing system, m
θ	Discharge angle of chopped material trajectory

Acknowledgment

Authors should express their acknowledgment toward Dr. S. M. Nassiri and Professor Jonathan Chaplin for their guidance and helps and also Shiraz University for its technical supports.

References

[1] Auernhammer H., Demmel M. and Pirro P.J.M. (1995) *Yield measurement on self propelled forage harvesters*. ASAE Paper No. 95-1757.
[2] Baganz K. (1991) *International Symposium on Locating Systems for Agricultural Machines*, Godollo, 18.
[3] Benjamin C.E., Mailander M.P. and Price R.R. (2001) *Sugar cane yield monitoring system*. ASAE 011189.
[4] Campbell R.H., Rawlins S.I. and Shufeng H. (1994) *Monitoring methods for potato yield mapping*. ASAE 94-1584.
[5] Cox G., Harris H. and Pax R. (1997) *Australian Society of Sugar Cane Technologists*, 38-43.
[6] Durrence J.S., Perry C.D., Vellidis G., Thomas D.L. and Kvien C.K. (1998) *Evaluation of commercially available cotton yield monitors in Georgia field conditions*. ASAE 983106.
[7] Durrence J.S., Hamrita U.T.K. and Vellidis G. (1999) *Precision Agriculture* 1, 301-317.
[8] Ehlert D. (1996) *Landtechnik* 51(H.1), 20-21.
[9] Ehlert D. (1999) *the 2nd European Conference on Precision Agriculture*, 797-804.
[10]Godwin R.J., Wheeler P.N., O'Dogherty M.J., Watt C.D. and Richards T. (1999) *Computers and Electronics in Agriculture* 23, 85-101.
[11]Hennens D., Baert J., Broos B., Ramon H. and De Baerdemaeker J. (2003) *Biosystems Engineering* 85 (4), 425-436.
[12]Kumhala P. and Prosek V. (2003) *Precision Agriculture* 4, 413 -419.
[13]Kumhála F., Kvíz Z., Kmoch J. and Prošek V. (2007) *RES. AGR. ENG., 53* (4), 149-154.
[14]Larsson K. (1994) *Teknik for Lantbruket*, 43, 7.
[15]Lee W.S., Schueller J.K. and Burks T.F. (2002) *ASAE Annual International Meeting*. 021165.
[16]Loghavi M. and Almaee M. (2009) *Agricultural Engineering International: the CIGR Ejournal*.
[17]Magalhaes P.S.G. and Cerri D.G.P. (2006) *Biosystems Engineering* 96 (1), 1-6.
[18]Martel H. and Savoie P. (2000) *Can. Agric. Eng*. 42,123-129.
[19]Meriyam J.L. and Kraige L.G. (2002) *Engineering Mechanics*, II: Dynamics. Fifth edition. John Wiley & Sons.
[20]Missoten B., Brooks B., Strubbe G. and De Baerdemaeker J. (1997) *The first European conference on Precision Agriculture*, Warwick University, 7-10, 529-536.
[21]Pagnano N.B. and Magalhaes P.S.G. (2001) *The Third European Conference on Precision Agriculture*, 839-843.
[22]Pelletier G. and Upadhyaya S.K. (1999) *Computers and Electronics in Agriculture* 23, 103-117.
[23]Perry C.D., Vellidis G., Wells N. and Kvien C. (2001) T*he Beltwide Cotton Conference*, ed: D.A. Richter, Memphis, Tenn.: National Cotton Council of America. 328-339.
[24]Perry C.D. and Vellidis G. (2005) *Cotton yield monitor instantaneous accuracy during steady-state and step-input conditions*. ASAE 051129.
[25]Persson A. (1998) T*he 1st International Conference on Geospatial Information in Agriculture and Forestry*, Ann Arbor, MI, 618-623.
[26]Pierossi M.A. and Hassuani S.J. (1997) *Semina´rio Copersucar de Tecnologia Agrono^mica* 7, Piracicaba, Brazil, 263-265.
[27]Rawlins S.L., Campbell G.S., Campbell R.H. and Hess J.R. (1995) *The 2nd Intl. Conf. Site-Specific Management for Agricultural Systems*.
[28]Roades J.P., Beck A.D. and Searcy S.W. (2000) *The Beltwide Cotton Conference*, 404-408.
[29]Sassenrath-Cole G.F., Thomson S.J., Williford J.R., Hood K.B., Thomasson J.A., Williams J. and Woodard D. (1999) *The Beltwide Cotton Conference*, 364-366.
[30]Savoie P., Lemire P. and Thériault R. (2002) *Applied Engineering in Agriculture* 18(4), 389-397.
[31]Sirvastava A.K., Georing G.E. and Rohrbach R.P. (1993) *Engineering Principle of Agricultural Machines*. ASAE, 601p.
[32]Sui R., Thomasson J.A., Mehrle R., Dale M., Perry C. and Rains G. (2004) *Computers and Electronics in Agriculture* 42, 149-160.
[33]Thomasson J.A., Pennington D.A., Pringle H.C., Columbus

E.P., Thomson S.J. and Byler R.K. (1999) *Applied Engineering in Agriculture* 15(1), 11-17.
[34]Thomasson J.A. and Sui R. (2000) *The Beltwide Cotton Conference,* 408-410.
[35]Thomasson J.A., Sui R., Wright G.C. and Robson A.J. (2006) *Applied Engineering in Agriculture* 22(6), 809-818.
[36]Upadhyaya S.K. (1999) *The Fourth International Conference on Precision Agriculture.*
[37]Van Canneyt T. and Verschoore R. (2000) *Agricultural Engineering into the Third Millennium,* 331-332.
[38]Vansichen R. and De Baerdemaeker J. (1993) *Journal of Agricultural Engineering Research,* 55, 1-10.
[39]Vellidis G., Perry C.D., Durrence J.S., Thomas D.L., Hill R.W., Kvien C.K., Hamrita T.K. and Rains G. (2001) *Transactions of the ASAE* 44(4), 775-785.
[40]Vellidis G., Perry C.D., Rains G., Thomas D.L., Wells N. and Kvien C.K. (2003) *Applied Engineering in Agriculture* 19(3), 259-272.
[41]Wallace T.P. (1999) *Computers and Electronics in Agriculture* 23, 1-8.
[42]Walter J.D., Backer L.F. and Hofman V.L. (1996) *Sugar beet yield monitoring for site-specific farming,* ASAE 961022.
[43]Walter J.D. and Backer L.F. (2003) *Precision Agriculture* 4, 421-31.
[44]Wild K. and Auernhammer H. (1999) *Computers and Electronics in Agriculture* 23, 119-132.
[45]Wilkerson J.B., Kirby S., Hart W.E. and Womac A.R. (1994) *Real-time cotton flow sensor.* ASAE 94-1054.
[46]Wilkerson J.B., Moody F.H. and Hart. W.E. (2002) *Applied Engineering in Agriculture* 18(2), 153-159.
[47]Wolak F.J., Khalilian A., Dodd R.B., Han Y.J., Keshlkin M., Lippert R.M. and Hair W. (1999) *The Belt wide Cotton Conference,* 361-364.

INSECTICIDAL EFFICACY OF MINTWEED, YAM BEAN AND CELERY SEED EXTRACTS ON *Aedes aegypti* L.

YONGKHAMCHA B. AND INDRAPICHATE K.*

School of Biology, Institute of Science, Suranaree University of Technology, Nakhon Ratchasima 30000, Thailand.
*Corresponding Author: Email- korakod@sut.ac.th

Abstract- Biological control of *Aedes aegypti*, the dengue virus vector, is needed to replace synthetic insecticides. This study demonstrated the insecticidal efficacy of water and ethanolic seed extracts of mintweed Hyptis suaveolens (MSE), yam bean Pachyrhizus erosus (YSE) and celery Apium graveolens (CSE) on controlling of Ae. aegypti larvae and adults. The cytotoxicity of the extracts was investigated by brine shrimp lethality assay (BSLA). Yam bean extracts were highly cytotoxic to brine shrimps. All seed ethanolic extracts were more toxic to mosquitoes than water extracts. The insecticidal efficacy of all extracts ranged as YSE/e > YSE/w > MSE/e > CSE/e > MSE/w > CSE/w. YSE/e was most toxic to Ae. aegypti 2nd instar larvae and adults with LC50 of 16.22 ± 0.20 μg/ml and 91.41 ± 0.49 μg/ml respectively and much higher than MSE/s and CSE/e. YSE in combinations produced strong synergistic effects to the other extracts. The insecticidal activities of MSE and CSE were mild. They could lessen the activity of YSE. This study demonstrates that the safe utilization of plant seed extracts, particularly of the very toxic yam bean seeds, as Ae. aegypti biological control agents are possible by either individuals or suitable combinations with other plant or seed extracts.

Key words- insecticidal efficacy, mintweed, yam bean, celery, *Aedes aegypti*.

Introduction

Aedes aegypti L. is the major vector of dengue virus, flavivirus, which transmits dengue hemorrhagic fever (DHF), the deathful disease in the tropical and subtropical regions. The outbreak of DHF appears to be cyclic and repeats every 2-3 years. In Thailand, DHF continues to be one of the major public concerns since its incidence had greatly increased from 46,829 to 115,845 cases during 2006-2010. Particularly in 2010, the incident rate of DHF was 182.36 per 100,000 and the mortality rate was 0.22 per 100,000 or 0.12% [1]. The re-emergence of DHF is primarily related to the Ae. aegypti reproduction. To avoid mosquito bites, mosquito nets are traditionally used, while physical catch and kill or uses of some plant products are also common. Plant products, such as pyrethrins, carvacrol, quassia, nicotine, allicin, anabasine, azadirachtin, d-limonene, and triptolide, are widely used in mosquito eradiation before the introduction of synthetically organic insecticides of pyrethroids and organophosphates [2]. However, the synthetic insecticides cause mosquito resistant and adverse effects to non-target organisms and environment [3-4]. Therefore, natural products from plants and bacteria have been researched to reduce the mosquito population and sustain the environment. Bacillus thuringiensis which produces crystal endotoxin has been introduced in the control of insect pests, including mosquitoes [5]. New phytochemicals as alternatives for controlling Ae. aegypti have been searched [6-8].

A few studies on insecticidal effects of plant seed products in controlling Ae. aegypti. Mintweed Hyptis suaveolens (L.) Poit is an aromatic weed possesses some medicinal properties and traditionally used to treat gastrointestinal infection, cramps and pain as well as to treat skin infections [9-10]. Mintweed seeds contained insecticidal activity [11-14]. Yam bean Pachyrhizus erosus L. is a tuber-root climbing plant. Its seeds possess a potent antitumor activity [15-16] and high toxic property to insects, fishes, rats, and fungi [3, 17]. Celery Apium graveolens L. is an aromatic and spicy vegetable. Its seeds provide a remedy for urinary tract infections and degeneration of joints [18]. Celery seeds were able to repel insects, including larvae, and adult mosquitoes [19]. This study aimed to explore insecticidal efficacy of the seed extracts of mintweed, yam bean and celery for controlling the second instar larvae and the adults of Ae. aegypti.

Materials and Methods

Seed Collection and Extract Preparation

Mintweed seeds were collected on Suranaree University of Technology campus. Yam bean and celery seeds were purchased from local farms in Mahasarakham and Nakhon Ratchasima, Thailand. Mintweed seeds were soaked in water and removed their mucilage before extraction. All seeds were cleaned, air dried and ground to powder. Ten grams of the seed powder were extracted in 150 ml of water or 70% ethanol in Soxhlet extractor (Buchi model B811, Germany). The extracts were evaporated, dried and stored at -20°C until used. Dimethylsulfoxide (DMSO) and pyrethroid were purchased from Sigma Chemical Co. (St. Louis, MO, USA).

Mosquito Rearing

Ae. aegypti was a gift from the Communicable Disease Control Centre, Zone 6, Ministry of Public Health, Khon Kaen, Thailand, and reared in the Center for Scientific and Technological Equipment, Suranaree University of Technology by the standard rearing techniques of World Health Organization (WHO). The eggs on a filter paper were placed on a pottery bowl containing water which was in a steel-net wire-cage (50x50x50 cm) and allowed to hatch at 26°C and 75 % relative humidity. The larvae were transferred into enamel trays (30x21x5 cm) and fed with ground rat food. The adults were maintained in a cage (80x80x80 cm). The males were fed with syrup containing vitamins and the females were fed with guinea pig blood.

Brine Shrimp Lethality Assay (BSLA)

Brine shrimp (*Artemia salina* L.) eggs were purchased from a local fish shop. They were hatched and reared in artificial seawater (120 g/l sea salt) under continuous light, at 25°C for 24 h [20]. Ten nauplii were transferred into a 24-well plate containing 200 ml of artificial seawater. The seed extracts of various concentrations were added and cultured for 12 h. DMSO, (0.1% v/v) and 1 µg/ml pyrethroid were used as controls. The numbers of dead (non-motile) nauplii were counted and calculated for mortality. The median lethal concentration (LC50) was analyzed by Probit analysis.

Larval Mortality Test

The extract powder was dissolved in water or 0.1 % DMSO. Fifty larvae, 2nd instar, were gently transferred into 10 ml of water in a-20 ml vial and treated with 0-100 mg/ml extracts for 24 h. Individual or combination (1:1, v/v) of the extracts were tested. DMSO and pyrethroid were used as controls. The tests were performed in triplicate and two repeats. The dead larvae were counted, calculated for mortality and analyzed for LC50 values.

Adult Mortality Test

Fifty mosquitoes, newly emerged, were transferred into a steel wire-net cage and the cage was covered with a plastic sheet. Five milliliters of extract solutions of different concentrations were firstly sprayed into the cage containing the mosquitoes. After 12 h, 5 ml of the extracts were secondly resprayed. The dead mosquitoes were counted after 12 h of the second spray. Individual and combination (pair wise; 1:1, v/v) of the extracts were applied. DMSO and pyrethroid were used as controls. The tests were performed in triplicate and two repeats. The mortality was calculated and analyzed for LC50 values. The percentage of mortality was corrected by Abbott's formula as following:

$$\%\text{Mortality} = \frac{\%\ \text{Test mortality} - \%\ \text{control mortality} \times 100}{100 - \%\ \text{control mortality}}$$

Data Analysis

All data were analyzed by analysis of variance (ANOVA) and completely randomized design (CRD) using Statistical Package for the Social Sciences (SPSS) program for Windows, v.17. The means were compared using the Duncan's Multiple Range Test (DMRT). The mortality was corrected by Abbott's formula [21]. The median lethal concentration, LC50, was determined by Probit analysis [22].

Results

Cytotoxicity

Cytotoxicity of the seed extracts of mintweed (MSE), yam bean (YSE), and celery (CSE) was investigated by brine shrimp lethality assay (BSLA). The toxicity of all extracts were concentration dependent, although the toxic magnitudes were broadly different due to the plant seeds and the extract solvents. The ethanolic extracts of all seeds showed much higher toxic than the water extracts (Table 1). The median lethal concentration, LC50 at 24 h, of the ethanolic extracts ranged as YSE/e > CSE/e > MSE/e and of the water extracts ranged as YSE/w > MSE/w > CSE/w. YSE/e produced highest cytotoxicity with LC50 of 0.02 ± 0.24 µg/ml and CSE/w was lowest toxic with LC50 of 3,134.45 ± 0.80 µg/ml. YSE/w had LC50 of 257.11 ± 0.29 µg/ml which was over 10^3 less toxic than YSE/e. In addition, YSE/e caused 100% mortality of the brine shrimps within 12 h. CSE/e cytotoxicity with LC50 of 72.30 ± 0.75 µg/ml was 43 fold of CSE/w. MSE/e cytotoxicity with LC50 of 506.23 ± 0.26 µg/ml was approximately 3 fold of MSE/w.

Larvicidal Effects

The efficacy of mintweed (MSE), yam bean (YSE) and celery (CSE) seed extracts on mortality of the 2nd instar larvae of *Ae. aegypti* determined by LC50 at 24 h, was depend upon plant seeds, solvents, and concentrations. All ethanolic extracts produced higher mortality of the larvae than water extracts. The efficacy compared as YSE/e > YSE/w > MSE/e > CSE/e > MSE/w > CSE/w (Table 2). YSE/e, the most potent, had LC50 of 16.22 ± 0.20 µg/ml and CSE/w, the least potent, had LC50 of 25,230.74 ± 0.12 mg/ml. YSE/e was 87 and 233 fold more potent in controlling of *Ae. aegypti* larvae than MSE/e and CSE/e respectively.

Adulticidal Effects

The *Ae. aegypti* adults were sprayed twice with various concentrations of the seed extracts of mintweed, yam bean and celery for 24 h. The effects of the seed products were also depend upon plant seeds, solvents, and concentrations. The efficacy, determined by LC50 values, compared as YSE/e > MSE/e > YSE/w > CSE/e > MSE/w > CSE/w (Table 3). YSE/e, the most effective, showed LC50 value of 91.41 ± 0.49 µg/ml and CSE/w, the least effective, showed LC50 value over 109 mg/ml. YSE/e was 32 and 275 fold more effective in controlling of *Ae. aegypti* adults than MSE/e and CSE/e respectively.

Effects of Extract Combinations

The effects of seed extract combinations (pair wise; 1:1, v/v) on the 2nd instar larvae and the adult mosquitoes were tested. The extract combinations enhanced the insecticidal activity against both 2nd instar larvae and adult mosquitoes as compare to the mortality caused by individual extracts, except the combination of YSE/e and CSE/e (Table 4, 5 and 6) CSE/e reduced YSE/e individual effect. The combination of MSE/e and YSE/e exhibited the highest efficacy on the control of the 2nd instar larvae with LC50 value of 11.79 ± 0.15 µg/ml and on the adult mosquitoes with LC50 value of 23.82 ± 0.20 µg/ml. The combination of MSE/w and CSE/w showed the least insecticidal activity on both larvae and adult mosquitoes with LC50 of 5,086.76 ± 0.10 µg/ml and 12,315.31 ± 0.23 µg/ml, respectively. It was noticed that YSE enhanced the activities of MSE and of CSE on both larvae and adults (Table 6). In addition, MSE enhanced only the activity of CSE with much less toxic magnitude than YSE did. The efficacy of the individual and the combination treatments of all extracts on *Aedes aegypti* larvae and adults are comparatively summarized in table 6.

Discussion

The biological control of *Aedes aegypti*, the important vector of the dengue hemorrhagic fever virus, in Thailand is very essential. Uses of natural products can avoid mosquitoes develop resistance to synthetic insecticides and safe to the environment and human beings as well. Plant products are possibly suitable alternatives and candidates as bio-insecticides for controlling of *Ae. aegypti*. Moreover, some plants are traditionally used for insect control in farms and are locally available. This study demonstrated the efficacy of three plant seed extracts which were mintweed *Hyptis suaveolens* (MSE), yam bean *Pachyrhizus erosus* (YSE) and celery *Apium graveolens* (CSE) in controlling of the 2nd instar larvae and the adult *Aedes aegypti*. Since *H. suaveolens* contained *b* - caryophyllene, 1,8-cineole, sabinene, *b*-selinene limonene, biclyclogermacrene, *b*-phellandrene, these phytochemicals could cause death to both larvae and adult mosquitoes [23-24]. There was a report that ethanolic extract of mintweed leaves could be used as larvicidal on *Ae. aegypti* larvae in Nigeria [11]. Petroleum ether extract of mintweed seeds was found against moth, *Plutell xylostella* [12]. However, mintweed extract at high dose, up to 500 mg/kg/day for 6 months, did not produce any toxic effects in rats [10, 14]. This could be certain that mintweed is safe to mammals, as confirmed by our investigation on brine shrimps. Celery seed extract (CSE) showed larvicidal and adulticidal activity against *Ae. aegypti* to some certain extend. Possibly, its constituents of limonene, selinene, glucosides and flavonoid may not that toxic to mosquitoes [25-26]. However, there was a report that the essential oil of whole celery was potent as an adulticide to *Ae. aegypti* [6].

It is clearly, the ethanolic seed extract of yam bean, YSE/e, was most toxic to *Ae. aegypti* larvae and adults. YSE/e possibly contained high toxic products as it was reported that yam bean seeds comprised of rotenone and cyanogenic glycosides which were known poisonous to human beings and animals [16, 27]. Rotenone is now widely used as an insecticide in agriculture and home gardening, in particularly, controlling of aphids, thrips, lices, ticks, moths, beetles, spider mites, and mosquito larvae [28-29]. Rotenone caused death to insects by inhibited NADH oxidation in mitochondria leading to critically blocking respiration and nerve conduction [29]. Having yam bean seed extract in combinations with the other seed extracts highly enhanced the insecticidal efficacy on *Ae. aegypti*.

Conclusion

In conclusion, Yam bean seed extracted with ethanol was highly cytotoxic to a higher animal like brine shrimps. In our study was the first investigation in the biological control of *Ae. aegypti* by mintweed, yam bean and celery seed extracts. It is concluded that the seed extracts of these three plants were safe to be used as botanical insecticides in controlling of *Ae. aegypti*, the dengue hemorrhagic fever mosquito. The insecticidal efficacy of these three seed extracts is very broadly differences. Yam bean seeds are the most potent candidate as mosquito plant-based larvicide and adulticide. However, yam bean seed extracts must be used with caution, since it is highly toxic, unless combining them with other appropriate plant extracts to lessen their toxicity for animal and human safety. The information provided here will be useful for further study for applying the very toxic plant seeds in controlling other kinds of mosquitoes, the important vectors of many tropical diseases, and other insect pests.

Acknowledgement

This work was financially supported by Suranaree University of Technology and Mahasarakham University, Thailand. *Aedes aegypti* was a gift from Communicable Disease Control Centre, Zone 6, Ministry of Public Health, Khon Kaen, Thailand. We thank Dr. Hatsachai Boonjung for his kind reviewing of statistical analyses.

References

[1] Bureau of the vector-borne Diseases (2011) Department of Disease control, Ministry of Public Health, Thailand. *http://www/thaivbd.org/uploads/dengue/2553/52/w eeks.pdf.*

[2] Wood A. (2010) Compendium of pesticide common names: Insecticides. *http://www.alanwood.net/pesticides/index.html.*

[3] Stenersen J. (2004) *CRC Press*; New York, U.K., 276.

[4] Mongkalangoon P., Grieco J.P., Achee N.L., Suwonkerd W., Chareonviriyaphap T. (2009) *J. Vector Ecol.*, 34, 217-24.

[5] Lee H.L., Chen C.D., Masri S.M., Chiang Y.F., Chooi K.H., Benjamin S. (2008) *Southeast Asian J. Trop Med Public Health*, 39, 601-9.

[6] Chaiyasit D., Choochote W., Rattanachanpichai E., Chaithong U., Chaiwong P., Jitpakdi A., Tippawangkosol P., Riyong D. and Pitasawat B. (2006) *Parasitol Res.*, 99, 715-721.

[7] Silva W.J., Dória G.A., Maia R.T., Nunes R.S., Carvalho G.A., Blank A.F., Alves P. B., Marçal R.M. and Cavalcanti S.C. (2008) *Bioresour Technol.*, 99, 3251-3255.

[8] Gillij Y.G., Gleiser R.M., Zygadlo J.A. (2008) *Bioresour Tech-*

nol., 99, 2507-15.
[9] Wulff R. (1987) *Can J Bot.*, 65, 2501-6.
[10]Attawish A., Chivapat S., Chavalittumrong P., Phadungpat S., Bansiddhi J., Chaorai B. (2005) *Songklanakarin J Sci Technol.*, 27, 1027-36.
[11]Amusan A.A.S., Idowu A.B., Arowolo F.S. (2005) *Tanzan Health Res Bull*, 7, 174-8.
[12]Keita S., Umoetok S.B.A., Smith J.G. (2006) *Agric J.*, 1, 11-3.
[13]Iloba B. N., Ekrakene T. (2006) *J Biol Sci.*, 6, 626-30.
[14]Niwatananun K., Niwatananun W., Lertprasertsook L., Okonogi S. (2006) *CMUJ*, 5, 369-75.
[15]Morales-Arellano G.Y., Chagolla-López A., Paredes-López O., Barba R.A.P. (2001) *J Agric Food Chem.*, 49, 1512-6.
[16]Hung Y.M., Hung S.Y., Olson K.R., Chou K.J., Lin S.L., Chung H.M., Tung C.N. and Chang J.C. (2007) *Intern Med J.*, 37, 130 -132.
[17]Barrera-Necha L.L., Bautista-Banos S., Bravo-Luna L., Garcia -Suarez F.J., Alaez-Sololano D., Reyes-Chilpa R. (2004) *Rev Mex Fitopatol.*, 2, 356-61.
[18]Boiling J. (2006) *http://everygreenherb.com/fromAtoz/celery.html.*
[19]Choochote W., Tuetun B., Kanjanapothi Rattanachanpichai E., Chaithong U., Chaiwong P., Jitpakdi A., Tippawangkosol P., Riyong D. and Pitasawat B. (2004) *J Vector Ecol.*, 29, 340-366.
[20]Briski E., Stappen G., Bossier P., (2008) *Aquaculture*, 282, 19 -25.
[21]Abbot W.S. (1925) *J Econ Entomol.*, 18. 265-267.
[22]Finny, D.J. (1971) *Cambridge University Press, England*, 333.
[23]Peerzada N. (1997) *Mol J.*, 1, 165-168.
[24]Campos I.F.P., Heleno D.F., Suzana C.S. (2001) *Phytochem.*, 57, 733-6.
[25]Fehr D. (1979) *Pharmazie*, 34, 658-62.
[26]Garg S.K., Gupta S.R., Sharma N.D. (1980) *Planta Med.*, 38, 363-5.
[27]Narongchai P., Narongchai S., (2005) *J Med Assoc Thai*, 88, 984-7.
[28]Dayan F.E., Cantrell C.L., Duke S.O. (2009) *Bioorg Med Chem.*, 17, 4022-34.
[29]Marrs T. C., Ballantyne B. (2004) *John Wiley and Sons*, West Sussex, England, 554.

Table-1 - Cytotoxic effects of seed extracts of mintweed (MSE), yam bean (YSE) and celery (CSE), determined by brine shrimp lethality assay (BSLA).

Conc.	% Mortality	LC50	Conc.	% Mortality	LC50
µg/ml	Mean ± SE	µg/ml, 24 h	µg/ml	Mean ± SE	µg/ml, 24 h
MSE/w			MSE/e		
10	1.67 ± 0.21^{a}	1,478.91 ± 0.90	1	1.67 ± 0.17^{a}	506.23 ± 0.26
50	3.33 ± 0.33^{a}		10	5.00 ± 0.22^{ab}	
100	5.00 ± 0.22^{a}		50	11.67 ± 0.17^{bc}	
1,000	16.67 ± 0.49^{b}		100	15.00 ± 0.22^{c}	
1,500	60.00 ± 0.26^{c}		1,000	68.33 ± 0.60^{d}	
YSE/w			YSE/e		
1	3.33 ± 0.21^{a}	257.11 ± 0.29	0.001	10.00 ± 0.26^{a}	0.02 ± 0.24
10	11.67 ± 0.31^{b}		0.01	40.00 ± 0.26^{b}	
50	21.67 ± 0.40^{c}		0.1	78.33 ± 0.31^{c}	
100	25.00 ± 0.43^{c}		1	100.00 ± 0.00^{d}	
1,000	85.00 ± 0.22^{d}		10	100.00 ± 0.00^{d}	
CSE/w			CSE/e		
1,500	5.00 ± 0.22^{ab}	3,134.45 ± 0.80	1	3.33 ± 0.21^{a}	72.30 ± 0.75
2,000	11.67 ± 0.31^{bc}		10	15.00 ± 0.50^{b}	
2,500	21.67± 0.48^{c}		50	35.00 ± 0.76^{c}	
3,000	35.00 ± 0.43^{d}		100	65.00 ± 0.72^{d}	
3,500	73.33 ± 0.42^{e}		1,000	100.00 ± 0.00^{e}	
Pyr (1 µg/ml)	100.00 ± 0.00^{d}	0.007 ± 0.18	DMSO (0.1%)	0.00	0.00

Each value is the mean ± standard error, n = 6. Numbers with different letters within the same column are significantly different (P £ 0.05). Pyr is pyrethriod

Table-2 - Effects of individual seed extracts of mintweed (MSE), yam bean (YSE) and celery (CSE) on the 2nd instar larvae of Ae. Aegypti

Conc.	% Mortality	LC50	Conc.	% Mortality	LC50
µg/ml	Mean ± SE	µg/ml, 24 h	µg/ml	Mean ± SE	µg/ml, 24 h
MSE/w			MSE/e		
6,000	8.33 ± 0.32^{a}	16,650 ± 0.46	800	40.33 ± 0.40^{a}	1,412.53 ± 0.82
8,000	14.33 ± 0.30^{b}		1,000	47.00 ± 0.43^{b}	
10,000	34.33 ± 0.48^{c}		2,000	56.00 ± 0.37^{c}	
20,000	51.67 ± 0.60^{d}		4,000	61.00 ± 0.43^{d}	
30,000	95.33 ± 0.33^{e}		6,000	70.00 ± 0.26^{e}	
40,000	98.67 ± 0.21^{f}		8,000	79.67 ± 0.31^{f}	
YSE/w			YSE/e		
100	4.33 ± 0.31^{a}	724.64 ± 1.25	5	12.33 ± 0.40^{a}	16.22 ± 0.20
200	8.00 ± 0.37^{b}		10	41.67 ± 0.31^{b}	
400	28.00 ± 0.58^{c}		15	53.33 ± 0.56^{c}	
600	37.33 ± 0.49^{d}		20	58.67 ± 0.49^{d}	
800	50.67 ± 0.42^{e}		25	61.00 ± 0.42^{e}	
1,000	70.00 ± 0.58^{f}		30	63.00 ± 0.76^{f}	
CSE/w			CSE/e		
6,000	5.33 ± 0.21^{a}	25,230.74 ±0.12	800	14.67 ± 0.21^{a}	3,778.39 ± 0.23
8,000	9.33 ± 0.33^{b}		1,000	29.00 ± 0.43^{b}	
10,000	21.00 ± 0.76^{c}		2,000	42.67 ± 0.67^{c}	
20,000	44.67 ± 0.66^{d}		4,000	9.00 ± 0.43^{d}	
30,000	59.00 ± 0.43^{e}		6,000	55.33 ± 0.33^{e}	
40,000	63.33 ± 0.49^{f}		8,000	67.67 ± 2.49^{f}	
DMSO (0.1%)	0.00	0.00	DMSO (0.1%)	0.00	0.00

Each value is the mean ± standard error, n = 6. Numbers with different letters within the same column are significantly different ($P \leq 0.05$). DMSO is dimethylsulfoxide.

Table-3 - Effects of individual seed extracts of mintweed (MSE), yam bean (YSE) and celery (CSE) on the adults of Ae. aegypti.

Conc.	% Mortality	LC50	Conc.	% Mortality	LC50
µg/ml	Mean ± SE	µg/ml, 24 h	µg/ml	Mean ± SE	µg/ml, 24 h
MSE/w			MSE/e		
20,000	13.38 ± 0.48^{a}	75,046.45 ± 0.12	1,000	26.69 ± 0.48^{a}	2,952.70± 0.11
40,000	20.74 ± 0.62^{b}		2,000	41.22 ± 1.00^{b}	
60,000	29.77 ± 0.45^{c}		4,000	54.73 ± 0.42^{c}	
80,000	54.85 ± 0.43^{d}		6,000	58.45 ± 0.43^{d}	
100,000	68.56 ± 0.56^{e}		8,000	77.70 ± 0.52^{e}	
200,000	84.62 ± 0.49^{f}				
YSE/w			YSE/e		
2,000	18.45 ± 0.76^{a}	9,600.51 ± 0.25	20	4.36 ± 0.43^{a}	91.41 ± 0.49
4,000	28.52 ± 0.43^{b}		40	10.74 ± 0.33^{b}	
6,000	40.27 ± 0.49^{c}		60	20.13 ± 0.56^{c}	
8,000	47.31 ± 0.47^{d}		80	46.64 ± 0.89^{d}	
10,000	55.03 ± 0.49^{e}		100	57.38 ± 0.48^{e}	
20,000	61.74 ± 0.26^{f}				
CSE/w			CSE/e		
20,000	4.02 ± 0.31^{a}	109,030.36 ± 17	10,000	15.88 ± 0.40^{a}	25,097.59 ± 0.12
40,000	12.38 ± 0.61^{b}		20,000	47.64 ± 0.91^{b}	
60,000	29.10 ± 0.56^{c}		40,000	61.49 ± 0.89^{c}	
80,000	44.15 ± 0.60^{d}		60,000	68.92 ± 0.42^{d}	
100,000	55.19 ± 0.76^{e}		80,000	97.97 ± 0.52^{e}	
200,000	63.55 ± 0.31^{f}				
DMSO (0.1%)	0.00	0.00	DMSO (0.1%)	0.00	0.00

Each value is the mean ± standard error, n = 6. Numbers with different letters within the same column are significantly different ($P \leq 0.05$). DMSO is dimethylsulfoxide

Table-4 - Effects of the seed extract combinations (pair wise) of mintweed (MSE), yam bean (YSE) and celery (CSE) on the 2nd instar larvae of Ae. aegypti.

Combined	% Mortality	LC50	Combined	% Mortality	LC50
Extract	Mean ± SE	µg/ml, 24 h	Extract	Mean ± SE	µg/ml, 24 h
MSE/w + YSE/w			MSE/e + YSE/e		
400	49.00 ± 0.43^{a}	225.43 ± 0.10	5	25.67 ± 0.31^{a}	11.79 ± 0.15
600	57.00 ± 1.00^{b}		10	53.00 ± 0.31^{b}	
800	67.67 ± 0.48^{c}		15	70.33 ± 0.56^{c}	
1,000	74.67 ± 0.33^{d}		20	95.67 ± 1.85^{d}	
2,000	80.00 ± 0.48^{e}		30	97.67 ± 0.40^{e}	
CSE/w + MSE/w			CSE/e + MSE/e		
2,000	36.00 ± 0.45^{a}	5,086.76 ± 0.10	40	21.67 ± 0.52^{a}	121.66 ± 0.10
4,000	43.33 ± 0.33^{b}		60	41.00 ± 0.43^{b}	
6,000	51.33 ± 0.67^{c}		80	50.00 ± 0.49^{c}	
8,000	56.00 ± 0.45^{d}		100	68.00 ± 0.84^{d}	
10,000	60.33 ± 0.40^{e}		200	77.33 ± 0.56^{e}	
YSE/w + CSE/w			YSE/e + CSE/e		
200	24.00 ± 0.52^{a}	930.65 ± 0.10	40	21.67 ± 0.91^{a}	82.04 ± 0.10
400	31.67 ± 0.31^{b}		60	41.00 ± 0.43^{b}	
600	36.00 ± 0.37^{c}		80	50.00 ± 0.37^{c}	
800	44.00 ± 0.37^{d}		100	68.00 ± 0.37^{d}	
1,000	55.33 ± 0.49^{e}		200	77.33 ± 0.42^{e}	
DMSO (0.1%)	0.00	0.00	DMSO (0.1%)	0.00	0.00

Each value is the mean ± standard error, n = 6. Numbers with different letters within the same column are significantly different (P £ 0.05). DMSO is dimethylsulfoxide.

KINETICS OF APRICOT THIN LAYER DRYING IN A MIXED AND INDIRECT MODE SOLAR DRYER

GHATREHSAMANI S.H.*, DADASHZADEH M. AND ZOMORODIAN A.

Department of Agricultural Engineering, Collage of Agriculture, Shiraz University, Shiraz, Iran.
*Corresponding Author: Email- Shirin63_saman@yahoo.com

Abstract- This paper presents mathematical modeling of a thin layer solar drying process of apricots. A forced convection mixed and indirect mode cabinet solar dryer (total collector surface area of $4m^2$ and total tray surface area of $0.7m^2$) was employed. Each experiment started from 9:00 to18:00 on sunny days in summer. The experiments were conducted at three air flow temperatures of 40, 50 and 60 °C with three air flow rates of 8.7, 6.3 and 3.8 m/s respectively for each indirect and mixed mode dryer types. The solar intensity was measured and recorded from 420 to 1000 W/m^2 during the course of experiments. In order to select the most appropriate drying model equation in each mode of drying, 10 different models were examined using the thin layer drying experimental data. The results of curve fitting were validated by comparing the proper statistical coefficients namely: r, χ^2, RMSE using non-linear regression analysis. According to the results, two-term drying model could satisfactorily describes the solar thin layer drying behavior of apricots with correlation coefficient (r) of 0.999 for both types of solar drying with $\chi^2=3.374*10^{-3}$, $RMSE=4.33*10^{-2}$ for the mixed-mode and $\chi^2=1.243*10^{-4}$, $RMSE=8.64*10^{-3}$ for the indirect mode respectively. The effect of drying operating conditions (air temperature and velocity) on the model constants and coefficients were also investigated using multiple regression analysis.

Key words- Solar drying, thin layer, drying models, Apricot

Introduction

Drying is one of the most common processes used to improve food stability, since it decreases considerably the water activity of the material, reduces microbiological activity and minimizes physical and chemical changes during its. Open-air sun drying has been used since the beginning of human life in the world to drying grain, fruits and other agricultural products as a means of preservation. However, this method has many disadvantages. It depends on climate conditions and requires a large surface and long time of exposure to the sun. This leads to deterioration of the dried product. In order to solve the problem, solar dryer have been proposed to utilize free, renewable and non polluting energy source provided by the sun [1]. The solar dryers could be an alternative to the hot air and open sun drying methods, especially in locations with good sunshine during the harvest season [2]. The introduction of solar dryers in developing countries can reduce crop losses and improve the quality of dried product significantly compared with traditional drying method [3]. Solar drying systems must be properly designed in order to meet particular drying requirements of specific crops and to give satisfactory performance with respect to energy requirements [4]. The improvement of the behavior of solar dryers passes through theoretical studies. Simulation models are helpful in designing new or in improving existing drying systems or for the control of the drying operation.

The drying kinetics of materials may be described completely by using their transport properties (thermal conductivity, thermal diffusivity, moisture diffusivity, and interface heat and mass transfer coefficients) together with those of the drying medium [5]. Different mathematical models for simulation were developed. The prediction of drying rate of the specific crops under various conditions is of importance for the design of the drying systems. In the development of thin layer drying models for agriculture products, generally the moisture content of the material at any time after it has been subjected to a constant drying air relative humidity and temperature condition is measured and correlated to the drying parameters [6]. Recently, there have been many researches on the mathematical modeling and experimental studies of the thin layer solar drying process of various vegetables and fruits. [7- 18]. Generally the model used to study a solar batch dryer, was based on the equation of heat applied to the product and on the equation of the drying rate [19]. In order to represent the moisture transfer, they have used Combes model [20].
This paper was mainly connected to the development of mathematical model of the thin layer solar drying for apricots using the mixed and indirect modes from a force convection solar dryer solar dryer under ecological conditions of Shiraz city, Iran.

Material and Experimental Set-Up

Solar Dryer

Fig.1 shows a schematic diagram of the solar dryer that was manufactured and installed in the college of agriculture, Shiraz University, Iran. It consists of two solar flat plate air heaters (4 m^2 surface area) and a drying cabinet that was made up of the available local materials and was used as an apparatus for the drying experiments. An aluminum plate (0.5 mm thickness) painted in black matt was used as an absorber plate for absorbing the solar radiation in the flat plate solar air heaters. Solar heaters and drying cabinet were assembly placed on a four wheeled iron frame and oriented 45° toward the south (local latitude is 30°). The frame of heaters was made of pressed wood and insulated by Styrofoam (5 cm thickness). A glass sheet was used as a cover on the heater to prevent the top heat losses. The drying cabinet was also constructed from pressed wood and kept by iron frame. It was also insulated by Styrofoam (5 cm thickness). The solar dryer was made in mixed-mode type; therefore, for letting the sun light pass through and hitting the drying products; the top of cabinet was covered by a glass sheet that was oriented towards the south as well at angle of 45°. The interior wall of the cabinet dryer was painted in black for absorbing solar radiation and increasing the temperature. This transparent glass sheet can be covered by an opaque and insulated sheet for indirect mode type of dryer application. On the back wall (north side) a lid was incorporated to let the operator for access to the drying cabinet. Two drying trays, The lower tray with 96×46 cm^2 in width and the upper one with 96×26 cm^2 (total tray surface area was 0.691 m^2) were placed inside the drying cabinet. Each tray made of wooden frames on which is fixed nylon net to facilitate the air flow. Collectors and drying cabinet were connected by a plenum camber. The outer surface of this chamber was insulated by Styrofoam. Heated air was flown between the trays by electric centrifugal fan (1400 RPM and 750W). To conduct the experiments at different air flow rates an adjustable damper was incorporated into the fan exhaust pipe. A long circular duct (15cm. in diameter with 150 cm. in length) was connected to the fan exhaust.

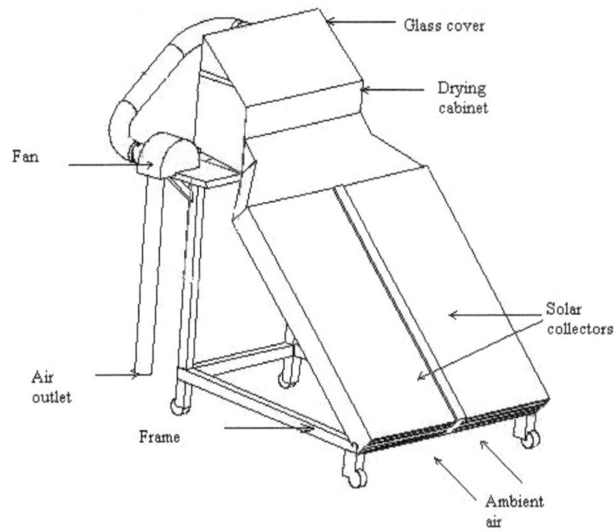

Fig. 1- schematic diagram of the solar dryer

Experimental Set-Up

Three sets of drying experiments were carried out during the periods of June to July 2007 in Shiraz, Iran. Each experiment started at 8:00 and continued till 18:00. During the drying experiments, the weather was generally sunny and no rain appeared. The air inlet and outlet temperatures in the solar collector and drying cabinet, the mass of sample trays containing representative apricots and the dryer, dry bulb and wet bulb temperature of ambient air and dryer exist air were measured at hourly intervals during the experiments. Moisture contents were reported on wet basis and then converted to kg water per kg dry matter. To prepare the fruit; apricots were washed; cut into half and removed the pits then the apricots were pre-treated with SO_2 to sulphur them before the drying operation [21]. Sulphured apricots were spread evenly in thin layers on the drying trays and then placed in the drying cabinet as undivided after having reached a suitable dryness level. The average weight of the sample used for the drying experiment was about 25 kg. The initial moisture content of the apricot samples were determined by using the vacuum oven method at 70 °C for 24 h [22]. These experiments were replicated thrice to obtain a reasonable average. Average moisture content was found to be 70% w.b. In order to investigate the effect of drying air velocity through the drying product, air velocity was set to 3.8, 6.3 and 8.7 m/s by an adjustable damper at the fan outlet. To examine performance of the device, a series of sensors and instruments were employed in the system. The drying air temperature was measured at the different points in the drying cabinet. The ambient and drying air temperature were measured by SMT-160 temperature sensors (±0.5°C) and air velocity by photoelectric type air velocity meter (0.01 m/s). The solar intensity was measured and recorded at the same time interval using a Casella Pyranometer (0 - 2000 W/m^2, 1mv=1W/m^2) placed beside the plane of the collector at the same inclination angle. All measurements were recorded by using a data logging system. The final moisture content of the product was determined while the product reaches the fixed weight in the drying cabinet.

Mathematical Modeling

In order to find the most appropriate thin layer drying model, the drying curves obtained were processed the most convenient on among 10 different expression defining drying rates given in Table 1 by several authors. However, the moisture ratio (MR) was simplified to M/M_0 instead of $(M-Me)/(M_0-Me)$ because relative humidity of the drying air continuously fluctuated in solar drying [23]. Regression analyses were carried out by using the Statistical methods. The correlation coefficient (r) was one of the primary criteria to select the best equation to account for variation in the solar drying curve of the dried samples [11; 24- 28]. In addition to r , the various statistical parameter such as; reduced chi-square (χ^2) as the mean square of the deviation, root mean square error (RMSE) between the experimental and calculated value for the models were calculated to determine the goodness of the fit experimental data. This parameter can be calculated as followings:

$$\chi^2 = \frac{\sum_{i=1}^{N}(MR_{\exp,i} - MR_{pre,i})^2}{(N-n)} \quad (1)$$

$$\text{RMSE} = \left[\frac{\sum_{i=1}^{N}(MR_{\exp,i} - MR_{pre,i})^2}{N}\right]^{1/2} \quad (2)$$

Table 1- Ten commonly used thin layer drying curve models

Model Name	Model
Newton	MR = exp(-kt)
Page	MR = exp(-kt^n)
Henderson and Pabis	MR = aexp(-kt)
Logarithmic	MR = aexp(-kt)+c
Tow Term	MR = aexp(-k$_0$t) + bexp(-k$_1$t)
Two term exponential	MR = aexp(-kt) + (1 - a) exp(-kat)
Wang and Singh	MR = 1 + at + bt^2
Approximation of diffusion	MR = aexp(-kt) + (1 - a)exp (-kbt)
Modified Henderson	MR = aexp(-kt) + bexp (-gt) + cexp(-ht)
Verma et al.	MR = aexp(-kt) + (1 -a)exp (-kgt)

Where $MR_{\exp,i}$ is the i th experimentally observed moisture ratio, $MR_{pre,i}$ is the i th predicted moisture ratio, N is the number of observations and n is the number of constant [5,10,29, 30].

The effects of initial and final moisture content, relative humidity, velocity and drying temperature on drying constants have been investigated by many researches [13; 20; 27; 31- 35].

Modeling the drying behavior of different agricultural products requires the statistical methods of regression and correlation analysis. Linear and non-linear regression models are important tools to find the relationship between different variables, especially those for which no established empirical relationship exists.

In the present work, the constants and coefficients of the best fitting model involving the drying variables such as drying air temperature and velocity was determined. The effects of drying parameters were also investigated by multiple linear regression analysis.

Result and Discussion

During the experiment, ambient air temperature ranged 25 to 30°C, ambient air humidity from 10 to 15%, drying air temperature from 40°C in air velocity of 8.7 m/s at indirect mode to 61°C in air velocity of 3.8 m/s at mixed-mode of solar drying and solar radiation from 420 to 1012 W/m2. The sulphured apricots of 70% (wb) average initial moisture content were dried to 16% (wb) by using air at different flow velocities (3.8, 6.3, 8.7 m/s) in the drying cabinet. The variation of moisture content versus drying time and the drying rates versus moisture content of apricots at each of two modes of drying are shown in Figs.2-5. The interruption of the lines in this figures represent the night periods of the drying operation. The changes in the moisture content of apricots after the sunset due to the thermal inertia of the drying system and apricots. It is apparent that in each mode of the solar apricot drying the constant rate drying period was absent and the drying process took place in the falling rate period. This result is in agreement with the earlier researches [5; 30; 33]. It can be said that the drying rate in the mixed-mode of solar drying is higher than the indirect mode of solar drying in every air flow velocity. The increase in the drying rate could be due to the values of higher temperature in the mixed-mode because of direct solar radiation on product in cabinet in addition to the heated air by solar heaters.

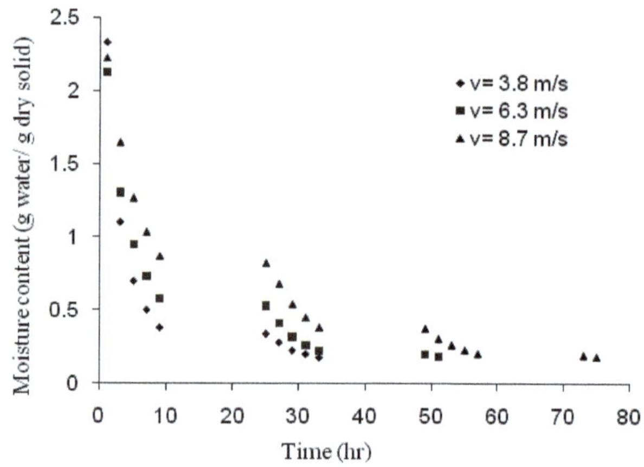

Fig. 2- Variation of moisture content versus drying time of apricots in indirect mode of solar drying

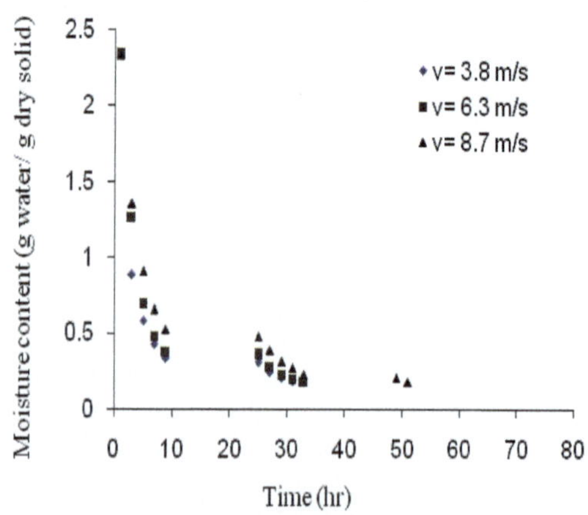

Fig. 3- Variation of moisture content versus drying time of apricots in mixed mode of solar drying

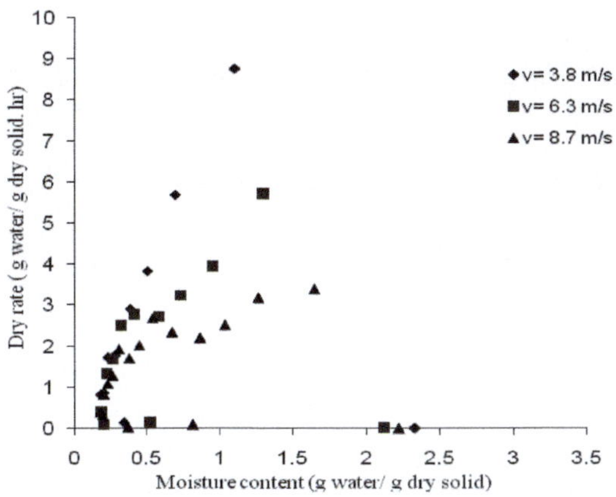

Fig. 4- Variation of drying rate versus moisture content of apricots in indirect mode of solar drying

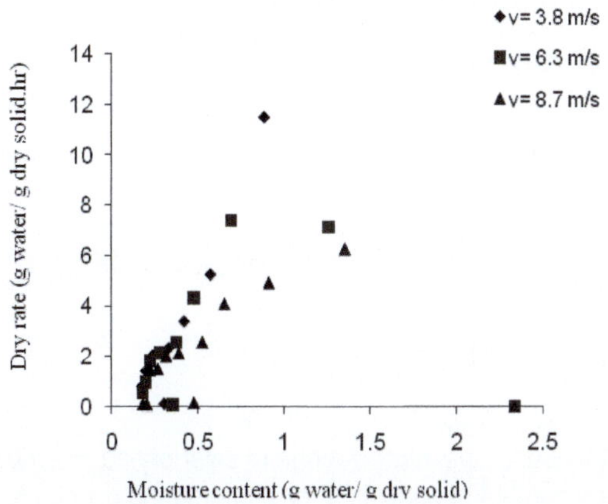

Fig. 5- Variation of drying rate versus moisture content of apricots in mixed mode of solar drying

Table 2- Estimated parameters and comparison criteria of moisture ratio for mixed mode solar cabinet drying

Model coefficients and constants	*r*	RMSE	x^2
$k = 0.2314$	0.951	$9.07*10^{-2}$	$9.27*10^{-3}$
$k = 0.2143$; $n = 1.0561$	0.951	$9.04*10^{-2}$	$10.5*10^{-3}$
$a = 1.2935$; $k = 0.3138$	0.976	$7.46*10^{-2}$	$7.16*10^{-3}$
$a = 1.4988$; $k = 0.5306$; $c = 0.111$	0.995	$2.58*10^{-2}$	$1*10^{-3}$
$a = 0.3432$; $k_0 = 0.09139$; $b = 1.6344$; $k_1 = 0.8673$	0.999	$4.61*10^{-3}$	$3.84*10^{-5}$
$a = 2.1045$; $k = 0.4233$	0.971	$8.98*10^{-2}$	$10.4*10^{-3}$
$a = -0.1594$; $b = 0.0065$	0.921	$10.8*10^{-2}$	$15*10^{-3}$
$a = 0.9837$; $k = 0.2596$; $b = -0.375$	0.968	$8.32*10^{-2}$	$10.4*10^{-3}$
$a = 0.5294$; $k = 0.6269$; $b = 0.4$; $g = 0.1068$; $c = 0.7144$; $h = 0.8$	0.996	$2.52*10^{-2}$	$1.91*10^{-3}$
$a = 0.9837$; $k = 0.2596$; $g = -0.0926$	0.976	$7.46*10^{-2}$	$8.35*10^{-3}$

Table 3- Estimated parameters and comparison criteria of moisture ratio for indirect mod solar cabinet drying

Model coefficients and constants	*r*	RMSE	x^2
$k = 0.1987$	0.965	$7.45*10^{-2}$	$6.16*10^{-3}$
$k = 0.1953$; $n = 1.01$	0.965	$7.44*10^{-2}$	$6.93*10^{-3}$
$a = 1.196$; $k = 0.2446$	0.98	$7.1*10^{-2}$	$5.04*10^{-3}$
$a = 1.306$; $k = 0.39$; $c = 0.1046$	0.996	$2.14*10^{-2}$	$6.59*10^{-4}$
$a = 0.3525$; $k_0 = 0.0833$; $b = 1.2343$; $k_1 = 0.6032$	0.999	$4.04*10^{-3}$	$2.73*10^{-5}$
$a = 0.461$; $k = 0.3088$	0.969	$7.18*10^{-2}$	$6.45*10^{-3}$
$a = -0.1399$; $b = 0.00506$	0.941	$9.21*10^{-2}$	$10.5*10^{-3}$
$a = 0.9836$; $k = 0.2225$; $b = -0.3592$	0.979	$6.66*10^{-2}$	$6.34*10^{-3}$
$a = 0.5142$; $k = 0.542$; $b = 0.466$; $g = 0.4365$; $c = 0.39$; $h = 0.058$	0.996	$6.72*10^{-2}$	$11.2*10^{-3}$
$a = 0.9837$; $k = 0.223$; $g = -0.08$	0.968	$6.67*10^{-2}$	$6.36*10^{-3}$

This moisture content data at the different drying air temperatures and velocities were converted to a moisture ratio, then fitted against the drying time were carried on the 10 drying models evaluated by the previous workers. These models and results of statistical analysis are given in Table 2 for the mixed-mode of apricots drying and Table 3 for the indirect mode of apricots drying.

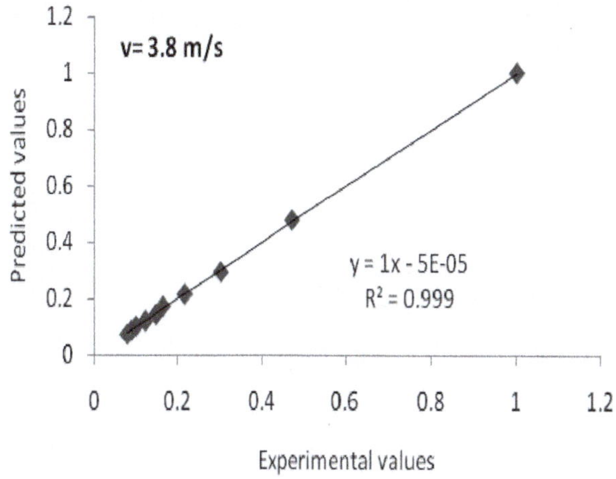

Fig. 6a-

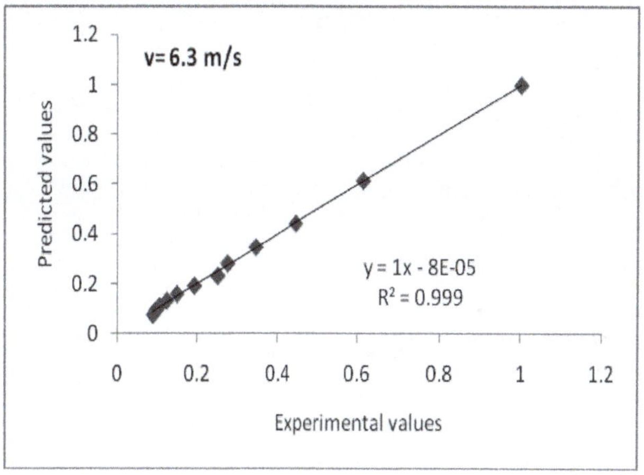

Fig. 6b-

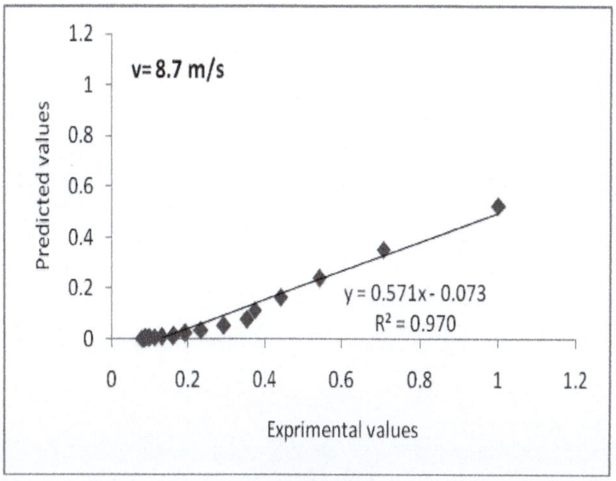

Fig. 6c-

Fig. 6- Experimental and predicted moisture contents values for various air flow rate in indirect mode of solar drying of apricots

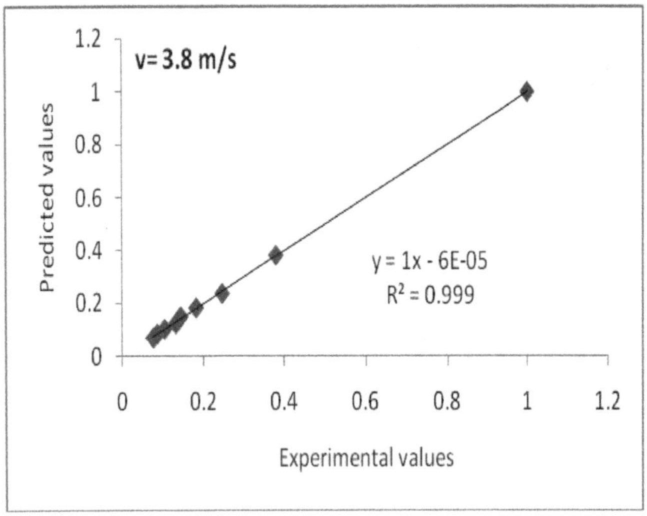

Fig. 7a-

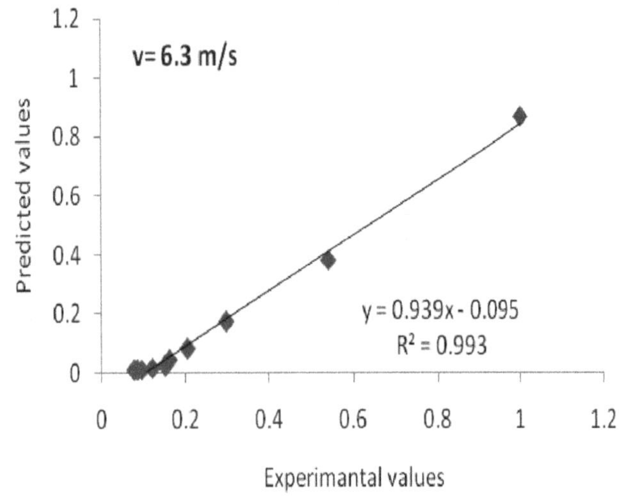

Fig. 7b-

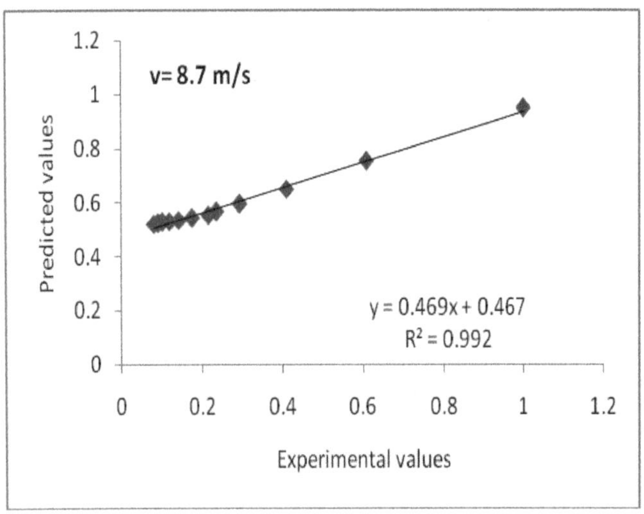

Fig. 7c-

Fig. 7c- Experimental and predicted moisture contents values for various air flow rate in mixed mode of solar drying of apricots

The two-term model gave a higher r and lower χ^2 was selected to present the thin layer behavior of apricots in the each of two modes of solar drying. It shown the solar drying curve of apricots was the same treat in the both of modes in this solar dryer. Therefore, using of mixed-mode for reduction of drying time is better than the indirect mode of drying.

The multiple combinations of the variables that gave the highest r were finally included in the final model. Based on the multiple regression analysis, the accepted model constants and coefficient were follows:

$$MR = M/M_0 = a\exp(-k_0t) + b\exp(-k_1t) \quad (3)$$

Where

$a= 16.908v+8.264T-559.732$ r =1.000

$k_0= 4.202v+2.034T-137.943$ r =1.000

$b= 707.38-21.453v-10.404T$ r =1.000

$k_1= 249.673-7.65v-3.662T$ r =1.000

Were determined for the mixed-mode of solar drying of apricots and

$a= 10.953v+5.398T-338.182$ r =1.000

$k_0= 6.268v+3.034T-190.606$ r =1.000

$b= 168.377-5.548v-2.656T$ r =1.000

$k_1= 184.642-6.059v-2.928T$ r =1.000

Were determined for the indirect mode of solar drying of apricots. The moisture content of apricots at any time during drying process in each of two modes could be estimated using these expressions with more accuracy. The consistency of the model and relationship between the coefficients and drying variables was evident with;

r =.999, $\chi^2=3.374*10^{-3}$, $RMSE=4.33*10^{-2}$ for the mixed-mode and r =.999, $\chi^2=1.243*10^{-4}$, $RMSE=8.64*10^{-3}$ for the indirect mode in this solar dryer.

Validation of the established model was evaluated by computed moisture content in any particular drying conditions with the observed moisture content.

The performance of the model at the different drying air velocities, drying air temperatures and sample areas is illustrated in Figs. 6 and 7. The predicted data generally banded around the straight line which showed the suitability of the two-term mathematical model in describing drying behavior of apricots.

Conclusions

The developed solar cabinet dryer can be used for dehydration of various agricultural products as well as the apricot under the climate conditions of Shiraz, Iran. The solar apricots drying process occurred in the falling rate period, starting from the initial to final moisture content of approximately 16% (wb). Drying rates of the apricot samples were affected by the temperature and velocity of the drying air in each mode of solar drying. Else, drying rate of the apricots in the mixed-mode is higher than in the indirect mode of solar drying. In order to explain the drying behavior and develop the mathematical modeling of apricots, 10 different thin layer drying models were applied to thin layer forced convection solar drying processes. Among these models, the two-term drying model could adequately describe the thin layer drying behavior of apricots in the mixed and the indirect modes of solar drying. When the effect of the drying air temperature and velocity on the constants and coefficient of the two-term model were investigated, resulting model gave an r of 0.999 and χ^2 of 3.374*10^{-3} for the mixed-mode and an r of 0.999 and χ^2 of 1.243*10^{-4} for the indirect mode of drying in this solar dryer. It is expected that this system will help reduce of drying time and increasing the quality of dried products. Further studies are ongoing.

Nomenclature

a,b, c,g,h,n	Empirical Constants in the Drying Models
k,k0,k1	Empirical Coefficients in the Drying Models
n	Number Constants
N	Number of Observations
MR	Moisture Ratio
MR_{exp}	Experimental Moisture Ratio
MR_{pre}	Predicted Moisture Ratio
M	Mean Moisture Content at t, %Dry Basis
M_e	Mean Equilibrium Moisture Content, %Dry Basis
M_0	Initial Moisture Content, %Dry Basis
r	Correlation Coefficient
RMSE	Root Mean Square Error
t	Time, min
T	Temperature, ^{0}C
V	Velocity, m/s
χ_2	chi-square

Reference

[1] Pangavhane D.R., Sawhney R.L. and Sarsavadia P.N. (2002) *Design, development and performance testing of a new natural convection solar dryer, Energy,* 27(6), 579-590.
[2] Sacilik K. (2007) *Journal of Food Engineering,* 79(1), 23-30.
[3] Muhlbaure W. (1986) *Energy in Agriculture*, 5, 121.
[4] Steinfeld A. and Segal I.A. (1986) *Drying Technology,* 4, 535-542.
[5] Togrul I.T. and Pehlivan D. *Journal of Food Engineering,* 65, 413-425.
[6] Midilli A., Kucuk H. and Yapar Z. (2002) *Drying Technology,* 20(7), 1503-1513.
[7] Simal S.A., Femenia M.C., Garau and Rosella C. (2005) *Journal of Food Engineering,* 66(3), 323-328.
[8] Menges H.O. and Ertekin C. (2006) *Energy Conversion and Management,* 47(15-16), 2337-2348.
[9] Dincer I. (1996) *Drying Technology,* 14, 1827-1838.
[10]Madamba P.S., Driscoll R.H. and Buckle K.A. (1996) *Journal of Food Engineering,* 29(1), 75-97.
[11]Sarsavadia P.N., Sawhney R.L., Pangavhane D.R. and Singh S.P. (1999) *Journal of Food Engineering,* 40, 219-226.
[12]Midilli A. (2001) *International Journal of Energy Research,* 25 (8), 715-725.
[13]Zomorodian A. and dadashzadeh M. (2009) *Science and technology journal of agricultural,* 11, 391-400.
[14]Akpinar E.A., Midilli b., Bicer Y. (2003) *Energy Conversion and Management,* 44, 1689-1705.
[15]Dandamrongrak R., Young G. and Mason R. (2002) *Journal of Food Engineering,* 55(2), 139-146.
[16]Doymaz I. (2004) *Journal of Food Engineering,* 61(3), 359-364.
[17]Lahsasni S., Kouhila M., Mahrouz M., Idlimam A. and Jamali A. (2004) *Energy,* 29(2), 211-224.
[18]Togrul I.T. and Pehlivan D. (2002) *Journal of Food Engineering,* 55, 209- 216.
[19]Youcef-Ali S., Messaoudi H., Desmons J.Y., Abene A. and Leray M. (2001) *Journal of Food engineering,* 45, 95-101.
[20]Daguenet M. (1985) *UNESCO.*
[21]Henderson S.M. (1974) *Transactions of the ASAE,* 17, 1167-1168.
[22]Association of Official Analytical Chemists (1990) *Official method of analysis* (934.06), Washington, DC, AOAC.
[23]Diamante L.M. and Munro P.A. (1993) *Solar Energy,* 51, 271-276.
[24]Yagcıoglu A. (1999) *Drying techniques of agricultural products.*
[25]Ozdemir M. and Devres Y.O. (1999) *Journal of Food Engineering,* 42, 225-233.
[26]Guarte R.C. (1996) *Modeling the drying behavior of copra and development of a natural convection dryer for production of high quality copra in the Philippines, Dissertation 287, Hohenheim University, Stuttgart, Germany.*
[27]Kassem A.S. (1998) *13th international congress on Agricultural Engineering,* 6, 2-6, Morocco.
[28]Werma L.R., Bukline R.A., Endan J.B. and Wratten F.T. (1985) *Transactions of the ASAE,* 28, 296-301.
[29]Pangavhane D.R., Sawhney R.L. and Sarsavadia P.N. (1999) *Journal of Food Engineering,* 39, 211-216.
[30]Martin M.B.S., Mate J.I., Fernandez T. and Virseda P. (2001) *Drying Technology,* 19(3&4), 681-690.
[31]Yaldiz O., Ertekin C. and Uzun H.I. (2001) *Energy,* 26, 47-465.
[32]Yaldiz O. and Ertekin C. (2001) *Drying Technology,* 19(3), 583 -596.
[33]Zhang Q. and Litchfield J.B. (1991) *Drying Technology,* 9, 383 -395.
[34]Tulasidas T.N., Raghaqvan G.S.V., Norris E.R. (1993) *Transactions of the ASAE,* 36, 1861-1865.
[35]Panchariya P.C., Popovic D. and Sharma A.L. (2002) *Journal of Food Engineering,* 52(4), 349-357.

CHARACTERISATION OF FIVE SIDEROPHORE PRODUCING ACTINOMYCETES FROM SOIL SAMPLES AND THE USE OF ANTIBIOTIC RESISTANCE TO DIFFERENTIATE THE ISOLATES

ISMINI NAKOUTI AND GLYN HOBBS*

School of Pharmacy and Biomolecular Sciences, Liverpool John Moores University, Byrom Street, Liverpool, L3 3AF, UK.
*Corresponding Author: Email- g.hobbs@livjm.ac.uk

Abstract- Organisms were isolated on their basis of survival in an iron-limited environment. The survivors of this treatment were largely actinomycetes. Of the viable cultures, most were found to produce siderophore like compounds. The most prolific producers as assessed by the Chromo Azuerol Sulphate assay were further characterised and found to belong to the genus *Streptomyces*. Attempts to taxonomically characterise these organisms illustrated the conserved nature of the 16S rRNA gene in this group of organisms. Physiological characterisation was undertaken and it was found that the resistance of these organisms to a range of antibiotics proved to be a useful discriminating factor .

Key words- Actinomycetes, iron chelation, siderophores, soil, streptomycetes, classification, resistance

Introduction

Iron is a fundamental element to microbial nature. Due to the evolution of the di-oxygen atmosphere in the earth's terrestrial environment, iron forms insoluble, bio unavailable ferric moieties at physiological pH [1]. In order to capture iron microorganisms have developed specific mechanisms involving the formation of siderophores [2]. The latter are low molecular weight compounds synthesised under iron starvation; they are produced extracellularly to bind iron and other ions forming a metal-siderophore complex. These complexes are then bound by the cells and transported intracellularly [3-6].

The actinomycetes are notorious environmental scavengers and have well-developed processes to capture ions in conditions depleted of elements; they are therefore an ideal target group for exploitation in ion binding technology [7-9]. Here we report the characterisation of four siderophore producing actinomycetes.

The taxonomy of streptomycetes has been extensively studied by many groups and comprehensive reviews are available in the literature [10-12]. Modern molecular taxonomical approaches involve the sequence analysis of the 16S rRNA gene, a particularly powerful tool in the classification of streptomycetes. Although 16S rRNA is a highly conserved area of the genome, it contains enough variations to allow us to investigate relationships at the genus, species and strain levels. Nevertheless it can be misleading due to intragenic variation, which highlights the need of genotypic and phenotypic observations [12]. One possible discriminatory factor is the resistance profile to antibiotics. This article describes data that uses antibiotic resistance to differentiate the siderophore producing isolates.

Materials and Methods

Purification of actinomycetes from soil:

A soil sample (1g) was suspended in 100ml of sterile distilled water and incubated 40^0 C for 24 hours. Actinomycetes were isolated from the soil suspension using a dilution plate technique on starch-casein agar (1.0 of starch, 0.4g of casein, 0.5 g of KNO_3, 0.1 g of $MgPO_4$, 0.2 g of K_2HPO_4, 0.1 g of $CaCO_3$ and 15g of agar per litre of distilled water [dH_2O]) supplemented with 150 mM 2,2'-dipyridyl (DIP). The agar plates were incubated at 27^{0}C for up to four weeks. Selective colonies were further sub-cultured and pure actinomycetes were isolated [13].

Siderophore production

Siderophore production was confirmed using the Chromo Azuerol Sulphate (CAS) assay (Renshaw et al. 2003). Arnow's [14] and Atkin's [15] assays confirmed the presence of a catecholic and hydroxamate compounds respectively.

Actinomycete characterisation

Actinomycete colonies were characterised according to Bergey's Manual of Systematic Bacteriology [16]; specific characteristics were observed by light and scanning electron microscopy.

The ability of the organisms to utilise different nutrient sources was tested using the SF-N2 (OXOID) and GP2 (OXOID) Biolog Microplates.

Antibiotic resistance of the strains was examined using MASTRINGS (Mast Diagnostics) and susceptibility discs (OXOID) performing the disc diffusion test. The model streptomycete, *Streptomyces coelicolor* (*S. coelicolor*) was also included in this study for comparison.

The strains were also tested for their ability to grow and sporulate on a range of liquid or solid media, including yeast extract-malt extract (YEME), Mannitol-Soya bean flower (MS), Muller-Hinton (MH), MH glucose, MH fructose and starch-casein agar [13].

16S rRNA amplification and taxonomic analysis

Genomic DNA was extracted using the 'Kirby mix procedure' [13]. Amplification of the 16S sequences were performed according to Rintala *et al.* 2001 [17].

The amplified products were purified using the QIAquick^{R} PCR purification kit (50) [Qiagen] and sent to Lark$_{TM}$ Technologies, Inc (United Kingdom) according to their requirements.

Pairwise sequence comparisons and retrieval of homologous sequences were conducted by the NCBI BLAST database (available online- http://www.ncbi.nih.gov/). Similar sequences, including an outgroup (ancestral sequence), were multiple aligned by BioEdit sequence alignment Editor 7.0.4.1 (Copyright 1997-2005, Hall, 1999). Taxonomic analysis was performed by the ClustalX software, version 1.83 and Bootstrap confidence values (1000) were provided by the Neighbour- Joining (NJ) algorithm. Finally the phylogenetic tree was presented using TreeView (Win 32) 1.6.6 (Copyright 2001) by Roderic. D. M. Page, available online: http://taxonomy.zoology.gla.ac.uk/rod/rod.html).

Results

Strain isolation

In an attempt to identify novel siderophore compounds we have isolated and purified 154 cultures from soil samples collected in Thailand (Tak province). The majority of strains are considered to be members of actinomycetes on the evidence of phenotypic examination. All the isolates were grown on starch-casein agar containing 150 mM DIP. This concentration of DIP was chosen as it reduced the number of surviving actinomycete colonies by approximately 99%. The remaining 1% was thought to be likely to have physiological attributes that enabled them to survive iron limitation. Twenty-two strains were confirmed CAS-assay positive with five strains (23F, 31B, 31C, 33D and 29C- GenBank accession numbers EF585403 - EF585407) exhibiting prolific siderophore production. Liquid based assays were conducted on culture supernatants in order to categorise the siderophore producers. The results classified the strains 31B, 31C, 33D and 29C as catechol producers, whereas 23F was categorised as a hydroxamate synthesiser.

16S rRNA gene sequence analysis and phylogenetic studies

16S rRNA gene sequence analysis was carried out to elucidate the taxonomic position of the isolates. Almost complete 16S rRNA sequences of the representatives were determined following the isolation of genomic DNA and sequencing of the amplified genes. The strains' phylogenetic positions are reported in figure 1. The levels of 16S rRNA sequence similarities between the isolates and actinomycetes species ranged from 97-99 %, confirming that the strains are members of the family, mainly streptomycetes.

The high percentage similarity values amongst species of the same genus are very common due to the highly conserved nature of the gene. For example although *Streptomyces macrosporus* and *Streptomyces megasporus* share 98 % similarities, they are clearly different on their phenotypic characterisation. A second paradigm includes *Streptomyces thermocarboxydovorans* (*S. thermocarboxydovorans*) and *Streptomyces thermodiastaticus* (99 %) or *S. thermocarboxydovorans* and *S. thermoviolaceous* (98 %) [18]. Moreover *Amycolatopsis albidoflavus* share 97.4 % similarities with *Amycolatopsis azurea* and *Amycolatopsis orientalis* [19].

Characterisation of the strains

All the streptomycete stains were further investigated in order to determine their morphological, physiological, growth and biochemical characteristics. Their ability to use a variety of nutrient compounds is described in Table 1.

Table 1- Nutrient utilisation of the five isolates

Nutrient utilisation of: (Measured using the Biolog system)	Isolates				
	23F	29C	31B	31C	33D
a-Cyclodextrin	+	-	+	-	+
b-Cyclodextrin	+	-	+	+	+
Dextrin	+	-	+	+	+
Glycogen	+	_	+	+	+
Inulin	+	_	+	+	+
Mannan	+	-	-	-	+
Tween 40	+	+	+	+	+
Tween 80	-	+	-	-	+
N-Acetyl-D-Glucosamine	+	-	-	-	+
N-Acetyl-b-D-Mannosamine	+	-	+	_	+
Amygdalin	-	-	-	-	-
L-Arabinose	+	-	+	-	+
D-Arabitol	-	-	+	-	-
Arbutin	+	-	+	-	+
D-Cellobiose	+	-	+	_	+
D-Fructose	+	+	+	-	+
L-Fucose	+	-	+	+	+
D-Galactose	-	-	+	-	+
D-Galacturonic Acid	-	-	+	-	+
Gentiobiose	-	-	+	-	+
D-Gluconic Acid	+	+	+	+	+
a-D-Glucose	+	+	-	-	+
m-Inositol	+	-	+	+	+
a-D-Lactose	-	-	-	-	+
Lactulose	-	-	-	-	+
Maltose	-	-	+	-	-
Maltotriose	-	-	+	-	-

Table 1 Continue

D-Mannitol	-	-	+	-	-
D-Mannose	-	-	-	-	-
D-Melezitose	-	-	+	-	-
D-Melbiose	+	-	+	+	+
a-Methyl-D-Galactoside	-	-	+	-	+
b-Methyl-D-Galactoside	-	-	-	-	-
3-Methyl Glucose	+	-	-	+	+
a-Methyl-D-Glucoside	+	-	+	+	-
b-Methyl-D-Glucoside	-	-	-	+	-
a-Methyl-D-Mannoside	-	-	-	+	+
Palatinose	+	-	+	+	+
D-Psicose	-	-	+	-	-
D-Raffinose	-	-	-	-	-
L-Rahamnose	-	-	+	-	-
D-Ribose	+	-	+	-	+
Salicin	-	-	-	-	+
Sedoheptulosan	-	-	-	-	--
D-Sorbitol	+	-	-	-	-
Stachyose	+	-	+	+	+
Sucrose	+	-	+	-	+
D-Tagatose	+	-	-	-	+
D-Trehalose	-	-	-	-	-
Turanose	+	-	-	+	-
Xylitol	-	-	-	-	-
D-Xylose	-	-	-	-	-
Acetic Acid	+	+	+	-	+
a-Hydroxybutyric Acid	-	+	-	-	-
b-Hydroxybutyric Acid	-	+	+	+	+
g-Hydroxybutyric Acid	-	+	-	-	+
p-Hydroxy-Phenylacetic Acid	-	-	-	-	-
a-Ketoglutaric Acid	-	+	+	+	+
a-Ketovaleric Acid	+	+	+	+	+
Lactamide	+	-	+	+	+
D-Lactic Acid Methyl Ester	-	-	+	-	+
L-Lactic Acid	-	+	-	-	-
D-Malic Acid	+	+	-	+	+
L-Malic Acid	+	+	-	-	+
Pyruvatic Acid Methyl Ester	+	+	+	+	-
Succinic Acid Mono-methyl Ester	+	+	-	+	+
Propionic Acid	+	+	+	+	+
Pyruvic Acid	+	+	+	+	+
Succinamic Acid	+	+	+	+	+
Succinic Acid	+	+	-	+	+
N-Acetyl-L-Glutamic Acid	+	-	-	-	+
L-Alaminamide	-	+	+	+	+
D-Alanine	-	+	-	+	+
L-Alanine	+	+	-	+	+
L-Alanyl-Glycine	-	+	+	_	-
L-lysine	+	+	+	+	+
L-Asparagine	+	+	+	+	+
L-Glutamic Acid	+	-	-	-	+
Glycil-L-Glutamic Acid	-	+	-	-	-
L-Pyroglutamic Acid	-	-	-	-	-
L-Serine	+	-	+	+	-
Putrescine	+	+	+	+	+
2,3-Butanediol	+	-	-	-	-
Glycerol	-	-	+	-	-
Adenosine	+	-	+	-	+
2'-Deoxy Adenosine	+	-	+	+	+
Inosine	-	-	+	+	+
Thymidine	+	-	+	+	+
Uridine	-	-	+	-	-
Adenosine-5'-Monophosphate	-	-	+	+	-
Thymidine-5'-Monophosphate	-	-	+	+	-
Uridine-5'-Monophosphate	-	-	-	-	-
D-Fructose-6'-Phosphate	+	-	+	+	+
a-D-Glucose-1-Phosphate	-	-	-	-	+
D-Glucose-6'-Phosphate	+	-	-	-	+
D-L-$_a$-Glycerol Phosphate	+	-	-	-	+

Table 1- Nutrient utilisation of strains 29C (GenBank accession number EF585407), 33D (GenBank accession number EF585406), 31B (GenBank accession number EF585404), 31C (GenBank accession number EF585405) and 23F (GenBank accession number EF585403).

Description of 29C (GenBank accession number EF585407):
Strain 29C is a Gram-positive, aerobic, white mycelia-forming organism that undergoes the typical streptomycete life cycle resulting in pink spirales spores. The spore surface was found to be smooth and this stage of the life cycle was only ever observed on starch-casein and YEME solid media. It did not sporulate on MS (Manitol-Soya agar) agar but it produced a diffusible pigment. It tolerated 150mM DIP and in starch-casein media it overproduced a very 'powerful' catechol that formed strong ferric complexes. It was found to produce a bactericidal agent against *Bacillus subtilis* (*B. subtilis*) and *Candida albicans* (*C. albicans*) under all conditions tested. In our hands it grew best in starch-casein media, pH 7.5 at 30^0 C.

Description of 33D (GenBank accession number EF585406):
It is a Gram positive, aerobic streptomycete species producing brown spirales spores. It grew rapidly on MS, starch-casein, yeast extract-malt extract (YEME) and slightly slower on Muller-Hinton-Fructose (MHF) or Muller-Hinton-Glucose (MHG). It was highly pigmented when grown on YEME and MS; nevertheless it did not sporulate well in the former cultivation media.
It grew poorly on MS, pH 5, and appeared to be unstable at pH 7. Optimum growth was on MS, pH 8 at 30^0 C. It produced an antimicrobial against *B. subtilis* and was found to be extremely tolerant to DIP (600mM), suggesting a well-adapted organism in its environment. It is worth noting that it overproduced a catechol on starch-casein agar, whereas it synthesised only a small amount of the specific siderophore on MS.

Description of 31C (GenBank accession number EF585405):
31C is a Gram positive, aerobic streptomycete producing rectiflexible white/grey spore chains. This strain only produced substrate mycelia on MHG or MHF agar, and on YEME solid media appeared to be extremely unstable forming heterologous colonies. It grew on MS agar, pH 5-7, with a preference for pH 7 where it only formed aerial mycelia. It solely sporulated on starch-casein media with an optimum growth at pH 7.5 at 30^0 C. It posed antibiosis against both *B. subtilis* and *C. albicans* but only in media where catechol production is negligible.

Description of 31B (GenBank accession number EF585404):
This isolate is a Gram positive, aerobic streptomycete producing rectiflexible spores. It was apparent that 31B had fastidious growth requirements and appeared unstable in our hands. It was also evident that its growth rate was extremely slow with a preference to starch-casein media, pH 7.5, at 30^0 C.

Description of 23F (GenBank accession number EF585403):
This strain is a Gram positive, aerobic actinomycete, which belongs to *Streptomyces* genus, and produced spores with an ornamented surface (spiny).
The organism was well adapted to its environment tolerating up to 600 mM DIP and growing in a range of pH (5-8) and temperature

(28^0 C- 37^0 C) conditions. It grew quickly on MS (3 days) agar, nevertheless much slower in starch-casein medium (approximately 10 days). It did not sporulate on YEME agar and on MHG and its optimum growth conditions involved cultivation on MS or starch-casein, pH 7, at 37^0C.

Antibiograms

Antibiotic resistance data offer valuable information in terms of strain characterisation and identification (Table **2**). In the specific small-scale experiment 14 out of 24 antibiotics demonstrated different patterns amongst the streptomycetes. Careful examination of the data revealed specific profiles for *Streptomyces coelicolor* (*S. coelicolor*) and strains 23F, 29C and 31B. For example sensitivity to amoxicillin, cefuroxim or oxaxillin was a unique characteristic of isolate 29C. Stain 31B was the only candidate that was resistant to vancomycin. A combination of growth in the presence of erythromycin and levofloxacin differentiated strain 23F from *S. coelicolor*. Culture 33D could be identifiable by its sensitivity against ampicillin and vancomycin. Finally *S. coelicolor* demonstrated a distinctive pattern of tetracycline (10 µg) and erythromycin resistance.

Table 2-Antibiograms of the siderophore producing strains. S. coelicolor is included as a reference

Antibiograms Resistant to:	Isolates 23F	29C	31B	31C	33D	S. coelicolor
Amoxycillin 2mg	+	-	+	+	+	+
Amoxycillin/clavulanic acid 3mg	+	+	+	+	+	+
Ampicillin 25 mg	+	+	-	+	-	+
Cefotaxime sodium 5 mg	+	+	+	+	+	+
Cefuroxime 5 mg	+	-	+	+	+	+
Cephalexin 30 mg	+	+	+	+	+	+
Chloramphenicol 50 mg	-	+	+	-	-	-
Ciprofloxacin 1 mg	+	-	+	+	+	-
Colistin sulphate 100 mg	-	-	-	-	-	-
Erythromycin 5 mg	+	-	-	-	-	+
Gentramicin 10 mg	-	-	-	-	-	-
Kanamycin 30 mg	-	-	-	-	-	-
Levofloxacin 1 mg	+	-	+	+	+	-
Nalidixic Acid 30 mg	+	+	+	+	+	+
Netilmicin 10 mg	-	-	-	-	-	-
Nitrofurantoin 50 mg	+	+	+	+	+	+
Nitrofurantoin 200 mg	+	-	-	+	+	+
Oxaxillin 1 mg	+	-	+	+	+	+
Penicillin G 1 unit	+	+	+	+	+	+
Piperacillin/ tazobactam 85 mg	+	+	-	-	+	+
Spectinomycin 25 mg	+	-	-	+	+	+
Streptomycin 25 mg	-	-	-	-	-	-
Tetracycline 10 mg	-	-	-	+	+	+
Tetracycline 100 mg	-	-	-	-	-	-
Trimethoprim 2.5mg	+	+	-	+	+	+
Vancomycin 5 mg	-	-	+	-	-	-

Table 2- Antibiograms of the siderophore producing isolates 29C (GenBank accession number EF585407), 33D (GenBank accession number EF585406), 31B (GenBank accession number EF585404), 31C (GenBank accession number EF585405) and 23F (GenBank accession number EF585403). *S.coelicolor*, the model streptomycete,is included as a reference organism.

Discussion

It is anticipated that the isolation and characterisation of new actinomycetes might lead to novel siderophores of significant pharmaceutical interest. Although 16S rRNA analysis is a very powerful tool in phylogenetic relationships, it does not provide enough evidence for classification of closely related species. Therefore a 'universal' approach, which will include the combination of modern molecular approaches with a number of morphological, physiological and biochemical characteristics, will facilitate the understanding of the genotypic and phenotypic behaviour of these precocious siderophore producers. Antibiotic resistance data provide useful information on the nature of the cells. A report by Hopwood reaffirms the importance of actinomycetes as a major reservoir of antibiotic resistance genes [20]. Clearly these characteristics can be a useful way of classifying strains. This could be a valuable evolutionary route to differentiate strains within a population. Given the fact that *Streptomyces* species have been found to have the capacity to produce multiple "antimicrobial" compounds [9, 21] and hence the ability to be resistant to these compounds, it will be interesting to develop a resistance fingerprint system for strains and to revisit phylogenetic analysis in the light of this information.

References

[1] Wandersman C. and Delepelaire P. (2004) *Annu Rev Microbiol* 58, 611-647.

[2] Gunter K., Toupet C. and Schupp T. (1993) *J Bacteriol* 175, 3295-3302.

[3] Neilands J.B. (1995) *J Biol Chem* 270, 26723-26726.

[4] Kalinowski B.E., Liermann L.J., Givens S. and Brantley S.L. (2000) *Chem Geol* 169, 357-370.

[5] John S.G., Christy E., Ruggiero L.E., Tung S. and Neu M.P. (2001) *Environ Sci Techol* 35, 2942-2948.

[6] Rodriquez M.G. and Smith I. (2003) *Molecul Microbiol* 47, 1485-1495.

[7] Meiwes J., Fiedler H.P., zahner H., Konetschny-Rapp S. and Jung G. (1990) *Appl Microbiol Biotechnol* 32, 505-510.

[8] Challis G.L. and Ravel J. (2000) *FEMS Microbiol lett* 187, 111-114.

[9] Bentley S.D., Chater K.F., Cerdeno-Tarraga A.M., Challis G.L., Thomson N.R., James K.D., Harris D.E., Quail M.A., Kieser H., Harper D., Bateman A., Brown S., Chandra G., Chen C.W., Colllins M., Cronin A., Fraser A., Goble A., Hilalgo J., Hornsby T., Howarth S., Huang C.H., Kieser T., Larke L., Murphy L., Oliver K., O'Neil S., Robbinowitsch E., Rajandream M.A., Rutherford B.G., Rutter S., Seeger K., Saunders D., Sharp S., Squares S., Taylor K., Warren T., Wietzorrek A., Woodward J., Barrell B.G., Parkhill J. and Hopwood D.A. (2002) *Nature* 417, 141-147.

[10]Williams S.T., Goodfellow M., Alderson G., Wellington E.M.H., Sneath P.H.A. and Sackin M.J. (1983) *J Gen Microbiol* 129, 1743-1813.

[11]Williams S.T., Goodfellow M., Wellington E.M.H., Vickers J.C., Alderson G., Sneath P.H.A, Sackin M.J. and Mortimer A.M. (1983) *J Gen Microbiol* 129, 1815-1830.

[12]Anderson A.S. and Wellington E.M.H. (2001) *Int J Syst Evol Microbiol* 51, 787-814.

[13]Kieser T., Bibb M.J., Buttner M.J., Chater K.F. and Hopwood D.A. (2000) *Streptomyces* genetics, 8, 168-169 & 342-343.

The John Innes Foundation, Norwich.

[14]Arnow L.E. (1937) *J Biol Chem* 118, 531-537.

[15]Atkin C.L., Neilands J.B. and Phaff H.J. (1970) *J Bacteriol* 103, 722-733.

[16]Locci R. (1989) *Streptomyces* and related Genera. In Bergey's Manual of Systematic Bacteriology ed Williams Wikins Company, Baltimore. 2451-1508.

[17]Rintala H., Nevalaines A., Ronka E. and Suutari M. (2001) *Mol Cell Probes* 15, 337-347.

[18]Kim C., Walter Lorenz W., Hoopes J.T. and Dean F.F.D. (2001) *J Bacteriol* 183, 4866-4875.

[19]Lee S.D. and Hah Y.C. (2001) *Int J Syst Evol Microbiol* 51, 645-650.

[20]Hopwood D.A. (2007) *Mol Microbiol* 63, 937-940.

[21]Omura S., Ikeda H., Ishikawa J., Hanamoto A., Takahashi C., Shinose M., Takahashi Y., Horikawa H., Nakazawa H., Osonoe T., Kikuchi H., Shiba T., Sakaki Y. and Hattori M. (2001) *Proc Natl Acad Sci USA* 98, 12215-12220.

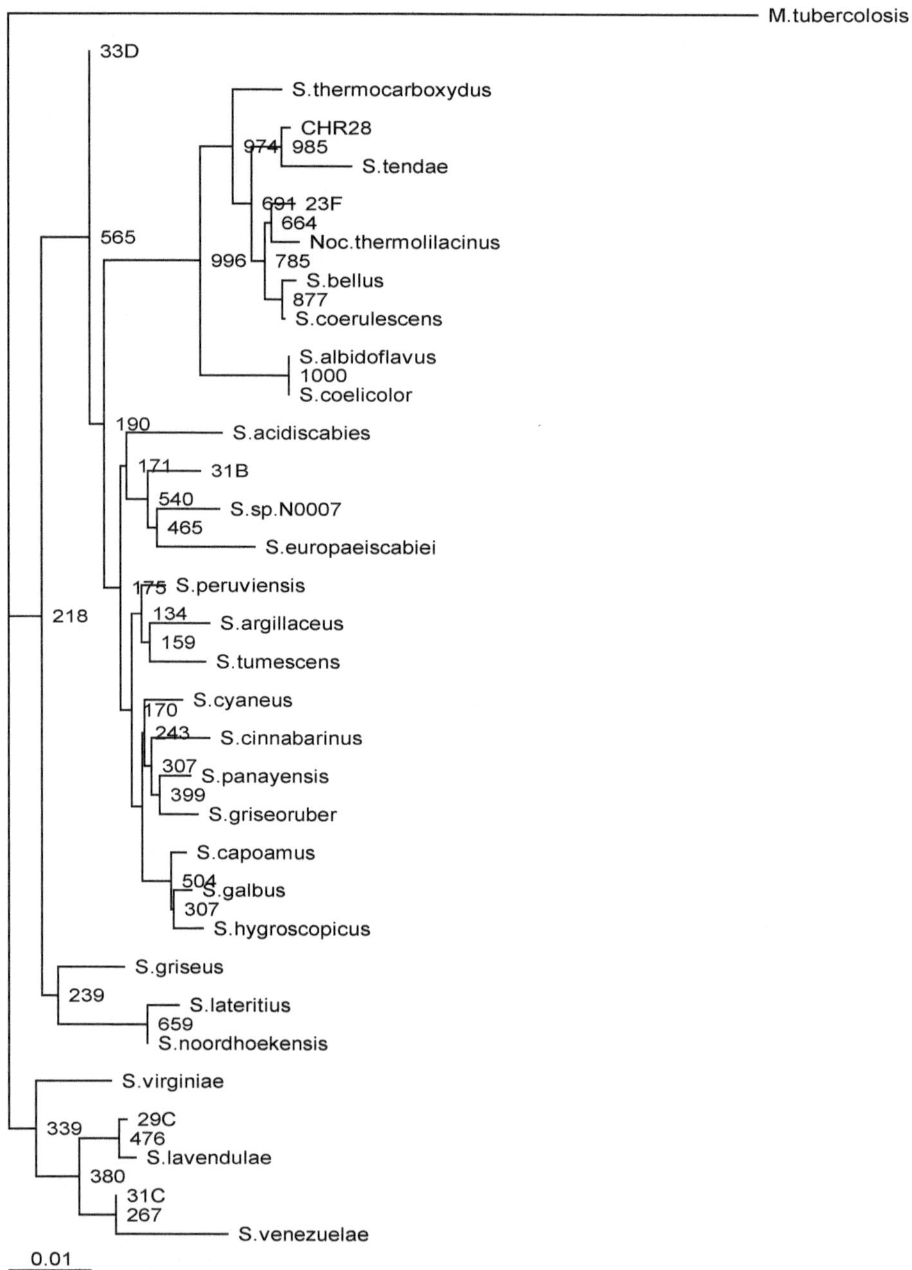

Fig. 1- Phylogenetic relationships of the isolates and related Actinomycetes based on the 16S rRNA sequence analysis. The tree is bootstrapped by 1000 times by Neighbour-Joining (N-J). M. tuberculosis was employed as an outgroup

21

EFFECT OF SEED STORAGE PERIODS, CONDITIONS AND MATERIALS ON SEED QUALITY OF SOME SOYBEAN CULTIVARS

KANDIL A.A., SHARIEF A.E. AND SHETEIWY M.S.*

Department of Agronomy, Faculty of Agriculture, Mansoura University, Egypt.
*Corresponding Author: Email- salahco_2010@mans.edu.eg

Abstract- In order to examine the effect of storage period, storage conditions and storage materials on seed quality of some soybean cultivars. A Laboratory experiment was conducted in Agronomy Department, Fac. of Agric., Mansoura Univ., Egypt during 2011 and 2012. The results showed that acid number, peroxide number and electrical conductivity (dSm-1)were increased with the period of ageingwere increased. The obtained results indicated that Giza 22cultivar exceeded other cultivars in electrical conductivity and moisture percentage. Giza 35 cultivar exceeded the other cultivars in acid number and peroxide number. Crawford cultivar exceeded the other cultivars in oil percentage. Storage under refrigerator conditions at 10°C ± 1 exceeded storage under ambient conditions in oil percentage and moisture percentage. Storage under ambient conditions exceeded storage under refrigerator conditions at 10°C ± 1 in acid number, peroxide number and electrical conductivity. Acid number and peroxide number significantly affected by storage material. Storage soybean cultivars as seeds or pods in plastic bags recorded lowest values of peroxide and acid number compared with other package materials. Results clearly showed that oil percentage, moisture percentage, acid number, peroxide number and electrical conductivity significantly affected by the varies interactions.

Keywords- Soybean, cultivars, storage, viability, deterioration, seed vigor and periods

Introduction

Both storage conditions and duration are important factors affecting soybean seed quality and the degree of biochemical changes in seed. In Egypt, soybean the most important crop and being the edible oil. Seed quality is an imperative parameter for commercialization and processing of the seed and can affect the value of the product and its derivatives [2]. Changes that occur in seed during ageing are significant in terms of seed quality among other things, also implies seed longevity [20]. Changes arising in seed during are very significant with regard to quality and longevity of seed. Characteristics of oily plants seed composition are relatedto specific processes occurring in seed during storage [14]. Seed quality during storage depend upon the initial quality of seed and the manner in which it is stored. The rapid seed deterioration of soybean is thought to be due to lipid peroxidation, subsequently resulting in loss of seed viability [27].

Changes in seed during ageing are significant in terms of seed quality, among other things, also implies seed longevity [20]. Balesevic-Tubicet, et al. [8] showed that defense of seed qualityhang onageing condition and duration. Khaliliaqdamet, et al. [19] suggested that variations occurred with the viability constants of a seed deterioration model between the soybean cultivars. Different periods of seed ageing, as well as storage conditions adversely affected the seed quality [30]. Similar inferences was reported by Sharma, et al [26], Simic, et al [29] and Vijay, et al [35].

Differences in oil affected by storage longevity were significant among tested crops and genotypes within crops. Storage longevity was negatively related with oil percentage. At storage conditions at 12°C/60%, decreasing of seed oil content than in storage conditions at 25°C/75% [28]. The amount of electrolytes in the soaked water increased as the period of time and the temperature of storage were increased. The temperature at which soybean seeds remain during storage may affect the amount of electrolytes in the soaked water and consequently the results of the electrical conductivity test [33]. Vijay, et al. [35] incubated soybean seeds at two different temperatures of 34°C and 40°C for varying times. They indicated that soybean viability decreased by increasing storage temperature.

A significant difference between soybean genotypes (Sb1, Sb2, Sb3, Sb4, Sb5 and Sb6) in oil percentage, acid number, peroxide number and moisture percentage. These differences between genotypes might be due to the genetic factors and seed chemical composition influence the expression of seed deterioration and vigor decline [30]. Khaliliaqdamet, et al. [19] reported that initial viability was same 100% among DPX and Sahar cultivars. Electrical conductivity had no significant difference between DPX and Sahar cultivars but in Shahar, were higher than DPX and EC in Sahar was lower. Similar conclusions were reported by Kalavathi, et al [18],

Simic, et al [29], Venkatareddy, et al [32].

In case of quality characters of soybean seed decreases all cases but the rate of deterioration is highest in cloth bag [22]. Seeds were stored in Metallized film bags and Aluminum foil bags observed highly standard quality and seed vigor, and keep water activity and seed moisture content in low level could delay seed quality deterioration followed by Polypropylene bags and woven bag [10]. Seeds were stored in aluminums foil bags observed highly seedling vigor and keep moisture content in low level could delay seed deterioration followed by polyethylene and wheat bags [31]. Sorghum seeds stored inside gene bank and freezer regardless of the packaging materials and type of accession retain their viability [24]. Similar results was obtained by [6,9,25]. Chuansin, et al. [10] found that soybean seeds were stored in Metallized film bags and Aluminum bafoil gs keep water activity and seed moisture content in low level could delay seed quality deterioration followed by Polypropylene bags and woven bag. The oil peroxide index differed significantly among samples due to the interaction between moisture content, temperature and storage period [3]. The objectives of this research were aimed to determine the effects of storage periods, storage conditions and materials on seed quality characters of some soybean cultivars.

Materials and Methods

A laboratory experiment was conducted at experimental seed testing laboratory of Agronomy Department, Faculty of Agriculture, Mansoura University, Egypt during 2011 and 2012. The objectives of this investigation were aimed to investigate the response of some soybean cultivars i.e. Giza 21, Giza 22, Giza 35, Giza 111 and Crawford to storage periods i.e. 3,6,9 and 12 months, storage conditions i.e. ambient conditions and refrigerator conditions at 10°C±1 as well as storage materials i.e. seed with cloth bags, seed with plastic bags, pods with cloth bags and pods with plastic bags on seeds quality characters.

Treatments and Experimental Design

Treatments were arranged in factorial experimental in completely randomize design, consisted of totally 160 treatments combinations involving tow storage conditions i.e. ambient conditions and refrigerator conditions at 10°C±1, four different kinds of package materials i.e. seed with cloth bags, seed with plastic bags, pods with cloth bags and pods with plastic bags of five soybean cultivars i.e. Giza21, Giza22, Giza35, Giza 111 and Crawford. Four storage periods i.e. 3,6,9 and 12 months were used. Soybean (*Glycine Max* L. Merr.) cultivars were harvested in October 2011, cleaned from dust and dirt then dried and processed for storage. Every three months, from a total 12 months of storage, electrical conductivity, acid number, peroxide number, oil percentage and moisture percentage were evaluated.

Studied Characteristics

The following seed quality characters were studied as following:

Oil Percentage (%)

Soxhelt apparatus was used for determination of ether extract percent, heating by electric heaters; cold water was used through the condenser. Petroleum ether (60-80°C) was preferred for extractions which continued for not less than eight hours (rate of siphoning was 6-7 /hr.) according to AOAC [1]. Moreover, oil percentage was calculated by multiplying ether extract percent × yield of dry matter (g).

Acid Number

Acid value was determined according to AOAC [1]. A known weight of sample (g) was dissolved in 20 ml ethylalcohol then the contents of the flask were shaken for one minute, then titrated with aqueous potassium hydroxide (0.1N) in the presence of phenolphthalein as indicator. The acid value was calculated according to the following equation [Eq-1]:

$$\text{Acid Value(A.V.)} = \frac{V \times N \times 56.1}{W} \quad (1)$$

Where, V = volume of KOH required to neutralize the free fatty acids, N = normality of KOH and W = weight of sample.

Peroxide Number

Peroxide value was determined as reported by AOAC [1]. A known weight of oil sample (g) was dissolved in 30 ml of glacial acetic acid:

Chloroform mixture (2:1, v/v), then one ml of freshly prepared saturated potassium iodide solution was added. The content of the flask were allowed to stand with occasional shaking for exactly 1 min and 30 ml of distilled water were added.

The liberated iodine was titrated with sodium thiosulphate solution (0.1N) until the yellow color had almost disappeared. The titration was continued after adding 1 ml of 1% starch solution as an indicator until the blue color has just disappeared.

The following equation [Eq-2] was used for calculation of peroxide value expressed as mill equivalents of oxygen / g oil:

$$\text{Peroxide Value} = \frac{V \times N}{W} \times 1000 \quad (2)$$

Where, V = volume of sodium thiosulphte used for sample, N = normality of sodium thiosulphate solution and W = weight of oil sample.

Electrical Conductivity of Seed Leachate (dSm^{-1})

Five grams of seeds in four replications were surface sterilized with $HgCl_2$ 1% solution and thoroughly washed in distilled water, later seeds were soaked in 25 ml distilled water and kept at 25±1°C for 24 hr. The leachate was decanted to another container and volume was made up to 25 ml by adding distilled water. The electrical conductivity of seed leachate was measured with electrical conductivity bridge and expressed in dSm^{-1} [16].

Moisture Percentage (%)

The moisture percentage of the seed was calculated according to the International Seed Testing Association (ISTA) rules [16] by hot air oven method maintaining 103°C ± 2°C for 17 hr. and then at 105°C till constant dry weight. The moisture content was calculated on wet basis and expressed in percentage by using the following formula [Eq-3]:

$$\text{Moisture Percentage (\%)} = \frac{(M2 - M3)}{(M2 - M1)} \times 100 \quad (3)$$

Where, M1 = Weight of empty moisture tin in (g), M2 = Weight of moisture tin and seed material before drying (g) and M3 = Weight of moisture tin and seed material after drying (g).

Result and Discussion

Storage Periods Effects

The results showed that oil percentage and moisture percentage were decreased as storage periods were increased as shown in [Table-1]. While, acid number, peroxide number and electrical conductivity were increased as storage periods were increased. Before storage treatment significantly exceeded the other storage periods in oil percentage and moisture percentage followed by storage for 3 months. While, storage for 12 months recorded lowest oil percentage and moisture percentage. It could be concluded that increasing storage periods from 6 and 12 months decreased oil percentage by 12.99 and 17.58%, respectively compared with before storage treatment. Increasing storage periods from 3, 6, 9 and 12 months decreased moisture percentage by 0.42, 1.38, 0.95 and 2.45%, respectively compared with before storage treatment. Increasing storage periods from 3, 6, 9 and 12 months increased electrical conductivity by 7.28, 11.57, 13.96 and 18.%, respectively compared with before storage treatment. Increasing storage periods from 6 and 12 months increased acid number by 14.39and 23.59%, respectively compared with before storage treatment. Increasing storage periods from 6 and 12 months increased peroxide number by 12.61 and 30.27%, respectively compared with before storage treatment. Decrease of seed quality is connected with biochemical changes in seeds of oil crops. These seeds had a rapid deterioration due to autoxidation of lipids and the increase of the content of free fatty acids during storage period. The longer seeds storage period increases intensity of seeds ageing [10]. These results are in good agreement with many researchers [5,7,12,13,15,17,19].

Storage Cultivars Effects

Results in [Table-1] revealed that Giza 35 cultivar surpassed other cultivars in acid number and peroxide number. Giza 21 cultivar recorded the lowest moisture percentage. Giza 22cultivar recorded the lowest peroxide number and acid number. Giza 35 cultivar recorded the lowest oil percentage. Giza 35cultivar surpassed Giza 21, Giza22, Giza 111 and Crawford cultivars in acid number by 3.31, 18.96, 10.68 and 4.05%, respectively. Giza 35 cultivar surpassed Giza 21, Giza22, Giza 111 and Crawford cultivars in peroxide number by 12.13, 25.10, 8.13 and 8.27%, respectively. In this respect, Tatic, et al [30] stated that there was significant differences between soybean genotypes (Sb1, Sb2, Sb3, Sb4, Sb5 and Sb6) in oil percentage, acid number, peroxide number and moisture percentage. These differences between genotypes might be due to the genetic factors and seed chemical composition influence the expression of seed deterioration and vigor decline. In addition, Doijoide [11] stated that the storability of different soybean cultivars is also regulated by initial seed quality, physical and chemical composition of seed as different cultivars possess different physical structure and chemical composition which determine the quality of seed in storage. Moreover, Gupta and Aneja [4] indicated that decreases in oil content between soybean cultivars may be due to age induced depletion phenomenon, improper storage environments and higher activity of storage fungi. These results are in good accordance with those obtained by Balesevic-Tubic, et al [7], Chuansin, et al [10], Erandes, et al [13] Jamro, et al [17] Khaliliaqdam, et al [19] Tatic, et al [30].

Storage Conditions Effects

The results in [Table-1] showed that a significant effect of storage conditions on average of oil percentage, acid and peroxide number and electrical conductivity. Storage under refrigerator conditions at 10°C surpassed ambient conditions in oil percentage by 4.71, 5.85, 11.67, 7.77 and 2.92%, respectively. While, storage under ambient conditions surpassed refrigerator conditions at 10°C in acid number, peroxide number and electrical conductivity by 26, 8.29 and 2.77%, respectively. In this regard, Arif [6] recorded that highest seed quality was in seed stored at 4°C than room temperature. Moreover, Vijay and Dadlani [35] incubated soybean seeds at two different temperatures of 34°C and 40°C for varying times. They indicated that soybean viability decreased by increasing storage temperature. These results are in good harmony with those obtained by Balesevic-Tubic, et al [7], Erandes, et al [13] Jamro, et al [17], Khaliliaqdam, et al [19], Mohammadi, et al [21], Vieira, et al [33].

Table 1- Averages of oil percentage (%), acid number, peroxide number, electrical conductivity (dSm-1) and moisture percentage (%) as affected by storage periods, soybean cultivars, storage conditions, package materials and their interactions.

Characters / Treatments	Oil (%)	Acid Number	Peroxide Number	Electrical Conductivity (dSm^{-1})	Moisture (%)
S- Storage periods (months)					
Before storage	22.86	4.34	5.32	1.91	9.38
3 months	-	-	-	2.06	9.34
6 months	19.89	5.07	6.47	2.16	9.25
9 months	-	-	-	2.22	9.29
12 months	18.84	5.68	7.63	2.27	9.15
LSD 5%	0.19	0.04	0.08	0.01	0.12
V-Soybean cultivars					
Giza 21	20.69	5.25	6.37	1.99	9.17
Giza 22	20.37	4.4	5.43	2.24	9.42
Giza 35	20.35	5.43	7.25	2.11	9.18
Giza 111	20.41	4.85	6.66	2.1	9.34
Crawford	20.82	5.21	6.65	2.2	9.28
LSD 5%	0.25	0.05	0.1	0.01	0.12
C-Storage conditions					
Ambient conditions	20.22	5.16	6.75	2.16	9.24
Refrigerator conditions (10°C±1)	20.83	4.9	6.19	2.1	9.32
F- test	**	**	**	**	NS
P-Package materials					
Pods in plastic bags	20.42	4.99	6.49	2.12	9.35
Pods in cloth bags	20.61	5.16	6.53	2.12	9.26
Seed in plastic bags	20.57	4.89	6.49	2.13	9.29
Seed in cloth bags	20.52	5.08	6.38	2.13	9.22
LSD 5%	NS	0.08	0.09	NS	NS
F-test Interactions					
V×S	**	**	**	**	NS
S×C	**	**	**	**	**
V×C	**	NS	NS	**	NS
V×C×S	NS	NS	NS	*	NS
S×P	NS	**	*	NS	NS
V×P	NS	**	**	**	NS
V×S×P	NS	**	*	*	NS
C×P	NS	NS	NS	*	NS
S×C×P	NS	NS	NS	NS	NS
V×C×P	NS	NS	NS	NS	NS
V×C×S×P	NS	NS	NS	NS	NS

Storage Package Materials Effects

The results in [Table-1]showed that a significant effect of package materials on the average of acid number and peroxide number. The results showed that highest value of acid number was obtained from storage soybean pods or seed in cloth bags. However, lowest value of acid number was obtained from storage soybean cultivars pods or seeds in plastic bags. Highest value of peroxide number was obtained from storage soybean pods in cloth bags. However, lowest value of peroxide number was obtained from storage soybean cultivars seeds in cloth bags. In this regard, Chuansin, et al [10] showed that seeds were stored in Metallized film bags and Aluminum foil bags observed highly standard quality and seed vigor, and keep water activity and seed moisture content in low level could delay seed quality deterioration followed by Polypropylene bags and woven bag. Moreover, Tatipata [31] stated that seeds were stored in aluminums foil bags observed highly seedling vigor and keep moisture content in low level could delay seed deterioration followed by polyethylene and wheat bags. These results are in harmony with those reported by Arefi and Abdi [5], Erandes, et al [13], Khaliliaqdam, et al [19], Tatipata [31].

Interaction Effects

Interaction between Soybean Cultivars and Storage Periods

Highest electrical conductivity was obtained from storage Giza 22 and Crawford cultivars for 12 months without significant differences between them. Lowest values of electrical conductivity were recorded from Giza 35 and Giza 21 cultivars before storage without significant differences between them [Fig-1]. The results clearly indicated that highest oil percentage was obtained from Giza 21 cultivar before storage followed by Giza 21 cultivar storage for 6 months. While, lowest oil percentage were recorded from storage Giza 21 and Giza 22 cultivar storage for 12 months without significant between them [Fig-2].

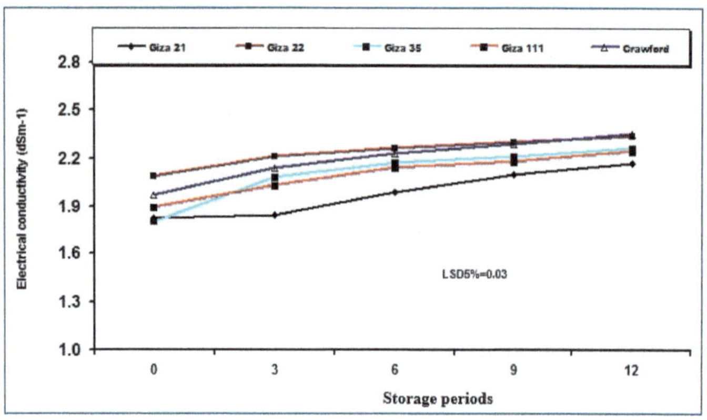

Fig. 1- Averages of electrical conductivity (dSm-1) as affected by the interaction between soybean cultivars and storage periods

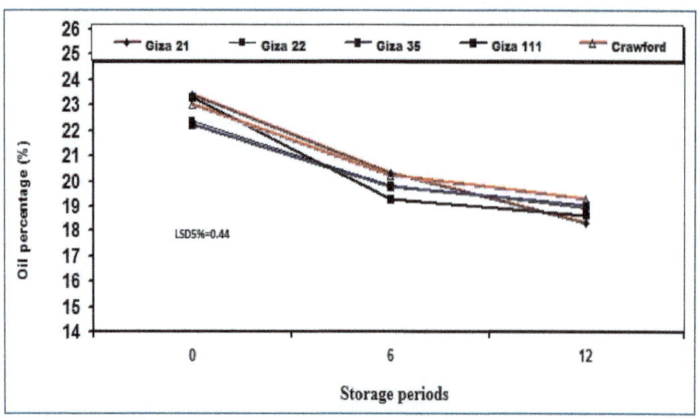

Fig. 2- Averages of oil percentage (%) as affected by the interaction between storage periods and soybean cultivars

The results clearly indicated that highest acid number was obtained from storage Giza 21 and Giza 35 cultivars for 12 months without significant differences between them. While, lowest acid number were recorded from Giza 22 before storage [Fig-3]. The results clearly indicated that highest peroxide number was obtained from storage Giza 21 and Crawford cultivars for 12 months without significant differences between them. While, the lowest peroxide number were recorded from before storage of Giza 22 [Fig-4]. These results in agreement with those reported by Arefi and Abdi [5], Balesevic-Tubic, et al [8], Chuansin, et al [10], Khaliliaqdam, et al [19], Muhammad [23], Tatipata [31].

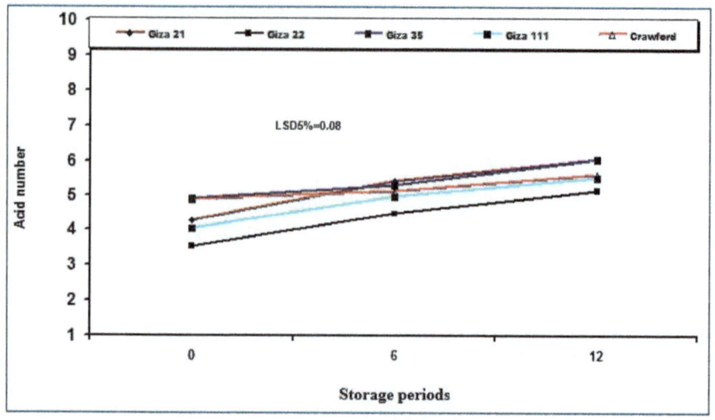

Fig. 3- Averages of acid number as affected by the interaction between storage periods and soybean cultivars.

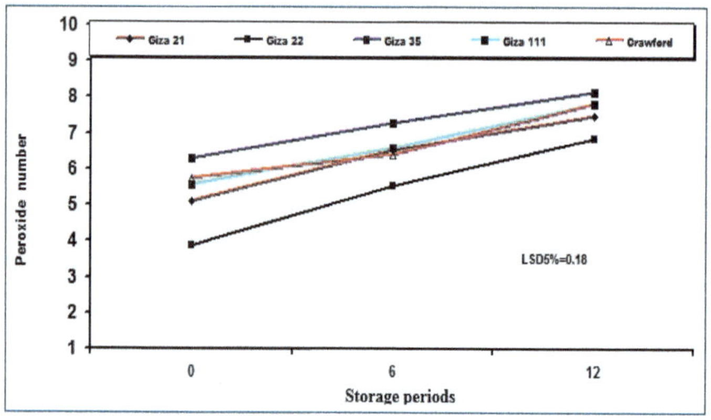

Fig. 4- Averages of peroxide number as affected by the interaction between storage periods and soybean cultivars.

Interaction between Storage Periods and Storage Conditions Effects

The results clearly showed that pre storage treatment recorded the lowest electrical conductivity followed by storage after 3 months under refrigerator conditions at 10°C [Fig-5]. Highest electrical conductivity was recorded from storage for 12 months under ambient conditions. The results clearly showed that pre storage treatment recorded highest moisture percentage followed by storage for 3 months under refrigerator conditions at 10°C [Fig-6]. Storage under ambient conditions for 12 months recorded the lowest moisture percentage. Pre storage treatment recorded highest oil percentage followed by storage after 3 months under refrigerator conditions at 10°C [Fig-7].

Storage under ambient conditions for 12 months recorded lowest oil percentage. The results clearly showed that pre storage treatment recorded lowest acid number followed by storage for 3 months under refrigerator conditions at 10°C [Fig-8]. Storage under ambient conditions for 12 months recorded highest acid number. The results clearly indicated that pre storage treatment recorded lowest peroxide number followed by storage for 3 months under refrigerator conditions at 10°C [Fig-9].

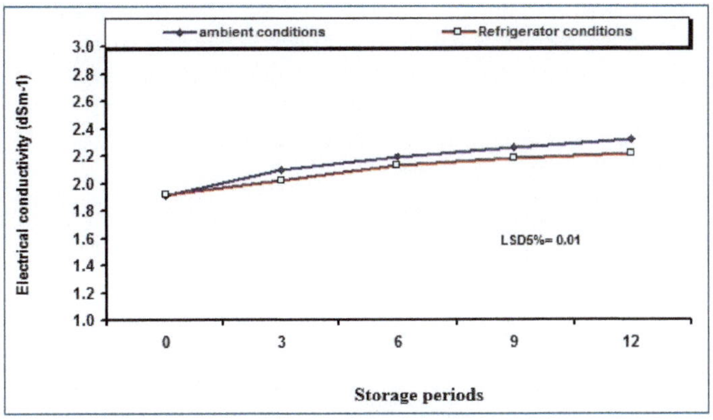

Fig. 5- Averages of electrical conductivity (dsm-1) as affected by the interaction between storage conditions and storage periods

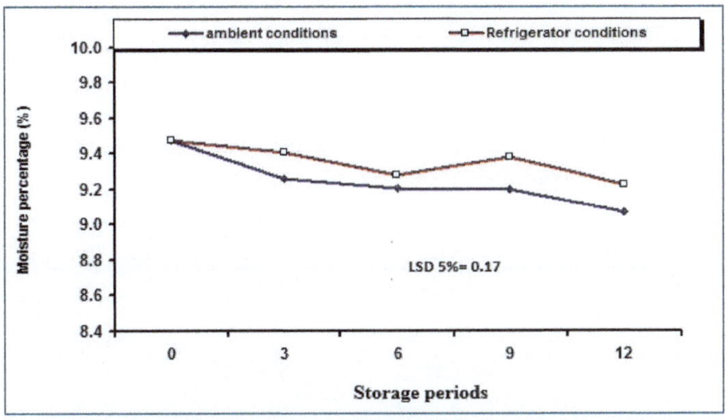

Fig. 6- Averages of moisture percentage (%) as affected by the interaction between storage conditions and storage periods.

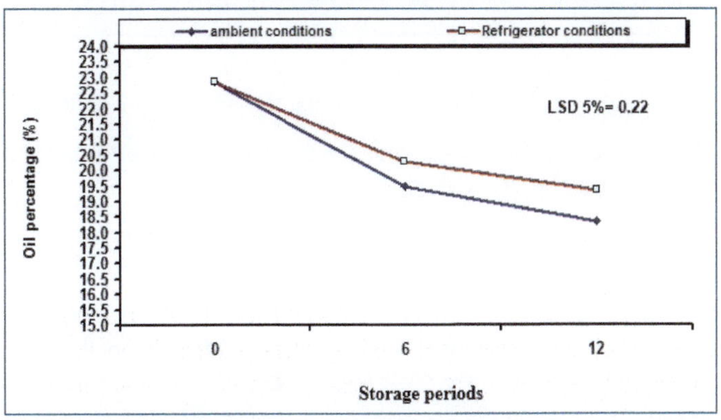

Fig. 7- Averages of oil percentage(%) as affected by the interaction between storage periods and storage conditions.

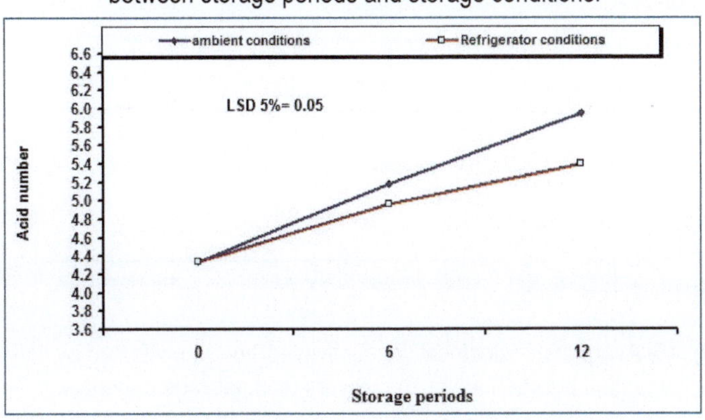

Fig. 8- Averages of acid number as affected by the interaction between storage periods and storage conditions.

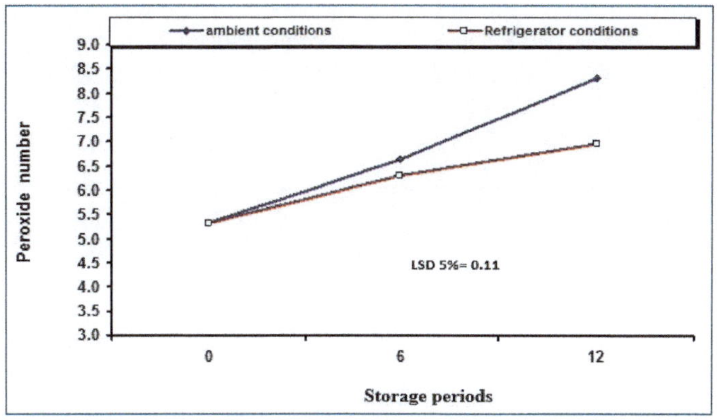

Fig. 9- Averages of peroxide number as affected by the interaction between storage periods and storage conditions

Storage under ambient conditions for 12 months recorded the highest peroxide number. The amount of electrolytes in the soaked water increased as the period of time and the temperature of storage were increased. In this respect Vieira, et al [33] showed that the temperature at which soybean seeds remain during storage may affect the amount of electrolytes in the soaked water and consequently the results of the electrical conductivity test. Moreover, Vijay, et al [35] incubated soybean seeds at two different temperatures of 34°C and 40°C for varying times. They indicated that soybean viability decreased by increasing storage temperature. These results are in conformity with those reported by Balesevic-Tubic, et al [7], Erandes, et al [13], Jamro, et al [17], Khaliliaqdam, et al [19], Mohammadi, et al [21], Vieira, et al [34].

Interaction between Soybean Cultivars and Storage Conditions Effects

Electrical conductivity significantly affected due to the interaction between soybean cultivars and storage conditions. Highest electrical conductivity was obtained from storage Crawford cultivar underambient conditions. The lowest electrical conductivity was obtained from storage Giza 21cultivar under refrigerator conditions at 10°C [Fig-10].

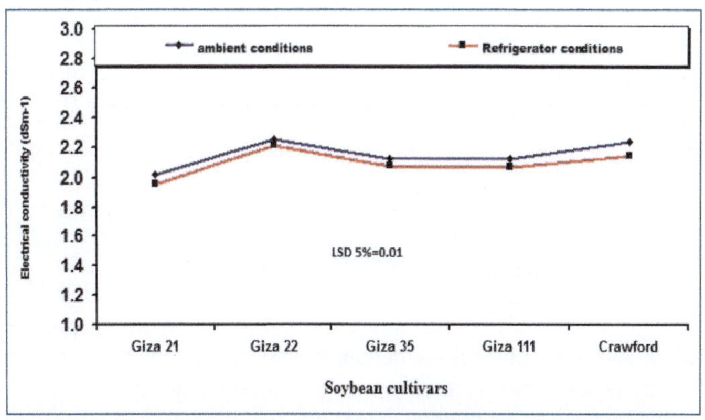

Fig. 10- Averages of electrical conductivity (dSm-1) as affected by the interaction between soybean cultivars and storage conditions.

The results clearly indicated that highest oil percentage was obtained from storage Crawford cultivar under refrigerator conditions (10°C±1). The lowest oil percentage was obtained from storage Giza 21 and Giza 35cultivars under ambient conditions without significant differences between them [Fig-11]. In this trend, Khaliliaqdam, et al [19] reported that initial viability was same 100% among DPX and Sahar cultivars. Electrical conductivity had no significant difference between DPX and Sahar cultivars but in Shahar, were higher than DPX and EC in Sahar was lower. These results are in conformity with those reported by Arefi and Abdi [5], Chuansin, et al [10], Erandes, et al [13], Khaliliaqdam, et al [19], Tatipata [31].

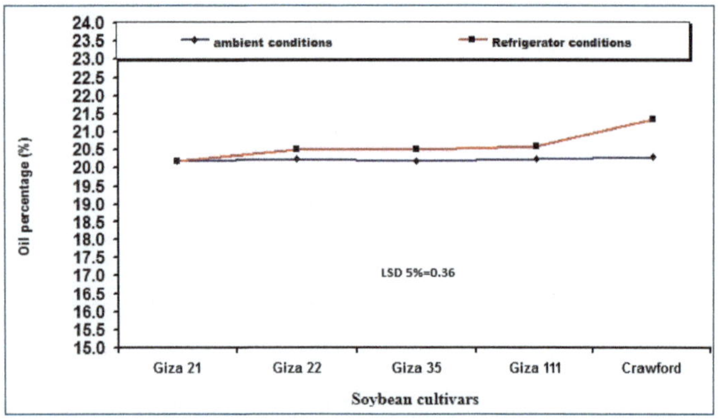

Fig. 11- Averages of oil percentage (%) as affected by the interaction between soybean cultivars and storage conditions.

Interaction between Storage Periods, Soybean Cultivars and Storage Conditions

The results clearly showed that highest electrical conductivity was obtained from storage Giza 22and Giza 111 cultivars underambient conditions for 12 months. The lowest electrical conductivity was obtained from pre storage of Giza 21 and Giza 35 cultivars [Fig-12]. In this regard, Tatic, et al [30] noticed that a significant differences between soybean genotypes (Sb1, Sb2, Sb3, Sb4, Sb5 and Sb6) in oil percentage, acid number, peroxide number and moisture percentage. These differences between genotypes might be due to the genetic factors and seed chemical composition influence the expression of seed deterioration and vigor decline. These results are in good agreement with those reported by Balesevic-Tubic, et al [7], Erandes, et al [13], Jamro, et al [17], Khaliliaqdam, et al [19], Mohammadi, et al [21], Vieira, et al [34].

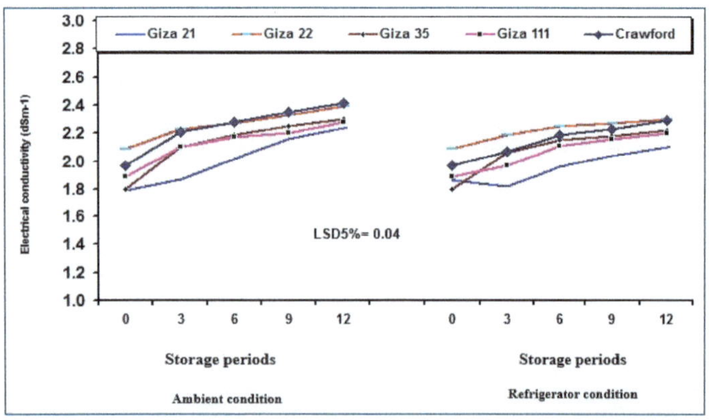

Fig. 12- Averages of electrical conductivity (dSm^{-1}) as affected by the interaction between storage periods, soybean cultivars and storage conditions.

Interaction between Storage Periods and Package Materials

Acid and peroxide numbers significantly affected due to the interaction between storage periods and package materials.

The results clearly showed that pre storage treatment recorded the lowest acid number followed by storage for3 months [Fig-13]. The highest acid number was recorded from storage soybean cultivars for 12 months seeds in cloth bags. Pre storage treatment recorded the lowest peroxide number followed by storage for 3 months [Fig-14]. The highest acid number was recorded from storage soybean cultivars for 12 months seeds in plastic bags. The oil peroxide index differed significantly among samples due to the interaction between moisture content, temperature and storage period [3]. These results are in conformity with those reported by Khaliliaqdam, et al [19], Muhammad [23].

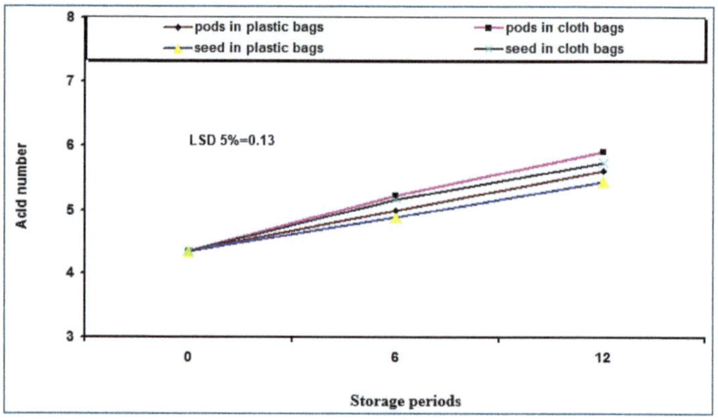

Fig. 13- Averages of acid number as affected by the interaction between storage periods and package materials.

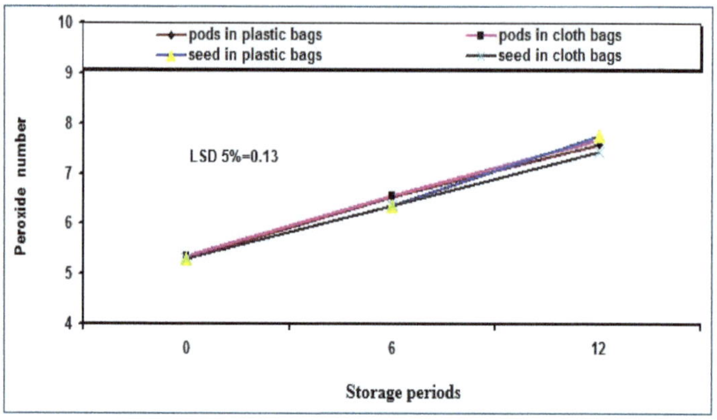

Fig. 14- Averages of peroxide number as affected by the interaction between storage periods and package materials.

Interaction between Soybean Cultivars and Package Materials

Highest electrical conductivity was obtained from storage Giza 22 cultivar pods storage in cloth bags or seeds in plastic bags. While, lowest values of electrical conductivity was obtained from Giza 21 cultivar storage as pods in plastic or cloth bags [Fig-15].

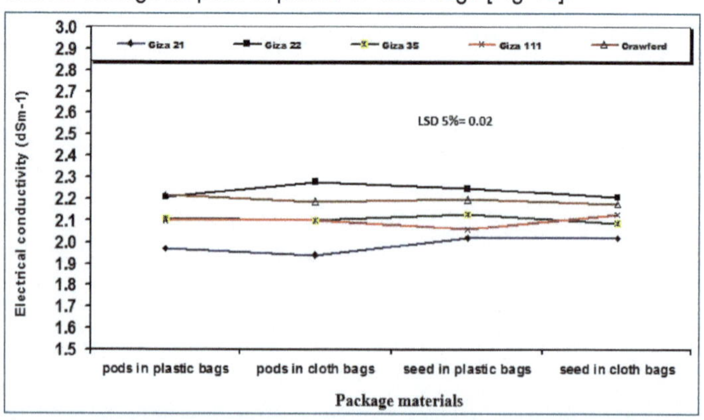

Fig. 15- Averages of electrical conductivity (dSm-1) as affected by the interaction between soybean cultivars fand package materials.

The results clearly showed that the highest acid number were produced from storage Giza 35cultivaras pods or seeds in cloth bags

[Fig-16]. The lowest acid number was obtained from storage Giza 22cultivarseeds in plastic or cloth bags without significant differences between them. Highest peroxide number was produced from storage Giza 35cultivaras pods in plastic or cloth bags [Fig-17]. The lowest peroxide number was obtained from storage Giza 22cultivarseed in cloth bags. These results are in conformity with those reported by Balesevic-Tubic, et al [7], Erandes, et al [13], Jamro, et al [17], Khaliliaqdam, et al [19], Mohammadi, et al [21].

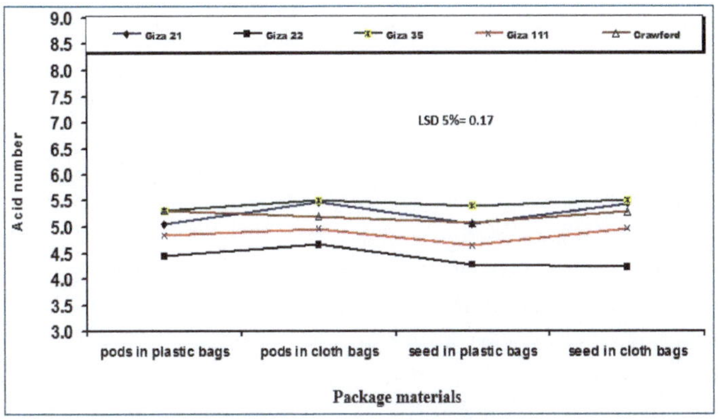

Fig. 16- Averages of acid number as affected by the interaction between soybean cultivars and package materials.

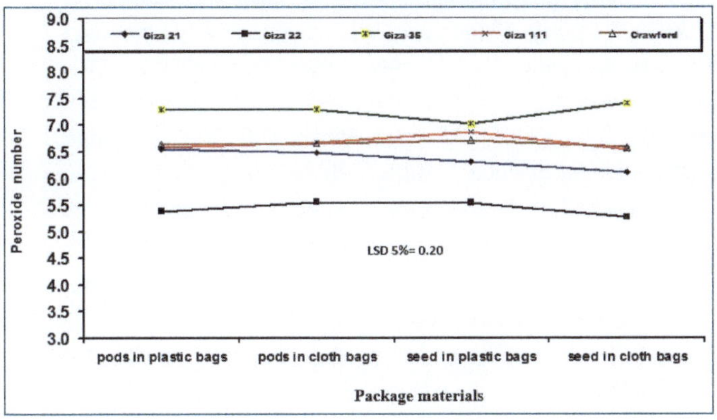

Fig. 17- Averages of peroxide number as affected by the interaction between soybean cultivars and package materials.

Interaction between Storage Periods, Soybean Cultivars and Package Materials

The results clearly showed that highest electrical conductivity was obtained from storage Giza 22cultivarseed in plastic bags for 12 months.

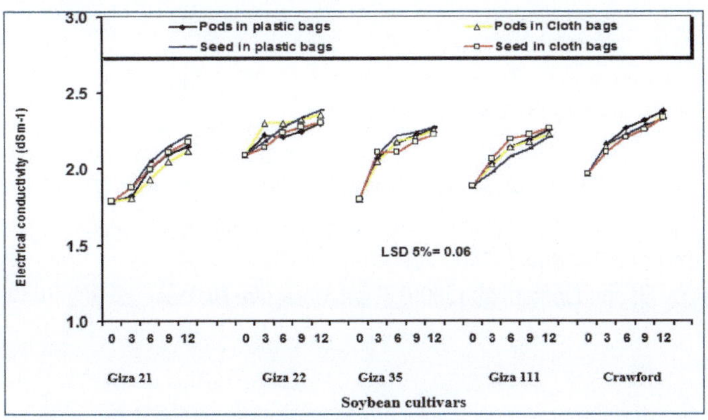

Fig. 18- Averages of electrical conductivity (dSm-1) as affected by the interaction between soybean cultivars, package materials and storage periods.

The lowest electrical conductivity was obtained from pre storage of Giza 21 and Giza 35 cultivars [Fig-18]. The results clearly showed that highest acid number was produced from storage Giza 35 cultivar as pods or seeds in cloth bags for 12 months [Fig-19]. The lowest acid number was obtained from pre storage of Giza 22 cultivar. The results clearly showed that highest peroxide number were produced from storage Giza 35 cultivar as seed in cloth bags for 12 months [Fig-20]. The lowest peroxide number was obtained from pre storage of Giza 35 cultivar. These results are in good agreement with those reported by Balesevic-Tubic, et al [7], Erandes, et al [13], Jamro, et al [17], Khaliliaqdam, et al [19], Mohammadi, et al [21], Vieira, et al [34].

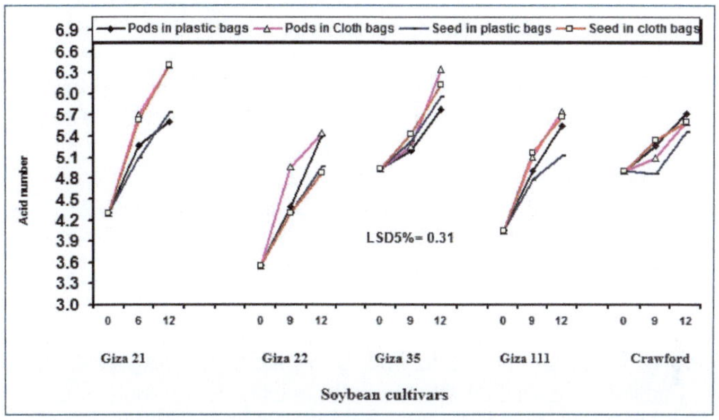

Fig. 19- Averages of acid number as affected by the interaction between storage periods, soybean cultivars and package materials.

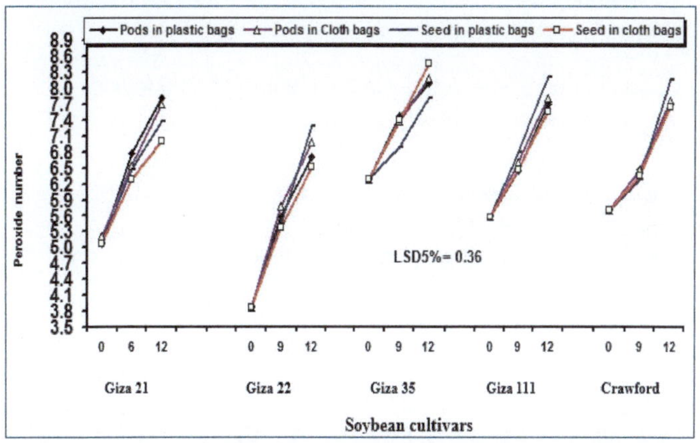

Fig. 20- Averages of peroxide number as affected by the interaction between storage periods, soybean cultivars and package materials

Interaction between Storage Conditions and Package Materials

The results clearly showed that highest electrical conductivity was obtained from storage soybean cultivars as pods in cloth bags under refrigerator conditions (10°C±1). The lowest electrical conductivity was obtained from storage soybean cultivars pods in cloth bags underambient conditions [Fig-21].

In this respect, Chuansin, et al [10] found that soybean seeds were stored in Metallized film bags and Aluminum foil bags keep water activity and seed moisture content in low level could delay seed quality deterioration followed by Polypropylene bags and woven bag. The oil peroxide index differed significantly among samples due to the interaction between moisture content, temperature and storage period [3]. These results in good harmony with those reported by Balesevic-Tubic, et al [7], Erandes, et al [13], Jamro, et al [17], Khaliliaqdam, et al [19], Mohammadi, et al [21].

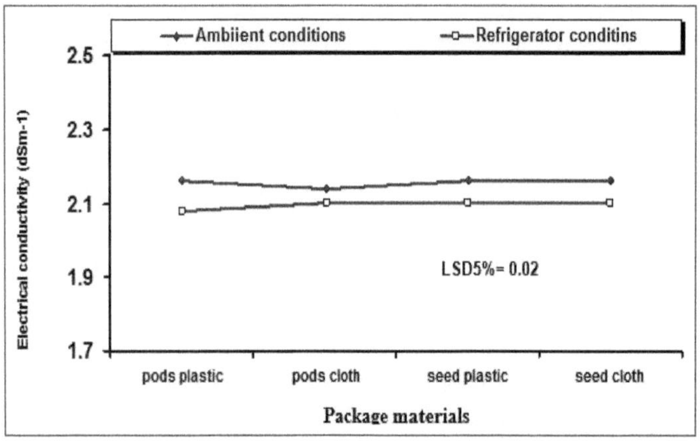

Fig. 21- Averages of electrical conductivity (dSm-1) as affected by the interaction between storage conditions and package materials

Conclusion

The results indicated that for maximizing soybean seed quality, it could be storage Giza 111 cultivar under refrigerator conditions (10°C±1) aspods or seeds in cloth bags for 3 months.

Acknowledgment

Sincere and gratitude thanks to Prof. Dr. Ali E. Sharief, Agronomy Dep., Faculty of Agriculture, Mansoura University, for possible help and extraordinary effort during the achievement of this investigation andhis guidance and encouragement during the extraordinary effort during this work.

References

[1] Association of Official Agric. Chem. (1996) *Official Methods of Analysis*, Washington, USA.

[2] Alencar E.R. and Faronil R.D. (2011) *Recent Trends for Enhancing the Diversity and Quality of Soybean Products*, 47-67.

[3] Alencar E.R., Faronil R.D., Peternellil A., Silva M.T.C. and Costa A.R. (2010) *Revista Bras. Eng. Agríc. Ambiental.*, 14(3), 303-308.

[4] Gupta A.A. and Aneja K.R. (2004) *Seed Research*, 32(1), 26-32.

[5] Arefi H.M. and Abdi N. (2003) *Iranian J. Rangelands and Forests Plant Breeding and Genetic Res.*, 11, 105-125.

[6] Arif M. (2006) *Khutab Al Nabi (Prophets Sermons) Collection, Compilation, Research, Interpretation, Explanation, Unraveling Strange Words & Determining the Date*, Ph.D. Dissertation, NWFP Agricultural University Peshawar, Pakistan.

[7] Balesevic-Tubic S., Malencic D., Tatic M. and Miladinovic J. (2005) *Hellia*, 28(42), 107-114.

[8] Balesevic-Tubic S., Tatic M., Dordevic V., Nikolic Z. and Dukic V. (2010) *Helia*, 33(52), 153-160.

[9] Charjan S. and Tararj L. (1992) *Indian Journal Agriculture Sciences*, 62, 500-502.

[10]Chuansin S., Vearasilp S., Srichuwong S. and Pawelzik E. (2006) *Conference on International Agricultural Research for Development*, University of Bonn, Bonn, Germany.

[11]Doijoide S.N. (1988) *Seed Research*, 1, 245-247.

[12]El-Abady M.I., El-Emam A.A.M., Seadh S.E. and Yousof F.I. (2012) *Research Journal of Seed Science*, 3, 1-11.

[13]Erandes R.D., Leda R.D.F., Luiz A. and Macro T.C. (2010) *Revista Brasileira de Engenharia Agricola*, 14 (3), 303-308.

[14]Ghasemnezhad A. and HonermeierB. (2007) *Int. J. Plant Prod.*, 3(4), 41-50.

[15]Hou H.J. and Chang K.C. (2004) *Journal of Agricultural and Food Chemistry*, 46, 4798-4805.

[16]International Seed Testing Association (1996) *Seed Science and Technology*, 24, 1-335.

[17]Jamro G.H., Hatam M., Jamali L.A. and Agha S.K. (2005) *Agri. Mech. in Asia, Africa and Latin America*, 36, 15-17.

[18]Kalavathi D.K. and Sundareswaran S. (1994) *Madras Agricultural Journal*, 81(11), 612-615.

[19]Khaliliaqdam N., Soltani A., Latifi N. and Gaderi F.F. (2012) *American-Eurasian Journal of Agriculture and Environments Sciences*, 12(2), 224-230.

[20]Milošević M. and Malešević M. (2004) *Naučni institut za ratarstvo i povrtarstvo, Novi Sad*, 110-114.

[21]Mohammadi H., Soltani A., Sadeghipour H.R. and Zeinali E. (2011) *International Journal of Plant Production,* 5(1), 65-70.

[22]Monira U.S., Amin M.H.A., Marin M. and Mamun M.A.A. (2012) *Bangladesh Research Publications Journal*, 7(4), 421-427.

[23]Muhammad A. (2008) *Effect of Planting Dates and Storage on Yield and Quality of Indigenous Land Races and Improved Varieties of Soyabean, Ph.D. Thesis*, Agri. Uni. Peshawar, Pakistan.

[24]Owolade O.F., Olasoji J.O. and Afolabi C.G. (2011) *African Journal of Plant Science*, 5(15), 873-877.

[25]Shanmugavel S.A. and Dadlani M. (1995) *Seed Research*, 23 (2), 61-66.

[26]Sharma S., Gambhir S. and Manshi S.K. (2007) *Asian Journal of Plant Sciences*, 6(3), 502-507.

[27]Shelar V.R., Shaikh R.S. and Nikam A.S. (2008) *Agric. Rev.*, 29 (2), 125-131.

[28]Šimić B., Popović R., Sudarić A., Rozman V., Kalinović I. and Ćosić J. (2007) *Agric. Conspectus Sci.*, 72(3), 211-213.

[29]Simic B., Sudaric A., Liovic I., Kalinovic I., Rozman V. and Cosic V. (2006) *9th International Working Conference on Stored Product Protection*, 59-63.

[30]Tatic M., Balesevic-Tubic S., Dordevic V., Miklic V., Vujakovic M. and Dukic V. (2012) *Helia*, 35(56), 119-126.

[31]Tatipata A. (2009) *Journal of Agricultural Technology*, 5(1), 51-64.

[32]Venkatareddy D.M., Bommegowda A. and Paramesh D.S. (1992) *Seed Tech. News*, 22(2), 41-46.

[33]Vieira R.D., TeKrony D.M., Egli D.B, Bruenning W.P. and Panobianco M. (2008) *Sci. Agric. (Piracicaba, Braz.)*, 65(5), 496-501.

[34]Vieira R.D., Tekrony D.M., Egli D.B. and Rucker M. (2001) *Seed Science and Technology*, 29, 599-608.

[35]Vijay D. and Dadlani M. (2009) *Plant Molecular Biology*, 22, 282 -291.

A STUDY ON PHYSICOCHEMICAL PROPERTIES OF SOILS UNDER DIFFERENT TEA GROWING REGIONS OF WEST BENGAL (INDIA)

RAY S.K. AND MUKHOPADHYAY D.*

Department of Soil Science and Agricultural Chemistry, Uttar Banga Krishi Viswavidyalaya, Pundibari, CoochBehar- 736165, WB, India.
*Corresponding Author: Email- dibsm107@gmail.com

Abstract- The physicochemical properties of soil (0-0.20 m) were analysed to evaluate the characteristics of tea growing soils under three different locations of West Bengal (India), considering the age and elevation of tea plantation. The soil samples of *Dooars* and *Terai* region of West Bengal were collected on the basis of age (young, medium and old) of the tea plants, while that from Darjeeling region as organic and non-organic tea growing soils. The organic and non-organic tea soils were collected on the basis of elevations of the sites. The soils of the *Dooars* region were clay to sandy loam in texture whereas, soils of *Terai* and Darjeeling were sandy loam in texture. The selected soils were strong to moderately acidic in reaction with low electrical conductivity (EC), Ca^{+2} and Mg^{+2} content. The organic carbon content of different regions was found medium to high, but very little variation was obtained with organic tea growing regions of Darjeeling. The soil available N and P content were low to medium in all the regions but higher available K content were found with the soils of *Dooars* and *Terai* regions. The cation exchange capacity (CEC) of soils varied from low to medium. The correlation study indicated that CEC, available N and K were influenced by soil organic carbon content, while the available P, Ca^{+2} and Mg^{+2} content by the soil pH. The available N, K, EC and CEC were negatively influenced by sand content of the soils.

Keywords- Physicochemical, electrical conductivity, texture and cation exchange capacity

Introduction

In West Bengal tea grows mostly in the districts of Darjeeling and Jalpaiguri. The major river (Jaldhaka, Tista, Torsha, Raidak, Sankosh) basin under North Bengal situation have large catchment area in the hills where the rivers cut down the ridge and mountain to come down to the plains of *Terai* and *Dooars* area [1]. In Darjeeling district, tea is mainly grown in regions of mountain slopes at an altitude of 2,000 metres and in Jalpaiguri district, it is grown mainly in *Terai* and *Dooars* regions. These sub-mountain tracts are characterized by deep steep sided valley, separated by terraced high lands, immediately to the south of the Himalayas. Tea naturally grows in tropical to temperate conditions, where the annual rainfall is more than 200 cm [2]. Tea performs best under acid soil conditions with well drained sloping ground and porous soil status for independent utilization and adsorption of nutrients. Tea grows well within the range of ideal physical and environmental conditions having heterogeneous soil and climatic conditions, for which tea growing regions/soils becomes fairly wide in character, that influence the soil physical and chemical nature. Evaluations of soil fertility status under tea growing soils are necessary to make a proper management and sound fertilizer recommendation for optimizing the yield of tea. Some of the areas in the lower foot hills below the Piedmont zone are prone to deposition of alluvial sediments and sands [3]. Hence, an attempt was made to study the important physicochemical properties of soils collected on the basis of age and elevation of the tea-plants under different tea growing regions of West Bengal.

Materials and Methods

Fifty four surface soil (0-0.2m) samples from eighteen tea gardens comprising six from each region were collected from two districts, namely, Jalpaiguri and Darjeeling of West Bengal. The soils of the Dooars and Terai region under Jalpaiguri districts were collected considering the age of the tea- plants (Young i.e., 10-20years, Medium i.e. 20-40 years and Old i.e., more than 50 years), while that from Darjeeling as organic and non-organic tea growing soils. The organic and non-organic tea soils were collected on the basis of elevations (low, medium and high) of the tea plantation. Soil samples were collected from Lakhipara, Haldibari, Nagrakata, Ghatia, Chalouni and Batabari tea gardens under Dooars region, while from the *Terai* region, Hansqua, Taipoo, Sayedabad, Mothidhar, Gayaganga and Lohagarh gardens were selected for collection of soil samples. Soil samples were collected from non-organic gardens of Darjeeling and under *Dooars* and *Terai* region. The

Sungma, Salimbong and Avongrove tea gardens were selected as organic and Glenburn, Bernesbeg and Namring as non-organic tea gardens. The important parameters (pH, EC, texture, available-N, P and K, CEC, exchangeable Ca^{+2} and Mg^{+2}) were determined following the standard methodology [4].

Statistical Analysis

Data were analysed statistically by using the software SPSS (Version 12.0).

Results and Discussion

Physicochemical Properties of Soils of *Dooars* Area

The soils (0-20 cm) were recorded to be strong to moderately acidic in reaction [Table-1] and non-saline in nature. The soils varied from clay to sandy loam in texture. Among the six tea gardens of Dooars regions, the soils of Nagrakata and Ghatia tea garden were predominantly clay to clay loam in texture, whereas, the soils of Lakhipara, Haldibari, Chalouni and Batabari were sandy loam to sandy clay loam in texture. The organic carbon varied from 19.4 to 28.4 g kg^{-1} in young, 14.9 to 28.4 g kg^{-1} in medium and 15.4 to 26.5 g kg^{-1} in old aged soils of *Dooars* region. The organic carbon was found relatively higher in young aged section of Ghatia and medium aged section of Chalouni tea garden. The cation exchange capacity (CEC) was low to medium and varied from 5.38 to 11.17 cmol (p^+) kg^{-1} in young, 5.11 to 13.8 cmol (p^+) kg^{-1} in medium and 4.39 to 11.56 cmol (p^+) kg^{-1} in old aged sections. The CEC of the medium aged sections of Nagrakata tea soil was higher, might be due to the presence of higher clay and least was in young aged sections of Batabari tea soil. The soil available N content varied from 78.56 to 203.55 kg ha^{-1} in the young, 78.56 to 245.55 kg ha^{-1} in medium and 95.96 to 202.59 kg ha^{-1} in old aged sections of Dooars. The available nitrogen status of all the soils were generally low to medium and this variation might be due to leaching loss of N from the surface soil. Available P content varied between 17.25 to 51.22 kg ha^{-1} in young, 19.22 to 56.73 kg ha^{-1} in medium and 12.56 to 56.73 kg ha^{-1} in old aged soils of Dooars. Higher P content was found with medium aged section of Lakhipara and the least was with old aged sections of Batabari. The available K content was found medium to high. Maximum potassium content was found with medium aged section of Nagrakata and the least was with young aged sections of Batabari tea soil.

Table 1- Important physicochemical properties of soils (0-20cm) of Dooars tea - growing region of Jalpaiguri Districts

Name of the Gardens	Age (years)	Clay (%)	Silt (%)	Sand (%)	Texture	pH (1:2.5)	EC (dSm^{-1})	Organic Carbon (g kg^{-1})	Available-N (kg ha^{-1})	Available-P (kg ha^{-1})	Available-K (kg ha^{-1})	CEC [cmol (p+)kg^{-1}]	Ca^{+2} [cmol (p+)kg^{-1}]	Mg^{+2} [cmol (p+)kg^{-1}]
	10-20	16.6	26.2	57.2	Sandy Loam	4.61	0.18	19.4	99.45	17.25	214.67	6.22	1.56	0.1
Lakhipara	20-40	16.6	20.2	63.2	Sandy Loam	4.97	0.15	19.8	102.56	56.73	261.33	8.56	0.15	0.12
	>50 yrs.	32.6	20.2	47.2	Sandy Clay	4.94	0.14	25.5	95.96	55.32	373.33	12.56	0.23	0.18
	10-20	16.6	18.2	65.2	Sandy Loam	4.93	0.25	19.5	78.56	51.22	289.33	7.12	0.12	0.91
Haldibari	20-40	18.6	26.2	55.2	Sandy Loam	4.84	0.12	14.9	78.56	43.04	153.07	8.54	0.23	0.91
	>50 yrs.	20.6	16.2	63.2	Sandy Clay loam	4.62	0.13	15.4	106.62	30.56	168	9.56	0.89	0.37
	10-20	30	29	41	Clay Loam	4.48	0.12	19.9	111.96	25.66	392	12.56	0.74	0.4
Nagrakata	20-40	35	39	26	Clay	4.6	0.11	22.7	116.25	30.12	709.33	13.8	1.11	2.29
	>50 yrs.	32.6	28.2	39.2	Clay	4.33	0.22	17	102.56	22.55	298.67	12.11	2.42	0.51
	10-20	39	36	25	Clay	4.62	0.09	28.4	159.94	29.34	599.83	10.75	1.22	3.02
Ghatia	20-40	38	38	24	Clay	4.42	0.09	15.1	89.56	23.12	401.33	11.47	2.56	0.61
	>50 yrs.	26	29	45	Clay Loam	4.46	0.05	26.5	202.59	25.22	476	8.56	1.25	0.4
	10-20	48.6	20.2	31.2	Clay	4.6	0.07	22.7	203.55	23.22	289.33	11.2	2.92	0.19
Chalouni	20-40	24.6	20.2	55.2	Sandy Clay loam	4.7	0.05	28.4	245.55	19.22	429.33	7.88	3.23	1.24
	>50 yrs.	26	34	40	Clay Loam	4.7	0.08	20.8	154.6	29.56	476	8.6	3.33	0.82
	10-20	19	27	54	Sandy Loam	4.68	0.12	20.8	127.95	33.25	130.67	5.38	1.22	0.09
Batabari	20-40	22	27	51	Sandy Clay Loam	4.45	0.11	17	127.95	26.25	181.07	7.44	2.66	0.27
	>50 yrs.	23	39	38	Clay Loam	4.4	0.09	17	109.94	12.56	143.73	7.23	2.56	0.09

The soil pH was positively correlated [Table-2] with available phosphorus (r = 0.855**) and sand content (r = 0.544*) and negative correlation was obtained with silt (r = -0.555*) and available calcium (r = -0.582*). The EC had a negative correlation with available nitrogen (r = -0.696**). The sand content had significant negative correlation with CEC (r = -0.631**) and potassium (r = -0.612*). Soil organic carbon had a significant positive correlation with available nitrogen (r = 0.711**) and available potassium (r = 0.629**), suggesting that soil organic carbon increased the availability of nutrients under tea growing soils. Strong and positive correlation was observed between cation exchange capacity and clay content (r = 0.748**) which indicated the cations adsorption at the exchange sites of the clay. The CEC had also positive correlation with available potassium content (r = 0.604*). The available calcium content had a significant positive correlation with available nitrogen (r = 0.543*) and negative correlation with available phosphorus (r =-0.742**). The significant positive correlation was observed between magnesium and available potassium (r = 0.744**) of the soil.

Table 2- Correlation matrix among soil physicochemical properties under Dooars region

Soil Properties	Clay	Silt	Sand	pH	EC	OC	N	P	K
Sand	-0.853**	-0.763**	1						
pH	NS	-0.555*	0.544*	1					
N	NS	NS	NS	NS	-0.696**	0.711**	1		
P	NS	-0.471*	NS	0.855**	NS	NS	NS	1	
K	0.545*	NS	-0.612**	NS	NS	0.629**	NS	NS	1
CEC	0.748**	NS	-0.631**	NS	NS	NS	NS	NS	0.604**
Ca^{+2}	NS	NS	NS	-0.582*	NS	NS	0.543*	-0.742**	NS
Mg^{+2}	NS	NS	NS	NS	NS	NS	NS	NS	0.744**

** Significant at 5% level; ** Significant at 1% level*

Physicochemical Properties of Soils Under *Terai* Region

Soils of *Terai* were strong to moderately acidic [Table-3] in reaction with low EC. Soils were sandy loam to clay in texture. The clay content varied from 7.6 to 20.0 percent in young, 6.6 to 40.4 percent in medium and 6.6 to 20.4 percent in old aged sections. The organic carbon content under *Terai* region varied from 15.1 to 41.2 g kg^{-1} in young, 17.4 to 44.6 g kg^{-1} in medium and 15.9 to 43.1 g kg^{-1} in old aged soils. The highest organic carbon was obtained with Lohagarh tea garden and the lowest in Hansqua tea estate. The CEC was low to medium in *Terai* soils under study and it varied from 4.20 to 9.79 cmol (p^+) kg^{-1} in young, 4.55 to 10.60 cmol (p^+) kg^{-1} in medium and 5.56 to 8.95 cmol (p^+) kg^{-1} in old aged sections.The highest CEC was obtained with medium aged sections of Mothidhar, might be due to higher clay content, and least with young aged sections of Sayedabad. The available soil N content varied from 47.04 to 210.74 kg ha^{-1} in the young, 40.32 to 226.12 kg ha^{-1} in medium and 65.86 to 286.00 kg ha^{-1} in old aged sections. Available P content varied between 3.91 to 41.08 kg ha^{-1} in young, 1.96 to 32.55 kg ha^{-1} in medium and 3.91 to 22.12 kg ha^{-1} in old aged soils and wide variation of soil available P might be due to the fixation of P in soil as insoluble aluminium and iron phosphate. Available K content under *Terai* region varied from the medium to high. Higher potassium content was found in both medium and young aged sections of Hansqua and Taipoo tea soils respectively. The Ca^{+2} and Mg^{+2} content were low, might be due to higher acidity and poor application of liming materials in soils.

Table 3- Important physicochemical properties of soils (0-20cm) under Terai tea - growing region of Jalpaiguri Districts

Name of the Gardens	Age (years)	Clay (%)	Silt (%)	Sand (%)	Texture	pH (1:2.5)	EC (dSm^{-1})	Organic Carbon (g kg^{-1})	Available-N (kg ha^{-1})	Available-P (kg ha^{-1})	Available-K (kg ha^{-1})	CEC [cmol $(p+)kg^{-1}$]	Ca^{+2} [cmol $(p+)kg^{-1}$]	Mg^{+2} [cmol $(p+)kg^{-1}$]
	10-20	8.6	26.2	65.2	Sandy Loam	4.35	0.05	17.3	47.04	6.25	448.00	7.89	0.25	0.41
Hansqua	20-40	6.6	27.2	66.2	Sandy Loam	4.70	0.03	18.9	105.23	22.56	485.33	8.89	2.16	1.22
	>50 yrs.	7.6	27.2	65.2	Sandy Loam	4.38	0.04	15.9	65.86	7.84	345.33	6.47	0.45	0.42
	10-20	7.6	25.2	67.2	Sandy Loam	4.41	0.05	15.1	98.56	9.22	485.33	9.79	0.77	0.33
Taipoo	20-40	7.6	19.2	73.2	Sandy Loam	4.42	0.06	17.4	40.32	8.92	235.44	4.55	1.22	1.22
	>50 yrs.	6.6	26.2	67.2	Sandy Loam	4.50	0.07	23.5	89.25	19.25	373.33	7.97	1.03	1.03
	10-20	18.4	19.2	62.6	Sandy Loam	4.38	0.02	15.6	108.45	41.08	242.67	4.20	0.57	0.33
Sayedabad	20-40	14.0	20.4	65.6	Sandy Loam	4.65	0.03	19.4	180.63	32.55	317.33	5.62	1.23	0.66
	>50 yrs.	12.0	22.8	65.2	Sandy Loam	4.36	0.05	17.5	125.45	3.91	164.27	6.90	0.23	1.26
	10-20	20.0	34.0	46.0	Clay Loam	4.75	0.09	27.6	126.50	24.45	189.23	8.99	1.55	2.01
Mothidhar	20-40	40.4	21.6	38.0	Clay	4.33	0.08	22.4	133.54	20.54	448.00	10.60	0.25	0.65
	>50 yrs.	20.4	21.6	58.0	Sandy Clay Loam	4.40	0.05	23.5	122.56	22.12	322.56	7.30	0.63	0.36
	10-20	14.0	28.8	57.2	Sandy Loam	4.13	0.02	17.9	98.26	14.67	420.00	6.10	0.44	0.23
Gayaganga	20-40	16.4	11.6	72.0	Sandy Loam	4.80	0.03	20.9	122.45	25.22	261.33	4.56	1.25	0.21
	>50 yrs.	20.0	26.4	53.6	Sandy Clay Loam	4.00	0.05	18.9	102.56	10.22	420.00	8.95	0.44	0.13
	10-20	16.0	17.6	66.4	Sandy Loam	4.30	0.06	41.2	210.74	3.91	298.67	7.80	0.36	0.37
Lohagarh	20-40	12.0	27.6	60.4	Sandy Loam	4.25	0.07	44.6	226.12	1.96	280.00	7.70	1.35	0.13
	>50 yrs.	12.0	14.4	73.6	Sandy Loam	4.10	0.02	43.1	286.00	7.25	2.13	5.56	0.58	0.26

The pH had a significant positive correlation [Table-4] with available P (r = 0.538*), Calcium (r = 0.677**) and Magnesium (r = 0.554*), which indicated that availability of P, Ca^{+2} and Mg^{+2} were dependent on soil acidity. Sand content had significant negative correlation with EC (r= -0.529*) and CEC (r= -0.609*).The EC was positively related to CEC (r= 0.616**) and Mg^{+2} (0.473*). Strong and positive correlation was observed between organic carbon and available N (r= 0.852**) and negative correlation with available K (r= -0.488*). The CEC had the significant positive correlation with silt (r= 0.557*), EC (r= 0.616**) and available K (r= 0.558*).

Table 4- Correlation matrix among soil physicochemical properties under Terai regions

Soil Properties	Clay	Silt	Sand	pH	EC	OC	N	K
Sand	-0.802**	-0.491*	1					
EC	NS	NS	-0.529*	NS	1			
N	NS	NS	NS	NS	NS	0.852**	1	
P	NS	NS	NS	0.538**	NS	NS	NS	
K	NS	NS	NS	NS	NS	-0.488*	-0.545*	1
CEC	NS	0.557*	-0.609**	NS	0.616**	NS	NS	0.558*
Ca^{+2}	NS	NS	NS	0.677**	NS	NS	NS	NS
Mg^{+2}	NS	NS	NS	0.554*	0.473*	NS	NS	NS

** Significant at 5% level; ** Significant at 1% level*

Physicochemical Properties Under Organic Tea Growing Soils of Darjeeling

Soils of organic tea growing gardens were found to be strong to moderately acidic in reaction, with low electrical conductivity (EC), might be due to leaching of soluble salts by heavy rainfall. Soils under study were sandy loam to sandy clay loom in texture and the clay content varied from 14.8 to 20.8, 14.8 to 18.8 and 16.8 to 22.8 percent at the low medium and high elevations respectively [Table 5]. Data pertaining to CEC ranged from 5.89 to 8.88, 5.97 to 8.96 and 5.89 to 13.89 cmol (p+) kg^{-1} in soils of low, medium and high elevations respectively.

The organic carbon content was high and ranged between 20.3 and 37.6, 19.1 and 29.9 and 21.3 and 28.6 g kg^{-1} of soils in low, medium and high elevations respectively and these variations might be due to the effect of management practices of these locations. The available N content varied from 40.14 to 189.63 kg ha^{-1} in low, 30.11 to 89.53 kg ha^{-1} in medium and 42.65 to 142.20 kg ha^{-1} in high elevations respectively. The lower available N in soil was found, might be due to less mineralization under low temperature conditions. The availability of soil P varied from 17.20 to 150.26 kg ha^{-1}, 18.14 to 22.80 kg ha^{-1} and 5.11 to 49.30 kg ha^{-1} under low, medium and high elevations respectively. The higher availability of phosphorus under lower elevation was observed. The available soil potassium content in organic tea growing areas of the Darjeeling regions was low to medium and the K content varied from 122.36 to 325.67 kg ha^{-1}, 105.23 to 380.37 kg ha^{-1} and 160.84 to 355.45 kg ha^{-1} in low, medium and high elevations, respectively. The soil Ca^{+2} and Mg^{+2} content were found low in all the elevations.

Soil pH had a strong and positive correlation [Table-6] with Calcium (r= 0.891**) and Magnesium (r= 0.807**) and negative correlation with available nitrogen (r= -0.484*).The EC had a significant negative correlation with organic carbon (r= -0.534*), phosphorus (r= -

0.591*) and available potassium (r= -0.658*). The CEC had a significant positive correlation with clay (r= 0.640*), silt (r= 0.633*) and potassium (r= 0.596*) and significant negative correlation with sand content (r= -0.775**). The strong and positive correlation was observed between Ca^{+2} and Mg^{+2} (r= 0.910**) under organic tea soil.

Table 5- Important physicochemical properties under organic tea growing soils (0-20cm) of Darjeeling

Name of the Gardens	Elevation (mt)	Clay (%)	Silt (%)	Sand (%)	Texture	pH (1:2.5)	EC (dSm^{-1})	Organic Carbon ($g\ kg^{-1}$)	Available-N ($kg\ ha^{-1}$)	Available-P ($kg\ ha^{-1}$)	Available-K ($kg\ ha^{-1}$)	CEC [cmol (p+)kg^{-1}]	Ca^{+2} [cmol (p+)kg^{-1}]	Mg^{+2} [cmol (p+)kg^{-1}]
	1137	14.8	16.2	69.0	Sandy Loam	4.56	0.02	37.6	189.36	150.26	322.25	5.89	0.03	0.04
Sungma	1367	18.8	24.2	57.0	Sandy Loam	4.97	0.06	29.9	89.56	19.53	380.37	8.96	0.33	0.19
	1498	19.8	27.2	53.0	Sandy Loam	4.68	0.05	24.8	104.25	49.30	355.45	13.89	0.30	0.12
	1214	17.8	15.2	67.0	Sandy Loam	5.21	0.02	23.7	40.14	27.44	325.67	8.88	0.25	0.09
Salimbong	1446	20.8	14.2	65.0	Sandy Clay Loam	4.96	0.05	19.1	51.43	18.14	145.23	6.80	0.30	0.10
	1534	16.8	22.2	61.0	Sandy Loam	5.32	0.06	21.3	42.65	11.16	204.31	5.89	0.43	0.27
	825	14.8	18.2	67.0	Sandy Loam	4.53	0.08	20.3	176.87	17.20	122.36	6.00	0.06	0.01
Avongrove	1149	14.8	18.2	67.0	Sandy Loam	4.31	0.09	21.0	30.11	22.80	105.23	5.97	0.02	0.03
	1501	22.8	22.2	55.0	Sandy Clay Loam	4.53	0.06	28.6	142.20	5.11	160.84	8.56	0.12	0.04

Table 6- Correlation matrix among soil physicochemical properties under organic tea growing regions of Darjeeling

Soil Properties	Clay	Silt	Sand	pH	EC	OC	N	K	Ca^{+2}
Sand	-0.716*	-0.889**	1						
OC	NS	NS	NS	NS	-0.534*	1			
N	NS	NS	NS	-0.484*	NS	0.593*	1		
P	NS	NS	NS	NS	-0.591*	0.749*	0.522*		
K	NS	NS	NS	NS	-0.658*	0.594*	NS	1	
CEC	0.640*	0.633*	-0.775*	NS	NS	NS	NS	0.596*	
Ca^{+2}	NS	NS	NS	0.891**	NS	NS	-0.528*	NS	1
Mg^{+2}	NS	NS	NS	0.807**	NS	NS	NS	NS	0.910**

** Significant at 5% level; ** Significant at 1% level*

Physicochemical Properties under Non-organic Tea Growing Soils of Darjeeling

The non-organic tea growing soils of Darjeeling districts were strong to moderately acidic in reaction with pH ranging from 4.51 to 5.33, 4.33 to 5.09 and 4.44 to 5.44 in low, medium and high elevations respectively [Table-7]. The EC of all the soils were low. The texture of the soil under study varied from sandy loam to clay but most of the soils were sandy loam in texture. The clay content varied from 14.8 to 35.2, 18.0 to 41.2 and 7.6 to 29.2 percent under low, medium and high elevations respectively. The organic carbon was medium to high, varying from 18.7 to 28.7, 16.5 to 27.9 and 18.2 to 30.2 g kg^{-1} under low, medium and high elevations respectively. The CEC varied from 4.12 to 8.76 cmol (p+) kg^{-1}, 4.18 to 10.22 cmol (p+) kg^{-1} and 3.80 to 9.50 cmol (p+) kg^{-1} in low, medium and high elevations respectively. The available soil nitrogen content was very low to low and it may be due to less inputs of fertilizer nitrogen and or organic amendments. The available P content of the soils varied widely from low to high. It was from 78.96 to 152.12 kg ha^{-1} in low, 18.61 to 167.93 kg ha^{-1} in medium and 62.35 to 180.03 kg ha^{-1} in high elevations respectively. The available K content in soils varied from 173.88 to 326.12 kg ha^{-1} in low, 122.56 to 445.57 kg ha^{-1} in medium and 86.94 to 225.13 kg ha^{-1} in high elevations respectively. These large variations in available K content might be due to the non-uniform application of K fertilizers. The Ca^{+2} and Mg^{+2} content in soils were low due to the surface leaching of basic cations by the excessive rainfall. The steep terrain physiography sustain quick run off of surface water.

Table 7- Important physicochemical properties under non-organic tea growing soils (0-20cm) of Darjeeling

Name of the Gardens	Elevation (mt)	Clay (%)	Silt (%)	Sand (%)	Texture	pH (1:2.5)	EC (dSm^{-1})	Organic Carbon ($g\ kg^{-1}$)	Available-N ($kg\ ha^{-1}$)	Available-P ($kg\ ha^{-1}$)	Available-K ($kg\ ha^{-1}$)	CEC [cmol (p+)kg^{-1}]	Ca^{+2} [cmol (p+)kg^{-1}]	Mg^{+2} [cmol (p+)kg^{-1}]
	444	21.2	25.6	53.2	Sandy Clay Loam	4.74	0.12	28.7	75.26	88.86	213.01	4.12	0.08	0.02
Glenburn	617	25.2	29.6	45.2	Clay Loam	4.59	0.11	27.9	66.48	167.93	122.56	7.60	0.17	0.02
	916	29.2	21.6	49.2	Clay Loam	4.44	0.14	24.1	69.23	62.35	86.94	8.12	0.04	0.03
	495	35.2	39.6	25.2	Clay	4.51	0.21	20.3	48.92	78.96	173.88	4.22	0.05	0.02
Bernesbeg	731	41.2	35.6	23.2	Clay	4.33	0.11	22.5	66.48	18.61	445.57	10.22	0.06	0.01
	849	7.6	32.0	60.4	Sandy Loam	4.53	0.09	30.2	112.36	180.03	225.13	9.50	0.05	0.06
	624	14.8	24.2	61.0	Sandy Loam	5.13	0.13	18.8	56.24	152.12	326.12	8.76	0.12	0.02
Namring	814	18.0	29.6	52.4	Sandy Loam	5.09	0.14	16.5	56.45	82.34	1.26.12	4.18	0.18	0.04
	1066	25.2	25.6	49.2	Clay Loam	5.44	0.11	18.2	58.96	103.73	122.36	3.80	0.17	0.23

Table 8- Correlation matrix among soil physicochemical properties under non-organic tea growing regions of Darjeeling

Soil Properties	Clay	Silt	Sand	pH	EC	OC	K
Sand	-0.922**	-0.724*	1				
pH	NS	NS	0.497*	1			
EC	NS	NS	NS	-0.681*	1		
OC	0.514*	NS	NS	-0.652*	NS	1	
N	0.633*	0.506*	-0.690*	NS	NS	0.776*	
K	0.648*	NS	-0.632*	NS	NS	NS	1
CEC	0.754*	0.561*	-0.804**	NS	NS	NS	0.539*
Ca^{+2}	NS	-0.533*	0.523*	-0.696*	-0.634*	NS	NS

** Significant at 5% level; ** Significant at 1% level*

The clay content had a significant positive correlation [Table-8] with OC (r= 0.514*), available N (r = 0.633*), available K (r= 0.648*) and CEC (r= 0.754**). The silt had a significant positive correlation with available N (r= 0.506*) and CEC (r= 0.561*) and negatively with calcium (r= -0.533*). The soil pH had a significant negative correlation with EC (r= -0.681**), organic carbon (r= -0.652*) and Ca^{+2} content (r= -0.696*). Strong and positive correlation was observed between organic carbon and available N (r= 0.776**), indicating the availability of nitrogen as influenced by the soil organic carbon.

Conclusion

It was observed that different pedogenic processes and related physiography of the tea-growing areas (differences in plant age, elevation, organic and non-organic inputs) had influences on the physicochemical characteristics of the soils (texture, nutrients availability, cation exchange capacity and organic carbon content of the soils). Therefore, this would help taking up appropriate measures on the problem of the tea soils of West Bengal. However, application of more labile organic inputs, liming materials and suitable inorganic fertilizers (N-P-K) would be effective for sustainable management and improving fertility status of the soils under tea-growing region.

References

[1] Jana M.M. (1994) *Workshop of River Behaviour in the North Bengal Rivers*, North Bengal University, India.

[2] Hadfield W. (1968) *Nature*, 219, 282-284.

[3] Jana M.M. (1997) *Journal of the Indian Society of Remote Sensing*, 25, 105-111.

[4] Page A.L., Millner R.H., Keeney D.R. (1982) *Method of Soil Analysis, Part 2: Chemical and Microbiological Properties*, *2nd ed.* ASA and SSSA, Madison, Wisconsin.

23

RESPONSE OF SOME RICE (*Oryza sativa L.*) CULTIVARS TO GERMINATION UNDER SALINITY STRESS

KANDIL A.A.[1], SHARIEF A.E.[1] AND NASSAR E.S.E[2]

[1]Department of Agronomy, Faculty of Agriculture, Mansoura University, Egypt.
[2]Central Administration for Seed certification (CASC), Ministry of Agriculture, Egypt.
*Corresponding Author: Email- sharief2005@yahoo.com

Abstract- The purpose of this study was found out to investigate the response of some rice (*Oryza sativa* L.) cultivars to germination parameters under salinity stress. This study was carried out to confirm rice seedling growth performance and to examine a range of genetic variability for salinity tolerance among ten rice cultivars under seven salinity levels. A laboratory experiment was accompanied at Giza Central Seed Testing laboratory (CASC), Egypt. The results showed that rice cultivars significantly varied in means of Final Germination Percentage (FGP), Germination Rate (GR), Germination Index (GI), Vigor Index (VI), Root Length, Shoot Length, Root Fresh Weight, Shoot Fresh Weight, Root Dry Weight, Relative Dry Weight and Seedling Height Reduction. Sakha 102 and Sakha 106 had positive effects in most of studied characters while, Giza 181 cultivar had negative effects in most of germination parameters under study. Salinity levels significantly varied in all germination parameters under study except shoot dry weight character. Increasing salinity concentration from 0 to 14% NaCl gradually reduced averages of germination and seedling characters. All studied characters were significantly affected by the interaction between cultivars and salinity stress, vice versa, shoot, root fresh weight and root dry weight didn't affect by interaction. It could be established that for exploiting rice germination percentage and seedling parameters under salinity stress are recognized by using rice Sakha 102, Sakha 106, Sakha 104 and Egyptian Yasmin cultivars with increasing salinity levels up to 14% NaCl. These cultivars were more tolerant to salinity and suggested to use in breeding program for attractive rice production in Egypt .
Key words- *Oryza sativa*, salinity concentration, germination characters, rice cultivars, seedling parameters

Introduction

Rice is considered as a major cereal crop for more than 50% of the world population. There is a great shortage in rice production more than 25% of the recent production may be needed [1]. Moreover, rice ranks second after wheat crop as the national income is concerned. The excessive increase in population in Egypt needs to increase the total yield of this crop in order to overcome this lack in production through its cultivation in the newly reclaimed lands especially under saline conditions of such soil. Rice is one of the most effective and commercial means of reclaiming hundreds of thousands of hectares of saline lands in Egypt. Growth characters including shoot height, fresh weight, dry weight and leaf area of salt-stressed seedlings were inhibited, depending on rice genotypes [2]. Rice varieties showed a great variation in germination due to salinity effects. Many investigators concluded that differences among rice cultivars in germination characters and seedling parameters[3,15,17,22].

Salinity is one of the major obstacles in enhancing rice production in growing areas in Egypt. Salinity decreases final germination percentage, speed of germination and led to reduction in shoot and root length and dry weight in all varieties and the magnitude of reduction increased with increasing salinity stress [11]. Therefore, development of salt tolerance varieties has been considered as one of strategies to increase rice production in saline prone areas or irrigation with mixed water at the end of water rivers. Salt stress decreased seed vigor of rice cultivars, LD a superior cultivar under all salt stress which can be suggested for cultivation under salinity condition was obtained. [12,14,17,20,22].

It is very important to study the interaction effects of variability of sodium chloride effects within studied rice varieties [3,11,17]. The objectives of this research were aimed to study the performance of cultivars and salinity stress and their interactions on germination parameters of rice plants under laboratory conditions.

Materials and Methods

A laboratory experiment was conducted at Giza Central Seed Testing laboratory of Central Administration for Seed Testing and Certification (CASC), Ministry of Agriculture, Egypt during the period of December 2011 to January 2012 to study the response of rice (*Oryza sativa* L.) cultivars to germinate it under salinity stress and to confirm the seedling growth performance to examine a range of genetic variability for salinity tolerance among rice cultivars.

Treatments and Experimental Design

The experiment was arranged in factorial experiment in Randomized Complete Block Design (RCBD) with four replications. The experiment included 10 different rice cultivars which were gotten from Rice Research Institute at Sakha, ARC, MOA, Egypt. Studied rice cultivars were stored under normal conditions in paper bags include 5 cultivars belong to Japonica type and 5 cultivars belong to Indicia type as showed in Table 1. Seven different salinity concentrations (NaCl) i.e. 0, 4%, 6%, 8%, 10%, 12% and 14% NaCl were used. Seeds of cultivars were surface sterilized by immersion for 5 minutes in sodium hypochlorite solution, then repeatedly washed with deionized water. Fifty seeds of uniform size in each treatment for each cultivar were allowed to germinate on a filter paper in 9 cm diameter petredishes. Each filter paper was moistened with a water solution at seven different NaCl concentrations. Thus, the whole experiment comprised 280 Petri dishes arranged in factorial experiment Randomized Complete Block Design (RCBD). The Petri dishes were placed in a growth chamber for 14 days at 28 ± C^0 for germination.

Table 1- Name, type, pedigree and year of release of studied cultivars

No	Name of Genotypes	Type	Pedigree	Year of release
1	Sakha 102	Japonica	GZ4096-7-1/GZ4120-2-5-2(Giza 177)	1997
2	Sakha 104	Japonica	GZ4096-8-1/GZ4100-9	1999
3	Sakha 105	Japonica	GZ5581-46-3/GZ4316-7-1-1	2010
4	Sakha 106	Japonica	Giza177/Hexi30	2010
5	Giza 178	Japonica	Giza175/Milyang49	1995
6	Giza 181	Indica	IR24/IR22	1988
7	Giza 182	Indica	Giza181/ir39422-161-1-3//Giza181	1999
8	Egyptian Yasmin	Indica	IR262-43-8-1 / NAHNG SARN	1992
9	Egyptian Hybrid 1	Indica	IR69625A× Giza178R	2005
10	Egyptian Hybrid 2	Indica	IR69625A×Giza181R	2007

Studied Characters

After 14 days ten seedlings were selected from each replicates and then seedlings were evaluated as follows:

- **Final Germination Percentage (FGP):** FGP was calculated according to the following equation [7,19].

$$(FGP) = \frac{\text{Number of Germinated Seed}}{\text{Total Number of Seed Tested}} \times 100.$$

- **Germination Rate (GR):** it was calculated according the following equation [6].

$$(GR) = \frac{\text{Number of Germinated Seeds}}{\text{Number of Germination Days}}$$

- **Germination Index (GI):** according to the following equation described by [16]. It was calculated according the following equation.

$$(GI) = \frac{\text{\%Germination Each Treatment}}{\text{\%Germination in the Control}}$$

- **Vigor index (VI):** it was calculated according to the following equation:

$$(VI) = (\text{Average Shoot Length} + \text{Average Root Length}) \times \text{Germination Percentage}$$

- **Root length (cm).**
- **Shoot length (cm).**
- **Root fresh weight (mg).**
- **Shoot fresh weight (mg).**
- **Root dry weight (mg).**
- **Shoot dry weight (mg).**
- **Relative dry weight (RDW):** It was calculated according to the following equation described by [12].

$$RDW\ (\%) = \frac{\text{Total Dry Weight under Saline Condition}}{\text{Total Dry Weight under Control Condition}} \times 100$$

- **Seedling height reduction (SHR):** It was calculated using the following equation described by [12].

$$SHR(\%) = \frac{\text{Plant Height at Control} - \text{Plant Height at Saline Condition}}{\text{Plant Height at Saline Condition}} \times 100$$

Statistical Analysis

All obtained data were statistically analyzed according to the technique of analysis of variance (ANOVA) for the split – plot design [9]. By using means of "MSTAT-C" computer software package. Least Significant Difference (LSD) method was used to test the differences between treatment means at 5% level of probability [21].

Results and Discussion

Performance of Cultivars

Results in Table 2 clearly showed that the ten tested cultivars of rice significantly varied for averages of root length (cm), shoot length (cm), root fresh weight (mg), shoot fresh weight (mg), root dry weight (mg), final germination percentage (FGP), germination rate (GR), germination index (GI), vigor index (VI), relative dry weight (RDW) and seedling height reduction (SHR). Sakha 102, Sakha106, Sakha104, Giza 178 and Egyptian Yasmin cultivars significantly exceeded the other studied rice cultivars in final germination percentage, without significant differences between them. It could be observed that both Sakha 102 and Sakha 106 cultivars more tolerant than other studied cultivars. In addition, Giza 178 and Egyptian Yasmin cvs moderate to salinity. The results clearly designated that Sakha 102, Sakha 104 and Giza 181 cultivars significantly surpassed the other studied cultivars in germination rate, without significant differences between them. Sakha102, Sakha 104, Giza 178 and Egyptian Yasmin cvs significantly surpassed other studied rice cultivars in germination index, without significant changes between them. Sakha 102, Sakha 104, Giza 178 and Egyptian Yasmin cvs was more tolerant to salinity up to 14% NaCl, while Sakha 105 was more sensitive to salinity.Sakha 106, Sakha 102 and Egyptian hybrid 1 CVs significantly surpassed the other studied cultivars in vigor index without significant differences between them. The differences between rice CVs in germination characters might be due to the genetical factors and heredity variation among the ten rice cvs under study which caused differed in germination characters. These results in good arrangement with those summarized that selected genotypes differently responded to salinity [7,12,18].

Table 2- Averages of final germination percentage (%), germination rate, germination index, vigor index, root length, shoot length, root fresh weight, shoot fresh weight, root dry weight, shoot dry weight, relative dry weight and seedling height reduction as affected by rice studied cultivars.

Cultivars Performance	FGP (%)	GR (days)	GI	VI	RL (cm)	SL (cm)	RFW (mg)	SFW (mg)	RDW (mg)	SDW (mg)	RDW (%)	SHR (%)
Sakha 102	98.92	3.54	99.25	11.81	6.69	5.17	0.039	0.112	0.01	0.016	61.88	38.2
Sakha 104	97.85	3.5	97.89	10.45	6.58	4.01	0.05	0.09	0.009	0.014	60.22	33.9
Sakha 105	80.71	2.78	82.71	8.82	5.91	4.03	0.02	0.084	0.007	0.014	64.47	36.4
Sakha 106	98.92	3.5	96.75	11.83	8.31	4.94	0.052	0.102	0.006	0.016	66.42	29
Giza 178	96.78	3.38	97.92	10.39	6.13	4.38	0.036	0.113	0.011	0.015	70.23	18.9
Giza 181	93.39	3.15	96	8.49	4.43	4.48	0.027	0.054	0.01	0.015	61.71	31.9
Giza 182	87.85	3.02	91.07	8.92	5.49	4.33	0.096	0.078	0.01	0.013	62.57	31.2
Egyptian Yasminn	97.85	3.44	98.21	7.43	4.21	4.03	0.055	0.079	0.008	0.014	62.54	33.3
Egyptian Hybrid 1	89.28	3.08	93.14	11.64	7.17	5.42	0.065	0.091	0.011	0.017	70.89	24.5
Egyptian Hybrid 2	77.32	2.81	87	7.74	4.36	5.27	0.018	0.077	0.01	0.017	63.43	28.8
F. test	**	**	**	**	**	**	**	**	**	**	**	**
LSD 5%	2.64	0.04	2.42	0.56	0.38	0.24	0.028	0.01	0.0013	0.0011	3.38	3.2
LSD 1%	3.47	0.05	3.18	0.74	0.5	0.32	0.037	0.02	0.0018	0.0014	3.64	4.2

*, ** *and NS indicate P< 0.05, P< 0.01 and not significant, respectively*

Sakha106 CV significantly outdone the other studied cultivars in root length in view of the fact that produced the tallest root length. Root length of Sakha 106 exceeded by 97.3% compared with Egyptian Yasmin CV and by 90.6% compared with Egyptian hybrid 2 and by 87.5% compared with Giza 181 CV. Egyptian Hybrid 1, Sakha 102 and Egyptian hybrid 2 CV significantly exceeded the other studied cvs without significant differences between them. Egyptian hybrid 1 CV exceeded Sakha 104 CV by 26.0%, Sakha 105 CV by 25.6%, Egyptian Yasmin CV by 25.6, Giza 181 CV by 17.3%, Giza 182 CV by 20.1%, Giza 178 CV by 19.1% and Sakha 106 CV by 8.8% in shoot length. Giza 182 significantly surpassed other studied cultivars in root fresh weight followed by Egyptian Hybrid 1 and Egyptian Yasmin, respectively. Giza 182 CV surpassed Egyptian hybrid 2 CV by 81.25%, Sakha 105 CV by 79.16%, Giza 181 CV by 71.8%, Giza 178 CV by 62.5%, Sakha 102 CV by 59.37%, Sakha 104 CV by 47.91%, Sakha 106 CV by 45.83%, Egyptian Yasmin CV by 42.7% and Egyptian hybrid 1 CV by 32.29% in root fresh weight. Giza 178 cultivar significantly exceeded the other studied CVs in shoot fresh weight (mg) in view of the fact that produced the highest shoot fresh weight, followed with statistical significance by Sakha 102 and Sakha 106 CVs in the second and third ranks, respectively. It could be discerned that Giza 178 cv surpassed Giza 181 cv by 52.2%, Giza 182 cv by 30.9%, Egyptian hybrid 2 cv by 31.8%, Egyptian Yasmin cv by 30.0%, Sakha 105 cv by 25.6%, Sakha 104 cv by 20.3%, Egyptian hybrid 1 cv by 16.9% and Sakha 106 cv by 9.7% in shoot fresh weight. Egyptian Hybrid 1, Egyptian hybrid 2, Sakha 102, Giza 178, Giza 181 and Giza 182 CVs significantly outdone other studied CVs in root dry weight, without significant differences between them. While, Sakha 106 CV recorded the minimum weight of root dry weight, However, Sakha 102, Giza 181 and Giza 182 CVs recorded intermediate weights of root dry weight compared to the other studied CVs. Egyptian Hybrid 1, Egyptian Hybrid 2, Sakha 102 and Sakha 106 CVs significantly surpassed the other studied rice CVS in shoot dry weight without significant differences between them. Egyptian hybrid 1 and Giza 178 CVs significantly surpassed other studied cultivars in relative dry weight without significant differences between them. Sakha 102 and Sakha 105 CVs significantly exceeded other studied cvs in seedling height reduction percentage without significant differences between them. Followed by Sakha 104, Egyptian yasmin, Giza 181 and Giza 182 CVs, without significant differences between them. The differences between rice CVs in seedling parameters might be due to salt level side effects and to genetically factors and heredity variation among the ten studied CVs which differences in seedling parameters. Many similar deductions were reported with conclusion that salinity affect growth process and physiological consequences at advanced growth stages of genotypes [4,11, 12,18].

Salinity Stress Effects

The results in Table 3 indicated that salinity stress treatments had significant effects on means of root length (cm), shoot length (cm), root fresh weight (mg), shoot fresh weight (mg), root dry weight (mg), final germination percentage (FGP), germination rate (GR), germination index (GI), vigor index (VI) and relative dry weight (RDW). Highest means of these fonts were recorded with control treatment. Highest salinity focus up to 14% NaCl recorded the lowest averages of these characters. Increasing salinity concentrations from 0 to 14% NaCl gradually decreased averages of root length (cm), shoot length (cm), root fresh weight (mg), shoot fresh weight (mg) and root dry weight (mg). Highest final germination percentage (96.87%) was recorded with the control treatment compared with other salinity concentrations, followed by 95.62 and 95.37% with 4% and 6% Nacl without significant differences between them. Highest salinity concentration i.e. 14% NaCl recorded the lowest final germination percentage which was 82.87%. Germination percentage not affected from increasing salinity levels from 0% to 10% NaCl, however, increasing it to 12 and 14% NaCl significantly decreased final germination percentage. Highest number of germination rate was recorded with control treatment compared with other salinity concentrations. Highest salinity concentration i.e. 14% NaCl recorded the lowest germination rate. Highest germination index was obtained from the control treatment and salinity at concentration of 4% NaCl compared with other salinity stress concentrations without significant differences between them, which were 98.7 and 97.15% respectively. Highest of salinity concentrations at 14% NaCl recorded the lowest germination index. Highest vigor index values was recorded with the control treatment compared with other salinity stress concentrations (13.88%) followed by salinity concentration of 4 and 6% NaCl as an intermediate value of salinity concentrations. Highest salinity levels 14% NaCl chronicled the lowest vigor index value. High levels of salinity affected on germination characters might be due to toxic effects of salinity stress. Salinity decreased germination typescripts of seeds which are directly related to the amount of absorbed water by the seed. [5,8,12,13]. Seed germination was quantitative trait controlled by several gens and strongly affect by salt stress [22].

Table 3- Averages of final germination percentage (%), germination rate, germination index, vigor index, root length, shoot length, root fresh weight, shoot fresh weight, root dry weight, shoot dry weight, relative dry weight and seedling height reduction as affected by salinity concentrations (NaCl).

Salinity Stress	FGP (%)	GR	GI	VI	RL (cm)	SL (cm)	RFW (mg)	SFW (mg)	RDW (mg)	SDW (mg)	RDW (%)	SHR (%)
O (control)	96.87	3.45	98.7	13.88	8.89	6.63	0.099	0.138	0.016	0.022	100	0
4% NaCl	95.62	3.37	97.15	13.14	8.19	6.07	0.063	0.123	0.013	0.021	90.36	9.7
6% NaCl	95.37	3.32	96.57	11.84	6.96	5.4	0.045	0.1	0.011	0.018	77.78	19
8% NaCl	93.12	3.26	95.42	10.66	6.41	4.8	0.042	0.085	0.009	0.015	63.65	27.7
10% NaCl	90.75	3.18	93.65	8.5	5.1	3.9	0.038	0.07	0.007	0.013	51.76	41.5
12% NaCl	88.62	3.07	91.2	6.06	3.42	3.19	0.017	0.057	0.005	0.01	40.11	51.4
14% NaCl	82.87	2.89	85.27	4.19	2.51	2.27	0.016	0.043	0.003	0.007	27.4	64.9
F. test	**	**	**	**	**	**	**	**	**	**	**	**
LSD at 5%	2.21	0.03	2.02	0.46	0.31	0.2	0.023	0.013	0.0011	0.0008	2.83	2.7
LSD at 1%	2.9	0.04	2.66	0.61	0.41	0.27	0.029	0.017	0.0014	0.001	3.71	3.6

**, ** and NS indicate $P< 0.05$, $P< 0.01$ and not significant, respectively*

Tallest root length was recorded with the control treatment compared with other salinity stress levels. Highest salinity level i.e. 14% NaCl reduced root length and recorded the lowest root length. The second salinity level i.e. 6% NaCl came in the second rank. It could be concluded that reducing salinity levels from 14%, 12%, 10%, 8%, 6% to 4% NaCl reduced root length by 71.7%, 61.5%, 42.6%, 28.0% 21.7% and 7.5% respectively compared with control treatment. Reducing salinity levels from 14%, 12%, 10%, 8%, 6% to 4% NaCl reduced shoot length by 65.7%, 51.88%, 41.17%, 27.6, 18.55% and 8.44%, respectively compared without salinity i.e. control treatment. Highest weight of root fresh was obtained with the control treatment compared with other salinity levels, followed by 4% NaCl and 6% NaCl as intermediate values. Highest salinity level i.e. 14% NaCl recorded the lowest root fresh weight. Reducing salinity levels from 14%, 12%, 10%, 8%, 6% to 4% NaCl reduced shoot fresh weight by 68.84%, 58.19%, 49.2%, 38.4%, 27.5, and 10.86% compared with the control treatment. Highest root dry weight was obtained from the control treatment compared with other salinity levels. Increasing salinity level to 4 and 6% NaCl came in the second and third rank after control. Highest salinity level 14% NaCl was recorded the lowest root dry weight. Highest shoot dry weight was obtained from the control treatment compared with other salinity levels, followed by 4 and 6% NaCl as intermediate values. Highest salinity level i.e. 14% NaCl recorded the lowest shoot dry weight. Highest relative dry weight was obtained from the control treatment compared with other salinity levels which was 100.0%, followed by salinity concentrations of 4 and 6% NaCl as an intermediate value of salinity levels. Highest salinity level of 14% NaCl recorded the lowest relative dry weight. Relative dry weight gradually decreased with increasing of salinity levels. Highest seedling height reduction was obtained from the salinity level of 14% NaCl treatment compared with other salinity concentrations, followed by salinity levels of 12 and 10% NaCl as an intermediate values of salinity levels. The control treatment recorded the lowest percentages of seedling height reduction value, followed by salinity level of 4 and 6% NaCl. The gradual decrease in seedling parameters with the increase in salinity as observed might be due to more inhibitory effect of NaCl salt to root growth compared to that of shoot growth and to sodium level in the shoots during early growth. In addition to the cellular toxicity of NaCl in rice seedlings and reduction of chlorophyll content under saline condition. [5,8,10]. Salinity decreased final germination percentage, speed of germination, germination energy percentage and led to reduction in shoot and root length and dry weight in all studied varieties and the magnitude of reduction increased with increasing salinity stress [11].

Interaction Effects

The interaction between rice cultivars and salinity levels significantly affected final germination percentage as shown in Fig. 1. Highest final germination percentage was produced from Sakha 102 and Sakha 106 cultivars with snowballing salinity levels up to 10% NaCl. Whereas, the lowest final germination percentages were resulted from sowing Sakha 105 and Egyptian Hybrid 2 cultivars at higher salinity levels of 14% NaCl without significant differences between them. Highest averages of germination rate were resulted from sowing Sakha 106, Sakha 104 and Egyptian Yasmin cvs and increasing salinity concentrations up to 10% NaCl without significant differences between them as shown in Fig. 2. The lowest average of germination rate was resulted from sowing Sakha 105 and Egyptian hybrid 2 cultivars at higher salinity level of 14% NaCl without significant differences between them. Uppermost germination index was recorded with Sakha 102 and Sakha 106 cultivars with increasing salinity level up to 10% NaCl without significant differences between them as shown in Fig. 3. On contrary, the lowest value of germination index was resulted from sowing Sakha 105 with higher salinity level of 14% NaCl. In addition, uppermost vigor index percentages were resulted from sowing Sakha 102 CV and the control treatment as shown in Fig. 4. On conflicting, the lowest percentage of vigor index was resulted from Sakha 104 or Giza 181 cultivar with highest salinity levels of 14% NaCl. Tallest root length was chronicled with Sakha 106 cultivar and the control treatment as shown in Fig. 5. Whilst, the shortest in root length were caused from sowing Sakha 104 or Giza 181 with high salinity concentration i.e.14% NaCl. Moreover, tallest shoot length was obtained from all studied cultivars at control treatment. Sakha 102 and Egyptian hybrid 1 at control treatment twisted tallest shoot length as shown in Fig. 6 .The shortest shoot length was obtained from Giza 181, Sakha 105 and Sakha 104 cvs without significant differences between them. Highest shoot dry weight was resulted from sowing Egyptian Hybrid 1,

Sakha 102, Sakha 104, Sakha 106 and Egyptian Yasmin cvs and control treatment without significant differences between them as shown in Fig. 7. The lowest shoot dry weight was resulted from sowing Giza 181, Giza 182 and Sakha 104 cvs and salinity concentration of 14% NaCl without significant differences between them. Highest relative dry weight percentages were resulted from sowing all studied cvs with control treatment as shown in Fig. 8. Snowballing salinity levels from 4% NaCl to 14% NaCl significantly reduced relative dry weight in all studied cvs. The lowest percentage of relative dry weight was resulted from sowing Giza 181, Giza 182 and Sakha 104 cvs with highest salinity levels of 14% NaCl. Highest percentages of seedling height reduction was occasioned from sowing Giza181 cultivar with salinity level of 14% NaCl treatment, followed by Sakha 102, Sakha 104 and Sakha 105 with salinity level of 14% NaCl without significant differences between them as shown in Fig. 9. The lowest percentages of seedling height reduction were resulted from sowing all studied cvs with control treatment. Differences between cultivars tolerance to salinity may be due to higher proline, protein and starch content under salt stress [3]. Salt stress decreased vigor of all studied rice cultivars [17] .

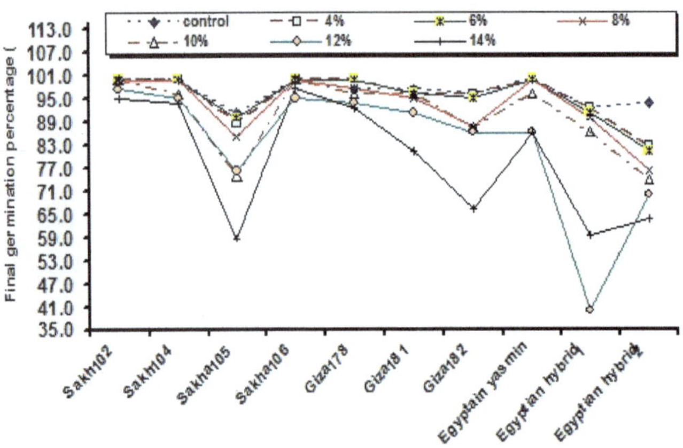

Fig. 1- Averages final germination percentage (%) as affected by the interaction between cultivars and salinity stress (NaCl %).

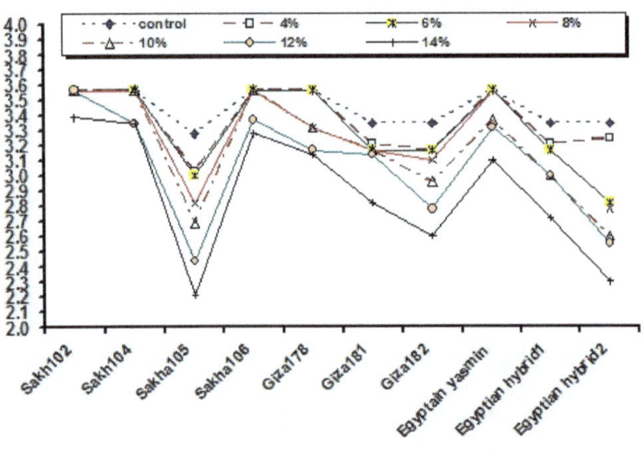

Fig. 2- Averages germination rate as affected by the interaction between cultivars and salinity stress (NaCl %).

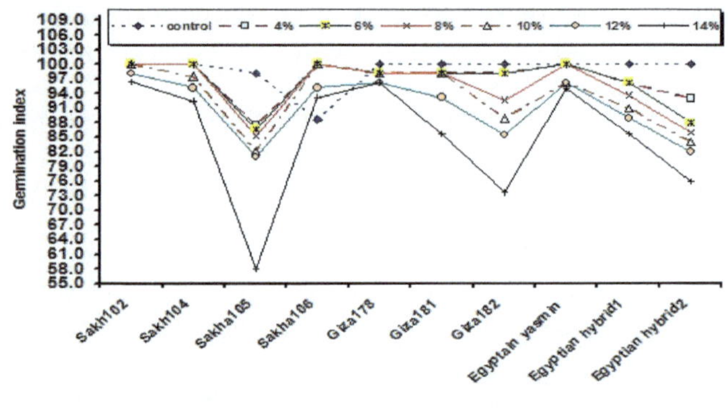

Fig. 3- Averages germination index % as affected by the interaction between cultivars and salinity stress (NaCl %).

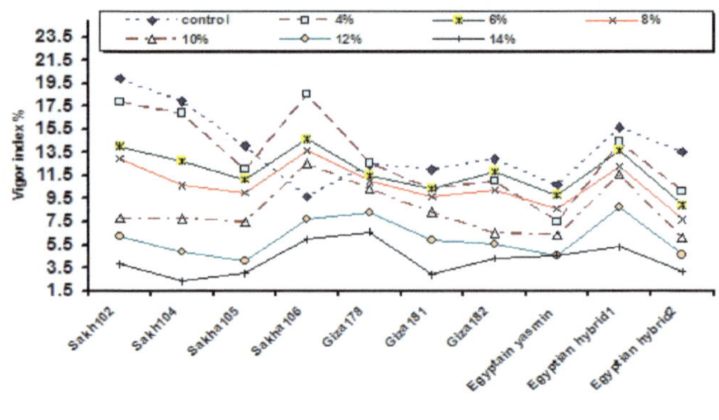

Fig. 4- Averages vigor index % as affected by the interaction between cultivars and salinity stress (NaCl %).

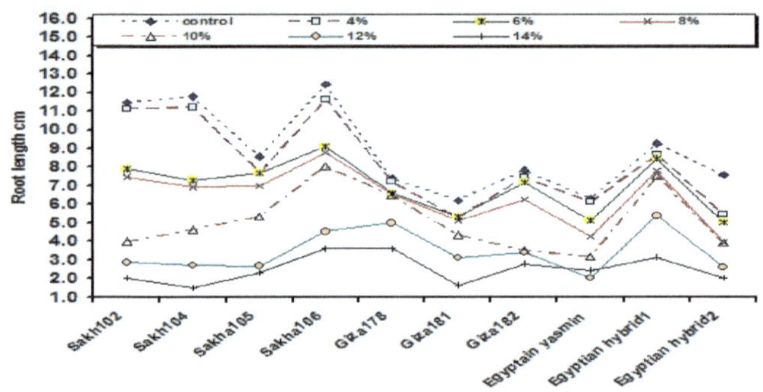

Fig.5- Averages root length cm as affected by the interaction between cultivars and salinity stress (NaCl %).

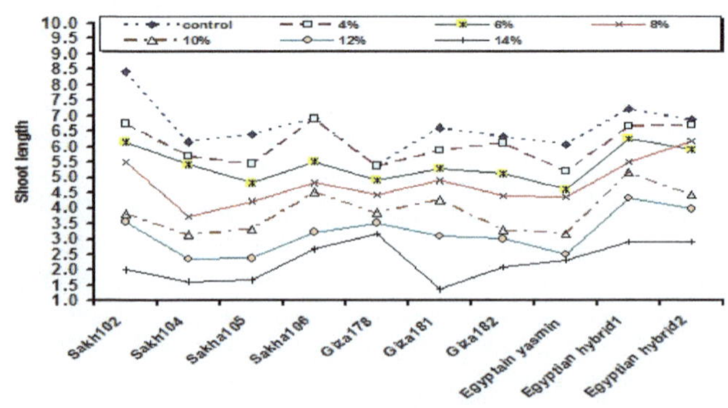

Fig. 6- Averages shoot length cm as affected by the interaction between cultivars and salinity stress (NaCl %).

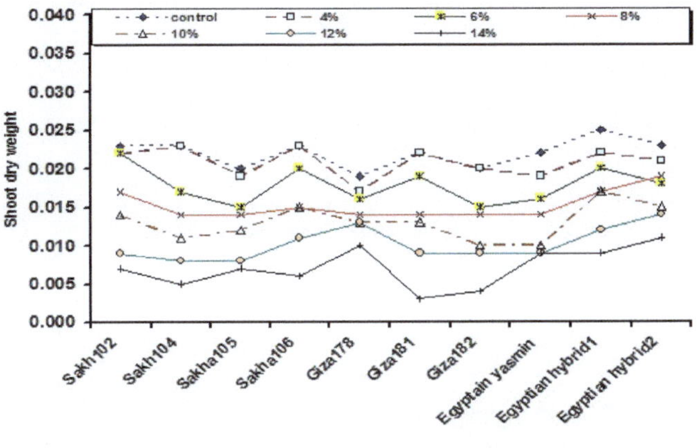

Fig7- Averages shoot dry weight mg as affected by the interaction between cultivars and salinity stress (NaCl %).

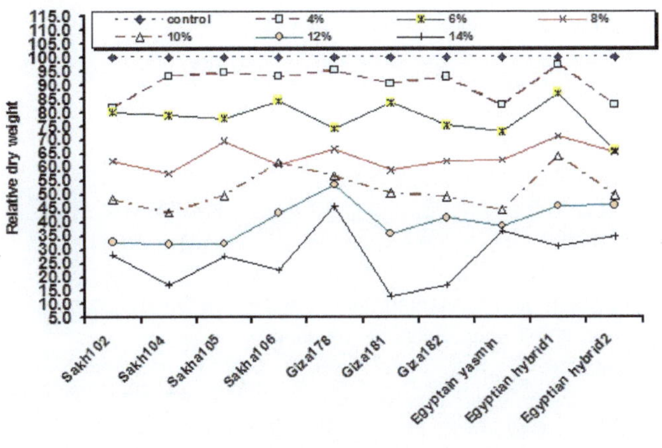

Fig.8- Averages relative dry weight % as affected by the interaction between cultivars and salinity stress (NaCl %).

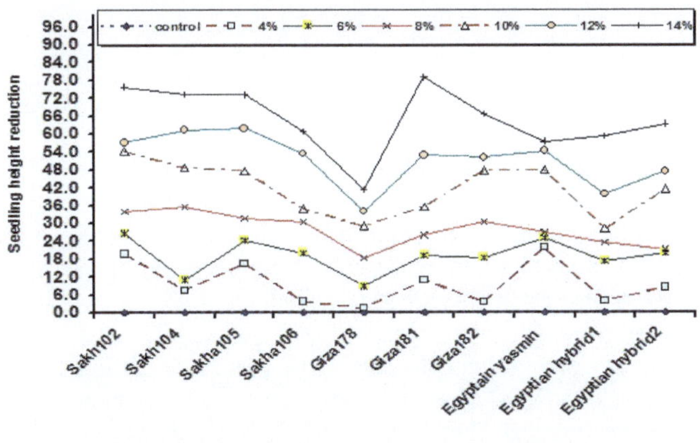

Fig. 9- Percentages seedling height reduction % as affected by the interaction between cultivars and salinity stress (NaCl %).

Conclusion

It could be concluded that for maximizing rice germination percentage and seedling parameters under salinity stress are producing by using rice Sakha 102, Sakha 106, Sakha 104 and Egyptian Yasmin cultivars under salinity levels level up to 14% NaCl. These cultivars were more tolerant to salinity and recommended to use them in breeding program for enhancing rice production in Egypt.

Abbreviations

- **FGP:** Final Germination Percentage
- **GR:** Germination Rate
- **GI:** Germination Index
- **VI:** Vigor Index
- **RL:** Root Length
- **SL:** Shoot Length
- **RFW:** Root Fresh Weight
- **SFW:** Shoot Fresh Weight
- **RDW:** Root Dry Weight
- **SDW:** Shoot Dry Weight
- **SHR:** Seedling Height Reduction

References

[1] Bhuiyan N.I., Paul D.N.R. and Jabber M.A. (2002) *National Workshop on Rice Research and Extension.*

[2] Cha-um S., Trakulyingcharoen T., Smitamana P. and Kirdmanee C. (2009) *Aust. J. Crop Sci.*, 3(4), 221-230.

[3] Danai-Tambhale S., Kumar V. and Shriram V. (2011) *J. Stress Physiol. Biochem.*, 7, 387-397.

[4] Djanaguiraman M., Senthil A. and Ramadass R. (2003a) *Madras Agric. J.*, 90(7-9), 506-510.

[5] Djanaguiraman M., Ramadass R. and Devi D.D. (2003b) *Madras Agric. J.*, 90 (1-3), 50-53.

[6] Ellis R.H. and Roberts E.H. (1980) *Towards a Rational Basis for Testing Seed Quality*, 605-635.

[7] Ellis R.H. and Roberts E.H. (1981) *Seed Sci. Technol.*, 9, 373-409.

[8] Golshani M., Pirdashti H., Saeb K., Babakhani B. and Heidarzade A. (2010) *World Applied Sci. J.*, 9 (2), 221-225.

[9] Gomez K.A. and Gomez A.A. (1984) *Statistical Procedures for Agricultural Research*, *2nd edition*, 680.

[10]Habib N., Muhammad A. and Ahmad M.S.A. (2010) *Pak. J. Bot.*, 42(6), 4071-4078.

[11]Hakim M.A., Juraimi A.S., Begum M., Hanafi M.M., Ismail M.R. and Selamat A. (2010) *Afric. J. Biotech.*, 9(13), 1911-1918.

[12]Islam M.M. and Karim M.A. (2010) *The Agricult.*, 8(2), 57-65.

[13]Jamil M. and Rha E.S. (2007) *Pak. J. Biol. Sci.*, 10, 4303-4306.

[14]Joseph B. and Jini D. (2010) *Int. J. Bot.*, 10, 456-464.

[15]Joseph B., Jini D. and Sujatha S. (2010) *Asian J. of Agric. Sci.*, 2(3), 99-105.

[16]Karim M.A., Utsunomiya N. and Shigenaga S. (1992) *J. Crop Sci.* 61, 279- 284.

[17]Kazemi K. and Eskandari H. (2011) *African Journal of Biotechnology*, 10(77), 17789-17792.

[18]Rahman M.S., Miyake H. and Takeoka Y. (2001) *Pak. J. Biol. Sci.*, 4(3), 351-355.

[19]Ruan S., Xue Q. and Thlkowska K. (2002) *Seed Sci. Technol.* 30, 451-458.

[20]Saeidzadeh F., Gurbanov E. and Dunyamaliyev S.A. (2011) *Son Yenilenme Pazar Ertesi.*, 4, 23-32.

[21]Snedecor G.W. and Cochran W.G. (1980) *Statistical Methods" 7th Ed.*, 507.

[22]Zhoufei W., Jianfei W., Yunyu B. and Hongsheng Z. (2011) *Euphytica.*, 178(3), 297-307.

POKA-A TRADITIONAL RICE WINE OF THE *GALO* TRIBE OF ARUNACHAL PRADESH, INDIA

BORA S.S.[1], LAHAN J.P.[1], MADHUMITA BAROOAH[1]* AND RUPJYOTI SARMAH[2]

[1]Department of Agricultural Biotechnology, Assam Agricultural University, Jorhat-785 013, Assam, India.
[2]ICAR Research Complex for NEH Region, Arunachal Pradesh Centre, Basar-791101, Arunachal Pradesh, India.
*Corresponding Author: Email- m17barooah@yahoo.co.in

Abstract- Arunachal Pradesh, the unique territory previously known as the North East Frontier Agency, is a mountainous region extending between the Brahmaputra Valley, whose eastern part it encloses like a horseshoe, Tibet to the north, Burma to the east, and Bhutan to the west. It is the home to 110 ethnic groups (sub-tribes) of great cultural diversity, but in many respects there is an overall uniformity. The *Galo* population estimated at 80,597 (2001 census) makes them the one of the most populous tribe of Arunachal Pradesh. *Poka*, a traditional rice wine plays an important role in the socio-cultural life of the *Galo* tribe of Arunachal Pradesh. It is consumed during most of the festive occasions and celebrations. This paper reports the traditional way of preparation of the wine with ethnobotanical observation.
Key words- *Poka,* Rice wine, *Galo*, Arunachal Pradesh, Agricultural, *Mopin,* starter, alcohol.

Introduction

Arunachal Pradesh is considered to be luxuriant in biodiversity and has been recognized as the 25th biodiversity hotspot in the world [1]. It is spreading over an area of 83,743 sq km with a variation in altitude from 150-6,500 m and unique climatic conditions. The original inhabitants of Arunachal Pradesh belong to 26 major tribes and 110 sub-tribes [2]. *Galo* is one of the prominent tribes occupying the southern part of Abor Hills, bounded by in Simen River east, Subansiri in the west, Siyom River Bayor Adi (Hill) in the north, and Lakhimpur district of Assam in the south. The maximum population of *Galo* is mainly concentrated on West Siang District. The tribe is composed of several groups inhabiting a compact area, comprising of many villages, and all are culturally and socially linked together [3]. The people are dependent on agriculture which is the backbone of their economy. They are socio-economically dominant in their area. *Donyi-Polo* religious traditions persist to a degree in most *Galo*, although Christianity is on the rise in recent years especially in the foothill areas [4]. With this traditional belief, they perform festivals that are mainly agricultural based. *Mopin* is the main festival of the tribe which is celebrated for four to five days in the month of April for social prosperity and wealth. Rice is their staple product; they produce it for food and preparation of rice-wine (*Poka*). They traditionally practice shifting cultivation (*Jhum kheti*) which involves intensive labor and often employs the entire village during cultivation and harvesting of the crops. *Galo*s have tremendous traditional knowledge to use natural resources to a great extent. They are well known for production of household liquor, *Poka,* which is associated with social, cultural socio-economical context. They offer it in their festivals like *Mopin*, at the onset of harvesting, their marriage and *Shhradha* ceremonies.

Methodology

The present study is based on Basar village (27°59'N 94°40'E.), which is situated in the West Siang District of Arunachal Pradesh. It has an average elevation of 578 metres above mean sea level and has a pleasantly cold climate. According to the 2001 India census, Basar has a population of 3,834 Galo (Tribe) people. Males make up 56% of the population and females form 44%. Basar has a mean literacy rate of 72%, where 61% of the males and 39% of the females are literate. 16% of the population is made up of children less than 6 years of age. It has a population density of 11 person per km [2]. Intensive field work was carried out during the 2010-2011. Investigations were made by interrogating the village local people mainly women, visits to local market and also by studying various literature. PICs (Prior Informed Consents) were taken during the investigation. Routine methods of botanical collection and herbarium techniques were followed. While collecting ethnobotanical aspects, standard approaches and methodologies were followed. Plant samples were identified by comparing their morphology recorded during field trips as well as laboratory studies, with local floras and monographs and with the help of plant taxonomists. The voucher herbariums were submitted to the Weed Herbarium of Assam Agricultural University, Jorhat.

Results

Starter Culture Preparation

The starter culture (locally known as *Apong Kusure*) is concocted by women folk with rice flour and several herbs (table1). Water is then added to form dough like mass with moisture content of 55-60% and then is inoculated using dry powdered starter from previous batches, followed by thorough mixing. The inoculated dough is shaped into small flattened or ball-shaped cakes about 4 cm in diameter and 1 cm thick. These are then covered with a thin layer of rice husks. The cakes are then sun-dried on a bamboo tray for 3-4 days.

Ethnobotanical Observation

The study recorded a number of six commonly used plants belonging to four families as enumerated below with photographs. The plants have been described in the table according to alphabetic order of botanical name along with their respective vernacular (Local) name (s) and the name of the families.

Table 1- Plants used in the preparation of starter cake

Local Name	Botanical name/family	Part Used	Uses
Oin	*Clerodendon viscosum* Vent/Verbanceae	Leaves	Antipyretic, laxative, vermifuge
Hibe/Onyor	*Debregesia longifolia* (Burm. f.) Wedd./ Urticaceae		Flavoring agent
Taka	*Diplazium esculentum* (Retz.) Sw./Athyriaceae		Laxative
Rare	*Pilea* sp./ Urticaceae		Antioxidant properties
Oyik	*Urtica hirta* Bl/Urticaeceae		Anti-diarrheic effect
Tita baigun	*Solanum kurzii* Brace ex Prain/Solanaceae	Fruit	Laxative

Clerodendon viscosum

Debregesia longifolia (Burm. f.) Wedd.

Diplazium esculentum (Retz.) Sw.

Pilea sp

Urtica hirta Bl

Solanum kurzii Brace *ex* Prain

Fig. 1- Some of the commonly used herbs for starter cake preparation

Method of preparation of *Poka*

Rice husks are stapled on *Tora* (*Alpinia allughas/* Zingiberaceae) leaves and are burnt slowly till they become black in colour, which may take upto 4-5 hours. Cooled ash is mixed rigorously in equal amount with warm boiled rice. Again, a heap of rice husk is laid on the pounded mass and is combusted slowly. Finally the ash is mixed thoroughly along with starter granules called *apong kusure* @ 2 cakes/1 kg of biomass. The inoculated pounded rice paste is then transferred into a rattan made traditional basket encapsulated with *Tora-pat* (*Alpinia allughas*). The container is sealed tightly and left to ferment for 15-30 days during summer at ambient temperature (in winter it may take up to 2-3 months). After the period of incubation (solid state fermentation), the mass is transferred to a conical shaped commercially available container which is encapsulated with a polythene sheet. The fermented content is eluted by pouring warm water slowly along the sides of the container. The product secreted out of the pore at the bottom of the container is collected and served. It is known as *Poka.* It is ash-colored and sweet in taste containing 4 to 8 percentage of alcohol.

Fig. 2a- A slowly burning heap of rice husk

Fig. 2b- Ash is collected & pounded rigorously with cooked rice

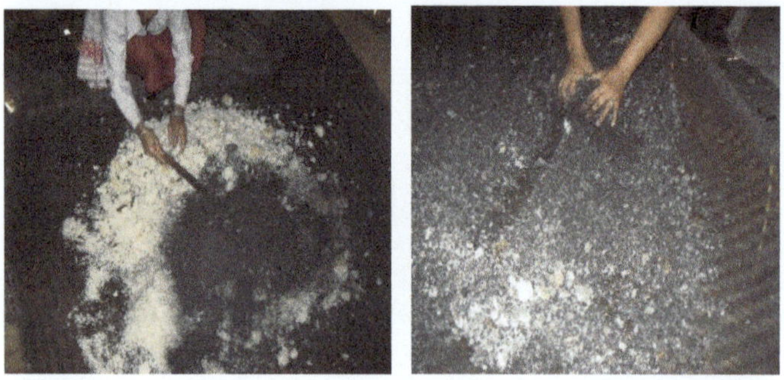

Fig. 2c- Ash is collected & pounded rigorously with cooked rice

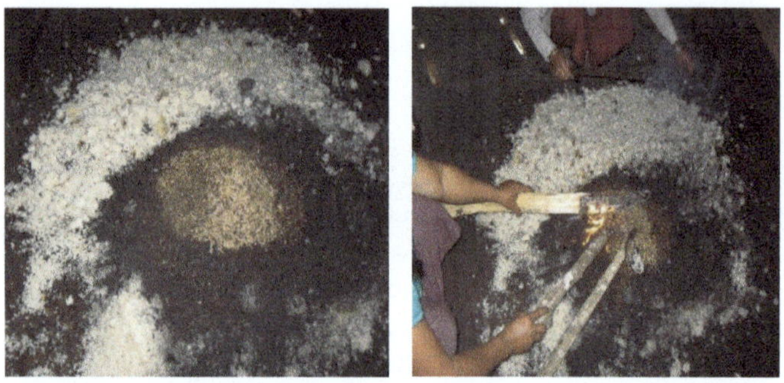

Fig. 2d- Second combustion of rice husk on rice and ash dough

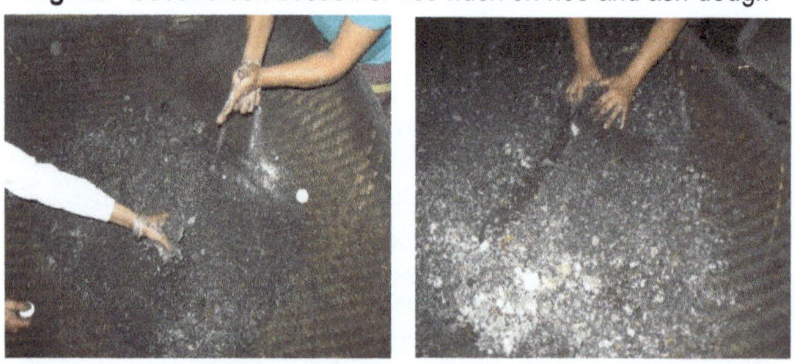

Fig. 2e- Mixing of starter cakes

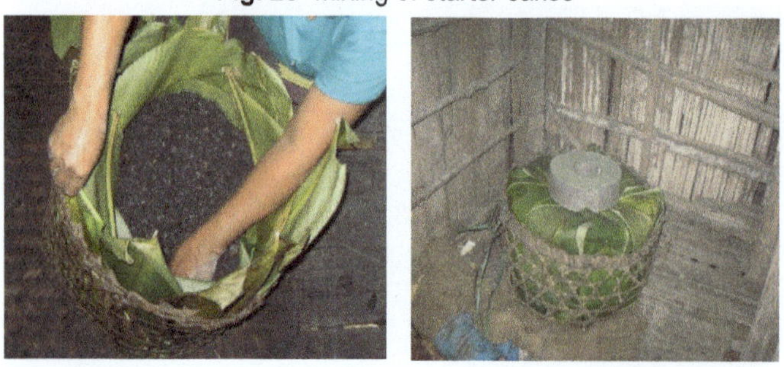

Fig. 2f- Transfer and storage of substrate for solid-state fermentation

Fig. 2g- Pouring of lukewarm water for collection of *Poka*

Discussion

The North-East of India is among the most cultural-linguistically rich and diverse regions of all Asia. This is due as much to its position at the cultural-geographical crossroads of East, South, and South-East Asia as to its varied topography and difficulty of travel and access, and consequent opportunities for populations to develop to a degree independently in relatively isolated niches [4]. Rice wine is one of the popular alcoholic beverages in South East Asian countries, postulated to spread as concentric cycle from the area of the middle of Chan Jiang River in the Yunggui area (Yunnan-Guizhou provinces) of China. Preparation of rice wine varies according to locations and traditional practices. Generally, rice wine fermentation can be categorized into submerged and solid state process. Submerged process involving saccharification of rice to liberate sugar and converted to ethanol by submerged fermentation of yeast in liquid medium occur with plenty availability of free water [5]. Solid state fermentation is another way of rice wine production. *Ruou nep* in Vietnam, *Ou* in Thailand and rice wine from Cambodia and Laos are produced using this method. Solid state fermentation is a process where microbial growth and product formation occurs on the surface of solid materials [6]. It was observed that the tribe under study also follows similar kind of processing.

Poka has been considered as a favorite drink among the *Galos*. All tribal women prepare it and men help them in pounding the cooked rice. It was noticed that the beverage is considered as a social drink which helps to bond and encourage people to help each other in every socio cultural aspects. Moreover, the respondents informed about the benefits of the drink in health problems such as dysentery, loss of appetite etc., though ethanol as a major product of alcoholic fermentation is well known for causing intoxication in human.

The preparation of fermented rice wine is an art of technology and is a family secret passed or inherited as a legacy. But, due to various anthropogenic activities, this traditional knowledge has been depleting. Again, knowledge about the significance of the presence of the metabolite and their chemical interaction in this beverage is still insufficient. Exploitation of traditional knowledge in microbes may not lead to destruction or eradication of biodiversity [7], but can contribute significantly in developing value added product for human welfare [8]. Such product with input from scientific and technological knowledge should be encouraged for sustainable utilization of locally available resources and traditional knowledge system. Moreover, people should also be encouraged to develop and capitalize entrepreneurship so as to use available resources for a low cost beverage with clinically documented health promoting properties.

Acknowledgement

Authors are thankful to all the faculty and staff members of the Department of Agricultural Biotechnology, Assam Agricultural University, Jorhat, Assam. Thanks are also due to local informants and village headmen for their valuable support and co-operation throughout the field work.

Reference

[1] Chowdhery H.J. (1999) *Floristic diversity and conservation strategies in India*, BSI, Kolkata.
[2] Srivastava R.C. (2009)*Indian Journal of Traditional Knowledge*, 8(2), 146-153.
[3] Medhi Birinchi K., Athparia R.P. and SVD K. Jose (2009) *Tribes of North-East India, Issues and Challenges*, New Delhi, 61-62.
[4] Post, Mark W. A Grammar of Galo. PhD Dissertation. Melbourne, La Trobe University Research Centre for Linguistic Typology (2007), 1.2.5.
[5] Law S.V., Abu Bakar F., Mat Hashim D. and Abdul Hamid A. (2011) *International Food Research Journal*, 18, 475-484.
[6] Ray R.C. and Ward O.P. (2006) *Microbial Biotechnology in Horticulture*, Enfield, New Hampshire, 1, 345-396.
[7] Aidoo K.E., Nout M.J.R. and Sarkar P.K. (2006) *Occurrence and function of yeasts in Asian indigenous fermented foods*, FEMS Yeast Research, 6, 30-39.
[8] Jeyaram K., Singh Anand Th, Romi W., Ranjita Devi A., Mohendro Singh W., H., N. Dayanidhi, Singh Rajmuhon and Tamang J.P. (2009) *Indian Journal of Traditional Knowledge.* Volume 8(1), 115-121.

MAPPING AND ASSESSMENT OF SOILS IN NAGARJUNASAGAR CATCHMENT, LINGASUGUR TALUK OF KARNATAKA

RAM R.L.[1]*, JHA P.[1], SHARMA P.K.[2], AHMED N.[3], KUMAR R.[4], SHARANAPPA[1], VASUDEO RAO D.A.L.[1] AND SINGH V.[1]

[1]Soil and Land Use Survey of India, Mrida Sarvekshan Bhavan, Rajendra Nagar, Hyderabad- 500030, AP, India.
[2]Department of Soil Science and Agricultural Chemistry, Institute of Agriculture Sciences, BHU, Varanasi- 221005, UP, India.
[3]Division of Soil Science and Agricultural Chemistry, IARI, New Delhi- 110012, India.
[4]Soil & Land Use Survey of India, IARI Buildings, New Delhi- 110012, India.
*Corresponding Author: Email- ramlakhanram08@gmail.com

Abstract- A detailed soil survey was carried out in the 4D5A2s, 4D5A2t, 4D5A2u and 4D5A3j Subwatersheds of Nagarjunasagar catchment, Lingasugur taluk of Raichur district, Karnataka, India covering an area of 22381 ha, spread over 47 villages. The survey area is represented by four major physiographic division namely subdued hills, upper pediplains, lower pediplains and alluvial plains developed over granite. Nine soil series were identified and mapped into 27 mapping units. The soils developed subdued hills is excessively drained, shallow, yellowish red to reddish brown, gravely sandy loam to gravelly clay, mapped into a single mapping unit and classified under LCC VI. The soils is spread over 0.7 percent of the total area and classified into Lithic Ustorthents. The soils of upper pediplains are well drained, shallow to deep, reddish brown to dark yellowish brown, gravelly sandy loam to gravelly clay and mapped into 14 mapping units. These soils belong to LCC II, III and IV and classified into Typic Haplustalfs, Typic Ustorthents and Typic Haplustepts which represent the 18.7 percent of the total area. The soils of lower pediplains and alluvial plains are moderately well drained to well drained, moderately deep to very deep, dark brown to dark grayish brown, calcareous, gravelly clay loam to clay, mapped under 12 mapping units which belongs to land capability classes II, III and V and classified into Typic, Sodic and Fluventic Haplustepts which represents the 78.04 percent of the total area.

Keywords- Subdued hills, pediplains, alluvial plains, detailed soil survey, land capability classification

Introduction

The modern agriculture requires precise information on various landuse and agro-climatic parameters like soil types and characteristics, rainfall, temperature, water resources etc. which influence directly the soil response to any specific use [1,2], however, the usefulness of a soil, is not solely due to its inherent qualities but the qualities which affect its capacity to respond to the inputs and management for a specific use or a combination of uses [3]. Soil survey is the only an effective array in the arsenal of planners and users to provides adequate information in terms of land form, terraces, vegetation as well as characteristics of soils (viz., texture, depth, structure, stoniness, drainage, acidity, salinity and so on) which can be utilized for the planning and development of modern agriculture. Likewise mapping and assessment of soils is the process of evaluation of a particular tract of land for specific purposes involving the execution and interpretation of data of natural resources and other related aspects of land in order to identify and make a comparison of the promising kind of land uses. To prepare a land resource inventory in terms of mapping units, its assessment and capability classification, a detailed soil survey work of the soils of Nagarjunasagar catchment, Lingasugur Taluk, Raichur district of Karnataka, India was undertaken by the Soil & Land Use Survey of India, Hyderabad centre, on 1:7920 scale.

Material and Methods

General Description of the Survey Area

The study area comprising 4D5A2s, 4D5A2t, 4D5A2u and 4D5A3j subwatersheds of Nagarjunasagar catchment, Lingasugur taluk of Raichur district, Karnataka lies between 16°03' to 16°21' N latitudes and 76°20' to 76°38' E longitudes and has a total geographical area of 22381 ha spread over 47 villages of entire taluka.

The physiography of the area has been divided into subdued hills, upper pediplains, lower pediplains and alluvial plains. The hills have strong to moderately steep (10-15%) side slope and have an altitude between 540 to 590 m above mean sea level whereas most of the area represents very gentle to gentle (1-5%) slopes of pediplains and alluvial plains and have an altitude between 480 to 540 m above mean sea level. All the subwatersheds are drained through Anehosur and Hirehalla stream which directly flowing into the Krishna River. The drainage pattern is dendratic to subdendratic with moderate to severe soil erosion.

Granite is the main geologic formation in the survey area and alluvium landscape is encountered around major stream bank. The granite rocks are medium to coarse textured, dark coloured, possessing distinct petrological characters with rhombic pyroxene hypersthenes and a high quality of the dark ferromagnesian compound which gives it dark colour.

The survey area is influenced by the semi-arid climate where evapotranspiration exceeds the precipitation for the major part of the year. The total annual rainfall is 740 mm, about 80.3 percent of which falling between June to October. The mean annual temperature is 30.5°C, whereas the mean summer soil temperature and mean winter soil temperature are 31.8°C and 26.5°C respectively. December is the coldest month (16.8°C) and April, May are hottest month (44.7°C). The soil moisture regime is *'ustic'* and soil temperature regime is *'isohyperthermic'* [4].

Soil Survey Methodology

Detailed soil survey of the area was carried out using cadastral map of 1:7920 based on grid system [5]. The soils of the area representing three physiographic divisions namely subdued hills, pediplains and alluvial plains were studied for their morphological characteristics [4] and the soil samples were collected from all the horizons and analyzed for their physico-chemical characteristics with standard procedures [6,7]. Mapping of soils [8] and land evaluation [9,10] procedure were adopted.

Result and Discussion

The area was represented by three major physiographic division namely subdued hills, Pediplains and alluvial plains developed over alluvium and granite. Nine soil series namely Bhupur, Chennampalli, Devergudda, Devsugur, Fatepur, Gabbur, Hembral, Mudgal and Tatikol were identified for mapping and assessment based on their physico-chemical and morphological properties.

Soil Morphology

Chennampalli series soils are excessively drained, shallow, yellowish red to reddish brown, gravely loamy sand to sandy loam representing pasture and grazing land use occur on strongly (10-15%) sloping, subdued hills [Table-1]. The surface layer is 18-22 cm thick, underlain by weathered granitic parent material. The lithic contact is encountered within 50 cm below the surface. Bhupur, Devergudda, Devsugur and Tatikol series soils of upper pediplains are well drained, shallow to deep, reddish brown to dark yellowish brown, gravelly sandy loam to gravelly clay occur on very gently to moderate (1-5%) sloping and suitable for rainfed crops. The surface layer is 7-25 cm thick which is underlain by 38-85 cm sub-surface soils and has weathered granitic parent material.

Fatepur, Hembral and Gabbur series soils of lower pediplains are moderately well drained to well drained, moderately deep to very deep, dark brown to dark grayish brown, calcareous, gravelly clay loam to clay occur on very gently to gently (1-5%) sloping and suitable for rainfed crops. The surface layer is 6-12 cm thick which underlain by 21-125 cm. cracks with 2-5 cm width and well developed slickensides are present in the soils of Hembral series. Mudgal series soils are moderately deep, very dark grayish brown, calcareous, sandy clay loam to gravelly clay occur on very gently to gently (1-5%) sloping, developed over alluvium. These soils are alkaline, suitable for pasture and grazing, but it can be utilized as agricultural lands adopting reclamation practices. The soil colour varies from dark brown (7.5 YR 4/4) to reddish brown (5YR 3/4) in subdued hills, reddish brown (5YR 3/4) to dark reddish brown (5 YR 3/3) and dark yellowish brown (10 YR 4/4) in upper pediplains and dark grayish brown (10 YR 4/2) to very dark grayish brown (10 YR 3/2) in lower pediplains and alluvial plains. The variation of colour in uplands and hill slope is due to prevalence of well drained conditions, admixture of organic matter and occurrence of iron in oxidized ferric form [11,12]. The soils of the taluk showed wide textural variation and have been classified into coarse loamy to very fine textural class. This wide variation in soil texture is caused by topographic position, nature of parent material, weathering, translocation of clay and age of soils [13]. The structure of the soils is weak, fine, subangular blocky in subdued hills, very fine, single grains to moderate, medium, subangular blocky in upper pediplains and moderate, medium, subangular blocky to angular blocky in lower pediplains and alluvial plains. The variation in soil structure is a reflection of physiographic position of the pedons [14]. The consistence of the soils occurring on the subdued hills and upper pediplains varied from loose, very friable, slightly sticky and slightly plastic to slightly hard, friable, sticky and plastic whereas the soils of lower pediplains and alluvial plains varied from slightly hard, friable, slightly sticky and slightly plastic to very firm, very sticky and very plastic. The wide variation in consistence is due to variation in clay content of pedons situated on different topographic position [12]. The boundary of the soils on subdued hills varied from abrupt smooth to diffuse smooth whereas the boundary of the soils on upper pediplains, lower pediplains and alluvial plains varied from clear smooth to gradual smooth. Strong effervescence with dilute HCl were observed in the soils of Devergudda, Devsugur series on upper pediplains and Fatepur, Gabbur, Hembral and Mudgal series on lower pediplains and alluvial plains. The vertic properties such as 2-5 cm wide cracks extended upto 50 cm depth and developed slickensides in lower horizons were observed in Gabbur and Hembral soil series of lower pediplains.

Soil Characteristics

Physico-chemical characteristics of the soils are presented in [Table -2]. The clay content varied from 33 to 34% in subdued hills, 8-63% in upper pediplains, 49-63% in lower pediplains and 25-45% alluvial plains. The high clay content in the soils of lower pediplains and alluvial plains as compared to upper pediplains and subdued hills was due to deposition of finer fractions in the lower pediplains and alluvial plains from the upper pediplains and subdued hills. Thus in a topo-sequence, the soil texture could be chiefly be attributed to the transportation of finer soil particles down the slope through runoff and their deposition in lower physiographic position [15]. The increase in clay content in lower horizons could be attributed to vertical migration of clay [16,17]. Silt content of all the pedons are ranges from 7 to 34%. Silt content in the soils of subdued hills and upper pediplains are almost increases with the depth whereas soils of lower pediplains and alluvial plains have irregular trends. Sand constitutes the bulk of mechanical fractions which could be attributed to dominance of physical weathering. Sand content in subdued hills and upper pediplains are higher (22 to 84%) than the lower pediplains and alluvial plains (21 to 37%). The bulk density of soils is highly dependent on texture and ranges from 1.31 to 1.85 Mg m^{-3}. The lower bulk density was noticed with clayey texture in lower pediplains and higher bulk density with Loamy texture in upper pediplains, which might be due to more compaction of finer particles in lower pediplains caused by overhead weight of the surface soils.

Table 1- Morphological characteristics of soils

Horizon	Depth (m)	Colour (moist)	Texture	Structure	Gravel (%)	Consistence			Cutans/ Slick-ensides/ cracks (cm)	Boundary	Pores	Roots	Efferves-cence
						Dry	Moist	Wet					
1. Subdued hills (10-15% slope)													
Chennampalli: Loamy skeletal, mixed, isohyperthermic, Lithic Ustorthents													
A1	0.00-0.08	7.5 YR 4/4	cl	sbk-1-f	25	ds	mfr	wss wps	-	as	c-vf-f	vf-f-p	-
Ac	0.08-0.18	5 YR 3/4	cl	sbk-1-m	25	dsh	mfr	wss wps	-	cs	c-vf-f	vf-f-p	-
C	0.18-0.48					weathered parent material of granite							
2. Upper Pediplains (1-5% slope)													
Bhupur: Fine, mixed, isohyperthermic, Typic Haplustalfs													
Ap1	0.00-0.07	7.5 YR 4/4	ls	sg-0-vf	-	dl	ml	wso wpo	-	cs	f-m	vf-f-f	-
A12	0.07-0.16	7.5 YR 4/4	sl	sbk-1-f	-	ds	mvfr	wss wps	-	gs	c-vf-f	vf-f-f	-
Bt1	0.16-0.31	5 YR 34/4	sc	sbk-1-m	-	dsh	mfr	ws wp	tn-py	gs	-	tn-py	-
Bt2	0.31-0.67	5 YR 34/4	sc	sbk-1-m	-	dsh	mfr	ws wp	tn-py	gs	-	tn-py	-
C	0.67-0.75					weathered parent materials of granite							
Devergudda: Fine loamy, mixed, isohyperthermic, Typic Haplustepts													
Ap1	0.00-0.14	5 YR 3/3	sl	sbk-1-m	8	dsh	mfr	wss wps	-	cs	c-vf-f	vf-f-f	-
Bw1	0.14-0.38	5 YR 3/3	scl	sbk-2-m	12	dsh	mfr	ws wp	-	cs	c-vf-f	vf-f-f	-
C	0.38-0.52					weathered parent material of granite							
Devsugur: Fine skeletal, mixed, isohyperthermic, Typic Ustorthents													
Ap1	0.00-0.09	10 YR 4/4	c	sbk-1-f	35	dsh	mfr	wss wps	-	cs	c-vf-f	vf-f-f	es
Ap2	0.09-0.18	10 YR 4/4	c	sbk-1-f	40	dsh	mfr	wss wps	-	cs	-	vf-f-f	ev
C	0.18-0.55					weathered parent materials of Granite							
Tatikol: Loamy skeletal, mixed, isohyperthermic, Typic Ustorthents													
Ap	0.00-0.07	5 YR 4/4	ls	sg-0-vf	35	dl	ml	wso wpo	-	cs	-	vf-f-f	-
Ac	0.07-0.18	5 YR 4/4	sl	sbk-1-f	37	ds	mvfr	wss wps	-	cs	c-vf-f	vf-f-f	-
C	0.18-0.55					weathered parent materials of granite							
3. Lower Pediplains (1-5% slope)													
Fatepur: Fine, montmorillonitic, isohyperthermic, Vertic Haplustepts													
Ap	0.00-0.08	10 YR 3/2	c	sbk-2-m	15	dsh	mfr	ws wp	-	cs	f-vf-f	vf-f-f	es
Bw1	0.08-0.16	10 YR 3/2	c	sbk-2-m	15	dsh	mfi	wvs wvp	pf	gs	vf-f	vf-f	es
Bwss2	0.16-0.27	10 YR 3/2	c	sbk-2-m	12	dsh	mfi	wvs wvp	ss	gs	vf-f	vf-f	es
C	0.27-0.68					weathered parent materials of granite							
Gabbur: Very Fine, montmorillonitic, isohyperthermic, Sodic Haplusterts													
Ap1	0.00-0.10	10 YR 3/3	c	sbk-2-m	8	dh	mfr	ws wp	-	cs	f-vf-f	vf-f-f	es
Bw1	0.10-0.32	10 YR 3/3	c	sbk-2-m	6	dvh	mvfi	wvs wvp	pf	gs	vf-f	vf-f	es
Bwss2	0.32-0.56	10 YR 3/3	c	sbk-2-m	6	dvh	mvfi	wvs wvp	ss	gs	vf-f	-	es
C	0.56-0.65					weathered parent materials of granite							
Hembral: Fine, montmorillonitic, isohyperthermic, Typic Haplusterts													
Ap1	0.00-0.08	10 YR 3/2	c	sbk-2-f	-	dh	mfr	ws wp	-	cs	f-vf-f	vf-f-f	es
A12	0.08-0.18	10 YR 3/2	c	sbk-2-m	-	dvh	mfi	wvs wvp	-	gs	f-vf-f	vf-f	es
Bw1	0.18-0.40	10 YR 3/2	c	sbk-2-m	-	dvh	mvfi	wvs wvp	pf	gs	f-vf-f	vf-f	es
Bwss2	0.40-0.81	10 YR 3/2	c	sbk-2-m	-	dvh	mvfi	wvs wvp	ss	gs	f-vf-f	-	es
Bwss3	0.81-1.10	10 YR 3/2	c	sbk-2-m	-	dvh	mvfi	wvs wvp	ss	-	f-vf-f	-	es
4. Alluvial plains (1-5% slope)													
Mudgal: Fine loamy, mixed, isohyperthermic, Fluventic Haplustepts													
Ap1	0.00-0.09	10 YR 3/2	scl	sbk-1-m	-	dsh	mfr	wss wps	-	cs	c-vf-f	vf-f-p	es
A12	0.09-0.22	10 YR 3/2	scl	sbk-2-m	-	dh	mfi	wv wp	-	gs	c-vf-f	vf-f-f	es
Bw1	0.22-0.41	10 YR 3/2	scl	sbk-2-m	-	dh	mfi	wv wp	-	gs	c-vf-f	vf-f	es
Bw2	0.41-0.95	10 YR 3/1	c	sbk-2-m	-	dh	mfi	wv wp	-	gs	c-vf-f	-	es
Bw3	0.95-1.05	10 YR 3/1	scl	sbk-2-m	-	dh	mfi	wv wp	-	-	c-vf-f	-	es

**Symbols used according to Soil Survey manual, (Soil Survey Staff, 2000 and AIS & LUS, 1970)*

The WHC of soils is highly depends on depth, clay and organic carbon content of the pedons. The minimum WHC (15.3%) was noticed in Bhupur series of upper pediplains to maximum (62.5%) in Gabbur series of lower pediplains.

The pH of the soils varied from 6.4 to 9.2 being higher in Devsugur series of upper pediplains and Fatepur, Gabbur, Hembral and Mudgal series of lower pediplains and alluvial plains. It is due to more accumulation of exchangeable bases removed from subdued hills and upper pediplains. The EC of the soils of all the series varied from 0.1 to 1.62 dS m^{-1}. The organic carbon content is low to moderately high which ranges from 0.12 to 1.55 percent respectively. The Chennampalli series of subdued hills represents pasture and grazing land use and are higher in organic carbon content whereas other series soils are low in organic carbon. It is due to addition of large amounts of leaf litter through leaf fall and also less intensive cultivation of field crops in subdued hills leading to low removal or depletion of organic carbon from the surface soils. The low CEC was noticed in the soils of Tatikol series of subdued hills followed by high CEC in the soils of Hembral series of lower pediplains which ranges from 2.5 to 47.1 cmol (P^+) kg^{-1}. The lowest ESP was noticed in Bhupur series of upper pediplains and highest in Mudgal series alluvial plains.

Table 2- Physico-chemical characteristics of soils

Soil	Mechanical Analysis (%)			Textural	pH	EC	OC (%)	$CaCO_3$	Exchangeable cations [cmol (p+) kg-1]				Sum	CEC	ESP	BS (%)	BD
Depth (m)	Sand	Silt	Clay	Class	(1:2.5)	(dsm^{-1})		(%)	Ex. Ca	Ex. Mg	Ex.Na	Ex. K					[Mg m^{-3}]
1. Subdued hills (10-15% slope)																	
Chennampalli: Loamy skeletal, mixed, isohyperthermic, Lithic Usthorthents																	
0.00-0.08	34	32	34	cl	7	0.1	1.55	3.2	6.2	6	0.5	0.4	13.1	14.6	3.4	89.7	1.34
0.08-0.18	33	34	33	cl	7.1	0.1	1.28	3.5	7.2	6.2	0.6	0.3	14.9	14.3	4.2	104.2	1.39
2. Upper Pediplains (1-5% slope)																	
Bhupur: Fine, mixed, isohyperthermic, Typic Haplustalfs																	
0.00-0.07	82	10	8	ls	6.7	0.2	0.15	1.5	2.5	3.3	0.3	0.1	6.2	6.3	4.8	98.4	1.81
0.07-0.16	80	10	10	sl	6.6	0.2	0.12	0.2	0.5	2.3	0.3	0.1	3.2	4.1	7.3	78	1.85
0.16-0.31	50	10	40	sc	6.5	0.1	0.3	2.5	4.3	4.5	0.2	0.3	9.3	11.3	1.8	82.3	1.39
0.31-0.67	47	8	45	sc	6.5	0.1	0.28	2.2	4	3.5	0.4	0.3	8.2	11.2	3.6	73.2	1.33
Devergudda: Fine loamy, mixed, isohyperthermic, Typic Haplustepts																	
0.00-0.14	81	7	12	sl	9.2	0.2	0.39	8.8	24	7.1	9.2	0.8	41.1	41.7	22.1	98.6	1.8
0.14-0.38	52	14	34	scl	9.1	0.2	0.32	4.6	24.2	8.8	11.5	0.8	45.3	45.8	25.1	98.9	1.44
Devsugur: Fine skeletal, mixed, isohyperthermic, Typic Ustorthents																	
0.00-0.09	24	15	61	c	9.2	0.2	0.39	8.8	24	7.1	9.2	0.8	41.1	41.7	22.1	98.6	1.34
0.09-0.18	22	15	63	c	9.1	0.2	0.32	4.6	24.2	8.8	11.5	0.8	45.3	45.8	25.1	98.9	1.33
Tatikol: Loamy skeletal, mixed, isohyperthermic, Typic Ustorthents																	
0.00-0.07	84	8	8	ls	7.1	0.1	0.12	0.5	1	0.7	0.2	0.2	2.1	2.5	1	84	1.42
0.07-0.18	68	14	18	sl	7	0.1	0.28	1.1	3.8	2.3	0.4	0.3	6.8	7	5.7	97.1	1.51
3. Lower Pediplains (1-5% slope)																	
Fatepur: Fine, montmorillonitic, isohyperthermic, Vertic Haplustepts																	
0.00-0.08	37	14	49	c	8.3	0.1	0.53	21.1	31.5	2.5	4.3	0.8	39.1	40.1	10.7	97.5	1.4
0.08-0.16	35	14	51	c	8.4	0.1	0.51	21.8	32.5	3	4.6	0.8	40.9	41.4	11.1	98.8	1.4
0.16-0.27	32	15	53	c	8.5	0.1	0.47	22.3	30.8	3.5	5.7	0.7	40.7	41.1	13.9	99	1.37
Gabbur: Very Fine, montmorillonitic, isohyperthermic, Sodic Haplusterts																	
0.00-0.10	23	18	59	c	8	0.1	0.37	14	32.5	5.3	5.4	0.8	44	42.8	12.6	102.8	1.43
0.10-0.32	27	12	61	c	9	0.1	0.45	13.8	28.8	6	6.5	0.8	42.1	41.9	15.5	104.7	1.37
0.32-0.56	21	16	63	c	9	0.2	0.39	13.7	30	5.8	9.8	0.8	46.4	46.8	20.9	99.1	1.34
Hembral: Fine, montmorillonitic, isohyperthermic, Typic Haplusterts																	
0.00-0.08	29	16	55	c	8.6	0.1	0.59	13.7	28.3	8	4.6	0.9	41.8	42.8	10.3	97.7	1.42
0.08-0.18	27	16	57	c	8.6	0.1	0.5	13.2	30	9	4.6	0.8	44.4	44.6	10.3	99.6	1.39
0.18-0.40	29	15	56	c	8.6	0.1	0.5	13.6	31.8	8.5	5	0.7	46	47.1	10.6	97.7	1.39
0.40-0.81	25	22	53	c	8.5	0.1	0.5	12.9	27.5	9.3	5	0.7	42.5	42.3	11.8	100.4	1.4
0.81-1.10	21	22	57	c	8.5	0.1	0.48	11.4	24.8	8.8	5.9	0.8	40.3	40.9	14.4	98.5	1.31
4. Alluvial plains (1-5% slope)																	
Mudgal: Fine loamy, mixed, isohyperthermic, Fluventic Haplustepts																	
0.00-0.09	51	18	31	scl	8.5	0.1	0.35	13	15.8	7	2.4	0.4	25.6	26.9	8.9	95.2	1.7
0.09-0.22	55	16	29	scl	8.6	0.1	0.21	14.7	13.5	7.8	3	0.3	24.6	25.6	11.7	96.1	1.72
0.22-0.41	57	16	27	scl	9.3	0.2	0.32	12.1	14.3	7	6.1	0.4	27.7	28.3	21.5	97.8	1.75
0.41-0.95	29	26	45	c	9.1	0.5	0.24	10.6	11.5	12	17.4	0.4	41.4	42.9	40.5	96.5	1.6
0.95-1.05	57	18	25	scl	8.6	1.6	0.12	8	10.5	15	10.5	0.3	36.3	38.2	27.4	95	1.7

Land Evaluation

Land evaluation classification stating land capability class, soil irrigability classification, land irrigability classification, soil suitability for paddy and hydrological soil groupings are given in [Table-3].

Soil Mapping

Chennampalli series soils of subdued hills are excessively drained, shallow, yellowish red to reddish brown and coarse loamy which mapped under C2hE3SR mapping unit and spread over 0.7 percent of the total area. Bhupur, Devsugur, Tatikol and Devergudda series soils of upper pediplains are well drained, shallow to deep, reddish brown to dark yellowish brown and fine which mapped into B4dB2, B4dC2, B4cC3, D2rB2, D2rC2, D2rC3, DG3dB2, DG3dC2, DG3kC3, T2dB2, T2dC2, T2dC2S, T2cC3SR and T2dD3SR mapping units and represents 18.7 percent of the total area. Fatepur, Hembral and Gabbur series soils of lower pediplains are moderately well drained to well drained, moderately deep to very deep, dark brown to dark grayish brown, calcareous, gravelly clay loam to clay which mapped into F3hB2, F3rB2, F3rC2, G4rB2, G4rC2, H5rB2, H5rC2 and H5rC2S mapping units and represents 74.3 percent of the total area. Mudgal series soils are moderately deep, very dark grayish brown, calcareous, sandy clay loam to gravelly clay occur on very gently to gently stream bank developed over alluvium. These soils are alkaline representing the 3.74 percent of the total area and mapped into M5kB2, M5kC2, M5rC2 and M5rC3 mapping units.

Land Capability Classification

The result on land capability classification shows that majority of soils belongs to class II (29.7), class III (32.0), class IV (8.3), Class V (26.7) and class VI (0.7) lands. Chennampalli series soils of subdued hills belong to land capability class VI and spread over 0.7 percent of the total area. These soils are suitable for permanent pasture and grazing. The surface layer of these soils are 18-22 cm thick which underlain by weathered granitic parent material. Soils of upper pediplains are well drained, shallow to deep, reddish brown to dark yellowish brown, gravelly sandy loam to gravelly clay belongs to land capability classes II, III and IV and represents the 18.7 percent of the total area. These soils are neutral to alkaline, suitable for agricultural practices with proper management whereas the soils of lower pediplains are moderately well drained to well drained, moderately deep to very deep, dark brown to dark grayish brown, calcareous, gravelly clay loam to clay belongs to land capability classes II, III and V and represents 74.3 percent of the total area. The Fatepur and Hembral series soils are suitable for intensive cultivation with proper management and Gabbur series soils are suitable for permanent pasture and grazing. Mudgal series soils are moderately deep, very dark grayish brown, calcareous, sandy clay loam to gravelly clay occur on very gently to gently slope, stream bank developed over alluvium. These soils are alkaline representing the 3.74 percent of the total area and belong to land capability class V. These soils are suitable for pasture and grassland. Similar studies were also made by [18,19].

Table 3- Mapping and evaluation of soils

Soil Mapping Unit	Description of mapping unit	Land Capability Classification	Soil irrigability Classification	Land irrigability Classification	Paddy soil group	Hydrological soil group
1. Subdued hills (10-15% slope)						
Chennampalli: Loamy skeletal, mixed, isohyperthermic, Lithic Ustorthents						
C2dE3SR	Chennampalli, shallow, gravely sandy loam, strongly sloping (10-15%) slope, severe erosion, stony and rocky.	VIes-1	E	6st	4	D
2. Upper Pediplains (1-5% slope)						
Bhupur: Fine, mixed, isohyperthermic, Typic Haplustalfs						
B4dB2	Bhupur, deep, sandy loam, very gently sloping (1-3%) slope, moderate erosion.	IIes-1	B	2st	2	C
B4dC2	Bhupur, deep, sandy loam, gently sloping (3-5% slope), moderate erosion .	IIIes-1	B	3st	3	C
B4cC3	Bhupur, deep, loamy sand, gently sloping (3-5% slope), severe erosion.	IVes-1	B	3st	3	C
Devergudda: Fine loamy, mixed, isohyperthermic, Typic Haplustepts						
DG3dB2	Devergudda, moderately deep, gravelly sandy loam, very gently sloping (1-3% slope), moderate erosion.	IIIes-2	C	3st	3	C
DG3dC2	Devergudda, moderately deep, gravelly sandy loam, gently sloping (3-5% slope), severe erosion.	IIIes-3	C	3st	3	C
DG3kC3	Devergudda, moderately deep, gravelly sandy clay loam, gently sloping (3-5% slope), severe erosion.	IVes-2	C	3st	3	C
Devsugur: Fine skeletal, mixed, isohyperthermic, Typic Ustorthents						
D2rB2	Devsugur, shallow, gravely clay, very gently sloping (1-3% slope), moderate erosion.	Ves-4	D	5st	4	D
D2rC2	Devsugur, shallow, gravely clay, gently sloping (3-5% slope), moderate erosion.	Ves-4	D	5st	4	D
D2rC3	Devsugur, shallow, gravely clay, gently sloping (3-5% slope), severe erosion.	Ves-5	D	5st	4	D
Tatikol: Loamy skeletal, mixed, isohyperthermic, Typic Ustorthents						
T2dB2	Tatikol, shallow, gravely sandy loam, very gently sloping (1-3% slope), moderate erosion	IVes-3	D	4st	4	D
T2dC2	Tatikol, shallow, gravely sandy loam, gently sloping (3-5% slope), moderate erosion.	IVes-3	D	4st	4	D
T2dC2S	Tatikol, shallow, gravely sandy loam, gently sloping (3-5% slope), moderate erosion, stony phase.	IVes-3	D	4st	4	D
T2cC3SR	Tatikol, shallow, gravely loamy sand, gently sloping (3-5% slope), severe erosion, stony and rocky phase.	IVes-4	D	4st	4	D
T2dD3SR	Tatikol, shallow, gravely sandy loam, moderately sloping (5-10%) slope, severe erosion, stony and rocky phase.	IVes-5	D	4st	4	D
3. Lower Pediplains (1-5% slope)						
Fatepur: Fine, montmorillonitic, isohyperthermic, Vertic Haplustepts						
F3hB2	Fatepur, moderately deep, clay loam, very gently sloping (1-3% slope), moderate erosion.	IIIes-2	C	3st	3	C
F3rB2	Fatepur, moderately deep, clay, very gently sloping (1-3% slope), moderate erosion.	IIIes-2	C	3st	3	C
F3rC2	Fatepur, moderately deep, clay, gently sloping (3-5%) slope, moderate erosion.	IIIes-3	C	3st	3	C
Gabbur: Very Fine, montmorillonitic, isohyperthermic, Sodic Haplusterts						
G4rB2	Gabbur, deep, clay, very gently sloping (1-3% slope), moderate erosion.	Ves-1	D	5st	4	D
G4rC2	Gabbur, deep, clay, gently sloping (3-5% slope), moderate erosion.	Ves-2	D	5st	4	D
Hembral: Fine, montmorillonitic, isohyperthermic, Typic Haplusterts						
H5rB2	Hembral, very deep, clay, very gently sloping (1-3% slope),	IIes-1	B	2st	2	D
H5rC2	Hembral, very deep, clay, gently sloping (3-5% slope), moderate erosion.	IIIes-1	B	3st	3	C
H5rC2S	Hembral, very deep, clay, gently sloping (3-5% slope), moderate erosion, stony.	IIIes-1	B	3st	3	C
4. Alluvial plains (1-5% slope)						
Mudgal: Fine loamy, mixed, isohyperthermic, Fluventic Haplustepts						
M5kB2	Mudgal, very deep, gravelly sandy clay loam, very gently sloping (1-3% slope), moderate erosion.	Ves-1	D	5st	3	D
M5kC2	Mudgal, very deep, gravelly sandy clay loam, gently sloping (3-5% slope), moderate erosion.	Ves-2	D	5st	3	D
M5rC2	Mudgal, very deep, gravelly clay, gently sloping (3-5% slope), moderate erosion.	Ves-2	D	5st	3	D
M5rC3	Mudgal, very deep, gravelly clay, gently sloping (3-5% slope), severe erosion.	Ves-3	D	5st	3	D

** Symbols used according to Klingebiel and Montgomery 1961, FAO 1976 and Soil Survey manual, (AIS & LUS, 1970).*

Soil Irrigability Classification

The soil irrigability classes are defined in terms of soil properties that express their degree of limitation for development and their requirement for irrigation management. Four soil irrigability classes are identified based on available WHC, effective soil depth, surface texture, infiltration, topography, drainage, quality of irrigation water and soil salinity and alkalinity [5]. Bhupur and Hembral soil series have moderate limitations for sustained uses under irrigation are grouped into "B" class. These soils are deep to very deep, moderately well drained, fine textured, moderately slow to slow permeable and have moderate erosion hazards. Devergudda and Fatepur soil series have severe limitations for sustained uses under irrigation are grouped into "C" class. These soils are moderately deep to very deep, moderately well drained, fine textured, moderately slow permeable and have moderate erosion hazards. Devsugur, Gabbur, Mudgal and Tatikol soil series are shallow, well drained, and loamy with moderate to severe erosion hazards. These soils are having severe limitations for sustained use under irrigation and grouped into class "D". Chennampalli series soils are shallow, coarse textured, excessively drained, developed over strongly sloping lands and having severe erosion hazards. These soils are not suitable for sustained uses under irrigation and grouped into class "E".

Land Irrigability Classification

The suitability of land for irrigation depends on physical and socio-economic factors in addition to the soil irrigability classes, quality of irrigation water, drainage pattern and other considerable factors. Bhupur, Devergudda, Fatepur and Hembral soil series have moderate to severe limitations of soils and topography for sustained uses under irrigation and classified under land irrigability sub-classes from 2st to 3st, whereas Tatikol (4st), Devsugur, Gabbur and Mudgal (5st) followed by Chennampalli (6st) with increasing limitations.

Soil Suitability for Paddy

Paddy soil group is a grouping of soils based on soil depth, pH, soil texture, puddling quality, permeability, slope and physiography. Insufficient precipitation for growing paddy needs irrigation in the survey area. Bhupur and Hembral soil series are highly suitable for paddy cultivation followed by Devergudda and Fatepur with increasing limitations. The paddy crops can be grown in the soils of Gabbur and Mudgal series adopting reclamation by gypsum. Chennampalli, Devsugur and Tatikol soil series are unsuitable for paddy cultivation.

Hydrological Soil Groupings

Hydrological Soil Groupings are suggestive of runoff generation capacity of soils from rainfall and useful for hydrological studies particularly in watershed management. Bhupur, Devergudda and Fatepur soil series have moderately high to high runoff potential and classified under class 'C' and rest are having high runoff potential and classified under class 'C'.

Soil Classification

Based on morphological, physical and chemical properties and climatic data, the soils were classified as Entisols, Inceptisols, Alfisols and Vertisols [4]. Soils of subdued hills and upper pediplains (except Bhupur and Devergudda series) do not have any diagnostic horizon, hence, classified as Entisols. Bhupur series soils were classified as Alfisols with argillic (Bt) sub-surface diagnostic horizon and Devergudda, Fatepur and Mudgal series soils having cambic (Bw) sub-surface diagnostic horizon were classified as Inceptisols. Due to wide cracks and slickensides (Bss) in sub-surface diagnostic horizons, Gabbur and Hembral series soils of lower pediplains were classified as Vertisols. Chennampalli series of subdued hills and Devsugur and Tatikol series of upper pediplains are shallow in depth and have gravelly texture, hence, grouped under Orthents at suborder level and Ustorthents at great group level due to '*ustic*' soil moisture regime. Further the Chennampalli series placed under Lithic Ustorthents at sub-surface level in the presence of lithic contact within 50 cm of the mineral soil surface whereas Devsugur and Tatikol series were placed under Typic Ustorthents in the absence of lithic contact within 50 cm of the mineral soil surface. The Bhupur series of upper pediplains showed the presence of argillic (Bt) sub-surface diagnostic horizon as evidenced by the fact that the illuvial horizon contained more clay than the eluvial horizon and also had base saturation more than 35% throughout the profile. However this series was classified as Ustalfs at sub-order level due the presence of '*ustic*' soil moisture regime. Bhupur series did not have duripan, plinthite, kandic, natric, or petrocalcic horizons and hence classifies as Haplustalfs at great group level and Typic Haplustalfs at sub group level. Devergudda and Fatepur series of lower pediplains and Mudgal series of alluvial plain were grouped under Ustepts at sub-surface level due to '*ustic*' soil moisture regime and Haplustepts at great group level because these soils did not have either duripan or calcic horizon. Further, Devergudda series did not have vertic properties and lithic contact within 50 cm from the surface, hence, this series was classifies as Typic Haplustepts at sub-group level. Fatepur series had vertic features like 2-3 cm wide cracks to a depth of 3-5 cm and slickensides in the lower horizons; hence, this series was classified as Vertic Haplustepts at sub-group level and due to irregular decrease or increase in organic carbon and clay content in lower horizons Mudgal series have been classified as Fluventic Haplustepts. Gabbur and Hembral series of lower pediplains were grouped under Usterts at sub-surface level due to '*ustic*' soil moisture regime Haplusterts at great group level. Further, Gabbur series were grouped into Sodic Haplusterts whereas Hembral series grouped as Typic Haplusterts.

Conclusion

Chennampalli series soils of subdued hills are excessively drained, shallow, yellowish red to reddish brown, gravely loamy sand to sandy loam mapped into one mapping unit which belongs to land capability class VI and spread over 0.7 percent of the total area. These soils are suitable for permanent pasture and grazing. Bhupur, Devsugur, Tatikol and Devergudda series soils of upper pediplains are well drained, shallow to deep, reddish brown to dark yellowish brown, gravelly sandy loam to gravelly clay mapped into 14 mapping units which belongs to land capability classes II, III and IV and represents the 18.7 percent of the total area. These soils are neutral to alkaline, suitable for agricultural practices with proper management. Fatepur, Hembral and Gabbur series soils of lower pediplains are mapped into 8 mapping units which belongs to land capability classes II, III and V and represents 74.3 percent of the total area. The Fatepur and Hembral series soils are suitable for intensive cultivation with proper management whereas Gabbur series soils are suitable for permanent pasture and grazing. Mudgal series soils are moderately deep, very dark grayish brown, calcareous, sandy clay loam to gravelly clay occur on very gently to gently slope stream bank developed over alluvium. These soils are alkaline representing the 3.74 percent of the total area and mapped into

4 mapping units which belongs to land capability class V and suitable for pasture and grassland. Though, the Gabbur and Mudgal series are alkaline and not suitable for agriculture but it can be improved by adopting soil reclamation measures.

Reference

[1] Ghosh S.P. (1991) *Agro-climate Zone Specific Research: India Perspective Under NARP*, Indian council of Agricultural Research, New Delhi.

[2] Hallmark Charles T., White G.N., Biswas S., Drees L.R., Biswas B., Panullah G.M., Shoaib J.U. Md., and Loeppert R. (2006) 18*th World Congress of Soil Science*, Philadelphia, Pennsylvania, USA.

[3] Ram R.L., Sharma P.K., Jha P, Das S.N. and Ahmad N. (2010) *Agropedology,* 20, 112-123.

[4] Soil Survey Staff (1998) *Keys to Soil Taxonomy, 8th ed.* SCS, USDA, Washington, D.C. USA.

[5] AIS and LUS (1970) *Soil Survey Manual,* All India Soil & Land Use Survey Organization, IARI, New Delhi.

[6] Black C.A. (1985) *Methods of Soil Chemical Analysis - Part* 2. American Society of Agronomy, Madison, Wisconsin, USA.

[7] Jackson M.L. (1979) *Soil Chemical Analysis*: *Advance Course*. University of Wisconsin, Madison and Wisconsin.

[8] Soil Survey Staff (2000) *Field Book for Describing and Sampling Soils*. National Soil Survey Centre, NRCS, USDA, Washington, USA.

[9] Klingebiel A.A. and Montgomery P.H. (1966) *Agricultural Handbook, No.* 210, USDA, Washington, USA.

[10]FAO (1976) *Soils Bulletin* 32, FAO Rome.

[11]Swaranam T.P., Velmurugan A. and Rao Y.S. (2004) *Agropedology,* 14, 114-122.

[12]Rao A.P.V.P., Naidu M.V.S., Ramawatharam N. and Rao G.R. (2008) *J. Indian Soc. Soil Sci.,* 56, 23-33.

[13]Nayak R.K., Sahu G.C., and Nanda S.S.K. (2002) *Agropedology,* 12, 1-8.

[14]Singh I.S. and Agrawal H.P. (2003) *Agropedology*, 13 11-16.

[15]Taha M. and Nanda S.S.K. (2003) *Agropedology,* 13, 30-3.

[16]Sarkar D., Baruah U., Gangopadhyay S.K., Sahoo A.K. and Velayutham M. (2002) *J. Indian Society of Soil Science*, 50, 196 -204.

[17]Hegde R., Anilkumar K.S., Rameshkumar S.C., Devaraju M. and Rudragouda (2008) *Karnataka J. Agric. Sci.*, 21(3) 373-378.

[18]Mini V., Patil P.L. and Dasog G.S. (2007) *J. Indian Society of Soil Sci.*, 55(3), 317-323.

[19]Siddhu S.S., Rana K.P.C., Lal T, Mahapatra S.K., Varma T.P. and Rao R.B.S. (2007) *J. Indian Society of Soil Sci.*, 55(3), 335-339.

SEEDLING PARAMETERS OF SOYBEAN CULTIVARS AS INFLUENCED WITH SEED STORAGE PERIODS, CONDITIONS AND MATERIALS

KANDIL A.A., SHARIEF A.E. AND SHETEIWY M.S.*
Department of Agronomy, Faculty of Agriculture, Mansoura University, Egypt.
*Corresponding Author: Email- salahco_2010@mans.edu.eg

Abstract- This investigation was elucidated seedling characters effects of soybean cultivars i.e. Crawford, Giza 111, Giza 21, Giza 22 and Giza 35 due to ageing (storage) in different packaging material i.e. seeds or pods in plastic or cloth bags for different periods i.e. 3, 6, 9 and 12 months under ambient and refrigerator conditions. Seedling parameters i.e. root and shoot length, seedling fresh and dry weight root/shoot ratio and seedling vigor index were decreased with the period of ageing. Soybean cultivars are stored in plastic bags were affected due to storage but the effects were more pronounced in the plastic bags as compared to cloth bags. It could be stated that Giza 111 exceeded the other cultivars in root length, shoot length and seedling vigor index. Giza 22 and Giza 21 cultivars exceeded the other cultivars in root/ shoot ratio. Giza 35 exceeded other cultivars in seedling dry and fresh weight. Storage under refrigerator conditions at 10$^{\circ C}$ ± 1 exceeded storage under ambient conditions in root length, shoot length, root/shoot ratio and seedling fresh and dry weight and seedling vigor index. Results clearly showed that root and shoot length, root/shoot ratio, seedling fresh and dry weight and seedling vigor index significantly affected by the varies interactions.

Keywords- Soybean, cultivars, storage, viability, deterioration, seed vigor and periods

Introduction

Soybean is one of the most important oil crops in the world agricultural economy as food for man. Seed deterioration is a serious problem in developing countries like Egypt where seeds are stored in places usually without a proper control of humidity and temperature. Seed storage conditions can regulate germination characteristics and vigor potential of seeds [14]. Reduction in seed vigor is due to decrease in seed quality, percentage, rate of germination can increase susceptibility to environmental stress [29]. Mc Donald [14] found that seed deterioration leads to decreases in seed quality, performance and stand establishment. Variations that occur in seed during aging are significant in terms of seed quality, among other things, also implies seed longevity [15]. The shoot and root length of seedling was lowest at the end of storage in cloth and the lowest seedling vigor. Balesevic-Tubic, et al. [4] showed that protection of seed viability depended on storage condition and duration. Khalilia-qdam, et al. [12] recommended that variations happened with the viability constants of a seed deterioration model between the soybean cultivars. Different periods of seed ageing and storage conditions adversely affected the seed vigor [27]. Arif, et al. [3] concluded seed viability gradually decreased from 64.5 to 39.2% as the time in storage increased, from 2 to 12 months Similar conclusions was reported by Sharma, et al. [24], Simic, et al. [25], Vijay, et al. [32].

Moreover, Arif, et al. [3], concluded that uppermost germination of seed stored at 40°C than ambient temperature. Similar results was indicated by Ghasemnezhad, et al. [8], Sharma, et al. [23], Simic, et al. [25], Vijay, et al. [32]. Mohammadi, et al. [16] incubated soybean seeds at two different temperatures of 34°C and 40°C for varying times. They indicated that seedling vigor and seedling parameters reduced by increasing storage temperature.

There were significant differences between soybean genotypes (Sb 1, Sb 2, Sb 3, Sb 4, Sb 5 and Sb 6) in seedling vigor index, shoot and root length. These changes between genotypes might be due to the genetic factors and seed chemical composition influence the expression of seed deterioration and vigor decline [28]. Khalilia-qdam, et al. [12] reported that preliminary viability was same (100%) among DPX and Sahar cultivars. Also, the vigor power i.e. electrical conductivity, seedling dry weight and seedling normal percentage had no significant difference between DPX and Sahar cultivars but in Sahar, values of all vigor trials except EC were higher than DPX and EC in Sahar was lower. Similar conclusions were reported by Kalavathi, et al. [10], Simic, et al. [25], Venkatareddy, et al. [31].

In case of dry weight per plant of soybean seed decreases all cases but the rate of weakening is highest in cloth bag [17]. Chuansin, et al. [6] reported that soybean seeds were stored in Metallized film bags and Aluminum foil bags observed highly typical germination and seed vigour, and keep water activity and seed moisture content in low level could delay seed quality deterioration followed by Polypropylene bags and woven bag. Tatipata, et al. [28] reported that soybean seeds were stored in aluminums foil bags observed highly seedling vigor and keep moisture content in low level could delay seed deterioration followed by polyethylene and wheat bags. Owolade, et al. [19] reported that sorghum seeds stored inside gene bank and freezer irrespective of the packaging materials and type of accession retain their viability to the tune of 90.67 to 100%. Whereas seeds stored at ambient temperature had low germination percentage (10.67 to 28.00%) except those stored in aluminum can (41.33%). Similar results was achieved by Arif, et al. [3], Charjan, et al. [5], Shanmugavel, et al. [22], Singh, et al. [26]. The objectives of this investigation was amid to study changes on seedling properties of some soybean cultivars under different storage condition, storage materials and periods and their interactions effect.

Materials and Methods

The experiment was conducted at experimental seed testing laboratory of Agronomy Department, Faculty of Agriculture, Mansoura University, Egypt during 2011 and 2012. The objectives of this investigation were aimed to study response of some soybean cultivars i.e. Giza 21, Giza 22, Giza 35, Giza 111 and Crawford to storage periods i.e. 3, 6, 9 and 12 months, storage conditions i.e. ambient conditions and refrigerator conditions at 10°C ±1 as well as storage materials i.e. seed with cloth bags, seed with plastic bags, pods with cloth bags and pods with plastic bags on seedling parameters.

Treatments and Experimental Design

The treatments were arranged in factorial experimental in completely randomize design, consisted of totally 160 treatments combinations involving tow storage conditions i.e. ambient conditions and refrigerator conditions at 10°C ±1, four different kinds of package materials i.e. seed with cloth bags, seed with plastic bags, pods with cloth bags and pods with plastic bags, four storage periods i.e. 3, 6, 9 and 12 months and five of soybean cultivars i.e. Giza 21, Giza 22, Giza 35, Giza 111 and Crawford. Soybean (*Glycine Max* L. Merr) cultivars were harvested in October 2011, cleaned from dust and dirt then dried and processed for storage. Every three months, from a total 12 months of storage, germination characters, electrical conductivity, acidity number, peroxide number, oil content and moisture content were evaluated.

Studied Characteristics

The following seedling parameters were studied as following:

1. **Root length (cm):** Ten normal seedlings were randomly selected in each petri dishes from final count and the root length was determined from the collar region to the tip of the primary root. The average root length of ten seedlings was computed and expressed in centimeter.
2. **Shoot length (cm):** Ten seedlings used for root length were used for shoot length measurement. The length between the collar region and the tip of the primary shoot was considered as shoot length and it was measured. The mean value of shoot length was calculated and expressed in centimeter.
3. **Root/ shoot ratio:** The root/shoot ratio was computed according to Harris, et al. [9], the root/shoot ratio is the ratio of the length of root divided on the length of the shoot.
4. **Seedling dry weight (g):** Ten seedlings used for measuring the seedling were also utilized for determining the dry weight of the seedling. Seedlings were dried in hot air oven maintained at 70±1°C for 24 hr. and then at 105°C until constant dry weight. After drying, seedlings were kept in desiccators for cooling and further weighed and expressed in grams.
5. **Seedling fresh weight (g):** Ten seedlings used for measuring seedling length were also utilized for determining the fresh weight of the seedling expressed in grams.
6. **Seedling vigor index:** Seedling vigor index was computed by adopting the formula as suggested by Abdul-baki, et al. [9] and expressed in whole number.

Vigor Index = Germination (%) x [(Shoot Length + Root Length)]

Results and Discussion

Storage Periods Effects

The results showed a significant effect of storage periods on the means of root length, shoot length, seedling dry weight, seedling fresh weight and seedling vigor index [Table-1]. The results showed that root length, shoot length, root /shoot ratio, seedling fresh and dry weight and seedling vigor index were decreased as storage periods were increased. Results revealed that before storage treatments significantly exceeded the other storage periods in root length, shoot length, root/shoot ratio, seedling fresh and dry weight and seedling vigor index followed by after 3 months. While, after 12 months from storage recorded lowest root length, shoot length, root /shoot ratio, seedling dry and fresh weight and seedling vigor index. It could be concluded that increasing storage periods from 3, 6, 9 and 12 months decreased root length by 7.73, 8.63, 13.34 and 22.08 %, respectively compared with root length before storage. Increasing storage periods from 3, 6, 9 and 12 months decreased shoot length by 5.94, 8.72, 13.27 and 21.99 %, respectively compared with shoot length before storage. Increasing storage periods from 3, 6, 9 and 12 months decreased seedling dry weight by 2.23, 3.91, 15.08 and 22.90 %, respectively compared with seedling dry weight before storage. Increasing storage periods from 3, 6, 9 and 12 months decreased seedling fresh weight by 5.01, 5.35, 9.23 and 13.34 %, respectively compared with seedling fresh weight before storage. Increasing storage periods from 3, 6, 9 and 12 months decreased seedling vigor index by 9.30, 14.24, 25.35 and 38.28 %, respectively compared with seedling vigor index before storage. In this regard, Manomani, et al. [9] reported that the decline in root length, shoot length, seedling dry and fresh weight and seedling vigor index with increase of storage periods might be due to their genetic differences age induced deterioration, inherent differences in seed structure and composition. In addition, Mohammadi, et al. [9] reported that seed deterioration results in decreased percentage of normal seedlings. Seedling growth and the fraction of seed reserve mobilization indicated a significant reduced with the advance of deterioration. Decrease of seed quality is connected with biochemical changes in seeds of oil crops. These seeds had a quick deterioration due to auto oxidation of lipids and the increase of the content of free fatty acids during storage period. The longer seeds storage period increases intensity of seeds aging [6]. These results are in good agreement with those reported by Arefi, et al. [2], Keshavulu, et al. [11], Khaliliaqdam, et al. [12], Sharma, et al. [24], Tatic, et al. [27].

Storage Cultivars Effects

Results revealed that Giza 111 cultivar significantly exceeded the other studied cultivars in root length, shoot length, root/shoot ratio and seedling vigor index. While, Giza 35 surpassed the other cultivars in seedling dry and fresh weight as shown in [Table-1]. Giza 21 cultivar recorded the lowest root length, shoot length, seedling dry and fresh weight and seedling vigor index. While, Giza 35 recorded the lowest root/shoot ratio. Giza 111 surpassed Giza 21, Giza22, Giza 35 and Crawford cultivars in root length by 33.96, 33.56, 21.40 and 9.54 %, respectively. Giza 111 surpassed Giza 21, Giza 22 and Giza 35 cultivars in shoot length by 36.39, 36.39 and 7.77 %, respectively. Giza 111 surpassed Giza 21, Giza 22, Giza 35 and Crawford cultivars in root/shoot ratio by 3.38, 3.38, 15.25 and 9.32 %, respectively. Giza 111 surpassed Giza 21, Giza 22, Giza 35 and Crawford cultivars in seedling vigor index by 40.20, 40.08, 16.22 and 12.19 %, respectively. Giza 35 surpassed Giza 21, Giza 22, Giza 111 and Crawford cultivars in seedling dry weight by 15.25, 11.29, 3.38 and 8.47 %, respectively. Giza 35 surpassed Giza 21, Giza 22, Giza 111 and Crawford cultivars in seedling fresh weight by 10.48, 5.76, 5.99 and 5.99 %, respectively. In this respect, Tatic, et al. [27] stated that there was significant differences between soybean genotypes. These differences between genotypes might be due to the genetic factors and seed chemical composition influence the expression of seed deterioration and vigor decline. Moreover, El-Abady, et al. [7] stated that assessment of some soybean cultivars seed quality during storage by monitoring germination and germination after storage percentages, in addition to seed and seedling vigor measurements may be reliable indicators for damages occurred after mechanical threshing method. These results are in good accordance with those obtained by Khaliliaqdam, et al. [12], Muhammad, et al. [18], Sharma, et al. [24].

Table 1- Averages of root length (cm), shoot length (cm), root/shoot ratio, seedling dry and fresh weight and seedling vigor index as affected by storage periods, soybean cultivars, storage conditions, package materials and their interactions.

Characters / Treatments	Root length (cm)	Shoot length (cm)	Root / shoot Ratio	Seedling dry weight (g)	Seedling fresh weight (g)	Seedling vigor index
S- Storage periods (months)						
Before storage	8.92	7.91	1.15	0.179	0.877	1601
3 months	8.23	7.44	1.14	0.175	0.833	1452
6 months	8.15	7.22	1.14	0.172	0.83	1373
9 months	7.73	6.86	1.14	0.152	0.796	1195
12 months	6.95	6.17	1.13	0.138	0.76	988
LSD 5%	0.22	0.1	NS	0.0042	0.013	153.6
V-Soybean cultivars						
Giza 21	6.57	5.4	1.22	0.15	0.777	1010
Giza 22	6.61	5.4	1.22	0.157	0.818	1012
Giza 35	7.82	7.83	1	0.177	0.868	1415
Giza 111	9.95	8.49	1.18	0.171	0.816	1689
Crawford	9.03	8.48	1.07	0.162	0.816	1483
LSD 5%	0.22	0.1	0.04	0.004	0.013	153.6
C-Storage conditions						
Ambient conditions	7.77	7.05	1.12	0.162	0.792	1252
Refrigerator conditions (10°C±1)	8.22	7.19	1.15	0.164	0.847	1391
F- test	**	**	*	NS	**	**
P-Package materials						
Pods in plastic bags	7.88	7.04	1.14	0.162	0.822	1273
Pods in cloth bags	8.17	7.25	1.14	0.163	0.821	1415
Seed in plastic bags	7.9	7.14	1.12	0.163	0.812	1322
Seed in cloth bags	8.03	7.05	1.16	0.166	0.822	1277
LSD 5%	0.17	0.11	NS	NS	NS	NS
F-test Interactions						
V×S	**	**	NS	**	**	NS
S×C	**	**	NS	**	**	NS
V×C	**	**	**	**	NS	NS
V×C×S	NS	**	NS	**	NS	NS
S×P	NS	**	NS	NS	NS	NS
V×P	NS	**	**	**	**	NS
V×S×P	NS	NS	NS	NS	NS	NS
C×P	**	NS	*	NS	NS	*
S×C×P	NS	NS	NS	NS	NS	NS
V×C×P	*	NS	NS	**	**	NS
V×C×S×P	*	NS	*	NS	NS	NS

Storage Conditions Effects

The results showed that a significant effect of storage conditions on the average of root length, shoot length, root/shoot ratio, seedling fresh weight and seedling vigor index [Table-1]. The results clearly indicated that storage under refrigerator conditions at 10°C surpassed ambient conditions in root length, shoot length, root/shoot ratio by 5.47, 1.94, 2.60, 5.50 and 9.99 %, respectively. Ageing process is affected by the genetic factors [30]. Meanwhile, Rai, et al. [21] suggest that the seed stored in jute bags enhances the storage life of maize seeds as compared to plastic bags. Seeds of inbred CM-138 showed better storability as compared with inbred CM-142. These results are in good harmony with those obtained by Arefi, et al. [2], Chuansin, et al. [6], Khaliliaqdam, et al. [12], Muhammad, et al. [18].

Storage Package Materials Effects

The results showed that a significant effect of package materials on the means of root length and shoot length [Table-1]. The results showed that root length significantly affected by package materials. Highest root length obtained from storage soybean pods in cloth bags followed by storage seed in cloth bags. However, the lowest root length obtained from storage soybean pods or seed in plastic bags without significant differences between them. Tallest shoot was obtained from storage soybean pods in plastic bags and seed in cloth bags. However, shortest shoot was obtained from storage soybean cultivars seed in cloth bags or pods in plastic bags. In this respect, Panobianco, et al. [20] stated that both the packaging materials, seedling parameters were found more affected in plastic bags as compared to jute bags in both the inbred lines of maize. Vigor percentage decreased with the period of storage. Inbred are stored in plastic bags were affected due to storage but the effects were more pronounced in the plastic bags as compared to jute bags. These results are in harmony with those reported by Arefi, et al. [2], Khaliliaqdam, et al. [12].

Interaction Effects

Interaction between Soybean Cultivars and Storage Periods

The results clearly indicated that pre storage of Giza 111 cultivar recorded the tallest roots followed by storage Giza 111 after 3 and 6 months without significant differences between them [Fig-1]. The lowest root length was recorded from storage Giza 21 and Giza 22 cultivars after 12 months without significant differences between them. The results clearly indicated that tallest shoots was produced from pre storage treatments of Giza 111 and Crawford cultivars followed by storage Crawford or Giza 111 cultivars for 3 months recorded highest shoot length. The shortest shoot was recorded from storage Giza 21 and Giza 22 cultivars after 12 months without significant differences between them [Fig-2]. The results clearly indicated that heaviest seedling dry weight was recorded from pre storage of Giza 111 cultivar followed with Crawford and storage Giza 35 after 3 months. The lowest seedling dry weight was recorded from storage Giza 21 cultivar after 12 months [Fig-3]. The results clearly indicated that highest seedling fresh weight was obtained from Giza 35 cultivar before storage followed by storage Giza 35 after 3, 6 and 9 months without significant differences between them. The lowest seedling fresh weight was recorded from storage Giza 21 cultivar after 12 months [Fig-4]. These results in agreement with those obtained by Arefi, et al. [2], Balesevic, et al. [4], Khaliliaqdam, et al. [12], Monira, et al. [17], Muhammad, et al. [18], Panobianco, et al. [20], Tatic, et al. [27].

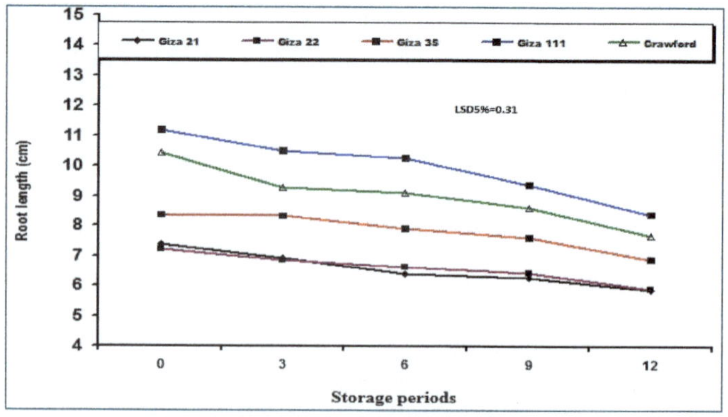

Fig. 1- Averages of root length (cm) as affected by the interaction between soybean cultivars and storage periods

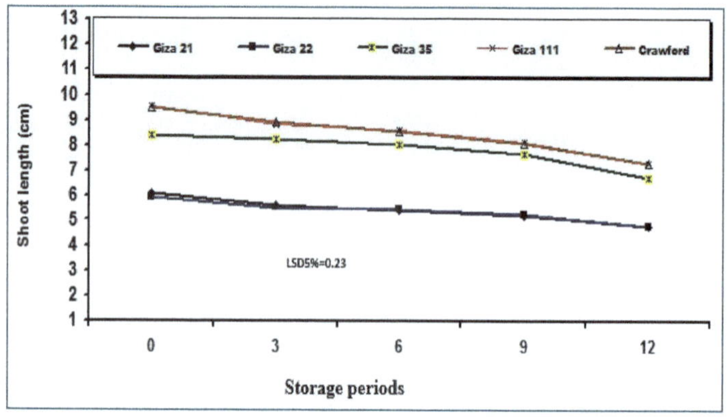

Fig. 2- Averages of shoot length (cm) as affected by the interaction between soybean cultivars and storage periods

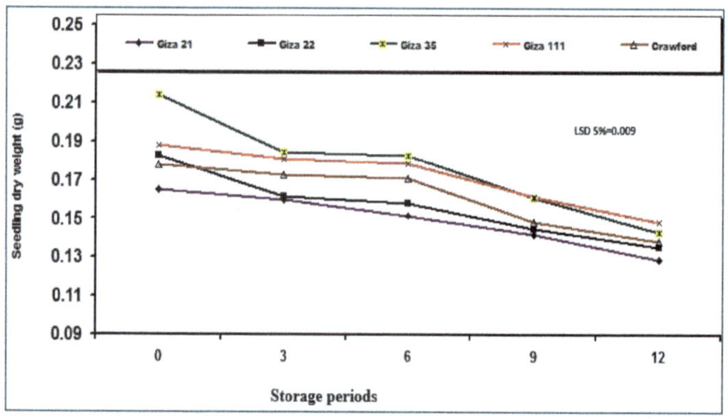

Fig. 3- Averages of seedling dry weight (g) as affected by the interaction between soybean cultivars and storage periods

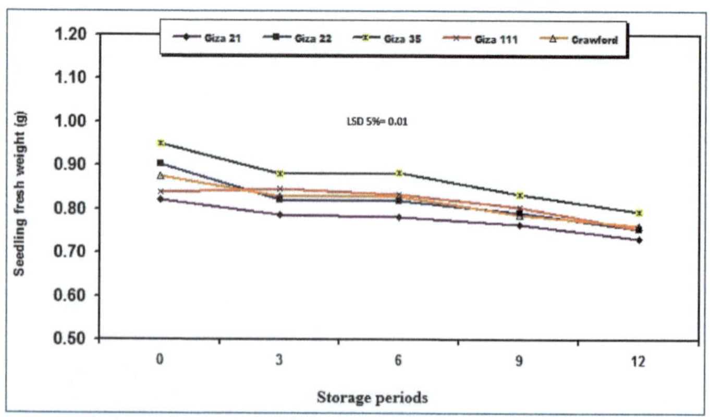

Fig. 4- Averages of seedling fresh weight (g) as affected by the interaction between soybean cultivars and storage periods

Interaction between Storage Periods and Storage Conditions Effects

The results clearly indicated that clearly indicated that root length significantly differed due to the interaction between storage periods and storage conditions. Storage under refrigerator conditions at 10° C for 3 or 6 months recorded highest root length without significant differences between them [Fig-5]. Storage under ambient conditions for 3 months came in the second rank. Shortest root was obtained from storage under ambient conditions for 12 months. The results clearly indicated that that shoot length significantly differed due to the interaction between storage periods and storage conditions [Fig-6].

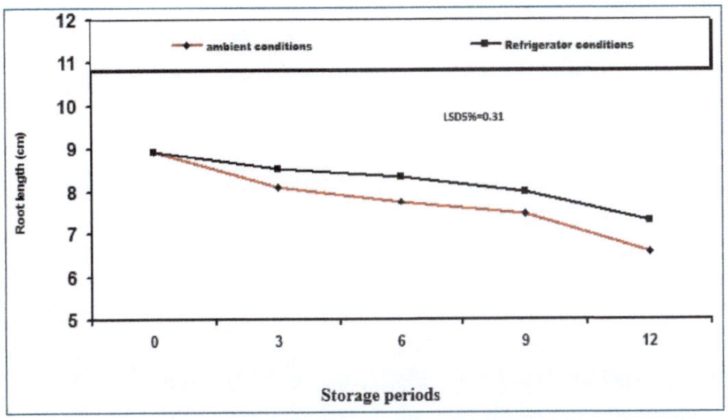

Fig. 5- Averages of root length (cm) as affected by the interaction between storage conditions and storage periods.

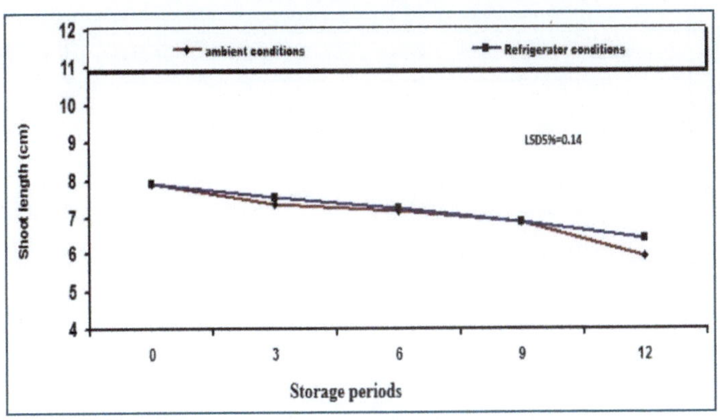

Fig. 6- Averages of shoot length (cm) as affected by the interaction between storage conditions and storage periods

It could be noticed that tallest shoots was produced from pre storage treatments followed by soybean storage under refrigerator conditions at 10°C for 3 months. The shortest shoot was obtained from storage under ambient for 12 months. The results clearly indicated that seedling dry weight significantly differed due to the interaction between storage periods and storage conditions [Fig-7]. It could be noticed that pre storage treatment recorded heaviest seedling dry weight followed by the storage under refrigerator conditions at 10°C for 3 months. Storage under ambient conditions for 3 months came in the second rank. The lowest seedling dry weight was obtained from storage under ambient for 12 months. The results clearly indicated that seedling fresh weight differed significantly due to the interaction between storage periods and storage conditions [Fig-8]. It could be noticed that pre storage treatment recorded highest seedling fresh weight followed by storage under refrigerator conditions at 10°C for 3 or 6 months without significant differences between them. Storage under ambient conditions for 3 months came in the second rank. The lowest seedling fresh weight was obtained from storage in ambient for 12 months. These results are in conformity with those reported by Arefi, et al. [2], Khaliliaqdam, et al. [12], Mohammadi, et al. [16], Panobianco, et al. [20], Sharma, et al. [24], Tatic, et al. [27].

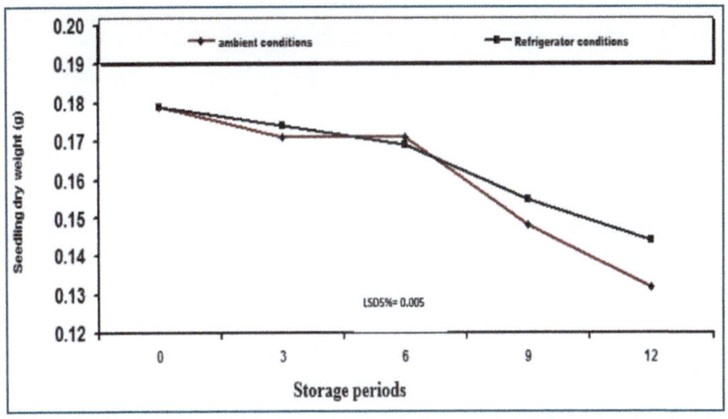

Fig. 7- Averages of seedling dry weight (g) as affected by the interaction between storage conditions and storage periods

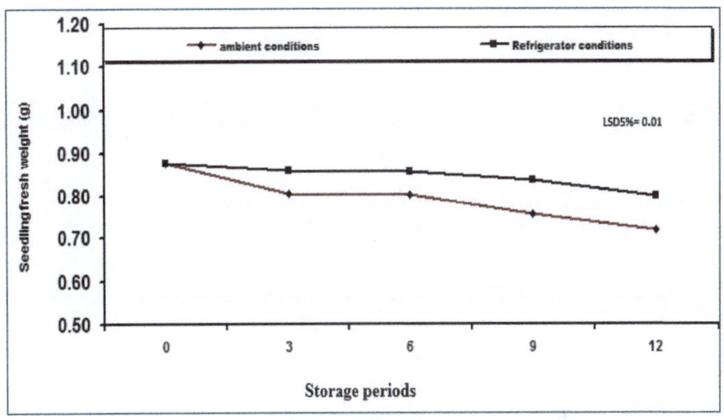

Fig. 8- Averages of seedling fresh weight (g) as affected by the interaction between storage conditions and storage periods

Interaction between Soybean Cultivars and Storage Conditions Effects

The results clearly indicated that root length significantly affected due to the interaction between soybean cultivars and storage conditions [Fig-9].

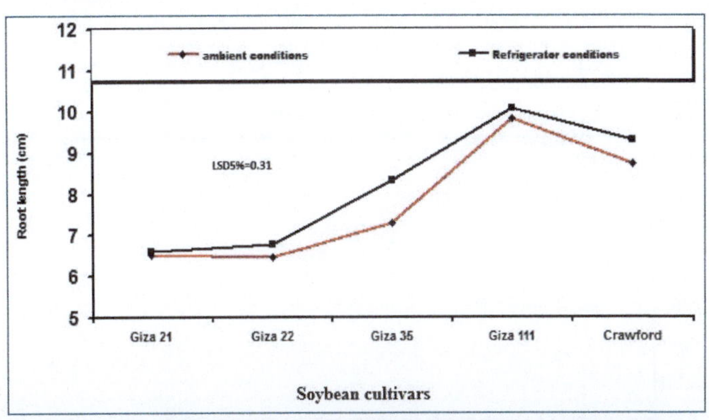

Fig. 9- Averages of root length (cm) as affected by the interaction between soybean cultivars and Storage conditions

The results clearly indicated that highest root length was obtained from storage Giza 111 cultivar in refrigerator conditions (10°C±1) followed by Crawford cultivar. The shortest root was obtained from storage Giza 21 and Giza 22 under ambient conditions without significant differences between them. The results clearly showed that shoot length significantly affected due to the interaction between soybean cultivars and storage conditions [Fig-10]. The results clearly showed that highest shoot length was obtained from storage Giza 111 cultivar in refrigerator conditions (10°C±1) followed by Crawford cultivar without significant differences between them. The shortest shoot was obtained from storage Giza 21 and Giza 22 cultivars in ambient conditions without significant differences between them. The results clearly indicated that root/shoot ratio significantly affected due to the interaction between soybean cultivars and storage conditions [Fig-11].

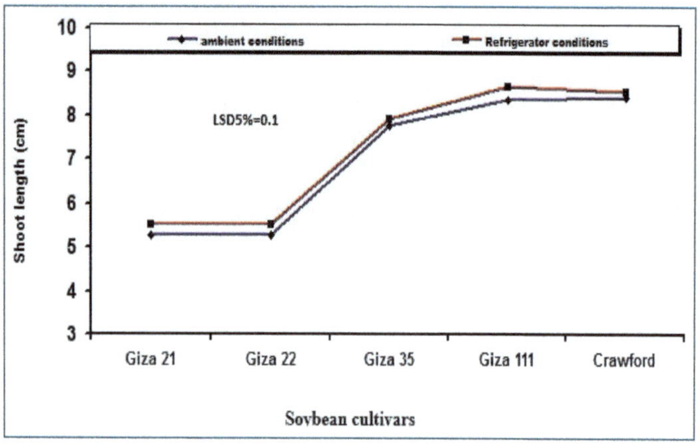

Fig. 10- Averages of shoot length (cm) as affected by the interaction between soybean cultivars and Storage conditions.

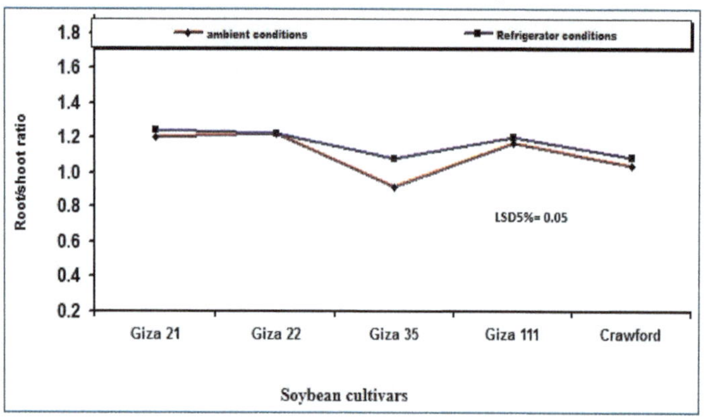

Fig. 11- Average root/shoot ratio as affected by the interaction between soybean cultivars and Storage conditions.

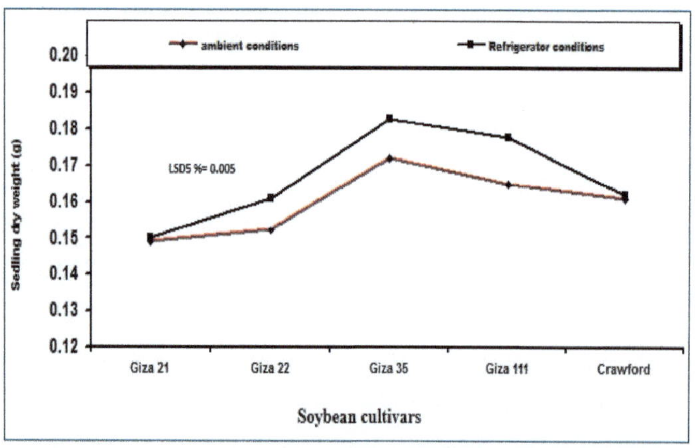

Fig. 12- Averages of seedling dry weight (g) as affected by the interaction between soybean cultivars and Storage conditions

It could be noticed that highest root/shoot ratio was obtained from storage Giza 21 in refrigerator conditions (10°C±1) followed by Giza 22 cultivars without significant differences between them. The lowest root/shoot ratio was obtained from storage Giza 35 under ambient conditions. The results clearly indicated that seedling dry weight significantly affected due to the interaction between soybean cultivars and storage conditions [Fig-12]. The results clearly showed that highest seedling dry weight was obtained from storage Giza 35 cultivar in refrigerator conditions (10°C±1) followed by Giza111 cultivar. The lowest seedling dry weight was obtained from storage Giza 21 and Giza 22 under ambient conditions without significant differences between them. These results are in conformity with those reported by Arefi, et al. [2], Khaliliaqdam, et al. [12], Mohammadi, et al. [16], Panobianco, et al. [20], Tatic, et al. [27].

Interaction between Storage Periods, Soybean Cultivars and Storage Conditions

The results clearly showed significant interaction between storage periods × soybean cultivars × storage conditions on shoot length [Fig-13]. It could be noticed that pre storage treatment recorded tallest shoot of Giza 111 or Crawford cultivar followed by storage Giza 111 and Crawford in refrigerator conditions at 10°C for 3 months. However, shortest shoots was produced from storage Giza 21 and Giza 22 cultivars under ambient or refrigerator conditions for 12 months without significant differences between them. The results clearly showed significant interaction on seedling dry weight between storage periods × soybean cultivars × storage conditions [Fig-14].

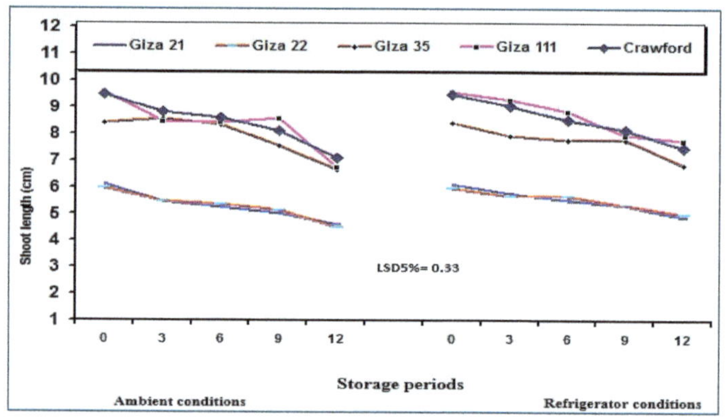

Fig. 13- Averages of shoot length (cm) as affected by the interaction between storage periods, soybean cultivars and Storage conditions.

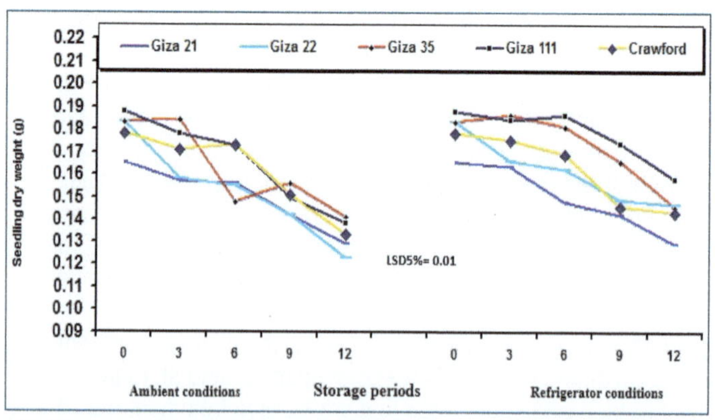

Fig.14- Averages of seedling dry weight (g) as affected by the interaction between storage periods, soybean cultivars and Storage conditions

It could be noticed that pre storage of Giza 111 cultivar recorded heaviest seedling dry weight followed by storage Giza 35 and Giza 111 under refrigerator conditions at 10°C for 3 months. However, the lowest seedling dry weight were produced from storage Giza 22 under ambient conditions for 12 months and or Giza 21 under ambient or refrigerator conditions at 10°C without significant differences between them. These results are in good agreement with those reported by Balesevic, et al. [4], Tatic, et al. [27].

Interaction between Storage Periods and Package Materials

The results clearly indicated that shoot length significantly affected due to the interaction between storage periods and package materials [Fig-15]. It could be noticed that tallest shoots was obtained from pre storage treatment followed by storage soybean cultivars seeds in plastic bags for 3 months. The shortest shoots was obtained from storage soybean pods in plastic bags for 12 months. These results are in conformity with those reported by Arefi, et al. [2], Khaliliaqdam, et al. [12], Mohammadi, et al. [16], Panobianco, et al. [20], Sharma, et al. [24], Tatic, et al. [27].

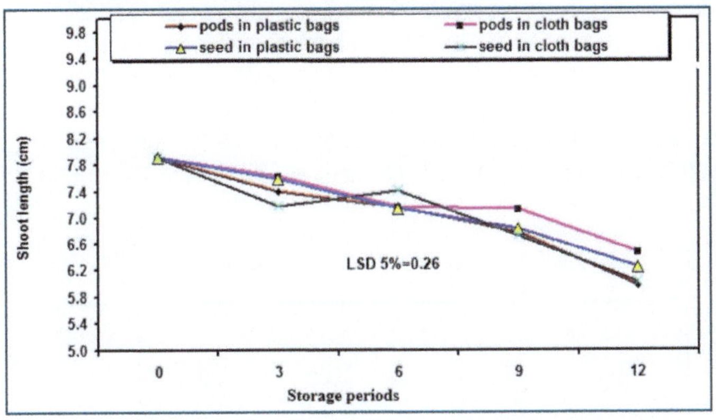

Fig. 15- Averages of shoot length (cm) as affected by the interaction between storage periods and package materials

Interaction between Soybean Cultivars and Package Materials

The results clearly indicated that shoot length significantly differed due to the interaction between soybean cultivars and packages materials [Fig-16]. The results clearly showed that the tallest shoots were produced from storage Giza 111 pods in cloth bags. The shortest shoots were obtained from storage Giza 21 or Giza 22 pods in plastic bags or cloth bags, as well as seed in plastic bags or cloth bags without significant differences between them. The results clearly indicated that highest root/shoot ratio were produced from storage Giza 22 seeds in plastic or cloth bags without significant differences between them [Fig-17]. The lowest root/shoot ratio were obtained from storage Giza 35 seeds or pods in plastic bags without significant differences between them. The results clearly indicated that highest seedling dry weight were produced from storage Giza 35 pods in plastic bags [Fig-18]. The lowest seedling dry weight were obtained from storage Giza 21 pods in plastic bags or in cloth bags and seed in plastic bags without significant differences between them. The results clearly indicated that highest seedling fresh weight were produced from storage Giza 35 seeds or pods in plastic bags without significant differences between them [Fig-19]. The lowest seedling fresh weight was obtained from storage Giza 21 pods or seeds in plastic or cloth bags without significant differences between them. These results are in conformity with those reported by Khaliliaqdam, et al. [12], Sharma, et al. [24], Tatic, et al. [27].

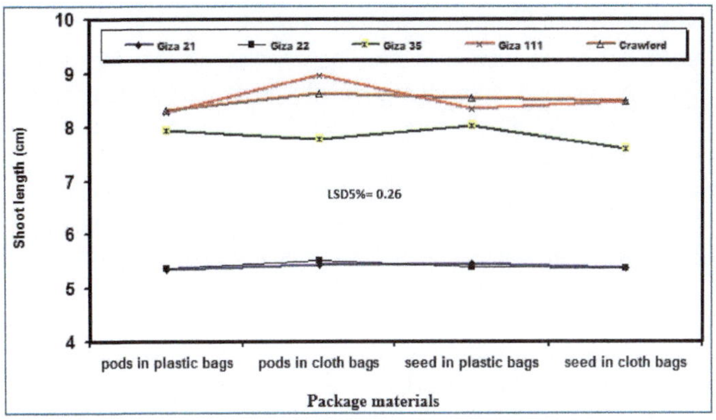

Fig. 16- Averages of shoot length (cm) as affected by the interaction between soybean cultivars and package materials.

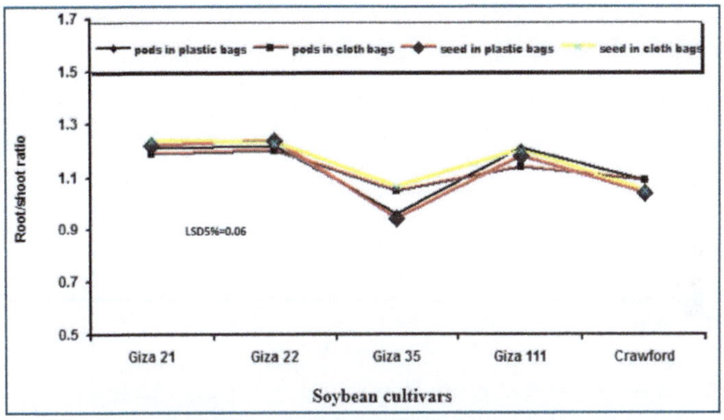

Fig. 17- Averages of root/shoot ratio as affected by the interaction between soybean cultivars and package materials.

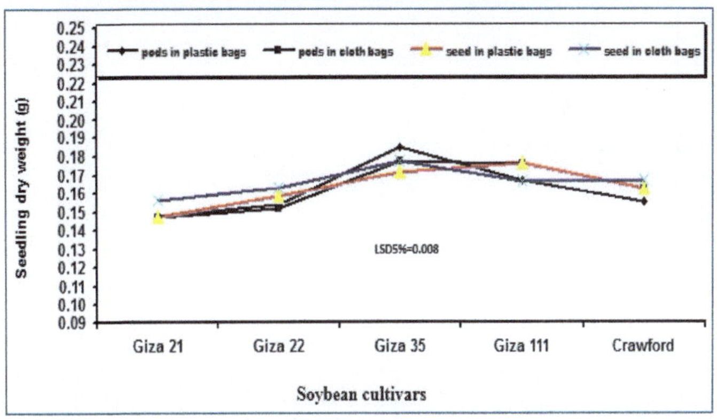

Fig. 18- Averages of seedling dry weight (g) as affected by the interaction between soybean cultivars and package materials.

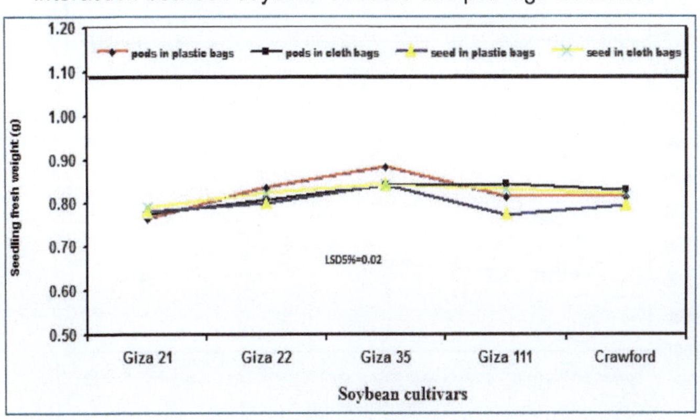

Fig. 19- Averages of seedling fresh weight (g) as affected by the interaction between soybean cultivars and package materials

Interaction between Storage Conditions and Package Materials

Root length due to the interaction effect between storage conditions and package materials showed significant effects on root length [Fig-20]. Highest root length was obtained from storage soybean pods or seeds in cloth bags under refrigerator conditions at 10°C without significant differences between them. However, the shortest root were obtained from storage soybean seeds in cloth bags and pods in plastic bags under ambient conditions without significant differences. The results clearly showed that highest root/shoot ratio were obtained from storage both soybean cultivars seeds or pods in cloth bags under refrigerator conditions at 10°C without significant differences between them [Fig-21].

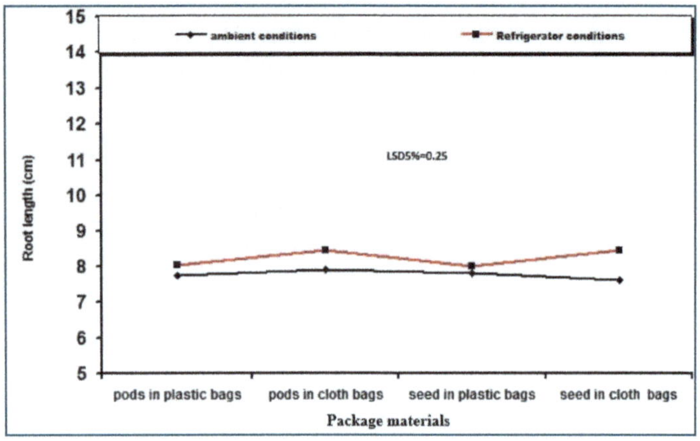

Fig. 20- Averages of root length (cm) as affected by the interaction between storage conditions and package materials.

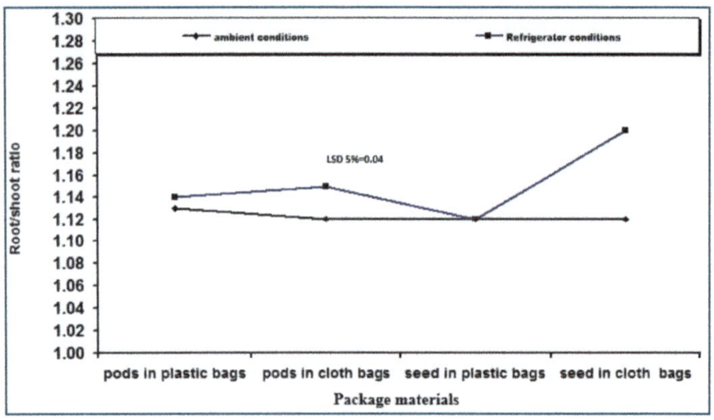

Fig. 21- Averages of root/shoot ratio as affected by the interaction between storage conditions and packages materials.

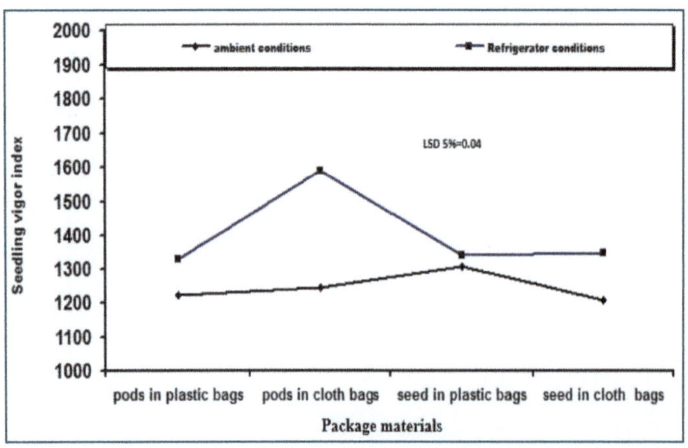

Fig. 22- Averages of seedling vigor index as affected by the interaction between storage conditions and packages materials

However, the lowest root/shoot ratio were obtained from storage soybean cultivars pods or seeds in cloth bags and seeds in plastic bags under ambient conditions without significant differences. The results clearly indicated that highest seedling vigor index was obtained from storage soybean pods in cloth bags under refrigerator conditions (10°C±1). However, the lowest seedling vigor index was obtained from storage soybean as pods or seed in plastic or cloth bags under ambient conditions without significant differences [Fig-22]. These results in good harmony with those reported by Khaliliaqdam, et al. [12], Muhammad, et al. [18], Sharma, et al. [24], Tatic, et al. [27].

Interaction between Soybean Cultivars, Storage Conditions and Package Materials

The results clearly indicated that tallest root was obtained from storage Giza 111 when storage in refrigerator conditions as pods or seeds in cloth bags followed by Crawford cultivar when storage in pods or seeds in cloth bags [Fig-23]. Shortest root was obtained from storage Giza 21 under ambient conditions or under refrigerator in pods or seeds in cloth bags or plastic bags followed by Giza 35 cultivar as pods or seed in plastic or cloth bags, in refrigerator conditions or in ambient conditions. The results clearly indicated that highest seedling dry weight was obtained from storage Giza 35 cultivar under refrigerator conditions pods in plastic or cloth bags followed by Giza 111 cultivar pods or seeds in Cloth or plastic bags [Fig-24].

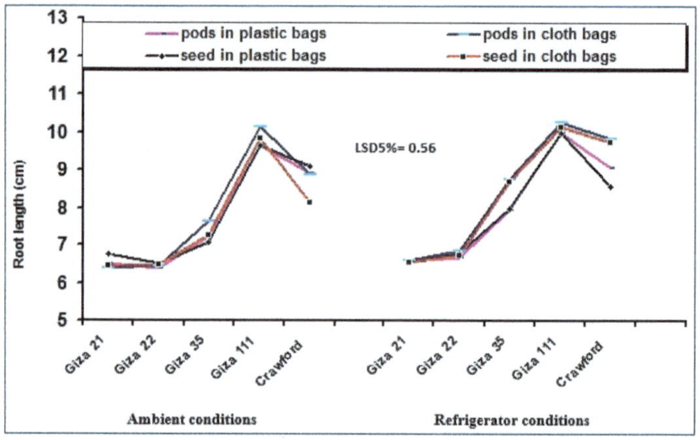

Fig. 23- Averages of root length (cm) as affected by the interaction between soybean cultivars, storage conditions and package materials.

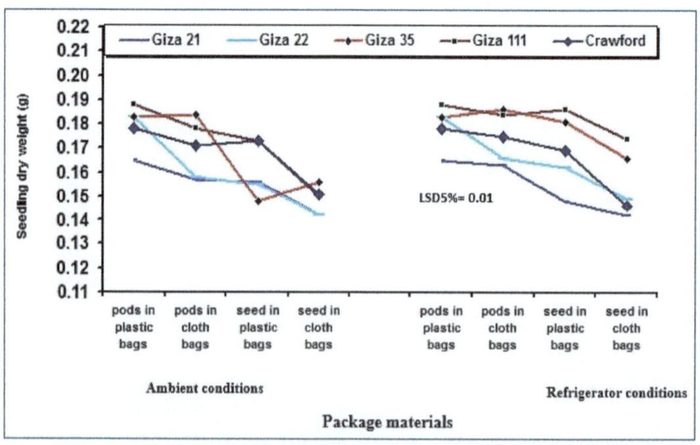

Fig. 24- Average seedling dry weight (g) as affected by the interaction between soybean cultivars, storage conditions and package materials.

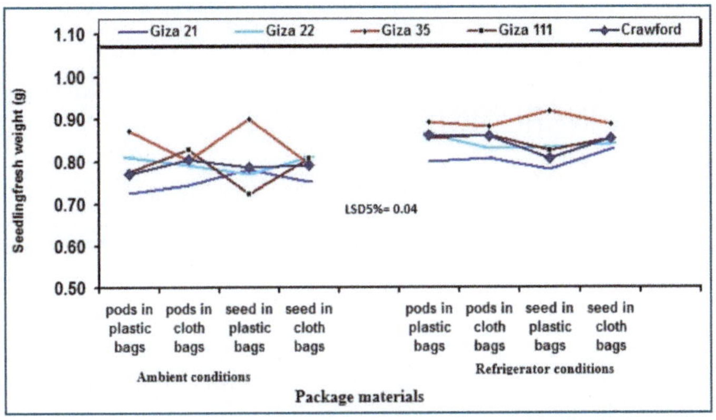

Fig. 25- Average seedling fresh weight (g) as affected by the interaction between soybean cultivars, storage conditions and package materials

Lowest seedling dry weight was obtained from storage Giza 21 cultivar under ambient conditions pods or seeds in plastic or cloth bags. The results clearly indicated that highest seedling fresh weight was obtained from storage Giza 35 cultivar under refrigerator conditions of soybean seed or pods in plastic or cloth bags without significant differences between them [Fig-25]. Lowest seedling fresh weight was obtained from storage Giza 21 cultivar under ambient conditions of soybean pods in plastic bags or soybean seed in plastic or cloth bags without significant differences between them. These results in agreement with those obtained by Arefi, et al. [2], Khaliliaqdam, et al. [12], Mohammadi, et al. [16], Muhammad, et al. [18], Tatic, et al. [27].

Conclusion

This investigation was revealed that for maximizing soybean viability and seedling parameters, it should be storage Giza 111 cultivar under refrigerator conditions (10°C±1) pods or seeds in cloth bags for 3 months.

Acknowledgment

Sincere and gratitude thanks to Prof. Dr. Ali E. Sharief, Agronomy Dep., Faculty of Agriculture, Mansoura University, for his help and extraordinary effort during the achievement of M. Sc. Thesis. Their guidance and encouragement during the extraordinary effort during writing this manuscript.

References

[1] Abdul-baki A.A. and Anderson J.D. (1973) *Crop Science*, 13, 630-633.

[2] Arefi H.M. and Abdi N. (2003) *Iranian J. Rangelands and Forests Plant Breeding and Genetic Res.*, 11, 105-125.

[3] Arif M. (2006) *Khutab Al Nabi (Prophets Sermons) Collection, Compilation, Research, Interpretation, Explanation, Unraveling Strange Words & Determining the Date*, Ph. D Dissertations, NWFP Agricultural University Peshawar, Pakistan.

[4] Balesevic Tubic S., Tatic M., Dordevic V., Nikolic Z. and Dukic V. (2010) *Helia*, 33(52), 153-160.

[5] Charjan S. and Tarar J.L. (1992) *Indian Journal Agriculture Science*, 62, 500-502.

[6] Chuansin S., Vearasilp S., Srichuwong S. and Pawelzik E. (2006) *Conference on International Agricultural Research for Development*, University of Bonn, Bonn, Germany.

[7] El-Abady M.I., El-Emam A.A.M., Seadh S.E. and Yousof F.I. (2012) *Research Journal of Seed Science*, 3, 1-11.

[8] Ghasemnezhad A. and Honermeier B. (2007) *International Journal of Plant Production*, 3(4), 41-50.

[9] Harris R.W. (1992) *Journal of Arboriculture*, 18(1), 39-42.

[10] Kalavathi D.K., Ramamoorthy A. and Sundareswaran S. (1994) *Madras Agricultural Journal*, 81(11), 612-615.

[11] Keshavulu K. and Krishnasamy V. (2005) *Seed Research*, 33 (2), 208-210.

[12] Khaliliaqdam N., Soltani A., Latifi N. and Gaderi F.F. (2012) *American-Eurasian J. of Agri. and Enviro. Sci.*, 12(2), 224-230.

[13] Manomani V. (2002) *Seed Research*, 30(1), 158-160.

[14] Mc Donald M.B. (1999) *J. Seed Sci. Technol.*, 27, 177-273.

[15] Milošević M. and Malešević M. (2004) *Semenarstvo. Naučni institut za ratarstvo i povrtarstvo,* Novi Sad, 110-114.

[16] Mohammadi H., Soltani A., Sadeghipour H.R. and Zeinali E. (2010) *International Journal of Plant Production.,* 5(1), 65-70.

[17] Monira U.S., Amin M.H.A., Marin M. and Mamun M.A.A. (2012) *Bangladesh Research Publications Journal*, 7(4), 421-427.

[18] Muhammad A. (2008) *Effect of Planting Dates and Storage on Yield and Quality of Indigenous Land Races and Improved Varieties of Soyabean*, Ph.D. Thesis, Agri. Uni. Peshawar, Pakistan.

[19] Owolade O.F., Olasoji J.O. and Afolabi C.G. (2011) *African Journal of Plant Science.*, 5(15), 873-877.

[20] Panobianco M. and Vieira R.D. (2007) *Revista Brasileira de Sementes*, 29(2), 97-105.

[21] Rai P.K., Kumar G. and Singh K.K. (2011) *Journal of Agricultural Technology*, 7(6), 1765-1774.

[22] Shanmugavel S., Varier A. and Dadlani M. (1995) *Seed Research*, 23(2), 61-66.

[23] Sharma S., Virdi P., Gambhir S. and Munshi S.K. (2006) *Journal Research Punjabi Agriculture University*, 43, 29-33.

[24] Sharma S., Gambhir S. and Manshi S.K. (2007) *Asian Journal of Plant Sciences*, 6(3), 502-507.

[25] Simic B., Sudaric A., Liovic I., Kalinovic I., Rozman V. and Cosic V. (2006) 9th *International Working Conference on Stored Product Protection*, 59-63.

[26] Singh P., Thakur C.P.D., Katiyar R.P. and Gupta P.K. (1998) *Seed Research*, 9(2), 75-80.

[27] Tatic M., Balesevic-Tubic S., Dordevic V., Miklic V., Vujakovic M. and Dukic V. (2012) *Helia*, 35(56), 119-126.

[28] Tatipata A. (2009) *Journal of Agricultural Technology*, 5(1), 51-64.

[29] TeKrony D.M., Egli D.B., Wickham D.A. (1989) *J. Crop Sci.*, 29, 1523-1528.

[30] Varghese K. and Rai P.K. (2005) *Variation in Growth and Performance of Maize Inbreds Stored under Ambient Conditions*, M.Sc. (Ag.) Thesis, Dept. of GPB and Seed Sci. and Tech, AAI (DU), Allahabad, UP, India.

[31] Venkatareddy D.M., Bommegowda A. and Paramesh D.S. (1992) *Seed Tech. News*, 22(2), 41-46.

[32] Vijay D. and Dadlani M. (2009) *Plant Molecular Biology*, 22, 282 -291.

EVALUATION OF *Pongamia pinnata* SEED EXTRACTS AS AN INSECTICIDE AGAINST AMERICAN BOLLWORM *Helicoverpa armigera* (HUBNER)

REENA[1], RAM SINGH[2] AND SINHA B.K.[3]

[1]Dryland Research Sub-Station, SKUAST-J, Dhiansar, Bari Brahmana- 183311, Jammu, J&K, India.
[2]Department of Entomology, CCS Haryana Agricultural University, Hisar- 125 004, India
[3]Division of Biochemistry and Plant Physiology, FOA, SKUAST-J, Chhatha- 180009, India
*Corresponding Author: Email- sinha_bk@indiatimes.com

Abstract

Background- Karanj has been reported to possess insecticidal properties; however most of the work done so far is restricted to its oil. Karanj seeds (both mature as well as immature) methanolic extracts (2.5, 5.0, 7.5 and 10.0%) were assessed for their insecticidal activities against *Helicoverpa armigera* (Hubner).

Results- The mature seed extract 5.0% exhibited more than 50% first instar larval mortality and more than 65% third instar larval feeding deterrence. Besides, a pronounced decrease in food consumption and utilization indices was noticed when fourth instar larvae were fed on extract treated chickpea pods. Adults when fed on the extract treated sucrose diet (0.5, 1.0, 1.5 and 2.0%) exhibited a marked reduction in oviposition and egg hatching. Extracts also deterred the adult females from egg laying when applied on to the oviposition substrate (no-choice, choice and without actual contact condition) at concentrations as low as 2.5%. Hatching inhibition of one-day old eggs on application of extracts was recorded. Mature seed extracts gave better results than the immature one, though both were effective.

Conclusion- Repeated application of mature karanj seed methanolic extract 2.5% may be incorporated in integrated pest management programmes to take care of *H. armigera* menace on crop plants.

Key words- *Helicoverpa armigera*, food utilization indices, methanolic extracts, oviposition deterrence, *Pongamia pinnata*
Short Title- Karanj seed extracts against *H. armigera*

Introduction

Plants are the richest source of organic chemicals. Many secondary plant metabolites are found to possess insecticidal properties. Plant products and their derivatives sometimes provide effective control of insect pests that have become resistant to other insecticides, are short- lived in the environment, pose relatively low risk to non-target and beneficial organisms and are naturally occurring or can be derived or manufactured with minimal technology and thus cost effective. Besides neem, the work on other botanical pesticides needs to be strengthened considerably [1]. *Pongamia pinnata* (L.) (Leguminosae and subfamily Papilionacae), commonly called Indian beach in English and popularly known as karanj, is a hardy, draught resistant, salinity tolerant, medium sized fast growing tree found in almost all parts of India as an avenue tree. Its different parts (*viz.*, leaves, root, bark, flowers and seeds) containing a number of furano-flavonoid compounds are known to possess pesticidal activity. The prinicipal furano flavonoids present in the seed oil are karanjin, pongamol, pongapin, glabrin, karanja chromene, karanjone and pongaglabrone [2]. The insecticidal potential of karanj has been aptly reviewed [3]. Most of the work done so far on karanj is restricted to its oil.

American bollworm, *Helicoverpa armigera* (Hubner) is cosmopolitan and is recognized worldwide as an international pest because of its polyphagous and multigenerational nature as well as high damage potential infesting more than 182 species of plants. A survey of insecticide use pattern to control this pest on cotton shows an increasing tendency of the farmers to undertake repeated applications of the insecticides with doses much higher than recommended. About 40% of the cost of cultivation in cotton accounts for chemical pesticides [4]. Hence, resistance development in this pest has been reported from different parts of the world *viz.*, Pakistan [5]; Australia [6]; India [7], etc.

The risks to human health and environmental side effects have forced to look for greener alternatives to chemical pesticides. Food production today must not only ensure food supply, but also human health and environmental safety, ecological viability as well as cost effectiveness [8]. Hence, in view of these considerations, it was considered worthwhile to assess the efficacy of karanj seeds methanol extract against *H. armigera*.

Experimental Methods

Pongamia pinnata immature pods were collected during 2nd and 3rd week of April while mature pods were collected during 1st week of May, 2002 from the trees grown on the campus, Chaudhary Charan Singh Haryana Agricultural University, Hisar, India. Seeds were separated by breaking the pods and allowed to shade-dry for a month, after which they were crushed into bits.

Methanolic extract of both mature as well as immature crushed seeds were prepared following the standard procedure of refluxing and distillation [9]. The crushed seeds (2.5 kg at a time) were taken in a round bottom flask (5 litre cap.) and methanol AR grade (2.5 litres) was added to it to immerse the bits. Refluxing was done by fitting the flask with a water condenser and boiling the set using a heating mantle for 6h. The extract was then filtered out of the flask and was concentrated by distillation process. This procedure of refluxing and distillation was repeated thrice for complete extraction of active ingredients from Karanj seeds. Likewise 20 kg seeds (10 kg of mature and immature each) were extracted to obtain 1650 and 1450 ml of methanolic extracts respectively. A stock concentration of ten per cent (w/v) was prepared in acetone and further dilutions were made using water to obtain different concentrations *viz.*, 2.5, 5.0 and 7.5% of methanolic extracts. Karanj oil (1.0%) was taken as a standard check. Suitable control (without extract) was also maintained for all the experiments.

Helicoverpa larvae were collected from CCS Haryana Agricultural University chickpea fields during February, 2003 and were reared on chickpea twigs bearing leaves, flower as well as pods as per the standard procedure [10]. All the experiments were conducted under laboratory conditions and were replicated thrice.

Estimation of Contact Larval Toxicity and Antifeedant Action

Contact toxicity was assessed against freshly hatched (<12 h old) first instar larvae using dry film technique. 0.5 ml of different concentrations of extracts and karanj oil (1.0%) prepared in acetone was taken in rimless test tube (20.0 X 2.0 cm) and evaporated by rotating the tube manually using flywheel. The larvae were allowed to remain in contact with dry films for four hours and mortality counts were taken thereafter. Moribund larvae were counted as dead.

Antifeedant Action

Deterrent activity of various extracts was assayed by offering treated and untreated weighed food (chickpea twigs bearing leaves as well as pods) to ten third instar larvae. The amount of unconsumed food was recorded after 24 h and per cent antifeedancy was calculated [11]. Weight loss in food due to drying / desiccation was adjusted by keeping suitable controls.

Effect on Consumption and Utilization Indices of Larvae

Chickpea pods dipped for 30 seconds in various concentrations, air-dried and weighed were provided to ten pre-weighed fourth instar larvae. Observations on increase in larval weight, uneaten food and faecal matter were recorded after 48 h of feeding. Various indices *viz.*, Consumption index (CI), Growth Rate (GR), Efficiency of Conversion of Ingested Food (ECI), Approximate Digestibility (AD) and Efficiency of Conversion of Digested Food (ECD) were calculated on wet weight basis [12].

Effect of Treated Diet of Adults on Egg Laying and Subsequent Hatching of Eggs

Different concentrations of extracts (0.5, 1.0, 1.5 and 2.0%) were prepared in 15 per cent sucrose solution. Four pairs of one-day old adults were released in a glass jar (21 X 15 cm) with its mouth covered with muslin. A cotton swab dipped in different concentrations of extracts was suspended inside the glass jar and clean muslin was provided as oviposition substrate [13]. Number of egg laid in 48 h and per cent hatching were recorded.

Assessment of Oviposition Deterrence Effects on Adults

Muslin treated with various concentrations of extracts were exposed to four pairs of adults in different situations *viz.*, choice (half of the oviposition substrate was treated), no-choice (whole of the substrate was treated) and without actual contact (jar was partitioned into two halves using a wire mesh and in one half treated muslin was kept while in other half adults were released along with untreated muslin). Number of eggs laid in 48 h was recorded under different situations [13].

Evaluation of Ovicidal Action

Muslin laden with one-day old *H. armigera* eggs (appox. 50) was taken from the general culture for this experiment. The eggs were then dipped for 5 seconds in various concentrations of extracts and per cent hatching was recorded to evaluate the ovicidal action.

Statistical Analysis

The data were subjected to analysis of variance (ANOVA) in a Completely Randomized Design after appropriate transformations. The difference of two means between treatments exceeding Critical Difference (CD) value is significant [14].

Results

Contact Larval Toxicity

Mature and immature seed extracts were on par with each other in the larval contact toxicity studies (Table 1) at all concentrations tested. With the increase in concentration from 5.0% to 10.0%, resultant mortalities did not differ significantly, though it was highest at 10.0% (83.33%). As compared to control, per cent mortality was significantly more in all the treatments. Further, the extracts were superior to karanj oil (36.67%) at all concentrations except at 2.5% where they were comparable with the oil. Acetone in itself was non-toxic as evident from per cent mortality (3.33%).

Table 1- Contact and antifeedant activity of Pongamia pinnata seed methanolic extract against Helicoverpa armigera

Extract	Conc. (%)	Larval mortality(%)[a]	Larval feeding deterrence(%)[a]
Mature seed methanolic extract	2.5	40.00 (39.23)	42.08 (40.44)
	5.0	53.33 (46.91)	52.73 (46.56)
	7.5	76.67 (61.12)	65.03 (53.75)
	10.0	83.33 (65.91)	73.77 (59.19)
Immature seed methanolic extract	2.5	36.67 (37.27)	31.69 (34.26)
	5.0	53.33 (46.91)	46.45 (42.96)
	7.5	73.33 (58.91)	60.11 (50.83)
	10.0	76.67 (61.12)	67.76 (55.40)
Karanj oil	1.0	36.67 (37.27)	36.34 (37.07)
Control (Water)		_	0.00 (4.05)
Control (Acetone)		3.33 (8.85)	0.57 (5.98)
SE(m)±		(2.69)	(0.65)
CD (P=0.05)		(7.92)	(1.89)

[a]*Figures in parentheses are arc-sine transformed values*

Antifeedant Action

The extracts treated chickpea pods largely deterred *H. armigera* larvae (Table 1). Minimum deterrence (31.69%) was recorded on food treated with 2.5% immature seed methanolic extract, while, the maximum (73.77%) in mature seed methanolic extract (10.0%) treated food, though all were superior over control. Feeding deterrence (%) showed a positive correlation with the extracts used. Even at 2.5% concentration mature seed methanolic extract showed 42.08% larval feeding deterrence. Mature seed extract proved better than the immature one at all concentrations and excluding immature seed extract (2.5%). Karanj oil (36.34%) was inferior to all other treatments.

Consumption and Utilization Indices

Food consumption, utilization, assimilation and growth efficiencies were significantly inhibited (Table 2) when the *H. armigera* larvae were fed on extract treated food, the inhibition being more at higher concentrations. The indices varied from 3.32-3.79 mg/mg/day (CI), 0.12-0.68 mg/mg/day (GR), 12.56-17.87% (ECI), 41.99-44.31% (AD) and 59.77-79.01% (ECD) at mature seed extract (10.0%) and immature seed extract (2.5%), while in control they were CI (3.99), GR (0.72), ECI (17.92), AD (45.18) and ECD (79.39). Karanj oil was the least effective except immature seed extract (2.5%). However, all the treatments effectively reduced the various indices as compared to control.

Effect of Treated Adult Diet on Egg Laying and Hatching

All the treatments greatly inhibited egg output as well as subsequent hatching of eggs in *H. armigera* adults offered with extract treated sucrose diet (Table 3). At concentrations higher than 2% methanolic seed extract, mortality of adults was noticed. So, the treatment concentrations were reduced accordingly for this experiment. Even at the lowest concentration (0.5%) of immature seed methanolic extract, egg laying was greatly reduced (76.00) as compared to control (166.00). Both mature as well as immature seed methanolic extract (1.0, 1.5, 2.0%) proved better than karanj oil (61.67 eggs). At higher concentration (2.0%) egg laying as well as their subsequent hatching was greatly hampered (6.33 eggs with 26.11% hatching and 13.00 eggs with 36.21% hatching in mature and immature seed extract respectively).

Table 2- Assessment of Pongamia pinnata seed methanolic extract on consumption and utilization indices of Helicoverpa armigera

Extract	Conc. (%)	Consumption index (CI)	Growth Rate (GR)	Efficiency of Conversion of Ingested Food (ECI)	Approximate Digestibility (AD)	Efficiency of Conversion of Digested Food (ECD)
Mature seed methanolic extract	2.5	3.62	0.59	16.23	43.91	73.94
	5.0	3.59	0.54	15.06	42.96	69.99
	7.5	3.44	0.49	14.16	42.71	67.12
	10.0	3.32	0.12	12.56	41.99	59.77
Immature seed methanolic extract	2.5	3.79	0.68	17.87	44.31	79.01
	5.0	3.62	0.58	16.10	43.85	73.47
	7.5	3.56	0.53	14.87	43.52	68.36
	10.0	3.33	0.17	14.12	42.29	66.77
Karanj oil	1.0	3.72	0.60	16.20	44.10	73.52
Control (Water)		3.99	0.72	17.92	45.18	79.39
Control (Acetone)		4.01	0.72	17.89	45.54	78.64
SE(m)±		0.04	0.01	0.29	0.34	1.67
CD (P=0.05)		0.11	0.03	0.88	1.00	4.92

Table 3- Effect of karanj seeds methanolic extracts treated sucrose diet on egg laying and per cent hatching of Helicoverpa armigera eggs

Extract	Conc. (%)	Mean no. of eggs laid in 48 h[a]	Hatching (%)[b]
Mature seed methanolic extract	0.5	62.33 (7.89)	72.43 (58.32)
	1.0	37.00 (6.08)	63.84 (53.03)
	1.5	14.33 (3.78)	48.35 (44.06)
	2.0	6.33 (2.52)	26.11 (30.73)
Immature seed methanolic extract	0.5	76.00 (8.72)	75.48 (60.32)
	1.0	51.67 (7.19)	64.45 (53.39)
	1.5	26.33 (5.13)	55.54 (48.18)
	2.0	13.00 (3.61)	36.21 (36.99)
Karanj oil	0.2	61.67 (7.85)	69.21 (56.29)
Control (15% sucrose solution in water)		166.00 (12.88)	86.15 (68.15)
Control (15% sucrose solution in acetone)		163.67 (12.79)	84.93 (67.15)
SE(m)±		(0.19)	(1.06)
CD (P=0.05)		(0.55)	(3.10)

[a]*Figures in parentheses are square-root transformed values*
[b]*Figures in parentheses are arc-sine transformed values*

Oviposition Deterrent Action

Under no-choice condition (Table 4) egg laying capacity of *H. armigera* adults was greatly suppressed, being only 12.33 eggs at 10.0% mature seed methanolic extract. While, the maximum was in 2.5% immature seed extract (83.67 eggs), which was too low as compared to control (168.33). Excluding 2.5% mature and immature seed extracts, karanj oil was inferior to all other treatments. Under choice condition (Table 4) number of eggs laid ranged from 16.67-70.33 on treated and 71.00-83.33 on untreated substrate. Maximum reduction over untreated substrate was recorded at highest concentration (10.0%) of mature and immature seed extract (76.39 and 75.81% reduction in egg laying on treated substrate over untreated respectively in mature and immature seed extract) while, the minimum was at the lowest concentration (2.5%). Except at 2.5% concentration, both mature as well as immature seed extracts showed greater suppression of oviposition than karanj oil. *H. armigera* adults laid fewer eggs (Table 4) in all the treatments under without actual contact condition of egg laying (varying from 87.33-156.67 eggs) than in control (171.00 eggs), though they were not in direct contact with the treated muslin. It followed the same trend as under no-choice condition, but the effect was less pronounced. All the treatments proved better than karanj oil except mature and immature seed extract (2.5%).

Table 4- Effect of extracts on mean no. of eggs laid by 4 females in 48 h under different conditions of egg laying

Extract	Conc. (%)	No-choice	Choice Condition			Without actual contact
			Treated	Untreated	% reduction over untreated	
		74.00[a]	72.33[a]	84.33[a]	14.25[b]	151.00[a]
	2.5	(8.60)	(8.51)	(9.18)	(22.18)	(12.29)
		42.00	52.33	80.67	37.19	139.67
		(6.48)	(7.23)	(8.98)	(37.58)	(11.82)
		23.67	28.67	79.33	63.78	114.67
	7.5	(4.87)	(5.35)	(8.91)	(52.99)	(10.71)
		12.33	16.67	71.00	76.39	87.33
Mature seed methanolic extract		(3.51)	(4.08)	(8.43)	(60.93)	(9.35)
		83.67	70.33	83.00	15.13	156.67
	2.5	(9.15)	(8.39)	(9.11)	(22.89)	(12.52)
		58.67	55.67	80.67	28.85	138.00
		(7.66)	(7.46)	(8.98)	(32.49)	(11.75)
		32.33	30.33	82.00	62.99	118.00
	7.5	(5.69)	(5.51)	(9.06)	(52.53)	(10.86)
		21.33	18.00	74.67	75.81	92.00
		(4.62)	(4.24)	(8.64)	(60.54)	(9.59)
		61.67	60.00	83.00	27.71	152.00
	1.0	(7.85)	(7.74)	(9.11)	(31.76)	(12.33)
		168.33	168.33	168.33	0.00	171.00
		(12.97)	(12.97)	(12.97)	(4.05)	(13.08)
		169.33	167.00	167.00	0.00	170.33
Immature seed methanolic extract		(13.03)	(12.92)	(12.92)	(4.05)	(13.05)
		(0.15)	(0.13)	(0.10)	(1.22)	(0.13)
		(0.45)	(0.39)	(0.30)	(3.57)	(0.37)
		(5.69	(5.51)	(9.06)	(52.53)	(10.86)
	10	21.33	18	74.67	75.81	92
		(4.62)	(4.24)	(8.64)	(60.54)	(9.59)
Karanj oil	1	61.67	60	83	27.71	152
		(7.85)	(7.74)	(9.11)	(31.76)	(12.33)
Control (Water)		168.33	168.33	168.33	0	171
		(12.97)	(12.97)	(12.97)	(4.05)	(13.08)
Control (Acetone)		169.33	167	167	0	170.33
		(13.03)	(12.92)	(12.92)	(4.05)	(13.05)
SE(m)±		(0.15)	(0.13)	(0.1)	(1.22)	(0.13)
CD (P=0.05)		(0.45)	(0.39)	(0.3)	(3.57)	(0.37)

[a]*Figures in parentheses are square-root transformed values*
[b]*Figures in parentheses are arc-sine transformed values*

Ovicidal Action

The mature seed extracts proved better over the immature one, though both were effective in inhibiting hatching in *H. armigera* eggs (Table 5) than control. Hatching percentage varied from 22.31-78.13 in mature seed extract (10.0%) and immature seed extract (2.5%) respectively. Significant reduction in hatching was noticed with the progressive increase in concentrations. Methanolic extracts of both mature as well as immature seeds (10.0%) proved better than karanj oil (1.0%).

Table 5- Ovicidal action of karanj seed methanolic extracts against one-day old Helicoverpa armigera eggs

Extract	Conc. (%)	Egg hatching (%)[a]
	2.5	70.37 (57.02)
Mature seed methanolic extract	5	51.48 (45.85)
	7.5	38.95 (38.62)
	10	22.31 (28.19)
	2.5	78.13 (62.12)
Immature seed methanolic extract	5	57.11 (49.09)
	7.5	45.03 (37.52)
	10	27.52 (31.64)
Karanj oil	1	36.83 (37.37)
Control (Water)		91.22 (72.76)
Control (Acetone)		84.33 (66.68)
SE(m)±		(0.48)
CD (P=0.05)		(1.42)

Discussion

The higher efficacy of mature seed extract of methanol compared to that of immature seed (Table 1-5) was probably due to more accumulation of active principles (responsible for insecticidal activity) in the mature karanj seeds.

Contact larval toxicity

Methanolic extracts exerted mortalities as high as 83.33% in *H. armigera* first instar larvae. More than 50.00% larval mortality was noticed at 5.0% of mature as well as immature seed methanolic extract. Though methanolic seed extract have not been studied so far for its contact action, aqueous karanj plant extract was found toxic to 4th instar *Spodoptera litura* larvae when applied topically [15].

Antifeedant Action

Both mature as well as immature seeds methanolic extracts had pronounced antifeedant effect on *H. armigera* third instars; the larval feeding deterrence ranging from 42.08-73.77% in mature and 31.69-67.76% in immature seed methanolic extract (Table 1). Study on this aspect of karanj seed extract is lacking. However, antifeedant properties of aqueous karanj extract have been reported against various pests [16].

Consumption and Utilization Indices

The concentration dependent reduction in dietary utilization indices (Table 2) relative to controls is directly related to the reduction in food consumption. This conclusion is corroborated by the results of direct bioassays for larval feeding deterrence, the treated food no-choice tests (Table 1), where substantial feeding inhibition on treated food was observed. This suggests that reduction in growth may result from both behavioural and physiological (post ingestive) effects [17]. The reduced ECI and ECD result from a reduction in the efficiency to convert foodstuffs into growth, perhaps by a diversion of energy from production of biomass into detoxification [18]. Likewise, rate of food consumption, assimilation and production by 5th instar *Euproctis fraterna* larvae, showed a negative correlation with the concentration of *P. pinnata* leaves acetone extract [19].

Effect of Treated Adult Diet on Egg Laying and Hatching

H. armigera adults fed on extracts treated sucrose diet laid fewer eggs and per cent hatching of eggs was also significantly reduced (Table 3). Decrease in egg laying and hatching was in a concentration dependent manner, the reduction being more at higher concentrations and vice versa. Adult mortality at concentrations higher than 2.0% shows the toxic effect of these extracts on *H. armigera* adults. Suppressed fecundity reveals the effect of adult diet on its various physiological activities. Further, reduced hatching (%) shows the carry over effect of these extracts. Likewise, *Helopeltis theivora* adults fed with karanj aqueous extract treated shoots exhibited reduced fecundity [20]. The resultant F1 adults obtained from the adults fed with treated shoots and further reared on untreated shoots also exhibited similar effect, but to a lesser extent.

Oviposition Deterrent Action

Oviposition as well as subsequent egg hatching was greatly hampered (Table 4) under all the three situations of egg laying. Reduced egg laying under no-choice condition shows the capacity of karanj seed methanolic extract to render the host plant unsuitable on application for egg laying by *H. armigera*. Moreover, given a choice between treated and untreated host plant, the *H. armigera* females will prefer the untreated one as clear from the choice experiment. Reduction in oviposition on the untreated substrate as well as egg hatching under choice and without actual contact conditions could be due to the deterrent / repellent action of the volatiles emanating from the treated substrate. Reduction in egg hatching especially on the treated substrate demonstrates the contact action of the extracts on eggs as proved in our earlier experiment with larvae (Table 1). Though karanj oil has been found to act as oviposition deterrent against several pests [21, 22], but in the present study, oil was found less effective than the extracts.

Ovicidal Action

The extracts adversely affected hatching in *H. armigera* eggs as evident form Table 5. Reduction in egg hatching demonstrates the contact action of extracts on one-day old *H. armigera* eggs even when exposed for just five seconds. Even at the lowest concentration tested (2.5%) the extracts reduced egg hatching to 70.37 and 78.13% in mature and immature seed methanolic extract, while in control it was 91.22%. Egg hatchability of *Corcyra cephalonica* was disrupted by the application of aqueous karanj seed extracts [22].

Conclusions

There are possibilities of incorporating mature karanj seed methanolic extract in integrated pest management programmes at 5.0% concentration. Repeated application of this extract at different stages of insect development can very well take care of *H. armigera* on crop plants like cotton, etc., where it is a menace. The results demonstrate the potential of karanj seed extracts for further development into a botanical insecticide against *H. armigera* under field conditions considering its effectiveness as well as ease of preparation.

References

[1] Jayaraj S. (2002a) *Modern Trends in Integrated Pest Management.*, 1-9.
[2] Bringi N.V. (1987) *Non traditional oil seeds and oils in India.*
[3] Kumar M. and Singh R. (2002) *Biol. Agric. Hort.*, 20, 29-50.
[4] Jayaraj S. (2002b) *Modern Trends in Integrated Pest Management*, 188-191.
[5] Ahmad M., Arif M.I. and Ahmad Z. (2003) *Crop Prot.* 22(3), 539-544.
[6] Gunning R.V. and Moores G.D. (2002) *The BCPC Conference Pests and Diseases,* 1&2, 793-798.
[7] Phokela A., Rao R.V.S. and Singh S.P. (2002) *Indian J. Ent.* 64(1), 108-110.
[8] Paroda R.S. (2003) *Sustaining our Food Security.*
[9] Kumar R., Ram Singh, Bharti Suhag P. and Kalidhar S.B. (2001) *J. Cotton Res. Dev.* 15, 156-161.
[10]Jaglan M.S., Khokhar K.S., Malik M.S. and Singh R. (1997) *J. Agric. Food Chem.* 45, 3262-3268.
[11]Facknath S. and Kawol D. (1993) *Insect Sci. Applic.*, 14, 571-574.
[12]Waldbauer G. (1968) *Adv. Insect Physiol.*, 5, 229-288.
[13]Gajmer T., Singh R., Saini R.K. and Kalidhar S.B. (2002) *J. App. Ent.*, 126, 238-243.
[14]Panse V.G. and Sukhatme P.V. (1978) *Indian Council of Agricultural Research, New Delhi.*
[15]Behera U.K. and Satapathy C.R. (1996) *Insect Envi.*, 2, 43-44.
[16]Rao N.S., Rajendran R. and Raguraman S. (2002) *J. Ent. Res.* 26, 233-238.
[17]Koul O. and Isman M.B. (1990) *J. Insect Physiol.* 37, 591-598.
[18]Koul O. and Isman M.B. (1991) *Insect Sci. Applic.*111, 47-53.
[19]Sridhar S. and Chetty J.S. (1989) *Proc. Indian Acad. Sci., Animal Sci.*, 98, 313-323.
[20]Deka M.K., Handique R., Singh K. and Hazarika L.K. (1998) *Crop Res.*, 16(1), 102-105.
[21]Morale R.S., Sarnaik D.N., Satapute U.S. and Sadawarte A.K. (2000) *Pestology,* 24(1), 26-28.
[22]Yadav J.P. and Bhargava M.C. (2002) *Indian J. Plant Prot.* 30 (2), 207-209.

EXPORT OPPORTUNITIES AND COMPETITIVENESS OF VEGETABLE CROPS IN GUJARAT

PAL V.*, JADAV K.S., DEVI G. & ZALA Y.C.

Department of Agricultural Economics, B.A. College of Agriculture, Anand Agricultural University, Anand - 388110, Gujarat, India.
*Corresponding Author: Email- vivekpal_167@yahoo.in

Abstract- The vegetable crops are priority in agriculture by virtue of its vast potential in improving the socio-economic conditions of the farmers due to its production potentiality in a short period. In this study tabular analysis and Nominal Protection Coefficient measures has been used to measure the export potentiality and competitiveness of different vegetable crops in Gujarat. Greater potentiality of vegetable export was found due to its increasing trend of production and productivity. Potato as well as onion crop was found greater potentiality of export from Gujarat. The export of different vegetables i.e. cabbage, potato, tomato and onion was found competitive to other countries and among these four major crops, cabbage was found to be highest export competitive crop. Therefore, cabbage crop needs special attention for promotion of export from Gujarat. Singapore, Indonesia and Qatar and Russia were found to be highest export competitiveness for cabbage, onion, tomato and potato, respectively.

Keywords- Export opportunities, competitiveness, vegetable crops, India

Introduction

Agriculture has been a way of life and continues to be the single most important livelihood of the masses. India is the second largest economy in Asia after China, as measured in terms of its GDP [1]. Indian agriculture has undergone rapid transformation in the past two decades. The policy of globalisation and liberalisation has opened up new avenues for agricultural modernisation. The Prime Minister's Economic Advisory Council (PMEAC), headed by C. Rangarajan, recommended promotion of high-value agriculture and reforms in agri-marketing policies including the APMC Acts for boosting further growth [2]. Trade in fruit and vegetables products has been among the most dynamic areas of international agricultural trade, stimulated by rising incomes and growing consumer interest in product variety, freshness, convenience and year round availability [3]. Agriculture is projected to grow at 4.8 per cent in 2013-14 as against 1.9 per cent in 2012-13. It said output of horticultural produce (fruits and vegetables) and animal husbandry (milk, eggs and meat) as well as that of fisheries are likely to remain in upward trajectory with annual growth of around 4 to 5 per cent in the current fiscal [2]. India's diverse climate ensures availability of all varieties of fresh fruits & vegetables. Presently, India is the second largest producer of fruits & vegetables in the world after China.

The cropping pattern in Gujarat has been shifting away from the low value traditional crops to high value commercial crops with business and export potential. Gujarat has substantial potential in the exports of agri-products. The export intensity of agri-products in Gujarat (11.9%) is substantially higher than the nation (3.2%), which makes the agri-exports from Gujarat positively sensitive to the exchange rate depreciation unlike the national exports of manufactures or agri -products [4]. The most urgent need today is to increase the production of nutritious food in a sustainable manner and improve farm income to ensure household food and nutritional security, while conserving the natural resource base. Vegetables are the vital sources of minerals, vitamins and dietary fibers and play an important role in supplying nutrition to human health. Vegetables require comparatively lesser quantities of agro-inputs to grow [5]. The state has ample potential to export various processed products of the vegetables as powder and dehydrated forms, canned vegetables, pickles, chutney (Sauce) & ketchup. Nearly half of the produce is being exported. The state has a variety of agro climatic conditions which allows one to produce diversified crops. Gujarat is one of the fastest transforming states in India. Farmers are moving towards high value crops and cash crops which allow them to earn greater income through export. In the backdrop of growing above potentialities, this study was conducted to analyse the opportunities and competitiveness for the export of different vegetables crops at global level.

Methodology

To achieve the specific objectives of the study, secondary data on area, production, productivity and export of different vegetable for TE 2010-11 were collected and compiled from the different official sources like various issues of Food and Agricultural Organisation (FAO) Trade Year Book, Export Statistics for Agro-Food Products (an Agricultural and Processed Food Exports Development Authority (APEDA) publication), Horticultural Database (a National Horticulture Board (NHB) publication, Statistical Abstract (Directorate of economic & statistics, Gandhinagar) and from Directorate General of Commercial Intelligence and Statistics, Ministry of Commerce & Industry, Govt. of India. The study also required information on transportation cost including cost of grading, packaging, loading, chemical material, fumigation etc. from Ahmadabad to export destination point (Kandla, Gujarat) during pre-export of different vegetable products, which is obtained from Greenways shipping, Rajendranagar, Ahmedabad. The Wholesale prices of different vegetables at Ahmadabad market recorded during the year of 2010-11 have been taken as domestic price of the crops.

The export competitiveness of different vegetable crops for the year 2010-11 was assessed by using a simple measure known as Nominal Protection Coefficient (NPC).This is the ratio of domestic price to border price.

Symbolically

$$NPC = Pd/Pb$$

Where, NPC = Nominal Protection Coefficient

Pd = Domestic price of commodity

Pb = Border price or reference price of commodity after taking care of transportation and marketing expenses.

NPC shows degree of divergence of domestic price from the international price and determines the degree of export competitiveness of the commodity in question *i.e.* if the NPC < 1 the commodity under consideration is of export competitive [6].

In order to identify the most favorable countries for export NPC value was computed for different importing countries of vegetable crops and classified into three categories *i.e.* Highly competitive (NPC < 0.5), Moderately competitive (0.5 < NPC < 1) and non competitive (NPC > 1).

Results and Discussion

The major vegetables producing countries in the world with respect to area, production and productivity for the average of three years *i.e.* 2008-09 to 2010-11 and the percentage share of area and production in total vegetable production in the world was shown in [Table-1]. The results revealed that highest proportion of area and production of vegetables in the world was found in China followed by India, USA, Turkey, and Egypt and lowest was found in Spain. In case of productivity, Spain was found highest (35.90 MT/ha) followed by USA (31.61 MT/ha), Italy (25.80 MT/ha), Iran (25.52 MT/ha), whereas India was found at lowest position with the productivity of 16.74 MT/ha. Greater proportion of area as compared to the proportion of its production in India shows greater potentiality to increase its productivity as well as its production in future.

[Table-2] shows that trend in percentage shares of area and production of Gujarat in total vegetable production in India. It was found that trend in area under vegetable crops in India as well as in Gujarat was increasing over a decade except during year 2002-03 and 2003-04 in India and during 2006-07 and 2008-09 in Gujarat. Further it is interesting to see that the trend in area under vegetables production in India has shown greater fluctuation over a decade as compared to Gujarat [Fig-1]. The trend in production under vegetable crops in India as well as in Gujarat was increasing over a decade except during 2002-03 in India and during 2006-07 and 2008-09 in Gujarat. In case of productivity Gujarat was found slightly higher as compared to India in most of the period in a decade. Almost similar trend of gradually increasing performance of productivity of vegetable crops has been observed in India as well as in Gujarat [Fig-2]. The trend in percentage share of Gujarat in total area and production of India has also been increased over a decade [Fig-3]. It is inferred that potentiality of export of vegetable crop was found reasonably higher in Gujarat as compared to India.

Table 1- Major vegetables producing countries in the world (TE 2010-11)

Country	Area	Per cent area (%)	Production	Per cent production (%)	Productivity
China	23069.1	42.47	462875	47.78	20.06
India	8153.6	15.01	136456	14.09	16.74
USA	1149.31	2.12	36325.8	3.75	31.61
Turkey	1099.77	2.02	26710.3	2.76	24.29
Iran	650.8	1.2	16608.7	1.71	25.52
Russian Fed.	796.17	1.47	13880.1	1.43	17.43
Egypt	899.98	1.66	17479.6	1.8	19.42
Italy	528.1	0.97	13627.4	1.41	25.8
Spain	353.67	0.65	12695.5	1.31	35.9
OTHERS	17613.5	32.43	232036	23.95	13.17
WORLD	54314	100	968694	100	17.84

(Area in 000 hectares/ Production in 000 MT/ Productivity in MT/ hectare)

Table 2- Area, production and productivity of total vegetables in India and Gujarat over a decade

Year	Area			Production			Productivity	
	India	Gujarat	Per cent of GJ (%)	India	Gujarat	Per cent of GJ (%)	India	Gujarat
2001-02	6156	236.7	3.85	88622	3299.13	3.72	14.4	13.94
2002-03	6092	248.16	4.07	84815	3515.16	4.14	13.9	14.16
2003-04	6082	325.38	5.35	88334	4579.82	5.18	14.5	14.08
2004-05	6744	331.43	4.91	101246	4866.98	4.81	15	14.68
2005-06	7213	380.18	5.27	111399	6308.24	5.66	15.4	16.59
2006-07	7581	366.05	4.83	114993	6062.63	5.27	15.2	16.56
2007-08	7848	411.78	5.25	128449	7402.92	5.76	16.4	17.98
2008-09	7981	394.89	4.95	129077	6807.07	5.27	16.2	17.24
2009-10	7985	406.81	5.09	133738	7255.49	5.43	16.7	17.84
2010-11	8495	515.92	6.07	146555	9379.47	6.4	17.3	18.18

(Area in 000 hectares/ Production in 000 MT/ Productivity in MT/ hectare)

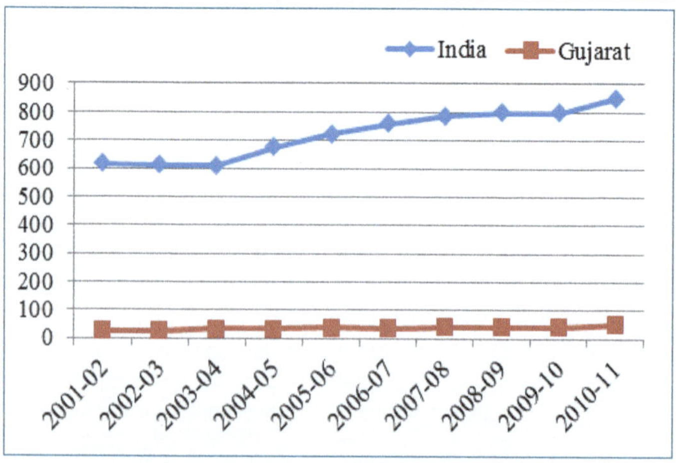

Fig. 1- Trend in Area ('000 ha) under Vegetables Production in India and Gujarat

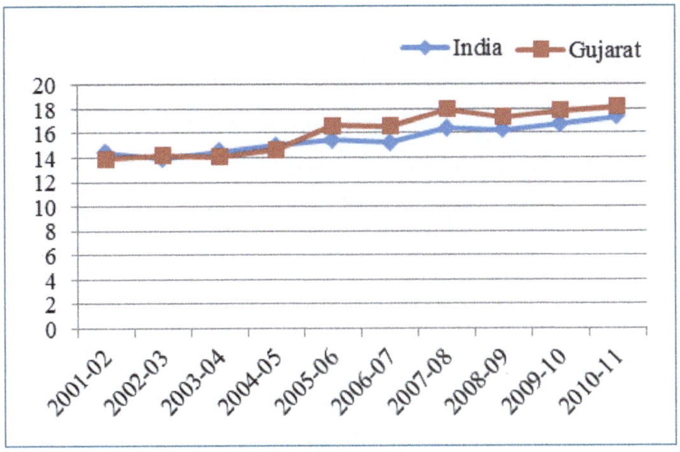

Fig. 2- Trend in Productivity (MT/ha) of vegetables in India and Gujarat

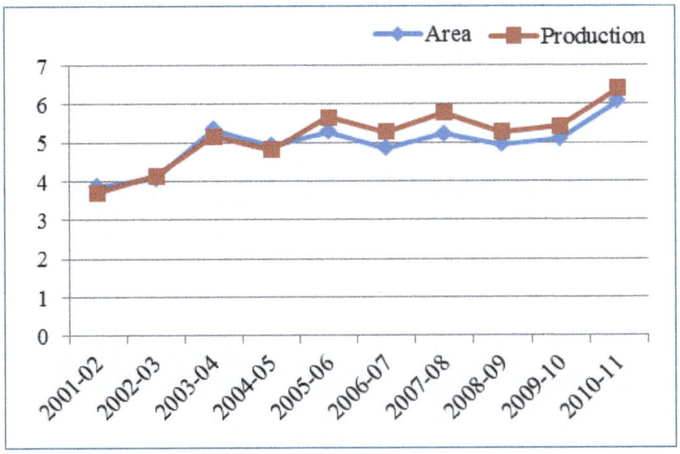

Fig. 3- Trend in percentage share of area and production of Gujarat in India

[Table-3] provides the estimates of area, production and productivity of different vegetable crops in India and Gujarat for the average of three years i.e. 2008-09 to 2010-11. It can be seen from the table that, area as well as production of potato crop was found highest among all the other vegetable crops in India as well as in Gujarat. In case of productivity, cabbage (21.87 MT/ha) and potato (27.35 MT/ha) was found highest among all other crops in India and Gujarat, respectively. Further it is observed from the table that, the productivity of all the vegetable crops including total vegetables was found higher in Gujarat as compared to India, except Cabbage crop. In case of production, the proportional share of onion crop in Gujarat was found highest (9.80 per cent) among all other crops including total vegetable in India. In Gujarat, potato crop was lowest proportional share both in case of areas (3.30 per cent) and production (4.40 per cent). Thus, potato as well as onion crops have greater potentiality of export from Gujarat whereas; potato crop needs to be more focused.

The perusal of [Table-4] revealed that Gujarat has placed on the seventh position in both area (5.69 per cent) as well as in production (6.06 per cent) of vegetables among different states in India. In case of productivity Gujarat has been ranked at eighth position (17.73 MT/ha) and the highest productivity was found in Lakshadweep (32.90 MT/ha) followed by Tamil Nadu (29.73 MT/ha), Kerala (22.47 MT/ha), Uttar Pradesh (20.83 MT/ha), Jammu & Kashmir (19.67 MT/ha), Punjab (19.63 MT/ha) and West Bengal (17.93 MT/ha).

Table 3- Area, Production and Productivity of different vegetables in India and Gujarat (TE 2010-11)

Vegetables	Area			Production			Productivity	
	India	Gujarat	Per cent of GJ (%)	India	Gujarat	Per cent of GJ (%)	India	Gujarat
Potato	1842	60.79	3.3	37769	1662.55	4.4	20.47	27.35
Tomato	699.33	34.37	4.92	13369.3	855.31	6.4	19.1	24.88
Onion	884.67	54.35	6.14	13614	1334.08	9.8	15.53	24.55
Cabbage	336.67	25.58	7.6	7366.67	477.29	6.48	21.87	18.66
Total	8153.67	439.18	5.39	136456	7814.01	5.73	16.73	17.79

(Area in 000 hectares/ Production in 000 MT/ Productivity in MT/hectare)

Table 4- Major vegetable producing States in India (TE 2010-11)

State	Area	Per cent area (%)	State	Production	Per cent Production (%)	State	Productivity
West Bengal	1325.33	17.18	West Bengal	23778.78	18.45	Lakshadweep	32.9
Uttar Pradesh	945.77	12.26	Uttar Pradesh	19688.41	15.28	Tamil Nadu	29.73
Bihar	835.97	10.84	Bihar	13974.23	10.84	Kerala	22.47
Orissa	640.17	8.3	Orissa	8407.03	6.52	Uttar Pradesh	20.83
Maharashtra	503.7	6.53	Tamil Nadu	8200.37	6.36	Jammu & Kashmir	19.67
Karnataka	451.83	5.86	Karnataka	7954.5	6.17	Punjab	19.63
Gujarat	439.17	5.69	Gujarat	7814.03	6.06	West Bengal	17.93
Andhra Pradesh	435.7	5.65	Andhra Pradesh	7513.77	5.83	Gujarat	17.73
Chhattisgarh	321.27	4.17	Maharashtra	6681.53	5.18	Karnataka	17.57
Haryana	315.23	4.09	Haryana	4176.57	3.24	Himachal Pradesh	17.53
Tamil Nadu	275.77	3.58	Jharkhand	3739.53	2.9	Pondicherry	17.03
Madhya Pradesh	275.37	3.57	Madhya Pradesh	3639	2.82	Chandigarh	17
Assam	251.8	3.26	Chhattisgarh	3630.3	2.82	Delhi	16.97
Jharkhand	237.9	3.08	Punjab	3506.2	2.72	Andhra Pradesh	16.93
Punjab	178.6	2.32	Kerala	3473.4	2.69	Bihar	16.7
Total	7712.17			128888.2			16.67

(Area in 000 hectares/ Production in 000 MT/ Productivity in MT/ hectare)

[Table-5] depicts the country wise export of cabbage from India for the average of 2008-09 to 2010-11.The table indicated that the total export of cabbage was 1014 MTs, whereas in terms of value it was Rs 84.77 lakhs. The highest export of cabbage was found to United Arab Emirates accounting for about 76.06 percentage share in total export quantity followed by Maldives (16.21 per cent), Nepal (2.12 per cent) and Singapore (0.04 per cent). The same pattern of export has been observed in terms of value.

Table 5- Export of Cabbage from India to different countries (TE 2010-11)

Country	Quantity	Per cent Quantity (%)	Value	Per cent Value (%)
United Arab Emirates (UAE)	771.23	76.06	58.5	69.01
Maldives	164.37	16.21	12.87	15.18
Nepal	21.5	2.12	1.1	1.3
Singapore	0.4	0.04	0.1	0.12
Others	56.5	5.57	12.2	14.39
Total	1014	100	84.77	100

(Quantity in MTs/ Value in INR Lakhs)

[Table-6] shows the country wise export of onion from India for the average of 2008-09 to 2010-11. [Table-6] indicated that the total export of onion was 1499527.10 MTs, whereas in terms of value it was Rs 196283.53lakhs. The highest export of onion was found to Bangladesh accounting for about 40.93 percentage share in total export quantity followed by Malaysia (19.12 per cent), UAE (10.29 per cent), Sri Lanka (8.80 per cent) and Pakistan (6.13 per cent). The same pattern of export has been observed in terms of value. [Table-7] revealed the country wise export of tomato from India for the average of 2008-09 to 2010-11. [Table-7] shows that the total export of tomato was 99554.17 MTs, whereas in terms of value it was Rs 11523.93 lakhs. The highest export of tomato was found to Pakistan accounting for about 60.00 percentage share in total export quantity followed by UAE (18.64 per cent), Bangladesh (15.37 per cent), and Nepal (2.43 per cent). The same pattern of export has been observed in terms of value.

Table 6- Export of Onion from India to different countries (TE 2010-11)

Country	Quantity	Per cent Quantity (%)	Value	Per cent Value (%)
Bangladesh	613700	40.93	81141.8	41.34
Malaysia	286754	19.12	30867	15.73
Sri Lanka	131992	8.8	17595.9	8.96
United Arab Emirates	154357	10.29	18689.2	9.52
Indonesia	21798.2	1.45	3893.33	1.98
Pakistan	91935.3	6.13	10650.1	5.43
Singapore	23267	1.55	2999.47	1.53
Nepal	36543.9	2.44	2783.33	1.42
Oman	16600.3	1.11	1929.7	0.98
Vietnam Social Republic	6403.37	0.43	895.73	0.46
Others	116175	7.75	24838	12.65
Total	1499527	100	196284	100

(Quantity in MTs/ Value in INR Lakhs)

[Table-8] revealed the country wise export of potato from India for the average of 2008-09 to 2010-11. It is clear from [Table-8] that the total export of potato was 154441.80 MTs, whereas in terms of value it was Rs 11449.43 lakhs. The highest export of potato was found to Nepal accounting for about 53.13 percentage share in total export quantity followed by Sri Lanka (11.53 per cent), Russia (3.31 per cent), and Maldives (2.66 per cent). [Table-9] revealed the country wise export of Dried & Preserved Vegetables from India for the average of 2008-09 to 2010-11. [Table-9] shows that the total export of Dried & Preserved Vegetables was 127549.43 MTs whereas in terms of value it was Rs 51515.37 lakhs. The highest export of Dried & Preserved Vegetables was found to United States accounting for about 16.92 percentage share in total export quantity followed by France (13.86 per cent), Russia (10.48 per cent), and Spain (9.87 per cent).

Table 7- Export of Tomato from India to different countries (TE 2010-11)

Country	Quantity	Per cent Quantity (%)	Value	Per cent Value (%)
United Arab Emirates	18557.9	18.64	3346.47	29.04
Bangladesh	15299.6	15.37	1523.87	13.22
Saudi Arabia	837.17	0.84	191.77	1.66
Oman	897.53	0.9	201.37	1.75
Pakistan	59729.9	60	5932.27	51.48
Nepal	2423.7	2.43	153.53	1.33
Maldives	1291.1	1.3	92.8	0.81
Bahrain	185.27	0.19	39.13	0.34
Kuwait	93.9	0.09	17.67	0.15
Qatar	21.77	0.02	4.87	0.04
Others	216.4	0.22	20.13	0.17
Total	99554.2	100	11523.9	100

(Quantity in MTs/ Value in INR Lakhs)

Table 8- Export of Potato from India to different countries (TE 2010-11)

Country	Quantity	Per cent Quantity (%)	Value	Per cent Value (%)
Nepal	82051.2	53.13	3947.87	34.48
Sri Lanka	17805.3	11.53	2439.03	21.3
Russia	5108.43	3.31	954.07	8.33
Malaysia	3323.23	2.15	351.33	3.07
Mauritius	3379.27	2.19	493.27	4.31
Maldives	4104.6	2.66	311.57	2.72
United Arab Emirates	3032.5	1.96	291.53	2.55
Singapore	1253.7	0.81	91.83	0.8
Vietnam Social Repbulic	185.83	0.12	26	0.23
Brunei	135.27	0.09	20.97	0.18
Others	34062.3	22.06	2522.07	22.03
Total	154442	100	11449.4	100

(Quantity in MTs/ Value in INR Lakhs)

Table 9- Export of Dried & Preserved Vegetables from India to different countries (TE 2010-11)

Country	Quantity	Per cent Quantity (%)	Value	Per cent Value (%)
Russia	13,366.17	10.48	5,000.67	9.71
Germany	4,665.53	3.66	4,119.90	8.00
Malaysia	2,686.23	2.11	1,431.73	2.78
United States	21,586.77	16.92	7,329.37	14.23
France	17,677.67	13.86	6,111.80	11.86
Spain	12,589.37	9.87	3,627.97	7.04
United Kingdom	3,919.37	3.07	2,720.57	5.28
Belgium	7,604.13	5.96	2,756.23	5.35
Poland	1,249.67	0.98	882.50	1.71
Brazil	1,272.13	1.00	995.63	1.93
Others	40,932.50	32.09	16,539.03	32.11
Total	127,549.43	100.00	51,515.37	100.00

(Quantity in MTs/ Value in INR Lakhs)

The data related to exports of different vegetable crops is used to work out the NPC which is presented in [Table-10]. It clearly shows that the export of all the different vegetables was found to be moderately competitive. Out of four vegetable crops cabbage was found most competitive with the lowest NPC (0.539), followed by potato

(0.712), tomato (0.786) and onion (0.843). The [Table-11] shows the country wise NPC for cabbage export from Gujarat during the year 2011. It indicates that export of cabbage was highest competitive to Singapore with the lowest NPC of 0.226, followed by UAE (0.392) .It was moderately competitive to the Maldives (0.706) whereas, it was found non competitive in case of Nepal (1.796). The country wise NPC for exporting of onion from Gujarat during 2010-11 is presented in the [Table-12]. It is clear from [Table-12] that Indonesia with NPC of 0.586 was found most competitive country for exporting of onion from India followed by Malaysia (0.735), Sri Lanka (0.818), Bangladesh (0.865), Singapore (0.908) etc. . The countries like Oman (1.022), Pakistan (1.094) and Nepal (1.627) was found non competitive.

Table 10- Crop wise analysis of vegetable export from Gujarat during 2010-11

Vegetable	Quantity (MT)	Value (INR Lakhs)	Wholesale Price (Pd) (INR/kg)	FOB price (INR/kg)	Other* charges (INR/kg)	Total Marketing expenses (INR/kg)	Reference price (Pb) (INR/kg) (FOB price – Total marketing expenses)	NPC (Pd/Pb)	Competitiveness
Cabbage	404.9	45.6	5.33	11.26	0.5	1.38	9.88	0.539	Moderately competitive
potato	184276.5	15424.1	4.86	8.37	0.66	1.54	6.83	0.712	"
Tomato	68183.7	11480.6	11.87	16.84	0.85	1.73	15.11	0.786	"
Onion	1163472.6	174155.4	11.41	14.97	0.55	1.43	13.54	0.843	"

Transportation cost (From Ahmedabad to Kandla): 0.88 INR/kg

*Other charges including grading, packing, handling & taxes in (INR/kg)

Table 11- Country wise analysis of cabbage export from Gujarat during 2010-11

Country	Quantity (MT)	Value (INR Lakhs)	FOB price (INR/kg)	Reference price (Pb) in INR/kg (FOB price - Total marketing expenses)	NPC (Pd/Pb)	Competitiveness
Singapore	1.2	0.3	25	23.62	0.226	Highly competitive
UAE	168.1	25.2	14.99	13.61	0.392	"
Maldives	214.9	19.2	8.93	7.55	0.706	Moderately competitive
Nepal	20.7	0.9	4.35	2.97	1.796	Non competitive

Wholesale Price (Pd): 5.33 INR/kg; Total marketing expenses: 1.38 INR/kg

Table 12- Country wise analysis of onion export from Gujarat during 2010-11

Country	Quantity (MT)	Value (INR Lakhs)	FOB price (INR/kg)	Reference price (Pb) in INR/kg (FOB price - Total marketing expenses)	NPC (Pd/Pb)	Competitiveness
Indonesia	43061.3	9000	20.9	19.47	0.586	Moderately competitive
Malaysia	273233	46343.5	16.96	15.53	0.735	"
Sri Lanka	114450	17599.2	15.38	13.95	0.818	"
Bangladesh	391550	57233.4	14.62	13.19	0.865	"
Singapore	20876.3	2920.6	13.99	12.56	0.908	"
Vietnam	15281.5	2077.5	13.59	12.16	0.938	"
UAE	126229	16627.7	13.17	11.74	0.972	"
Oman	17988.6	2266.2	12.6	11.17	1.022	Non competitive
Pakistan	45689.1	5417.6	11.86	10.43	1.094	"
Nepal	29977.6	2530.4	8.44	7.01	1.627	"

Wholesale Price (Pd): 11.41 INR/kg; Total marketing expenses: 1.43 INR/kg

Table 13- Country wise analysis of tomato export from Gujarat during 2010-11

Country	Quantity (MT)	Value (INR Lakhs)	FOB price (INR/kg)	Reference price (Pb) in INR/kg (FOB price - Total marketing expenses)	NPC (Pd/Pb)	Competitiveness
Qatar	19.6	8	40.82	39.09	0.304	Highly competitive
Oman	2203.6	532	24.14	22.41	0.53	Moderately competitive
Saudi Arebia	2261.5	539.4	23.85	22.12	0.537	"
Kuwait	125.9	29.9	23.75	22.02	0.539	"
Bahrain	442.3	102.2	23.11	21.38	0.555	"
UAE	33902.9	7468.6	22.03	20.3	0.585	"
Bangladesh	19711.9	2048.2	10.39	8.66	1.371	Non competitive
Pakistan	3708.9	340.6	9.18	7.45	1.593	"
Nepal	3266.6	227.3	6.96	5.23	2.27	"
Maldives	2447.7	169.8	6.94	5.21	2.28	"

Wholesale Price (Pd): 11.87 INR/kg; Total marketing expenses: 1.73 INR/kg

Table 14- Country wise analysis of potato export from Gujarat during 2010-11

Country	Quantity (MT)	Value (INR Lakhs)	FOB price (INR/kg)	Reference price (Pb) in INR/kg (FOB price - Total marketing expenses)	NPC (Pd/Pb)	Competitiveness
Russia	15325.3	2862.2	18.68	17.14	0.284	Highly competitive
Brunei	254.7	40.2	15.78	14.24	0.341	"
Mauritius	4871.2	758.5	15.57	14.03	0.346	"
Vietnam	448	66.9	14.93	13.39	0.363	"
Sri Lanka	34744.8	4687.5	13.49	11.95	0.407	"
UAE	3389.7	416.4	12.28	10.74	0.452	"
Malaysia	8144.4	892.3	10.96	9.42	0.516	Moderately competitive
Maldives	6019	478.8	7.95	6.41	0.758	"
Singapore	1716.5	134	7.81	6.27	0.776	"
Nepal	107838	4948.3	4.59	3.05	1.594	Non competitive
Wholesale Price (Pd): 4.86 INR/kg; Total marketing expenses: 1.54 INR/kg						

The values in [Table-13] represent the country wise NPC for tomato export from Gujarat during 2010-11. The Qatar was highly competitive with NPC of 0.304 for export of tomato whereas the countries like Oman (0.530), Saudi Arebia (0.537), Kuwait (0.539), Bahrain (0.555) and UAE (0.585) were found moderately competitive. Further, the export of tomato was found non competitive for Bangladesh (1.371), Pakistan (1.593), Nepal (2.270) and Maldives (2.280). During the year 2010-11, [Table-14] reveals that export of potato was found highly competitive for countries like Russia (0.284), Brunei (0.341), Mauritius (0.346), Vietnam (0.363), Sri Lanka (0.407) and UAE (0.452). It was found moderately competitive for Malaysia (0.516), Maldives (0.758) and Singapore (0.776). Nepal with NPC of 1.594 was found non competitive for export of potato from Gujarat.

Conclusion

Though the increasing trend of production of vegetable crops in Gujarat was observed during recent years, the productivity of vegetables was yet found lower as compared to advanced countries. Therefore, it may be implied that there has been scope to increase the production of vegetable crops in Gujarat. As government has been trying to promote the export of horticultural crops, lot of opportunity may be apparent in export of vegetable crops from Gujarat. Gujarat has greater potentiality to compete in export of vegetables among different countries in the world.

Potato crop was found highest area, production as well as productivity among all other vegetables in Gujarat. Potato as well as onion crop was found greater potentiality of export from Gujarat. The countries like UAE, Bangladesh, Pakistan, Nepal and United State account for major share in export of cabbage, onion, tomato, potato and dried and preserved vegetables, respectively from India.

Overall, the export of all these vegetables *i.e.* cabbage, potato, tomato and onion was found competitive and among these four major crops, cabbage was found to be highest export competitiveness. Therefore, cabbage crop needs special attention for promotion of export from Gujarat. Singapore, Indonesia and Qatar and Russia were found to be highest export competitiveness for cabbage, onion, tomato and potato, respectively.

Conflicts of Interest: None declared.

References

[1] Khandare V., Misal D. & Kharat D. (2012) *International Journal of Humanities and Applied Sciences*, 1(5), 138-141.

[2] The Economic Times (2013) *PMEAC pegs farm growth at 4.8% in FY'14 on good monsoon*, Sep 13, 2013.

[3] Siddayya & Atteri B.R. (2011) *Millennial Asia*, 2(1), 23-42.

[4] Dholakia R.H. (2003) *Vikalpa*, 28(4), 41-52.

[5] Singh K. (2004) *Vegetable Research in India: Some Issues, in the proceedings if Impact of vegetable research in India*, National Centre for Agricultural Economics and Policy Research, New Delhi, India.

[6] Anonymous (2011-12) *Export competitiveness of cotton crops from Gujarat*, Research report for eighth Meeting of Agricultural Economics, Agricultural Statistics and Extension Education Sub -Committee of the Research, JAU, Gujarat.

PHYSICAL AND MECHANICAL PROPERTIES OF *Citrus aurantiifolia* Swingle VAR. Pramalini (KAGZI-LIME)

PAWAR S.G.[1*], SHINDE G.U.[2] AND KHODKE S.U.[3]

[1]Department of Agricultural Process Engineering, Dr. Punjabrao Deshmukh Krishi Vidyapeeth, Akola - 444 104, MS, India.
[2]Department of Farm Machinery and Power Engineering, Vasantrao Naik Marathwada Krishi Vidyapeeth, Parbhani - 431 402, MS, India.
[3]Department of Agriculture Process Engineering, Vasantrao Naik Marathwada Krishi Vidyapeeth, Parbhani - 431 402, MS, India.
*Corresponding Author: Email- savitapawar100@gmail.com

Abstract- Kagzi- lime (*Citrus aurantifolia)* is a citrus fruit variety in Maharashtra, Andhra Pradesh and Karnataka state of India available in small to medium size, its pulp is juicy greenish with strong adherence to the yellowish, green, thin or papery and shiny. It is mainly used for processing purpose. Physical and frictional characteristics were determined in laboratories. It includes, linear dimensions, mass, bulk density, Sphericity, rolling angle, static co-efficient of friction and dynamic co-efficient of friction were determined for the Kagzi-lime. Polar diameter of Kagzi-lime was in the range of 3.33 cm to 4.48 cm whereas equatorial diameter was in the range of 3.34 cm to 4.47 cm. The maximum thickness value was in the range of 3.34 cm to 4.35 cm whereas weight of Kagzi-lime was in the range of 21.90 gm to 46.32 gm. The bulk density of Kagzi-lime was in the range of 551.80 kg/m^3 to 585.83 kg/m^3 and sphericity varied between 97.5 to 99.1 per cent, respectively. The highest rolling angle was found on plywood surface (8.05 degree) and lowest on aluminum surface (6.65 degree) was recorded for the Kagzi-lime fruit.

Keywords- Kagzi-lime, Physical properties, frictional properties

Introduction

Lemons are less poplar than the lime in India. Acid-lime stands second next to Nagpur-santra in Western Vidarbha region of Maharashtra state. Acid-lime is tropical in its climatic requirement. Maharashtra produces 12% of the total production of lime/lemon in the country and is the third largest producer in the country.The state produces 0.26 m MT of lime/lemon from an area of 0.04 m. ha. with productivity of 6.0 MT/ha [1] . A well drained soil with a pH of 6.5 to 7.0 is ideal for better growth and yield. In Maharashtra, *Pramalini, Vikram* are two varieties identified by clonal selection. These have been released for commercial cultivation as they are *Canker-free* and *Prolific-bearer*. The improved varieties are *Pramalini, Vikram, Chakradhar, PKM1, Seedless lime* and *Tahiti lime*. A good acid lime plant (7-year old) may yield 2000-5000 fruits annually. The average yield being 3000 to 3,500 fruits/tree. Harvesting of lime and lemons differs with different species, verities and regions of cultivation. The peak season of harvesting is from March to April and lean season May to June in Maharashtra. The storage life of lemon fruit is between 6 to 8 weeks at temperature 9 to10^{0}C and 85 to 90 % relative humidity [2].

The scientific classification of *Citrus Aurantifolia* is from Kingdom: Plantae, Order: Sapindales, Family: Rutaceae, Genus: *Citrus*. The important species are *Citrus aurantifolia- Key lime, Citrus maxima-Pomelo, Citrus medica-Citron, Citrus reticulate- Mandarin orange.* .Kagzi- lime (*Citrus aurantifolia)* is a citrus fruit variety available in small to medium size, its pulp is juicy greenish with strong adherence to the yellowish, green, thin or papery and shiny. Lime and lemons are harvested with shades of green and yellow colour which are not acceptable in the market. Therefore, harvested fruits are degreened. Degreening of citurs fruit in India is achieved with the aid of calcium carbonate crystals. They are harvested with a pole harvester having an iron hook and a net at packing sheds as soon as possible. They should never be allowed to stay under the sun for a long time. Kagzi-limes are oval or spherical in shape with smooth porous skin. In the ripening chambers calcium carbide releases ethylene gas which changes green colour and allows the development of yellow colour without affecting the quality of fruits [3].

Per capita consumption of the lime in world is around 1.30 kg/year on national map [4]. The fruit of acid lime are available at throw away prices due to glut in the market at time in the month of August and September. The fruit has valued not only for its nutritional and medicinal qualities but also extensively used for the preparation of value added products, like squashes, syrup, cordials, pickles, manufacture of citric acid and for culinary uses in the daily diet of Indians. Per day requirement of ascorbic acid for an adults to maintain the skin and teeth, to resist stress and infection is 70-90 mg [5]. The importance of flavonoids of citrus fruit is well documented including the ability to prevent red blood cell clumping and capillary fragility, anti-cancer and anti- viral activity. The lime peel oil has wide application as flavoring agent in food industry. The products and by-products developed from the fruit of Kagzi-lime are comparatively well known on commercial scale.

There are two main principle season of production of fruit i.e. February to April and October to December. The fruits are harvested during November to March are round in shape and good test. The rainy season fruits are oblong and have inferior test. The method of harvesting and stage of maturity at harvesting time are vital factors in grading and packing. Therefore, the post harvest management of fruits is of prime importance in order to sustain higher production.

Cleaning, sorting and grading are the important post harvest operations related to the market. It provides basics for buying and selling. In market, size of any fruit has been considered as the main criteria for selling followed by fruit maturity. In post harvest-handling, conveying and grading are two important operations responsible for mechanical injury of fruits. Fresh crop and damage free post-harvest handling of fruits and vegetables were considered basic requirements to increase the farmer's profit.

Singh, et al [6] said that the data availability of physical and frictional properties for Citrus fruit is limited. Tirthakar, et al [7] evaluated the fruit quality of acid lime orchard from fourteen locations of Akola district of Maharashtra. They measured physical parameters, in that length and breadth of fruit ranged between 3.52 to 5.12 cm and 3.41 to 4.93 cm, respectively. The weight and volume of fruit ranged between 32.34 gm to 48.20 gm and 33.79 cc to 48.6 cc, respectively Sharifi, et al [8] measured some physical properties of Orange. FAO [4] measured the physical properties of Apricot kernels as a function of moisture content varies from 2.86 to 13.03 per cent (wet basis).

Materials and Methods

Local variety of Kagzi-lime *(Pramalini)* fruits was procured from local market, Parbhani. Random samples were drawn from a freshly harvested lot of Kagzi-lime. Twenty number of Kagzi-lime in each three size grades: first (large), second (medium) and third (small) were taken as (measurement of physical and frictional properties) study samples. Physical and frictional characteristics were determined in laboratories. It includes, linear dimensions, mass, bulk density, Sphericity, rolling angle, static co-efficient of friction and dynamic co-efficient of friction were determined for the Kagzi-lime. Mean and standard deviation were computed according to the standard methods.

Determination of Physical Characteristics of Kagzi-lime

Linear Dimensions

There are two categories of Kagzi-lime fruit diameter, polar diameter and equatorial diameter. Polar diameter is the distance between the fruit crown and the point of root attachment of the Kagzi-lime. Equatorial diameter is the maximum width of the Kagzi-lime in a plan perpendicular to the polar diameter [9]. Equatorial diameter (De), polar diameter (Dp) and thickness (T) of each twenty fruits were measured for each grade using Vernier calliper with Least count of 0.05 mm.

Bulk Density

The bulk density of sample was computed by the mass of fruit (gm) and volume of fruit (cm^3) [10]. The mass of bulk fruit was determined by using a precise weight balance with accuracy 0.001 gm and volume was computed as length x breadth x thickness of the box used for measuring the bulk weight of fruits.

Sphericity

The fruit shape was expressed in terms of its sphericity. The sphericity index was calculated based on recommendation of Mohsenin, et al [10].

$$\text{Specificity (\%)} = \frac{(Dp \times De \times T)1/S}{Dp} \times 100 \qquad (1)$$

Where,

Dp - Polar diameter (mm).

De - Equatorial diameter (mm).

T - Thickness of kagzi lime (mm).

Determination of Frictional Characteristics of Kagzi-lime

Rolling Angle

To determine the rolling angle, the Kagzi-lime was kept at the centre of the working surface, (horizontal platform) in the most stable position (on their base) to prevent toppling over (top upwardsss). The platform was inclined until the Kagzi-lime begins to roll. When the rolling of Kagzi-lime was started, the position of platform was noted by protractor. For the next test, the platform was brought to the initial horizontal position [11].

In this experiment, the rolling angle was measured in three different platforms such as plywood, aluminum and mild steel. Rolling angle was measured with mild steel platform was considered in the design of Kagzi-lime grader.

Co-efficient of Static Friction

The co-efficient of static friction is the tangent of the slope angle of the table measured with a protractor [12]. It was determined on three surfaces i.e. plywood, aluminum and mild steel for Kagzi-lime fruits. A frame made with a rectangular wooden were placed on the surface. The frame was filled with fruits. The table was tilted slowly manually until movement of the whole fruit mass.

Co-efficient of Dynamic Friction

The co-efficient of dynamic friction was the ratio of the force required to slide the fruit over a surface divided by the normal force pressing the fruit against the surface [12]. It was measured for the same material i.e. plywood, aluminum and mild steel for Kagzi-lime fruits. The frame was filled with fruits. The force required to pull that frame was noted in term of weight.

Results and Discussion

The average value of each parameter for Kagzi-lime was measured and presented in [Table-1] and [Table-2]. Based on the average diameter values of Kagzi-lime, the size of fruits are classified into following three grades-

1. Grade-I – above 41mm.
2. Grade-II- between 40 to 36 mm.
3. Grade-III- below-35 mm

[Table-1] shows that, polar diameter of Kagzi-lime was in the range of 33.3 mm to 44.8 mm whereas equatorial diameter was in the range of 33.4 mm to 44.7 mm. The maximum thickness value was in the range of 33.4 mm to 43.5 mm whereas weight of Kagzi-lime was in the range of 21.90 gm to 46.32 gm.

The bulk density of Kagzi-lime was in the range of 551.80 kg/m^3 to 585.83 kg/m^3 and sphericity varied between 97.5 to 99.1 per cent, respectively.

The frictional properties of Kagzi-lime depend on the moisture content of fruit peel [13]. Initial moisture content of Kagzi-lime peel was 43.93 per cent (w.b.). [Table-1] shows the frictional properties of Kagzi-lime on three different surfaces (plywood, aluminum and mild steel). The highest rolling angle was found on plywood surface (8.05 degree) and lowest on aluminum surface (6.65 degree) was recorded for the Kagzi-lime fruit.

Table 1- Physical properties of Kagzi-lime

Particular	Large (G-I)		Medium (G-II)		Small (G-III)	
	Mean	SD	Mean	SD	Mean	SD
Polar diameter (mm)	44.8	3.4	38. 1	0.96	33.3	1.47
Equatorial diameter (mm)	44.7	2.6	38.1	0.95	33.4	1.12
Thickness (mm)	43.5	2.6	37.5	0.85	33.4	1.06
Weight (gm)	46.32	7.9	30.11	1.92	21.9	3.22
Bulk density (kg/m^3)	551.8	0.03	552.99	0.02	585.83	0.04
Sphericity (%)	97.5	3.1	98.2	0.7	99.1	1.2

Table 2- Frictional properties of Kagzi-lime

Particular	Plywood		Mild steel		Aluminum	
	Mean	SD	Mean	SD	Mean	SD
Rolling angle (degree)	8.05	0.8	6.75	0.78	6.65	0.78
Static co-efficient of friction	0.57	0.02	0.64	0.03	0.62	0.02
Dynamic co-efficient of friction	0.45	0.03	0.54	0.13	0.51	0.01

The static co-efficient of friction value varies from 0.57, 0.65, and 0.62 for Kagzi-limes on plywood, mild steel and aluminum, respectively. The dynamic co-efficient of friction value was found highest on mild steel (0.54) and lowest on plywood (0.45) for Kagzi-lime. Approximate values for these properties were noted by Singh, et al [6].

Evaluation of the Regression Models

The equations were derived through regression analysis and on the basis of the polar diameter as the first and only independent variable the fruit mass was estimated (coefficient of determination = 0.935).

$$y = 22.452x-53.505 \quad R^2 = 0.935 \quad (2)$$

A coefficient of determination of 0.969 was achieved when the equatorial diameter was also added to the model.

$$y = 24.232x - 60.394 \quad R^2 = 0.969 \quad (3)$$

The thickness of was also added to the model with a coefficient of determination of 0.953.

$$y = 26.48x - 67.58 \quad R^2 = 0.953 \quad (4)$$

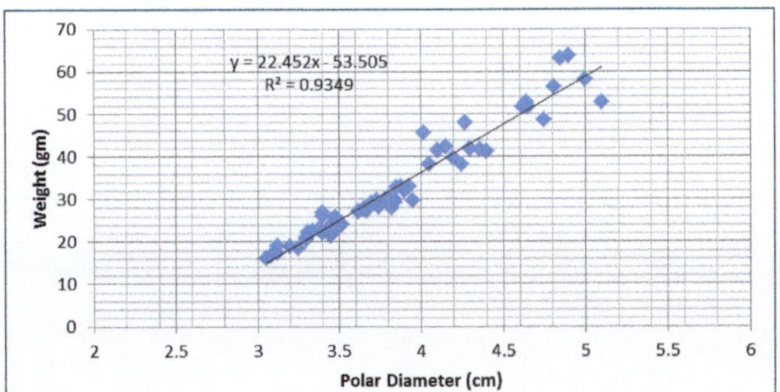

Fig. 1- Relationship between mass of Kagzi-lime with polar diameter

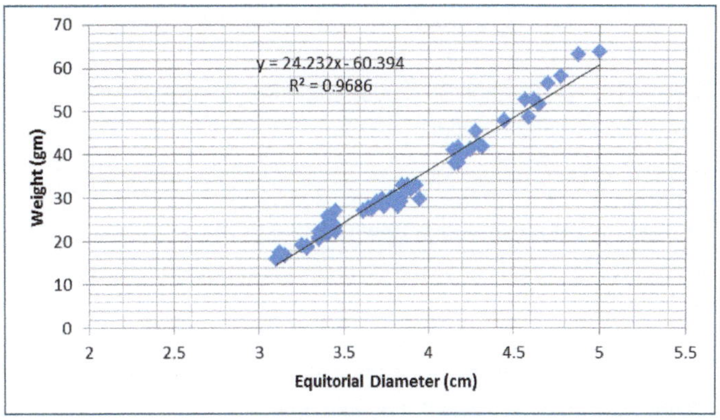

Fig. 2- Relationship between mass of Kagzi-lime equatorial diameter

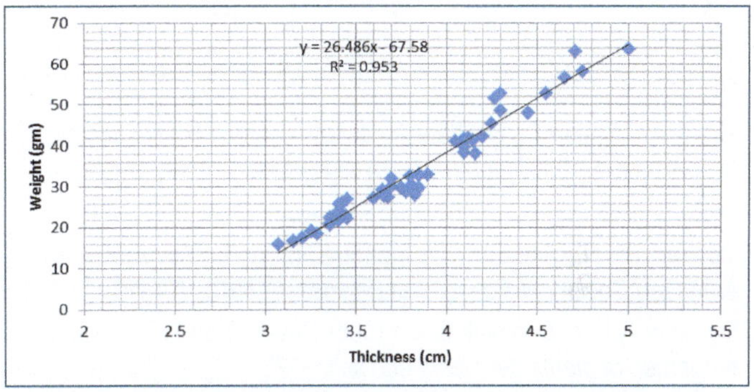

Fig. 3- Relationship between mass of Kagzi-lime with thickness

A linear equation was found to be more responsive to estimate the mass of kagzi-lime based upon any of the independent variables of dimension with respect to any of the polar diameter, equatorial diameter and thickness. These are demonstrated in [Fig-1], [Fig-2] and [Fig-3] in order of their coefficients of determination.

Any of the above variables is in significant correlation with Kagzi-lime mass and can be employed in development of the linear regression for estimation of the Kagzi-lime mass.

Conclusions

The following conclusions were found during experimentation:

The sphericity of fruit was found between 97.5 to 99 percent; therefor shape of fruit is spherical.

The highest rolling angle was found on plywood surface (8.05 degree) and lowest on aluminum surface (6.65 degree) was recorded for the Kagzi-lime fruit.

The static co-efficient of friction value was found highest on mild steel (0.64) and lowest on plywood (0.57) and the dynamic co-efficient of friction value was found highest on mild steel (0.54) and lowest on plywood (0.45) for Kagzi-lime.

A linear model of the Kagzi-lime mass was developed on the basis of the independent variables of polar, equatorial diameters and thickness. The most recommended regression model to fit Kagzi-lime mass was the one based upon the equatorial diameter of lime.

Conflicts of Interest: None declared.

References

[1] Anonymous (2015) *National Horticulture Mission, Ministry of Agriculture, Joint inspection Team report June 2013*, Depart-

ment of Agriculture & Cooperation Krishi Bhawan, New Delhi, India.

[2] Mistry N.C. & Singhal B. (2008) *National Horticulture Board, Indian Horticulture database*, 51-65.

[3] Chadha K.L. (2001) *Handbook of Horticulture, Directorate of information and publication of Agriculture*, ICAR, Krishi Anusandhan Bhavan, Pusa, New Delhi - 110 012.

[4] FAO (2005) Food and Agricultural Organization, Statistics Rome, 225.

[5] Tetens I. (2013) *Scientific Opinion on Dietary Reference Values for vitamin C*, EFSA Panel on Dietetic Products, Nutrition and Allergies, European Food Safety Authority.

[6] Singh K.K., Reddy B.S., Varshney A.C. & Mangraj S. (2004) *Applied Engineering in Agriculture*, 20, 821-825.

[7] Tirthakar S.S., Kuchonwar O.D., Sarode P.C. & Wagh S.P. (2004) *Panjabrao Krishi Vidhpith Research Journal*, 28(2), 158-161.

[8] Sharifi M., Rafiee S., Keyhani A., Jafari A., Mobli H., Rajabipour A. & Akram A. (2007) *International Agrophysics*, 21(4), 391-397.

[9] Arora S. & Kumar S. (2005) *Journal of Agriculture*, 86, 58-60.

[10]Mohsenin N.N., Jindal V.K. & Manor A.N. (1978) *Transaction of American Society of Agricultural Engineers*, 21(3), 594-600.

[11]Buyanov A.I. & Voronyuk B.A. (1985) *Physical and mechanical properties of plants, fertilizers and soils*, 528.

[12]Oje K. and Ugbor E.C. (1991) *Journal of Agricultural Engineering Research*, 50, 305-313.

[13]Paksoy M. & Aydin C. (2004) *Journal of Food Engineering*, 65 (2), 225-231.

[14]Anonymous (2001) *Physico-chemical analysis of Seedless lime*. Annual report of research work of horticultural crops, Marathwada Agricultural University, Parbhani, India.

[15]Bahnasawy A.H., El-Haddad Z.A., El-Ansary M.Y. & Sorour H.M. (2004) *Journal of Food Engineering*, 62(3), 255-261.

[16]Fathollahzadeh H., Mobli H., Beheshti B., Jafari A. & Borghei A.M. (2008) *Agricultural Engineering International: CIGR Journal*, X, FP 08 008.

[17]Mohsenin N.N. (1986) *Physical properties of plant and Animal Materials*, Gordon and Breach Science Publications, New York.

[18]Owolarafe O.K. & Shotonde H.O. (2004) *Journal of Food Engineering*, 63(3), 299-302.

EVALUATION SOME NEW HYBRIDS FOR CULTIVATION IN CONVENTIONAL SYSTEM IN SPRING CLIMATIC CONDITIONS OF ROMANIA

DRAGHICI E.M. AND PELE M.*
University of Agronomical Sciences and Veterinary Medicine, Faculty of Horticulture, Bucharest, Romania
*Corresponding Author: Email- mpele50@yahoo.com

Abstract- Considering the fact that in Romania, the vast majority of tomatoes harvest comes from the farmers who have relatively small areas of solariums and greenhouses, between 100 and 2000 sqm, farmers are always looking for new hybrids with the special productive performance in low temperature conditions corresponding to beginning of spring, with resistance to diseases and pests.

Aim of this study was to test and recommend farmers from the south of Romania's new tomatoes hybrids with outstanding performance for the culture in solar of the tomatoes in conventional system (culture on soil) with early production.

In this context, it has studied 11 hybrids the Israeli provenance tomatoes.

Among hybrids analyzed, the best results in relation to the tomatoes productive potential per $1m^2$ were obtained for 3319 hybrid noted V8 with a production of 16.667 kg/m^2 under the conditions of the early spring with a day maximum temperature of 20°C. In addition, although this hybrid germinated least (only 87% of seeds) and later towards others tested (after 11 days) had the highest early production 9.060kg/m^2 at first harvest from July 1.

In terms of storage conditions at 9°C, the best behavior has had the hybrid 3351 whose firmness was kept undisturbed for 2 weeks.

Such meticulous tests concerning the overall development of hybrids of germination, growth, flowering and production showed the importance of these experiments to identify the most suitable hybrids for envisaged climatic conditions and culture. In this specific, case the hybrid 3319.

Keywords- tomatoes hybrids, climatic conditions, solar, productivity

Abbreviations- IF: fruit shape index, cm: centimeter, mm: millimeter, %: percent, g: gram, kg: kilogram, cm^2: square centimeter, m^2: square meter, MU: Measuring unit, DL: degrees of their liberty, Mt: control, O: significant negative, N: insignificant, *: significant, ***: very significant

Introduction

Tomatoes are today one of the most respected and consumed vegetables worldwide. In 2010 the area planted with tomatoes was 4,338,834 ha and production of 145,751,507 tons. Actually, tomatoes dominate the production of vegetables in the EU. Area planted with tomatoes Europe represented approximately 13% of the world total area and production achieved was 14% of the world [1]. In addition, the statistical data we noticed that in 2010 the area planted with tomatoes decreased by about 5% compared with 2005, but total production was about 11% higher than the average production in 2005.The main countries tomato producers are Italy and Spain, which assume 2/3 of the total production. In 2012, Romania area cultivated with vegetables accounted for about 9% of the area planted with tomatoes in Europe, and the average production was at half of the average production in European countries. Therefore the annual production of vegetables in Romania varies around 750 000 tones [2].

For example from official the statistical data published by FAO in 2012 is noted that both Israel and the Netherlands, the main countries supplying tomato hybrids for Romania, areas planted with tomatoes are very small compared to Romania. However, yields obtained on one ha are very high in Holland 479411 kg/ha compared with Israel 85883 kg/ha and especially in Romania of 15446 kg/ha in 2012 [1].

The content of tomatoes in sugars, vitamins, minerals, amino acids and organic acids, which play a large role in the human body's physiological balance, determine also the therapeutic value of these products. Although in Europe are cultivated of only around 130 years, have come to be consumed both those reached the physiological maturity as well as green. The range of ways in which they are consumed is extremely varied: fresh in simple salads or with other vegetables, pickled, cooked in various broths, soups, processed foods and industrial pasty, broth, canned, regular or spicy juices etc. [3].

In addition to the nutrients listed, tomatoes contain a compound "miracle" for medicine, lycopene. Numerous medical studies have showed the direct action of this pigment on free radicals in the body.

Free radicals are considered particularly strong oxidative molecules that attack cell membranes of various tissues of the body, attacking also human DNA inducing replication defective processes that lead to rapid aging of the body [4].

These properties determine their wide consumption. Thus, in the world of all of the tomatoes produced annually, about 80% are sold in various forms processed and only 20% are consumed fresh.

To meet the needs of Tomatoes, the official lists of varieties of the world or the EC as well as in Romania try annually new tomato hybrids with resistance to temperature variations, diseases and pests, with a very high productivity and high capacity storage.

In Romania, the largest tomato production is obtained from small farmers who are particularly interested in newly created varieties. It is known that hybrids may respond differently in terms of production in particular to the environmental conditions. [5,6]. Therefore testing new hybrid tomato concerning their productivity is a priority for most researchers in order to prescribe to farmers the most appropriate hybrid both for the climate [7-11] as well as for the culture system [7].

Due to high demands of tomatoes during the entire year, they are cultivated not only in the field but also in greenhouses and solariums. Exigencies of the consumers concerning the quality of tomatoes have known a strong orientation towards the sensory seeking and appreciating as much as taste and flavor tomatoes even at the expense of a commercial aspect of the fruit perfectly. The biological material choice for tomato cultures setting up is an important sequence of vegetable culture technology under shelter [12-15]. Most of the extra early hybrids introduced in Romania over time were hybrids of Dutch origin.

From officially the statistical data published by FAO in 2012 is noted that in the Netherlands yields obtained on one ha are very high compared with those in Romania [1]. Not all hybrids that are marketed for cultivation in greenhouses and solariums shows adaptability and resistance to all pathogens that attack tomato plants or climate variations in Romania. This may be one of the causes of differences in production per hectare between the two countries. Tomato growers trying to overcome these disadvantages by choosing the hybrids suitable for their culture system offered on the market.

In examining of different varieties are taken into consideration a minimum of characters that tomatoes must cover according to the protocol of Community Plant Variety Office (CPVO)-TP 44/3 of 21.03.2007 [16], namely concerning the seedlings, stem, plant, flowering, during flowering and fruit maturation period.

In this context, it is necessary that tomato market of Romania to be enriched with hybrids adapted to the conditions in this country. Varieties that besides a high yield to be developed conveniently and in the conditions enabling energy savings through early development even at lower temperatures.

Thus in this paper are analyzed several hybrids of Israel with marketing perspective. These hybrids are supposed to have increased resistance to specific environmental conditions during the Romanian spring (low temperature specific first part of spring, days shorter, high nebulosity, high relative humidity), to pests and diseases but which on the other hand ensures high yields and high quality of early tomatoes respectively their appearance, taste and a long shelf life. In addition the introduction of new hybrids in production is envisaged and diversification of early tomatoes. It also takes into consideration the need to assess the various hybrids under protected culture in order to develop recommendations on their suitability for cultivation in solariums.

Material and Methods

The experiments were placed in the teaching and experimental field, namely in the unheated solarium of the Department of Vegetable Growing. The experiments were conducted during February to August 2011.

In experiments were used 11 hybrids of Israeli origin: V1-3349; V2-3350; V3-3330; V4-3326; V5-3323; V6-3335; V7-3351; V8-3319; V9-3352; V10-3348; and V11-2821.

Sowing was carried out in greenhouse multiplier, providing a constant optimum temperature for germination of 20°C. For sowing and transplant we used professional substrate Plantaflor stamp to produce seedlings purchased from the Hortigala company, Germany, with the following composition: 30% red peat and 70% black peat with 0-7 mm granulation and pH (H_2O) of 5.5-6.0. The level of fertilization was of 0.8 kg/m^3.

Culture was established in the solarium, on 04.04.2011. Planting scheme was as follows: the distance between rows 70cm and 40cm between plants in the row. Culture was mulched with biodegradable foil. Plant density was 3.5714 plants/m^2 respectively 35,714 plants per one ha.

The care works seedlings were applied: daily watering, thinning plants when they grew; 2 treatments with Mospilan 0.2% for greenhouse whiteflies; two preventive treatments with Merpan 0.2% when seedlings had 4 true leaves and 2 days before planting with Topsin 0.2%. The care works cultures were applied. Dripping watering plants whenever it was necessary. Plants were trellising after about 12-14 days after planting. The management and guiding plant growth around the string; pinch off the shoots as they arose when they were about 5-10cm. Growth limited in height after 5-7 inflorescences. Basal leaves defoliation. Phytosanitary treatment with Topsin 0.2%, 0.2% Merpan and 4 treatments for greenhouse whiteflies and spider with Mospilan 0.2% and Nisorun.

The following analysis and determinations were made:

At the seedlings: dynamics of seed germination, dynamic growth in height of tomato seedlings, dynamics of the number of leaves.

In culture: dynamic growth in the solarium of plants height, the dynamics of the inflorescences, the number of leaves per plant, the number of inflorescences, the number of flowers and fruits in the inflorescence, the average weight of fruits, the early and total production, fruits characteristics concerning average weight, shape index, fruits colour and the characteristics concerning the fruits keeping of under controlled conditions.

The measurements concerning of plants height growth were made at an interval of 10 days. The number o leaves appeared were regularly recorded.

Harvested tomato fruits have been counted, weighed individually on inflorescences and was determined fruit shape index (fruit shape index = IF).

Statistical calculations were performed to highlight differences between experimental variants.

Results and Discussion

Among the 11 hybrids tested differences have appeared even from the germination. Thus, hybrids V3, V6 and V7 have sprung up fastest namely 4 days after sowing. Variants V1, V2, V5, V9, V10 and V11 have sprung up to 5 days, but two hybrids were germinated after 9 days (V4) respectively 11 days (V8). It can be noticed also that not all seeds germinate. The highest percentage of germination is 97% at the variant V5. V4 and V8 variants have sprung up at the latest and the germination percentages were also the lowest of 92% and 87% [Fig-1].

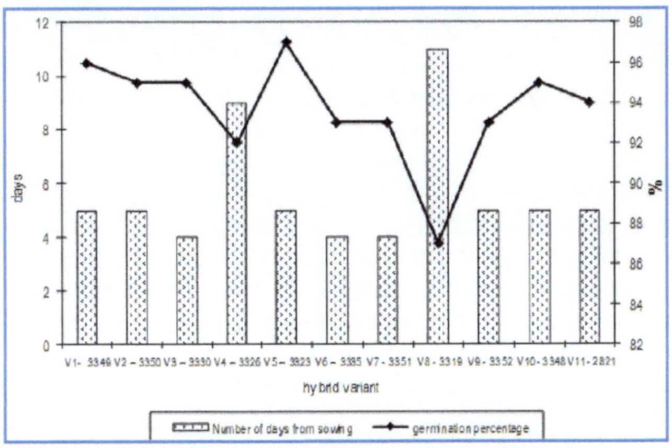

Fig. 1- Number of days from sowing to sprouting and the percentage of seeds sprung

The data presented in [Table-1] noted that seedling height determined on 25. 03. 2011 ranged from 6.33cm to 12.33cm for V8 respectively for V11. It seedling height was expected to be the smallest for V8 considering that this variant has sprouted at the latest. Nevertheless, it is interesting however, that the variant V11, which has sprouted at five days, has a height greater than the variants that arose after 4 days. This shows a higher growth capacity of this hybrid.

From statistical point of view, the differences concerning the height between the seedlings at planting in solar were insignificant.

Another factor chased it was to determine the number of leaves in seedlings. The average results are shown in [Fig-2]. Can be noticed that in general the number of leaves of seedlings was about 5. Variant V1 is distinguished with an average number of leaves 7 and variant V11 with an average number of 6 leaves. The lowest number of leaves had the variant V8 with an average of 4.33 leaves.

Table 1- Dynamic of growth in height of tomato seedlings

Variants	Measuring unit (MU)	Determination of the Height of Seedlings at: March 25 2011	April 04 2011	The difference cm	%	The Significance	The daily average of growth rate cm/day
V1-3349	cm	9.5	14.25	-0.52	96.5	N	0.317
V2-3350	cm	10.33	14.33	-0.44	97.04	N	0.318
V3-3330	cm	10.67	15.67	0.9	106.12	N	0.348
V4-3326	cm	7.67	12.67	-2.1	85.8	N	0.295
V5-3323	cm	8	16	1.23	108.35	N	0.356
V6-3335	cm	10	15	0.23	101.58	N	0.333
V7-3351	cm	10	14.85	0.08	100.57	N	0.33
V8-3319	cm	6.33	12.33	-2.44	83.5	N	0.308
V9-3352	cm	10.33	15.33	0.56	103.82	N	0.341
V10- 3348	cm	11.67	15.67	0.9	106.12	N	0.348
V11- 2821	cm	12.33	16.33	1.56	110.59	N	0.363
The average		9.71	14.77	0	100	Control	

DL = degrees of their liberty

DL5%= 2.740 DL5% in %= 18.5557

DL1%= 3.900 DL1% in %= 26.4114

DL01%= 5.640 DL01% in %= 38.1949

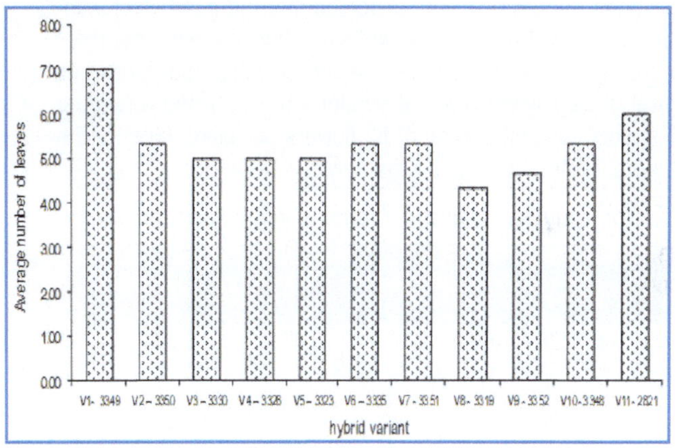

Fig. 2- Average number of leaves per hybrid at planting

According to UPOV characteristics, an important criterion for differentiating of cultivars is also the anthocyanin coloration of the plantlets hypocotyl in the seedling stage. At V1, V2, V9, V10 and V11 variants were noticed a purple colour of hypocotyls and for the rest of variants was green coloration. Dynamics of growth in height of tomato plants after planting was observed in the solar measuring the height of thereof at intervals of 10 days. Thus, the evaluation of plant growth dynamics of tomatoes in solar of the hybrids tested is represented in [Fig-3]. As it is observed, the increase rate was more than 2.2cm (centimeter) per day to all hybrids. Have increased the fastest variants V1 followed by V2 and V6. The variants with belated germination have had the smallest growth rate namely V8 the smallest with 2.21cm/day and V4 with 2.29cm/day.

A heeded character to knowing the hybrids recommended for high tunnel and greenhouse cultures is also leaf length and size of leaflets. Between hybrids analyzed variants 3, 4, 7 and 11 presented rich foliage with the length of leaves between 37cm and 45cm. This is important for determining distances of planting and carrying out of some works to remove some leaves with the aim for better circulation of the air among plants but also for better lighting the basal part.

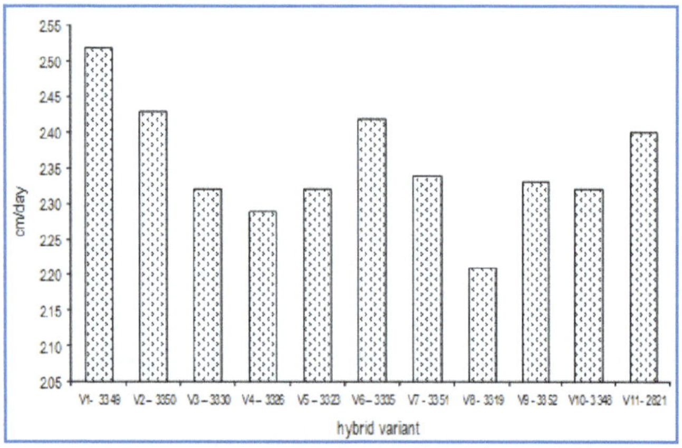

Fig. 3- Average daily rate of growth of different hybrids in cm/day

From the point of view of tomato growers, the most important factors are obviously the number and amount of fruit formed by each hybrid. In this context was observed the evolution of inflorescences formation, distance between them and fruits formation.

To the hybrids analyzed, it was observed that the number of inflorescences varied between 4 to 7 with a number of flowers per plant formed between 22 and 52. Thus, the variant 7 had four blossoms and a total of 22 flowers while at variant 1 have formed a number of 7 inflorescences with a total of 52 flowers per plant. Most of the flowers were formed at variant 2, namely 53 [Table-2].

Table 2- Total number of flowers formed in inflorescences and on the plant

Variants	Inflorescences							Total number of flowers/plant formed
	I	II	II	IV	V	VI	VII	
V1- 3349	6	9	9	7	7	7	7	52
V2-3350	11	9	6	9	9	9		53
V3-3330	7	8	6	8	8			37
V4-3326	9	9	9	9	5			41
V5-3323	7	7	7	9	7	5		42
V6-3335	7	7	7	6	7			34
V7-3351	6	6	4	6				22
V8-3319	7	7	7	7	6	3		37
V9-3352	6	6	6	6	6			30
V10-3348	5	6	6	4	5			26
V11-2821	9	6	6	6	6	6		39

Determining the average distance between inflorescences is important to assess the productive capacity of the hybrid. Distance between inflorescences was different from one hybrid to another, in average it was 19.36cm per variants. Shorter distances between inflorescences were recorded at variants 1 (17cm), 8 (17cm) and 11(17.5cm). The greatest distance between inflorescences, 24.5cm was recorded at the variant 7 [Table-3].

From the Statistical point of view the variant 7 has made a very significant difference compared to the average of variants. Differences significant negative were at the variants 1 & 8. Variant 6 also showed a significant positive difference. The others hybrids did not show significant differences towards average of experiences.

Table 3- The average distance between inflorescences (at inflorescences 1-4)

Variants	Distance between inflorescences	Difference		Significance
	cm	cm	%	
V1- 3349	17	-2.36	87.79	O
V2-3350	18	-1.36	92.96	N
V3-3330	20	0.64	103.29	N
V4-3326	19	-0.36	98.12	N
V5-3323	17,5	-1.86	90.38	N
V6-3335	22	2.64	113.62	*
V7-3351	24,5	5.14	126.53	***
V8-3319	17	-2.36	87.79	O
V9-3352	21,5	2.14	111.03	N
V10-3348	19	-0.36	98.12	N
V11-2821	17,5	-1.86	90.38	N
Average	19.36	0	100	Mt

*DL5%= 2.230, DL1%= 3.170, DL01%= 4.590, O= significant negative, N= insignificant, *= significant, ***= very significant, Mt= control.*

To all the hybrids analyzed, the inflorescences were simple, unbranched. One aspect observed only at one hybrid (V10-3348) it was issuance in large numbers of adventitious roots overall stalk.

Number of fruits produced in inflorescences and per total plant was between 17 fruits at V7 and 41 to V2. This data and the percentual assessment of fruits produced per inflorescence, per floors and per total plant are presented in [Table-4].

The lowest number of fruits per inflorescence it was seen at V7 namely an average of 4.25 fruits and the highest number at V2 of 6.83 fruits.

An interesting behavior had variant 1. Thus, we can notice that the overall percentage of fruits produced per plant, was the lowest at this variant of only 61.54%. However, this hybrid almost had the highest average flowers, formed per plant. It can be noticed that the lowest percentage of binding in accordance with the period when have been developed the inflorescences it was for one inflorescence I of only of only 33.33% and 28.57% for inflorescence VII. Therefore, it was the first and last inflorescence. Thus can be assumed that in this case extreme temperatures or lower or very high sunstroke would cause the low percentage of binding.

The best percentage of binding per plant was observed at the variant 8 (no. 3319) of 91.89%. It likewise remarks at this hybrid, which in general on inflorescences percentage was maximum at inflorescences 1, 2, 5 and 6. In addition, this percentage was high and the other inflorescences. Thus to inflorescence 3 was 85.71% and 71.43% in inflorescence 4 [Table-4].

Fruits evaluation was performed depending on their number and the weight of each fruit.

Fruits with small masses between 30-68 g were obtained only for three hybrids namely 33 to the variant V4, 11 to the variant V2 and 2 to the variant V6. Fruits with higher average weights between 141 g -170 g were in number of 5 to V1, V3 and V8, Of 10 respectively 24 fruits to variants 6 and 9.

The average weight of fruits was generally different from one inflorescence to another. Thus in the [Table-5] are presented the average values of tomatoes weights per inflorescence for each hybrid.

Table 4- The average number and the percent of fruits produced per inflorescence and plant

Variants	Inflorescences							The average number of fruits per inflorescence	Total number and percentage of fruits per plant
	I nr/ %	II nr/ %	III nr/ %	IV nr/ %	V nr/ %	VI nr/%	VII nr/ %	Chunks number	nr/%
V1- 3349	2/33.33	8/88.89	7/77.78	6/85.71	4/57.14	3/42.86	2/28.57	4.57	32/61.54
V2-3350	11/100	7/77.78	5/83.33	6/66.67	3/33.33	9/100	-	6.83	41/77.36
V3-3330	7/100	6/75.00	6/100	2/25.00	2/25.00	-	-	4.6	23/62.16
V4-3326	9/100	8/88.89	6/66.67	6/66.67	4/80.00	-	-	6.6	33/80.49
V5-3323	5/71.43	6/85.71	6/85.71	2/22.22	6/85.71	4/80.00	-	4.83	29/69.05
V6-3335	7/100	4/57.14	3/42.86	5/83.33	7/100	-	-	5.2	26/76.47
V7-3351	6/100	4/66.67	3/75.00	4/66.67	-	-	-	4.25	17/77.27
V8-3319	7/100	7/100	6/85.71	5/71.43	6/100	3/100	-	5.67	34/91.89
V9-3352	6/100	5/83.33	4/66.67	5/83.33	4/66.67	-	-	4.8	24/80.00
V10-3348	3/60.00	6/100	5/83.33	4/100	4/80.00	-	-	4.4	22/84.62
V11-2821	8/88.89	5/83.33	6/100	4/66.67	4/66.67	5/83.33	-	5.33	32/82.05

In [Table-6] are presented total masses on inflorescences and on plant harvested.

Table 5- The average weight of fruits per inflorescence (g)

Variants	Inflorescence						
	I	II	III	IV	V	VI	VII
V1-3349	122.5	82.78	103.43	122.33	124	157.33	162.5
V2-3350	49.66	93.86	108.2	104.5	117.66	82.66	
V3-3330	90.37	99.33	103.33	140.5	143.5		
V4-3326	48.44	51.38	55	53.67	56.25		
V5-3323	121.8	103	103.71	133.5	103.14	124.5	
V6-3335	98.71	121.25	123.67	120.4	83.71		
V7-3351	113.17	148	171.66	156.25			
V8-3319	134.86	136.43	137.33	140	135.33	144	
V9-3352	153.67	153.4	173.75	159.4	187		
V10-3348	121.67	115.67	120.4	122.5	119		
V11-2821	81.88	94.6	95.67	96	95	95.2	

Table 6- Total mass of fruits produced on inflorescences and on total plant

Variants	Total harvested per inflorescence, g							Total harvested per plant in g
	I	II	II	IV	V	VI	VII	
V1-3349	245	737	724	734	496	472	325	3733
V2-3350	585	664	546	633	356	753		3537
V3-3330	716	596	620	286	292			2510
V4-3326	44.5	51.38	55	53.66	56.25			1724
V5-3323	609	715	720	267	716	498		3525
V6-3335	691	485	371	602	586			2735
V7-3351	679	592	515	625				2411
V8-3319	944	955	824	700	812	432		4667
V9-3352	922	767	695	797	748			3929
V10-3348	365	694	602	490	476			2627
V11-2821	655	473	574	384	380	476		2942

At V1, the lowest fruits with an average of 82.78 g were obtained the second inflorescence. Within these inflorescences were obtained fruits even of 55-69 g. From inflorescences 1, 3, 4 and 5 were harvested fruits over 100 g. From inflorescences, 6 and 7 fruits were fewer on inflorescence but the average weight of over 157 g. The average weight of fruits harvested per plant was of 3733g.

From V2 were obtained smallest fruits in inflorescence 1. Within that inflorescence were formed 11 fruits. Also at inflorescences, 2 and 6 fruits were below 100 g. The fruits from inflorescences 3, 4 and 5 had a little over 100 g. This variant has formed the most irregular fruits. Per plant were harvested on average 3537 g tomatoes.

The fruits harvested from the variant 3 were on average around 100 g. The inflorescences 4 and 5 were formed only 2 each fruits and their average weight was about 140 g. Per plant were harvested on average 2510 g tomatoes.

From variant 4 were harvested fruits between 48.44 g average mass from inflorescence 1 and 56.25 g from inflorescence 5. Fruits generally small but uniforms characterize this hybrid. Total average weight per plant was harvested 1724 g.

From the variant 5 were obtained fruits with average weight over 100 g. Of over 120 g fruits were harvested from inflorescences 1, 4 and 6. At the 4th inflorescence were obtained of over 130 g fruits but were only two fruits. In total 3525 g were harvested tomatoes on the plant.

The fruits harvested from V6 were below 100 g in the case of inflorescences 1 and 5 but were produced a large number of fruits per inflorescence. In this situation could have been limit the number of fruits per inflorescence so that the fruit to grow in weight. Otherwise, fruits were uniform, with average weights about 120g. This variant gave an average of 2735 g per plant tomatoes.

Large fruits characterize the variant 7. Inflorescence 1 presented fruits with average weight of about 113 g. The rest of the inflorescences had fruits were over 140 g, and to inflorescence 3 the fruits were over 170 g. The total mass of fruits harvested per plant was 2411 g.

At the variant V8 were harvested fruits averaging about 135 g each. Their size was generally uniform for all inflorescences. From this variant was harvested largest amount fruit on plant, namely 4667 g.

In the case of variant 9 the fruits had higher weights over 153 g. At this variant was observed that the inflorescences were broken under the weight of fruits. This variant gave 3929 g fruits per plant.

At the variant V 10 were obtained generally uniform fruits over about 115 g. Most fruits have had about 120 g. From the inflorescences were harvested from 365 g to inflorescence 1, from 694 g to inflorescence 2. A total of 2627 g per plant were harvested.

Small Fruits, averaging about 80 g to about 96 g were collected from V11. Fruits were generally uniform within inflorescences. The large number of fruits in inflorescences determined to obtain on inflorescence between 380 g and 655 g and 2942 g per plant.

There have been a number of 4 harvests. To variants 4, 5, 10 and 11 there has been no harvesting on 1 July 2011 as result these hybrids are later.

The data obtained concerning plant production are recorded in [Table-7].

Table 7- The graph of tomatoes harvest and average production obtained per plant

Variants	Total production g/plant	1st harvest Jul-01 g/plant	2nd harvest Jul-15 g/plant	3rd harvest Jul-23 g/plant	4th harvest Aug-01 g/plant
V1-3349	3733	779	865	1150	939
V2-3350	3537	1120	780	975	662
V3-3330	2510	784	577	380	769
V4-3326	1724	0	830	450	444
V5-3323	3525		1650	985	890
V6-3335	2735	1463	530	375	367
V7-3351	2411	709	1320	382	0
V8-3319	4667	2537	650	875	605
V9-3352	3929	1695	1275	630	329
V10-3348	2627		1157	878	592
V11-2821	2942		1415	875	652

Analyzing hybrids in terms of the early production was found that the highest percentage of 54.36% of total production was registered at hybrid V8. The data recorded can appreciate the production capacity of hybrids analyzed per m^2. These assessments are presented in [Fig-4]. It can notice that in this case too hybrid V8 recorded the highest yield at 1 m^2 in a period of only three months of vegetation. The hybrid V4 showed the lowest production, of only 6.1 kg/m^2 but also the most belated.

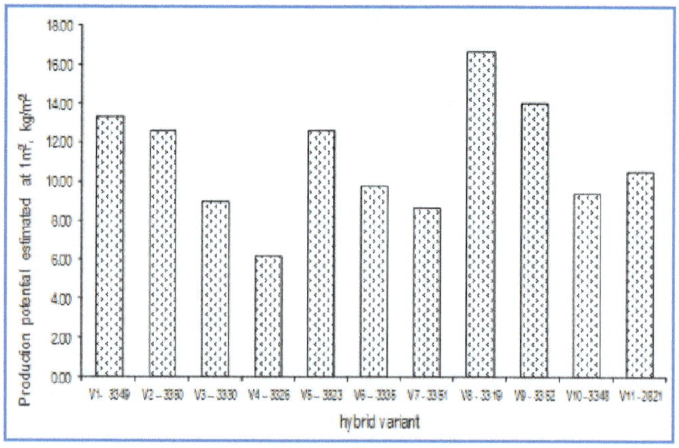

Fig. 4- Production potential estimated at one m^2

In [Table-8] the characteristics of tomato fruits are presented to all hybrids analyzed. Thus index of form, relationship between the height and the diameter of the fruit, showed that the fruits have had generally spherical shape flattened on most variants except of variant 11 where the fruits were elongated, the index being 1.25.

In cross section was observed that the number of the seminal loggias was lower at variants 11 and 8 (2-3 loggias).

Immature fruits showed a uniform green color to the variants 1,2,3,5,6,8 and 11 and only at the variants 4, 7, 9 and 10 was noted the presence of dark green with collaret. At the physiological maturity the fruits have had deep red color except of variant 11 where the fruits were red-orange easily.

Table 8- The characteristics of tomato fruits to hybrids tested

Variants	Shape of the fruit IF	Number of seminal loggias	Unripe fruit color	Physiological maturity of the fruit color	Fruit appearance	Pericarp thickness, cm	Size of the pestle area, mm	Peak shape
V1-3349	0.85	4-5	Uniform green	Red uniform	Smooth	0.7	2-3 mm	Without peak
V2-3350	0.79	4	Green	Red uniform	Fruit coated prone to cracking easily	0.8	4-5 mm	Without peak
V3-3330	0.84	4	Green	Red uniform	Smooth	0.5	2-3 mm	Without peak
V4-3326	0.75	3	Green with collaret	Red uniform	slightly rippled	0.6	1-2 mm	Without peak
V5-3323	0.84	3-4	Green	Red uniform	easily coated	0.7	2-3 mm	Without peak
V6-3335	0.73	3-4	Green	Red uniform	easily coated	0.7	4-5 mm	Without peak
V7-3351	0.67	4	Green with collaret	Red uniform	Smooth	0.9	1-2 mm	Without peak
V8-3319	0.85	2-3	Green	Red uniform	Smooth	0.8	1-2 mm	Without peak
V9-3351	0.82	4	Green with yellow collaret	Red uniform	Smooth	0.7	2-3 mm	Without peak
V10-3348	0.85	4	Green with collaret	Red uniform	Smooth	0.7	2-3 mm	Without peak
V11-2821	1.25	2	Green	Red-orange	Smooth	0.5	absent	Without peak

Table 9- Fruits firmness and dry substance content on 7-19-2011

Variants	Firmness: At the temperature of 9°C kgf/cm²	Firmness: At ambient temperature kgf/cm²	Dry substance: At the temperature of 9°C %	Dry substance: At ambient temperature %
V1	1.79	4.7	4.7	5.1
V2	1.96	5.4	5.4	5
V3	1.19	5.5	5.5	5
V4	2.07	5.4	5.4	4.6
V5	2.32	5.3	5.3	4.7
V6	1.74	4.6	4.6	4.6
V7	2.11	5.7	5.7	6
V8	1.59	5.2	5.2	4.9
V9	3.7	5.4	5.4	5.4
V10	1.9	5.3	5.3	4.6
V11	1.17	5.6	5.6	4.8

The thickness of pericarp was small only to V11 meaning of 0.5cm.

Another characteristic was the size of pestle area that most of the times when this is large can be characterized as a defect because the fruit can crack this area. Most fruits showed a smooth appearance except variants 2.4, 5 and 6, where appearance was slightly coated. All variants showed no peak fruits appearance being rounded.

Fruit firmness at harvest was between 4.6 and 5.7 kg/cm^2 to 6 kg/cm^2 V to V7. In terms of 9°C temperature, storage for 14 days it was observed that fruit firmness was lowers [Table-9].

Conclusion

The analysis of assessment of those 11 hybrids from Israel in early spring climatic conditions of Romania led us to some interesting conclusions. Thus between the seeds those 11 hybrids the lowest germination rate have had hybrids V4-3326 and V8-3319. While most of hybrids were germinated in 4-5 days hybrid V4-3326 hybrid germinated after 9 days and V8-3319 after 11 days. In addition, they had the lowest germination percentage namely 92% for V4 and 87% for V8.

In terms of seedlings developing, has been an increase between 0.295cm/day and 0.363cm/day. The lowest rate of growth of seedlings had the variant V4 and the highest the variant V11.

At planting tomato seedling with the fewest leaves was hybrid noted V8 with an average of 4.33 leaves while the most for V1 with an average by 7 leaves/plant. In addition at planting seedlings of 1,2,9,10 and 11 variants showed hypocotyls of purple color.

As it was found until, planting the variant V8 showed the worst performance followed by the variant 4. Even in the solar, the lowest height has had a hybrid V8-3319.

Pursuing formation of inflorescences and then fruits is one of the most important issues for horticulturists. Thus, plants had a different number of inflorescences respective flowers. Ratio between the number of flowers and the fruit produced was also different. Thus, although hybrid V1 has had one of the largest numbers of flowers of which only 61.54% have produced fruit, being one of the weakest production hybrids. At the same time the hybrid that look the weakest growth during seedling production (v8) produced the highest percentage of fruit connected, namely 91.89%.

In terms of fruits size the smallest fruits with an average of 52.24 g/fruit were produced by the hybrid V4-3326 and the highest with an average of 163.7 g/fruit were produced by the variant V9-3352.

Evaluating the production capacity per plant and the potential production per 1m^2 has found that the variant V4 has the lowest potential of 6.1571 kg/m^2. It is very interesting that the greatest potential productive it has the variant 8 (hybrid 3319) of 16.667 kg/m^2, the one that was seen from the start as the variant with the lowest capacity of germination and growth. This is extremely significant because this hybrid not only has a large production capacity but it is the earliest of all hybrids tested. Thus even from the first harvest gathered 54.4% of the total production at this hybrid.

As regards of fruits firmness in storage conditions for 2 weeks at 9° C most resistant fruits proved those provided by the hybrid V7-3351.

These results lead to the suggestion of the experiments to improve germination capacity of the hybrid V8 and recommendations for early crops in solar, in climate conditions in Romania.

Acknowledgements

This paper is supported by a contract funded by Executive Unit for Financing Higher Education, the Research, Development and Innovation Grant PN-II-PT-PCCA-2011-3.2-135, CNCSIS-UEFISCSU-Partnerships Program, Bucharest, Str. D.I. Mendeleev nr. 21-25, sector1, Romania.

References

[1] Bleekera P.M., Mirabellaa R., Diergaardeb P.J., VanDoornb A., Tissierc A., Kantd M.R., Prinsb M., de Vosb M., Haringa M.A., Schuurinka R.C. (2012) *Proc. Natl. Acad. Sci. USA*, 1-6.

[2] Draghici E.M. (2009) *Producerea Semintelor si Materialului saditor la Specii Legumicole*, Romania, 35-42.

[3] Ciofu R., Stan N., Popescu V., Chilom P., Apahidean S., Horgoş A., Berar V., Lauer K.L. and Atanasiu N. (2004) *Tratat de legumicultură.*

[4] Mayeaux M., Xu Z., King J.M., Prinyawiwatkul W. (2006) *J. Food Sci.*, 71, C461-464.

[5] Berova M., Stoeva N., Zlatev Z., Ganeva D. (2008) *Scientific Journal In Agriculture,* 9(4), 723-732.

[6] Suchindra R., Sankaranarayanan R., Nainar P. (2012) *Plant Archives,* 12(1), 257-258.

[7] Ali W., Jilani M.S., Naeem N., Waseem K., Khan J., Ahmad M.J., Ghazanfarullah (2012) *Sarhad J. Agric.,* 28(2), 207-212.

[8] Pandey Y., Pun A., Upadhyuy K. (2006) *Nepal Agric. Res. J.,* 7, 11-15.

[9] Suchindra R., Sankaranarayanan R., Nainar P. (2012) *Plant Archives,* 12(1), 245-247.

[10]Ahmad F., Khan O., Sarwar S., Hussain A., Ahmad S. (2007) *Sarhad J. Agric.,* 23(3), 581-585.

[11]Kacjan Maršić N., Osvald J., Jakše M. (2005) *Acta Agriculturae Slovenica,* 85(2), 321-328.

[12]Drăghici E.M., Hoza G. (2009) *Bulletin of Agricultural Sciences and Veterinary Medicine Cluj-Napoca,* 66(1).

[13]Drăghici E.M., Dobrin E. (2011) *Lucrări Stiinţifice USAMV Bucureşti, Seria B,* LV, 89-93.

[14]Lushi I., Balaj N., Haxhinasto L., Zhitia O. (2009) *Acta. Hort.,* 830, 163-168.

[15]Mejia L., Teni R.E., Vidavski F., Czosnek H., Lapidot M., Nakhla M.K., Maxwell D.P. (2005) *Acta. Hort.,* 695, 251-256.

[16]Commission Directive (2008) *Official Journal of the European Union.* L 219/55-L 219/57.

DECISION SUPPORT SYSTEMS IN A GLOBAL AGRICULTURAL PERSPECTIVE - A COMPREHENSIVE REVIEW

MIR S.A.[1]*, QASIM M.[2], ARFAT Y.[3], MUBARAK T.[1], BHAT Z.A.[4], BHAT J.A.[5], BANGROO S.A.[1] AND SOFI T.A.[1]

[1]Krishi Vigyan Kendra (Farm Science Centre), Kulgam, SKUAST Kashmir, J&K, India.
[2]Krishi Vigyan Kendra (Farm Science Centre), Poonch, SKUAST Jammu, J&K, India.
[3]Krishi Vigyan Kendra (Farm Science Centre), Narkura Srinagar, SKUAST Kashmir, J&K, India.
[4]Mountain Research Station Khudwani, SKUAST Kashmir, J&K, India.
[5]Krishi Vigyan Kendra (Farm Science Centre), Ganderbal, SKUAST Kashmir, J&K, India.
*Corresponding Author: Email- mir.shabir972@gmail.com

Abstract- In global agricultural, decision making at all levels are increasing rapidly due to unreliable and uneven distribution of combined experience of complex processes associated with the food production. Besides flaws in human initiative judgement have far-reaching consequences for the sustainable agriculture development. Therefore, aiding the reliability and deficiencies of human judgement and quality decision-making has been a major focus of research throughout the history particularly with the development of Decision Support Systems (DSSs). DSS is interactive computer-based system that help decision makers solve unstructured problems under complex and uncertain conditions by providing access to data through procedures and analytical reasoning. These systems have been designed to address complex tasks involving multiple disciplines of agriculture, enabling us to mimic the necessary requirements of crop with respect to the land characteristics so that the optimum objectives specified by the user is obtained. Experimental use of these systems has resulted in improved integrated crop production and management with restricted use of the natural resources together with improved farm income and reduced negative environmental impact. With the advent of powerful computation, efficient database management system, on-line analytical processing (OLAP), improved knowledge representation, efficient searching techniques, data warehousing and mining, Geographical Information System (GIS) and Remote Sensing (RS); DSS are being extensively used world-wide using state of art technologies, integrating multiple models for sustainable agriculture development from farm to fork, despite "Problem of Implementation". Once the barrier is crossed, it will give a leading role to DSS in global sustainable agriculture.

Keywords- Decision Support System, Sustainable Agriculture, Crop modelling, Management Information systems, Agricultural Systems Research, Integrated pollution control, Land use planning, knowledge representation and searching

Introduction

Agriculture is considered as a complex process of air, water, weather, soil, plant, animal, micro-organism, which are again unreliable and uneven in distribution, covering many disciplines and interacting at various scales [1, 2]. In order to manage different crops one has to generate alternatives to make choice, supported by estimation, evaluation and/or comparison. There is a substantial amount of empirical evidence that human initiative judgment and decision making can be far from optimal and it deteriorates further with added complexity and stress [3]. Also decisions made by the farm managers are irreversible and have far-reaching consequences for the crop being managed [4]. Therefore, aiding the deficiencies of human judgement and quality decision making has been a major focus of research to develop less intensive and integrated farming systems with lower inputs of fertilizers and pesticides and with restricted use of the natural resources (water, soil, energy, etc.) with the broader objective to maintain crop production in both quantitative and qualitative terms, maintain or preferably improve farm income, and at the same time reduce negative environmental impacts as much as possible[5].

For making rational decisions, disciplines such as statistics, economics and operational research have developed various methods, which have recently been enhanced by various techniques originating from Information Science, Cognitive Science, Artificial Intelligence and Pattern recognition. These methods have been implemented in the form of computer programs either as stand-alone system or complex computing environments for complex decision making using common name of decision support systems (DSSs).

Regarded as response to complexity, Decision Support System (DSS) is an interactive computer-based system or subsystem intended to help decision makers to use information and communication technologies, data, knowledge and/or models to facilitate formal and informal communication, mining knowledge, and building knowledge repositories [6]. DSS improves personal efficiency; speed up the process of decision making; increases organizational setup and control; encourages exploration and discovery of unknown data; speeds up problem solving capabilities; facilitates interpersonal communication; promotes learning and training abilities, provides new evidences in support of a decision and reveals new approaches to think about the problem more specifically through automation of managerial process.

DSSs are used by an organization or group of users to support and enhance decision making [7] when faced with ill-structured situations [8,9] for instance in natural resource issues [10-13]. They can assist in making better decisions by integrating information into a more useable form. They alter production systems, enhances management skills and reduces cost of production by applying modelling, hypothesis and optimization techniques. The current DSS industry is one of the most buoyant areas of investment despite the IT downturn of the early to mid 2000s. Recently, many attempts have been made at building decision support systems by employing mathematical models and databases [14].

Global agriculture has seen an increasing interest in the potential use of DSSs in recent years. The application of these systems to agriculture seems natural considering widespread use of extension agents in and out of the field. Status of soil, selection and yield characteristics of a variety, likely interaction between the field and variety, disease presence in the area, yield, weather conditions, current market and prices and related information have become important. For crop production, DSSs provides hypothetical environment for researchers to integrate all possible information from different sources to recommend the most appropriate action or choices [15]. Expert knowledge, management models and timely data are key elements of DSS and are used to assist producers with both daily operational and long-range strategic farm decisions [16].

Use of DSS in agriculture for providing information and recommendation on efficient utilization of fertilizer [17], reducing herbicide use [18], plant protection [19], variety-specific information [20], management of environment risks [21], Integrated Nutrient Management [22], Forest Management [23], Crop disease control [24], Agricultural practices and extension [25], Farm mechanization [26], Seasonal Climate Prediction[27], Integrated Pollution Control [28], labour requirements and land use planning [29], profit maximization and risk minimization [30] etc., have been designed and implemented successfully mainly for improving economic returns, changing farming practices or minimizing environmental risks. Developments in climate downscaling have helped to bridge the spatial and temporal resolution gaps between what climate modellers are currently able to provide and what impact assessors require [31]. In addition, tools that integrate climate and plant growth modeling will continue to be developed to help assess future impacts and adaptation options [32,33]. Future climate-related DSS are therefore likely to provide information on the current farming season, and potential changes in climate and farming risks for future seasons

Advancement in hardware and mathematical modelling, artificial intelligence techniques, data warehousing and mining, OLAP (examines and manipulates large amounts of detailed and consolidated data interactively in real time with rapid response from many perspectives), enterprise resource planning (ERP) systems, intelligent agents are adding new capabilities to agricultural DSS research and development. Moreover, the space technologies such as Remote Sensing (RS), Geographical Information System (GIS) are also providing support to agricultural DSS. Furthermore, with the rapid development of internet/intranet technology, Web technology is growing up to a new branch in the development of Automated Decision Support System or ADSS opening new opportunities for DSS design and implementation [34].

This paper chronicles basic concept of DSSs, their characteristics, capabilities, architecture, classification and applications in global agricultural perspective. Further it provides knowledge representation, searching mechanism, issues and future projections.

What are Decision Support Systems?

Decision Support System (DSS) has been used and defined in various ways depending upon the author's point of view [9, 35, 36] defines it as "useful and inclusive term for many types of information systems that support decision making". For [37], it is "a computer-based system that aids the decision making process". [38] has defined it specifically as "an interactive, flexible, and adaptable computer based information system, especially developed for supporting the solution of a non-structured management problem for improved decision making". [39] define DSS as "model-based set of procedures for processing data and judgments to assist a manager in decision-making". [40] Define DSS as "Computer-based support for management decision making". [41] define it as "extensible systems capable of supporting adhoc data analysis and decision modelling, used in future planning". [8] described DSS as "interactive Computer-based systems that help decision makers to solve unstructured problems using data and/or models". [42] is not of the opinion to give DSS a precise definition. For him "there can be no definition of DSS". He further adds that DSSs have typically quantitative output and place emphasis on the end-user for final problem solution. Often ES is developed around very specific and highly detailed "Domains" and thus tends to be narrow in their range of knowledge [43]. In order to avoid exclusion of any of the existing types of DSSs, we define them roughly as interactive computer based systems that aid in making a quality decision.

Characteristics and Capabilities

- Decision Support System comprises of vast set of characteristics and capabilities of DSS. The key characteristics and capabilities of DSS as reported by [38] are as follows:
- Ability to support in semi-structured and unstructured problems, including human judgment and computerized information.
- Ability to support managers at all levels.
- Ability to support individuals and groups.
- Ability to present knowledge on ad hoc basic in customized way.
- Ability to select any desired subset of stored knowledge for presentation or derivation during problem solving.
- Ability to support for interdependent or sequential decisions.
- Ability to support intelligence, design, choice and implementation. viii. Ability to support variety of decision processes and styles.
- Ability to support modelling and analysis.
- Ability to support data access.
- Benefits must exceed cost.
- Allow modification to suit needs of user & changing environment.
- Support quick decision-making using standalone, integration or web-based fashion DSSs having maximum number of these key characteristics and capabilities can be more useful and adoptable.

Major Fields of DSS

- **Personal Decision Support Systems (PDSS):** usually small-scale systems that are developed for specified managers;
- **Group Support Systems (GSS):** they use combination of DSS technologies to facilitate the effective decision process;
- **Negotiation Support Systems (NSS):** Here primary focus remains on negotiation between opposite perceptions;
- **Intelligent Decision Support Systems (IDSS):** It uses artificial intelligence techniques to facilitate decision;
- **Knowledge Management-Based DSS (KMDSS):** They provide knowledge storage, retrieval, transfer using organizational memory and inter-group knowledge access;
- **Data Warehousing (DW):** systems that provides the large-scale data infrastructure in multiple formats for decision support;
- **Enterprise Reporting and Analysis Systems:** enterprise focused DSS including executive information systems (EIS), business intelligence (BI), and corporate performance management systems (CPM). BI tools access and analyze data warehouse information using business intelligence software, query and analysis tools [44].

Classification of Decision Support Systems

There is no universally accepted taxonomy of DSSs, as different authors propose different classifications [Table-1]. However, important types of Decision Support Systems are:

- **A Model-driven DSS:** Model-driven DSS provides access to and manipulation of various underlying models by using data and parameters provided by users to assist decision making process. Dicodess[45] is an example of an open source model-driven DSS generator.
- **A Communication-driven DSS** allows more than one person to work on a one task. Microsoft's NetMeeting or Groove [46] is an example of communication driven DSS.
- **A Data-driven DSS or data-oriented DSS** emphasizes access to and manipulation of internal or external data. Example of such DSS is OLAP [47].
- **A Document-driven DSS** utilizes unstructured information in a variety of electronic formats for manipulation, retrieval and management ; example includes Google Search Engine.
- **A Knowledge-driven DSS** stores facts, rules, procedures and structures for expertise problem solving purposes. Mycin [48] is an example of knowledge driven DSS.

Table 1- Classification of Decision Support Systems given by different authors

Author	Classification	Features	Example	Criterion
Bhargava & Power [62]	Model-driven DSS	Emphasizes on access to and manipulation of statistical, financial, optimization or simulations model.	Dicodess [45]; production planning management Decision Support[49].	Mode of assistance
	Communication- driven DSS	Emphasizes communications, collaboration, and shared decision making support.	Microsoft's Net Meeting or Groove [46]; Basic Group Decision Support System [50].	
	Data-driven DSS	Emphasizes on access to and manipulation of time series data.	Data Driven DSS with OLAP [47].	
	Document-driven DSS	Manages, retrieves and manipulates unstructured information in a variety of electronic formats	Search engine [9]; On-line Analytical Processing [47].	
	Knowledge-driven DSS	Specialized problem solving expertise stores as facts, rules, procedures etc.	MYCIN [48]; EXSYS [51]; DENDRAL.	
Power [58]	Enterprise-wide DSS	Linked with large data warehouse, which serves many managers	Web-based DSS [52].	Scope
	Desktop DSS	Single user, small system that runs on managers Personal Computer	Visible calculator	
Haettenschwiler [59]	Active DSS	It aids the process of decision without bringing out explicit suggestions or solutions	Walmart	User relationship
	Passive DSS	Brings out explicit suggestions as Well	Exsys[53]	
	Cooperative DSS	modifies, completes or refines the decision suggFestions	Co-op [41]	
Holsapple & Whinston [60]	Text-oriented DSS	Works on text as input	Management, planning and control[54]	Type of Inputs used for decision making
	Database oriented DSS	Has a database in the back end for Inputs	ONVAREF[2]	
	Spread sheet oriented DSS	Uses spread sheet as inputs e.g. Excel	Optimization solver add-in for Microsoft Excel	
	Solver Oriented DSS	Mainly designed for solving problems. e.g. Linear Equations	Brandaid [55]	
	Rule oriented DSS	Uses inputs in the form of rules based on reasoning	NuMaSS[56]	
	Component Oriented DSS	Hybrid System that includes two or more of basic five structures described by Holsapple & whinston, 1996	GRAM++[57]	
Hackathorn & Keen [61]	Personal Support	Supports only one user	ONVAREF[2]	Scope
	Group Support	Supports group of user	Mindsight; Group Systems ; SAMM; PLEXSYS	
	Organizational Support	Supports an organization as a Whole	EXPRESS	

Components of DSS

Like classification, different authors have identified different components in a DSS and have proposed different architecture [Table-2]. Every DSS does not fits neatly into one category, but a mix of two or more architectures. The most general architecture of DSS can be divided into four subsystems viz., Database Management Subsystem, Knowledge-based Management Subsystem, User Interface subsystem and the User. Now-a-days as most of the data comes from external sources such as weather, satellite etc., and separate subsystems are used for model representation and transaction processing so the typical architecture of the DSS can be divided in to six subsystems viz., data management subsystem, model management subsystem, knowledge engine, the user interface, DSS architecture and network and the user [Fig-1].

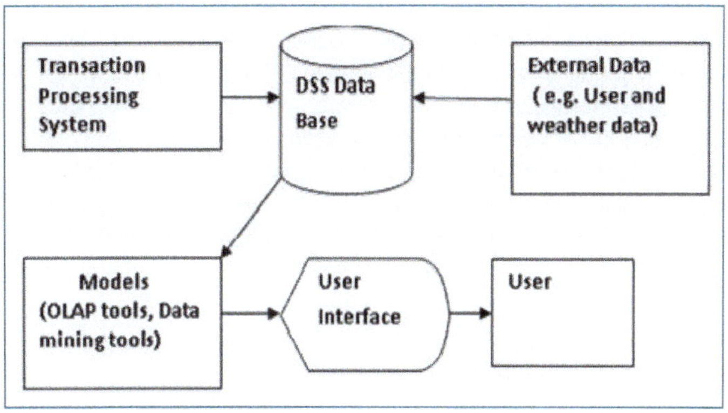

Fig. 1- Typical architecture of DSS

Table 2- Architectural components of DSS given by different authors

Author	Architectural Component	Description of component
Sprague & Carlson [63]	Database Management System (DBMS) Model Base Management system (MBMS) Dialog generation and management system	Stores information Integrates models Provides user interfaces to manage system
Haag et. al. [64]	Same as above but describes them in detail	Stores information (that can be further sub divided into organizations traditional data repository, form external source such as internet or from experience of individual user Using various kinds of models it, handles representation of events, facts, or situations. Integrates models and provides user interface. Participates different roles or functions in the data management process
Haettenschwiler [59]	User Decision context Target system Knowledge base	Specifically defined decision rules Describes majority of the preferences External data sources, knowledge databases, working databases, data warehouse, metadata bases, models, methods, integrates search engines to responding system
Marakas [65]	Database Management System Memory Management System Knowledge Engine User Interface User	Stores, manages and provides access to the data Organizes memory efficiently Inference procedure or control structure for utilizing the knowledge Allows user to interact with the system One who uses the system
Power [9]	The user interactive The database interactive The model and analytical tool The DSS architect and network	Interacts with the user over a command line Interacts with a single or a group of users using a database for heuristics Model designed for analysis Interacts with the other DSS or database server.

Knowledge Representation in DSS

Agricultural based DSS primarily represents knowledge in three broad categories viz., rules based, semantic networks and frames, depending on the type of DSS and domain for which it is designed.

Rule based knowledge representation are used for correlation of simple facts with a decision. The principle of rules is based on the correlation of one or more with one or more results that are activated or considered as true, when all treaties are true. Rules are constructed from a sequence of "if then" statements illustrated as follows:

Rule 1: *If* temperature is above 300 and relative humidity is between 50-70 and month is September

Then it is likely to spray Nuvan for stem borer

Rule 2: *If* temperature is above 300 and relative humidity is between 50-70 and month is August

Then it is likely to spray Kohiban for Leaf Hoper

The Semantic Networks

Often used as a form of knowledge representation, a semantic network represents semantics relations between concepts. It is a directed or undirected graph consisting of vertices, which represent concepts and edges. Representation of knowledge in Semantic Network [Fig-2].

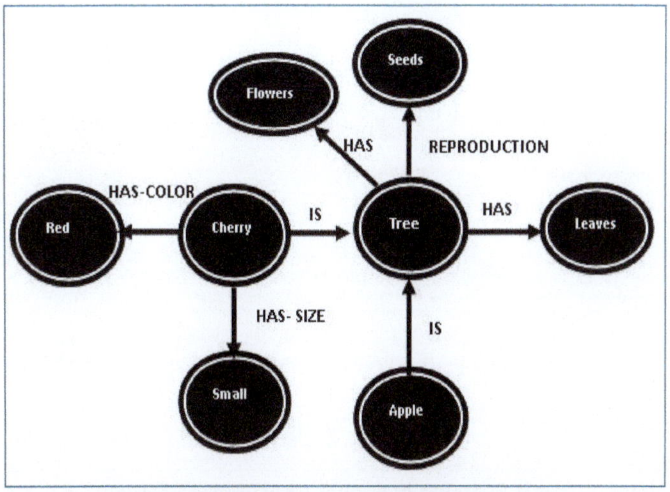

Fig. 2- Representation of a Semantic Network

Representation of Semantic Network in Lisp language

```
(defun *database* () ((apple (is-a tree)
(has flowers) (size big))
(cherry (is-a fruit) (color red))
(tree (is-a big plant) (has-part leaves) (reproduction by-seeds))))
```

The Frames

A frame contains information that accompanies an object. That information contains values or indicators to different frames or rules or processes, from where the values correspond to information and relevant with the object are drawn, is placed in slot. The frame contains information on how to use the information contained frame, next possibilities and course of action when desired expectations are not met. Some information in the frame is generally does not change in contrary within terminals. Different frames may share the same terminals. Simple structure of a frame is given in [Table-3].

Each piece of information about a particular frame is held in a slot. The information contains:

Facts or Data

Values (called facets)

Procedures (also called procedural attachments)

IF-NEEDED : deferred evaluation

IF-ADDED : updates linked information

Default Values

For Data

For Procedures

Other Frames or Subframes

Searching Strategies in DSS

Searching through huge networks, decision-trees and data bases is mostly used for document driven DSS in which unstructured information in a variety of electronic formats is retrieved, manipulated and managed. Some important search strategies used in agricultural DSS aids are explained as under:

Depth-first Search: The search backtracks and tries another path from B. A Depth-first strategy probes a knowledge-tree or a decision -tree by asking the detailed questions first in a limb of the tree (top downward) as the first path through the tree. In other words it explores a path all the way to a leaf before backtracking and exploring another path. For example, after searching parent node A it probes left siblings and goes deeper and deeper till the leaf node is encountered. Node are explored in the order A B D E H L M N I O P C F G J K Q. N will be found before J [Fig-3](a).

Breadth-first Search: A Breadth-first strategy, in contrast, searches all nodes at the same depth and then proceeds to the next lower level of nodes. In other words it explores nodes nearest the root before exploring nodes further away. For example, after searching parent node the search probes all the nodes at one level before shifting to another level. Nodes are explored in the order A B C D E F G H I J K L M N O [Fig-3](b).

The Best-first Search: It combines best features of the Depth-first and the Breadth-first strategy. A heuristic, or specific knowledge, guides the search to choose either a Depth-first or a Breadth-first strategy at any particular node, depending on the knowledge. In other words it explores a tree by expanding the most promising node chosen according to a specified rule [Fig-3](c).

Table 3- Structure of a frame

Slot	Value	Type
BOY	–	(This Frame)
ISA	Person	(parent frame)
SEX	Male	(instance value)
AGE	Under 12 yrs.	(procedural attachment - sets constraint)
HOME	A Place	(frame)
NUM_LEGS	Default = 2	(default, inherited from Person frame)

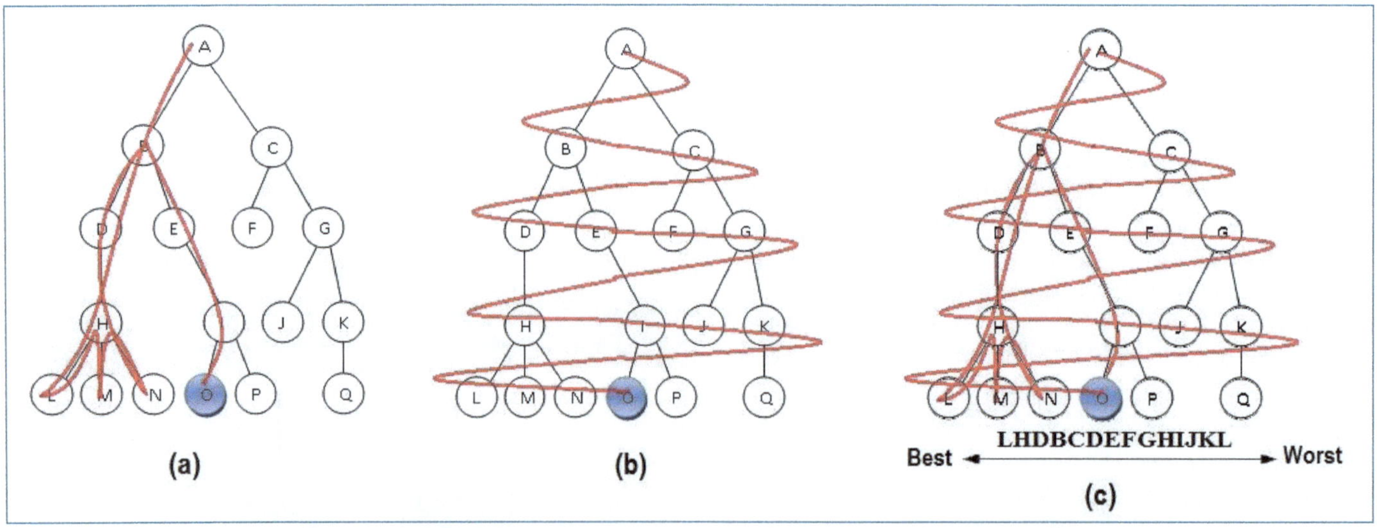

Fig. 3- Decision trees illustrating depth first, breadth first and best first searching

Decision Support Systems in Global Agricultural Perspective

Major DSS Projects

For global sustainable agriculture, pre-requisition is to maintain crop production in both quantitative and qualitative terms, maintain or preferably improve farm income, and at the same time reduce negative environmental impacts as much as possible [66,67]. DSS have proved to be an effective tool in this direction by development of Integrated Production [68] and Integrated Farming (IF) [69], which have been developed as holistic concepts that involve all crop and farming activities and shape these activities according to the individual site and farm. Globally large scale multipurpose projects on agricultural DSS have been successfully developed, covering diverse fields, incorporating different specialty to highest ever precision.

During 2002-10, Thailand and Cambodia have created a TRF-DSS research and development network, which consists of 59 research projects, involving 12 universities, two line agencies in Thailand, and one line agency in Cambodia. More than 20 DSS tools were developed and implemented for variety of users engaged in short and long-term planning and management. Most of these tools were designed to integrate biophysical and socioeconomic data as well as the decision support modules for alternatives evaluation and analysis; supporting dynamic simulation and multi-criteria analyses for model base software development to allow users evaluate various alternatives in agricultural and natural resource management [70].

The University of Tokyo in collaboration with National Agriculture and Food Research Organization, National Institute of Agro-Environmental Sciences Ishikawa, Prefectural University, Toyama Prefecture and Fukui Prefecture have developed decision support system for optimal agricultural production under global environment changes. It contains crop growth models for rice, wheat and barley; soil and water models for local water resource and whether models for locally useful high resolution data [4].

University of Stirling, UK has developed a user friendly PC-based expert system/decision support system CeserDSS for countermeasure implementation in agricultural food production systems in Scotland. It is regarded as a decision-aiding tool in the planning of long-term remediation at the level of a farm or small agricultural area [71].

In United States, the USDA-ARS Agricultural Systems Research Unit (ASRU), in collaborative effort with Colorado State University (CSU), developed the Great Plains Framework for Agricultural Resource Management (GPFARM) DSS. The general purpose of GPFARM is to serve as a whole-farm/ranch DSS for strategic agro-ecosystem sustainability evaluation across the U.S. Great Plains. The GPFARM DSS is primarily composed of six major components: 1) a Microsoft® Windows-based graphical user interface; 2) Microsoft® Access databases containing soil, crop, weed, climate, equipment, chemical, and economic parameters; 3) an object oriented modelling framework and science simulation model; 4) a stand-alone economic analysis tool; 5) a set of analysis tools, multi-criteria decision support system, an output visualization module and summary report tables and graphs; and 6) a web-based information system. Thus, GPFARM is unique in that it brings together the above suite of decision support tools integrated across a whole-farm/ranch system [72].

Also University of Nabaraska has developed GreenLeaf Project (Google-like Infrastructure for Climate Data), which is a revolutionary collection of decision-support tools and methodologies linked to provide dramatically new and unique views of the data critical to supporting producers, decision markers and researchers in analyzing climate -associated risk events and mitigating their effects. Recently The U.S. Department of Agriculture (USDA) and the National Aeronautics and Space Administration (NASA) signed a Memorandum of Understanding (MOU) to strengthen future collaboration to assimilate NASA's Moderate Resolution Imaging Spectroradiometer (MODIS) data and products into an existing decision support system (DSS) operated by the International Production Assessment Division (IPAD). Integration of MODIS data and derived products into the IPAD, DSS provides better characterization of land surface conditions at the regional scale and enables monitoring of changes in the key agricultural areas in a more timely fashion and at a higher resolution.

In Kenya, the Regional Centre for Mapping of Resources for Development (RCMRD) in collaboration with European and African partners developed a decision support system (DSS) for sustainable development of dry lands in three countries of eastern Africa (Ethiopia, Kenya and Tanzania). In this endeavour, GIS modelling was used to support decision-making on agricultural product development and distribution, natural resource management, environmental rehabilitation, and institutionalization of indigenous environmental knowledge [73].

India under National Agricultural Innovative Projects (NAIP) is working on Precision Farming Technologies based on Microprocessor and Decision Support Systems for Enhancing Input Application Efficiency in Production Agriculture with the aim to develop sensor-based systems for assessing spatial and temporal variability of soil, moisture, crop and weed parameters and creating databases by adapting available components; precision applicators for seed, fertilizers, herbicides and water to address site-specific variability; and integrate variable rate input application technology with the available decision support systems (DSS) for higher use efficiency.

Australian agricultural has great experience with using DSS for Agriculture, which assist farmers with information and procedures to management decisions with regard to crop forecasting and management (*Whopper Cropper; Yield Prophet*), livestock production and management (*GrazFeed*), pasture growth and management(*GRASP, GrassGro*); weather and climate forecasting (*Rainfall Reliability Wizard; Water and the Land*), crop growth, irrigation and nutrient management[11,74,29]. CottonLOGIC DSS is considered to have influenced the adoption of sustainable farming systems in the cotton industry [75].

Recently Australian National University Fenner School of Environment and Society has developed EXCLAIM2 -A tool for assessing climate change impacts on natural resources at a regional scale. It contains underlying models that links climate, hydrology and ecology to predict the impacts to ecological health. It helps to explore plausible scenarios of climate change and helps to explore how plausible climates scenarios impact on: hydrology –river flow, water allocation; water quality –turbidity, nutrients, algal blooms; ecological health –habitat conditions for vegetation and water birds [76].

Area Specific DSS

Nutrient Management

Fertilizers and lime are increasingly expensive but are commonly needed to grow high-yielding and good-quality crops. However, unnecessary use is wasteful, reduces farm profits and increases the risk of diffuse nutrient pollution. To maximize profits and avoid waste, farmers need to plan their use of nutrients for each field crop in each year. Organic manures (farmyard manure, sewage sludge, slurries, etc.) contain large quantities of nutrients which can often mean that large reductions are possible in the need for inorganic fertilizers. Nutrient management can play an important role in many of the regulatory and non-regulatory duties of farm-related management, and can protect, restore and enhance the status and diversity of all surface water ecosystems and ensure the progressive reduction of groundwater pollution.

For Nutrient management, different DSSs have been designed to recommend site-specific and need-based parameters that result in an optimized fertilizer management strategy. One example of such system is CERES, which simulates the whole soil crop system [77]. Another example is rice fertility DSS, which provides recommendation on efficient utilization of fertilizer for the production of flooded rice in Arkansas [17]. A DSS for reduction in potential nitrogen (N) losses to the environment has been developed, which saves fertilizer expenditure [78]. There is an ongoing trend to develop Nutrition Management Decision support tools to make them available to the farmers through World Wide Web. These facilities are enabling farmers to use the service of these tools irrespective of computer owner-ship, which is being reported as one of the reason for low adaptation of DSS among farmers. Haifa Nutri-Net is an example of such system [79]. It is a comprehensive crop Nutrition DSS, operated over the web, assisting growers to formulate their crop nutrition programs and irrigation schemes by integrating virtually all relevant cultivation parameters. It is based on comprehensive databases of crop nutrition, irrigation, soil and climate, covering all most every growth environment. FarmN is another web based DSS providing INM recommendations [80].

Most of the existing DSSs are based on very specific aspects, for example, Nutrient management. One system that addresses all the major manure management systems has been identified [81]. DSS for Planning Land Applications of Nutrients for Efficiency and the Environment (PLANET)[82], provides best management practice tool for farmers and their advisors to adopt in the use of organic manure and fertilizers. Fertilizer recommendations for field are calculated based on the precious cropping fertilizer and organic manure application. To encourage maximum uptake of DSS by the farming community, the logic to generate fertilizer recommendations based on input data was developed and made available to commercial agriculture software developers for integration within their systems, which are being widely used by farmers. In India number of DSS have been designed mainly for nutrient and micronutrient management in field crops [83-86].

Insect and Pest Management

Plant protection is definable as the reasoned application of different methods, products as well as chemicals to allow optimal productive factors, with the objective to satisfy farm worker, consumer and safeguard environment. The concept of computerized DSSs for pest management is not new. DSS models have been developed for diseases that could expand very rapidly or those that should be controlled regularly. Development of weather-related DSS during 1990s resulted in a lower risk of crop damage by diseases and pests and minimal use of other input dosage [87]. [88] Reported 67 different DSSs for plant protection. Out of these, 30 deal with insect pest problems, 20 with plant disease and 17 with weeds. Most of the use systems have been developed in North America, particularly United States. [89] reported DSS for pesticide use management. For detection and tackling of air-borne herbicides, a DSS has been reported by [90]. Another model known as CLIMEX is used in 20 countries to examine the distribution of insect, plants, pathogens and vertebrates. It was developed by CSIRO Entomology in 1985 and has had numerous applications under practical conditions [91]. An insect pest forecasting tool, known as *SOPRA*, has been developed with the objective of optimizing timing of monitoring, management and control measures related to fruit orchards in Switzerland [92]. It uses local weather data, and simulates the age structure of the pest populations and crucial events for management activities are provided. Growers in alpine valleys and north of Alps can use it for decision support for the eight major insects and pests of fruit orchards.

In Europe, numerous DSSs exist to predict appropriate pesticide application times for various crops including potato. Of these, the most common are *Negfry, Prophy, Plant Plus* and *Simphyt,* which mainly focus on reducing fungicide inputs [93]. These four late-blight advice systems have the advantage of being in direct competition with each other, resulting to be useful for farmers. The Thematic Strategy on the Sustainable Use of Pesticides adopted in 2006 by the European Commission aims to establish minimum rules for the use of pesticides in the Community so as to reduce risks to human health and the environment from the use of pesticides. IPM creates synergies by integrating complementary methods drawing from a diverse approaches including biocontrol agents, plant genetics, biotechnologies, cultural and mechanical methods and information technologies, together with some pesticides that are still needed to control the most problematic pests and to manage critical situations [5].

Agricultural Land Use and Planning

With the rise in human population and their aspirations, land becomes an increasingly scarce resource – a scenario calling for land use planning. Land use planning is defined as a systematic assessment of land and water potential, alternative land use choices for better economic and social conditions. It has become essential to mitigate the negative effects of land use and to enhance the efficient use of resource with minimal impact on future generations. Land use planning is becoming complex and multidisciplinary as planners face multiple problems that need to be addressed within a single planning framework. These includes non-point-source deforestation, urbanization, pollution, ecosystem deterioration, water allocation, global warming, poverty and employment, deterioration of farmland and low economic growth. For land use planning it is increasingly necessary to recognize the complex tradeoffs between the multiple objectives of stakeholders. This is particularly apparent where outcomes of scales above the land management unit are considered important (e.g. water quality, biodiversity and land use planning). Many different DSS tools for land use related decision-making have been designed for agricultural planning [94], sustainable watershed management [95], forest planning [23], environmental planning [2], site selection [96], species protection plans [97] and conservation preserves planning [98]. A conceptual framework and a spatial DSS for rural land use planning have been developed for supporting decision making on selected area for different watershed management schemes for conservation planning [99]. The system provides suggestions and warnings for land use.

Linear programming approach based decision support have been employed variously covering wide range of including land use planning. The first linear programming models applied to land use planning were single objective problems [100,101]. However, because of the complexity of agricultural planning, multi-objective models are becoming increasingly more common. Within these models, goal programming is one of the techniques most frequently applied. [102] developed a goal programming addressing land use and the cropping system, maximization of gross margin and the minimization of risk. [29] reported continuous development of the DSS with multiobjective land use planning tools. Linear Programming model and Goal Programming-based DSS for farm regions in Greece have been designed, having development possibilities of agricultural sector in relation with the agricultural processing industries of the region [96,103-106]. It aims at the development of farm regions through a better utilization of available agricultural resources and agricultural industries.

Trends in modern land use planning increased with involvement of stakeholders in the planning process, which causes the need for interactive programming to exchange information between the decision-maker and the system. Interactive multiple objective learner programming has been successfully applied to agricultural development policy analysis [107], land use strategy evaluation [95] and land resource utilization [108]. DSS for sustainable land use planning to address conservation of land, improving soil quality and fertility, and local water balance with minimization of soil and nutrient translocation into surface water bodies and downstream fields have been designed keeping in view the optimized benefits for farmers as well as for the society [98].

Global Environment Change and Forecasting

Now-a-days global environment change is happening. Human activities related to the production, supply and consumption of food are responsible for changing global climate and giving rise to other globally and locally environment changes [109]. These include a change in fresh water supplies, carbon and nitrogen cycling, biodiversity and land cover and soils [110]. Ever increasing demand for food will further degrade the environment [111] the consequences of these changes must be estimated well in advance to minimize their impact for sustain ability of global agriculture [112]. DSSs for agricultural application of climate forecast can determine where, when and which sections of society are at most risk. They can construct future scenarios from food security point of view and reduce the vulnerability of food systems to Global Environment Change (GEC). Apparently, developed and deployed agricultural DSSs have the capacity to encapsulate scientific, practitioner and stakeholder knowledge and to present the consequences of alternative land use scenarios, to achieve an appropriate balance between economic, social and environmental outcomes [29].

An integrated pollution control DSS has been developed for rapid estimation of the extent and impact of pollution in a given situation [113]. Another model based on geographical information system (GIS), remote sensing and precision agriculture has been developed for Minland East Asia [114]. This model has a value in prioritizing and targeting of agriculture research in developed countries [115]. Decision Support System for Agro-technology Transfer (DSSAT) has been designed to answer "what if" questions frequently asked by policy-makers and farmers concerned with sustaining an economically sound and environmentally safe agriculture [116]. DSS for metrology has been designed by [117], which locates and analyzes historical situations of interest. It is used mostly by agronomists to forecast and provide recommendations to the growers. Climate forecasting based DSS are estimating the potential economic values for farm scale management decisions. They are helping in management and adjustment of different crops between specified regions having different climatic conditions [12]. They are also improving dissemination of grassland technology by predicting impact of climatic risks [118].

Water and Drought Management

Nearly one billion people worldwide are malnourished. The majority of these people live in developing countries, where increasing water scarcity complicates efforts towards food self-sufficiency. Huge amounts of water are needed to produce more food and eradicate hunger among increasing populations [119]. The current limited approaches to increasing demands of water will not be enough to eradicate hunger, especially in areas with growing populations and amidst dry climates in most developing countries. The central issue is how to manage water for all the different functions for which it is needed. With the advent of agronomic models that show how vegetation is likely to respond to climatic stress, with remote sensing to monitor vegetations conditions from airborne and space borne platforms, and with GIS to display spatial and temporal data in more comprehensible ways, it is now feasible to more accurately assess the impacts of drought [120]. Different DSSs have been developed to tackle with the problems related to water and drought management, [121], TEMPEST allows to model water flow both saline and fresh and predict the responses to each facet of the landscape to management [122]. Aussie GRASS, developed by the Queensland Department of Natural Resources, provides timely estimation of the extent of severity of drought [69]. A DSS developed in Vietnam formulates the plans for sustainable watershed management, using a combined approach of linear programming, goal programming and GIS for deriving the sustainable watershed management plan [95].

In Thailand, the collaborative project between the Department of Geography, Faculty of Liberal Arts, Thammasat University, and the Remote Sensing and Geographic Information System field of study at the School of Engineering and Technology, Asian Institute of Technology, aims to assess the effects of climate variability (especially droughts or dry spells) on rice production in rainfed environments and to develop a DSS tool that might help to properly anticipate and adapt farming to maximize agricultural production [70].

Other Applications

TropRice, an integrated rice management system being used by researchers, extension workers and some farmers has been used in Asia for irrigated rice areas [123]. It provides some generic and some site-specific information for rice cultivation. It is currently being used in many countries like China, India, Indonesia, the Philippines, Thailand and Vietnam. It is being translated or localized by national collaborators for local conditions in these countries. Other applications of DSS in agriculture range from conserving soil, local water balance, agronomical practices, canopy management cropping system analysis, and conservation of forest and computer multimedia instruction [98,124,125,126,127,128]. Web- based DSSs are assisting in forest management, resource management, and, agricultural emergency response [57, 129, 23]. DSS for farm mechanization using GIS based on linear programming provides machinery selection and planning to minimize farmer mechanization has been developed by [26]. The system includes the natural factors (climate and soil conditions), plot geographic site and the crop and machinery data. The GIS is part of the system to carry out a spatial analysis of the farm results to make machinery grouping.

Issues

Despite of vast applications of DSS in different areas of agriculture, their acceptability and adoptability has been constantly questioned throughout design, development to implementation, which has limited their application [2,10,11,19,63,91,129]. In other words they are facing problems of 'implementation' issues in comparison with the other DSS e.g., Medical DSS, which have revolutionised healthcare. Most of the researchers have pinpointed DSS issues viz., crisis for professional relevance, failure to support more than the exceptional circumstance, system complexity, user participation, ease of use, low adaptation, failure to show cost-benefits, complexity with user inputs, relationship between developer and user, regret avoidance, distrust for output, lack of field testing, lack of integration among heterogeneous components, success measurements, non-involvement of end user before and after development stages and under-definition of end users [10,130-132]. Evaluation of every DSS at post-implementation stage based on usability, portability, extendability, usefulness, operateability and adoptability investigates beyond user involvement and tends to monitor developed tool besides user [133]. Moreover, design, development and implementation of DSS through participatory approach in recent years has enhanced uptake of DSS in agricultural [134,135].

Future Trends

Researchers are optimistic about the future of DSS. This optimism continues to produce products and contributions to literature. A host of new tools and technologies are adding new capabilities to DSS. They include hardware and mathematical software development, artificial intelligence techniques, data warehousing and mining, OLAP enterprise resource planning, ERP, intelligent agents and World Wide Web (WWW) [136]. Separated from operational databases and optimized for decision support, data warehousing is an integrated, time-variant and non-volatile collection of a relational or multidimensional database (MDDB). It organizes data as an n-dimensional cube so that users deal with multidimensional data views such as crop, region, yield and area, with speedy query response time. Also known as knowledge Data Discovery, Data Mining refers to discovering hidden pattern from data, not known before. It attempts automatic extraction of knowledge from the large databases like data warehouse, spreadsheets, weather observatories, text documents etc.

Intelligent agents research is an emerging interdisciplinary research area involving researches from such fields as ESs, DSS, cognitive science, psychology and databases. Intelligent agent's research has contributed to the emergence of a new generation of active and intelligent DSS. This approach will enable us to integrate simulation models, GIS and multimedia with ESs, giving DSS a dominant role to play in modern agriculture. Development of domain-specific tasks will help in knowledge sharing and reuse. Sophisticated user interfaces for different media types are expected to be designed based on the users expertise and need.

World Wide Web (WWW) is becoming an infrastructure for the next generation of DSSs and groupware applications. There is also a trend to develop tools and techniques that could facilitate the dissemination of ESs through WWW. High bandwidth, reliable internet connectivity and carefully prepared underlying data will be keys to the future success of web-based decision tools. ERP, a new generation of information system, is integrating information and information-based processes within and across functional areas in an organization. The extensive databases created by the ERP system provide the platform for decision support using data warehouse, data mining and executive support systems. Global DSSs are emerging as the new frontiers in MIS area. Over the next decade, DSS will focus on large scale decision making involving groups, teams within distributed and decentralized structures [137]. In future, DSS will be a small tool for aiding farmer's tactical decisions, a versatile simulator as a consultant's tool, a core of a facilitated learning & training and a formal frame work that supports regulatory objectives in constraining and documenting farming practice [131]. DSS integrated with precision agricultural equipments, GIS and site-specific farming are changing the realm of modern agricultural practices. Future developments may include the possibility of implementing a number of DSS models into a GIS, which will support precision agriculture by providing adjusted spraying advice based on plot-specific characteristics [87].

In future, design and development of DSS is expected to get advantage from promising technologies like data warehousing and mining, agent-based approach, intelligent agents and enterprise resource planning besides advancement in hardware and software technologies. These technologies shall facilitate easier design of more complex DSSs. Agricultural is expected to get maximum benefits out of these as well as new milestones laid by the technologies like modelling, hypothesis, simulations and projections. Continued progress in system modelling combined with increasing growth in computer power, improvements in Remote Sensing, Geographical Information Systems, Precision Agriculture, new developments in the data extraction like data warehousing and date mining with new concepts of data exchange over the Internet should all contribute to expanded use of DSS for cropping systems in the future. Also there has been renewed interest in search strategies that can exploit the rapidly expanding information base on the Internet. These strategies may make qualitative information much more accessible to computer based reasoning systems to give new spin in DSS research and development.

Conclusion

DSS practice, research and technology continues to evolve though its history covers only a relatively brief span of years. There is no general account of classification and architecture of DSS. But it is possible to reconstruct the history from retrospective accounts considering published and unpublished material and redefine classification and architecture of DSS through in-depth research and literature contribution. Agriculture is in the midst of powerful changes influenced by industrialization and modernization, farm consolidations, environmental limitations, land use conflicts and overall increased risk; the availability, accessibility and application of contemporary expert agricultural information in the form of portable DSS is of high priority for farmers, technicians and researchers. With these changes and demands, different useful, scientifically valid and user-specific models have been implemented successfully. Many scientific and academic institutions have turned to computerized DSSs as a means of packaging biological and technical information to make the information more easily accessible and useful for various intending users in a short span of time.

In the last one decade researchers have reviewed the use of computer models in land use planning, forecasting, agronomical practices, water resources and emphasized the need for DSSs to make these models more useful. Incorporation of simulation and optimization models with interactive graphical capabilities is encouraging the acceptance of techniques related to literature development in practice. Now we have interactive, user-friendly computer systems which are becoming the rule, rather than the exception. Perhaps the greatest challenge to DSS development is how they are delivered to and used by industry and policy-makers. In order to achieve these goals we must not only learn from experience but be ready to accommodate new technologies and research areas. In addition, adoption to long-term climatic change, integrating process planning, natural hazards mitigation, matching new scientific information with local and indigenous knowledge are being addressed for sustainability of vulnerable areas.

Conflicts of Interest: None declared.

References

[1] Bennet A. & Bennet D. (2008) *Handbook on Decisions Support Systems*, Springer-Verlag, Berlin, Heidelberg.

[2] Shim J.P., Warkentin M., Courtney J.F., Power D.J., Sharda R. & Carlsson C. (2002) *Decision Support Systems*, 33, 111-126.

[3] Adams M.L., Cook S.E. & Corner R. (2000) *Prec. Agric.*, 2(1), 39-54.

[4] Moore J.H. & Chang M.G. (1980) *ACM SIGMIS Database*, 12(1 -2), 8-14.

[5] Rossi V., Meriggi P., Giosue S., Caffi T. & Bettati T. (2010) *A web-based decision support system for managing durum wheat crops*, INTECH Open Access Publisher.

[6] Courtney J.F. (2001) *Decision Support Systems*, 31, 17-38.

[7] Panigrahi S.K., Sahu A. & Pattnaik S. (2011) *International Journal of Computer Science Issues*, 8(1), 226-232.

[8] Sprague Jr.R.H. & Carlson E.D. (1982) *Building Effective Decision Support Systems*, Prentice-Hall, Inc., Englewood Cliffs, NJ.

[9] Power D.J. (2002) *Greenwood/Quorum*, Westport, CT.

[10]Cox P.G. (1996) *Agricultural Systems*, 52, 355-381.

[11]McCown R.L. (2002) *Agricultural Systems*, 74, 179-220.

[12]Jones J.W., Hansen J.W., Royee F.S. & Messian C.D. (2000) *Agricultural, Ecosystems & Environment*, 83, 169-184.

[13]Mysiak J., Giupponi C. & Rosato P. (2005) *Environmental Modelling & Software*, 20, 203-214.

[14]Tan C.H., Teo H.H. & Benbasat I. (2010) *Information Systems Research*, 21(2), 305-326.

[15]Agrios G.N. (2005) *Plant Pathology*, 5th ed., Academic Press, New York.

[16]Sonka S.T., Bauer M.E. & Cherry E.T. (1997) *Precision Agriculture in the 21st Century: Geospatial and Information Technologies in Crop Management*, National Academy Press, Washington, D.C.

[17]Chai K.L., Costello T.A., Wells B.R. & Norman R.J. (1994) *App-eng-agric.*, 10, 849-855.

[18]Coble H.D. (1994) *Proceedings of the annual meeting of the Northeastern Weed Science Society*, 48, 155-159.

[19]Jørgensen L.N., Noe E., Langvad A.M., Jensen J.E., Ørum J. E. & Rydahl P. (2007) *EPPO Bulletin*, 37(2), 374-377.

[20]Detlefsen N.K. & Jensen A.L. (2004) *Agricultural Systems*, 81, 55-72.

[21]Bazile D., Le Page C., Dembélé S. & Abrami G. (2005) *Perspectives of modelling the farmer's seed system for in situ conservation of sorghum varieties in Mali*, 5th Conference of the European Federation for Information Technology in Agriculture, Food and Environment and 3rd World Congress on Computers in Agriculture and Natural Resources. Universidade de Trásos-Montes e Alto Douro, Vila Real, Portugal.

[22]Mosseddaq F., Dnidane S. & Lahlou M. (2007) *Integrated Wheat N Nutrition Management in Morocco: A Decision Support Model.* Progress of Information Technology in Agriculture--Proceedings of the 4th International Symposium on Intelligent Information Technology in Agriculture (ISIITA).

[23]Riberir R.P. & Borges J.G. (2005) *MetaForest- a web-based decision system to support forest management involving multiple-ownership*, EFITA/WCCA, Vila Real, Portugal.

[24]Ibragimov T.Z., Ibragimova I.T. & Sanin S.S. (2005) *Decision support systems for cereal crop disease control*, EFITA/WCCA, Vila Real, Portugal.

[25]Sodtke R. (2005) *Multifunctionality of Landscapes - Analysis, Evaluation and Decision Support*, International Conference, Univ. Giessen, 193.

[26]Suarez de Cepeda M., Recio B. & Rubio F. (2005) *Decision Support System for Farm Mechanization*, 5th Conference of the European Federation for Information Technology in Agriculture, Food and Environment and 3rd World Congress on Computers in Agriculture and Natural Resources, Vila Real, Portugal.

[27]Hansen J.W. (2002) *Agricultural Systems*, 74, 305-307.

[28]Kseniya L. (1999) *Pollution Management Discussion Note*: In focus, 3

[29]Matthews K. & Buchan K. (2003) *Proceedings of the Modelling and Simulation Society of Australian and New Zealand*, 4, 1534-1539.

[30]Meinke H., Baethgen W.E., Carberry P.S. Donatelli M., Hammer G.L., Selvaraju R. & Stockle C.O. (2001) *Agricultural Systems*, 70, 493-513.

[31]Wilby R.L., Dawson C.W. & Barrow E.M. (2002) *Environmental Modelling and Software*, 17, 147-159.

[32]Betts R.A. (2005) *Philosophical Transactions of the Royal Society B: Biological Sciences*, 360, 2049-2065.

[33]White J.W., Hoogenboom G., Kimball B.A. & Wall G.W. (2011) *Field Crops Research*, 124, 357-368.

[34]Fangquan M. (2003) Review of China Agricultural Science and Technology, 5, 13-17.

[35]Druzdzel M.J. & Flynn R.R. (1999) *Encyclopedia of Library and Information science*, A. Kent, Marcel Dekker, Inc.

[36]Power D.J. (1997) *The On-Line Executive Journal for Data-Intensive Decision Support*, 1(3).

[37]Finlay P.N. (1994) *Introducing Decision Support Systems*, Blackwell Publishers, Oxford, UK; Cambridge, MA.

[38]Turban E. (1995) *Decision support and expert systems: management support systems*, Prentice Hall, Englewood cliffs, NJ.

[39]Little J.D.C. (1970) *Management Science*, 16(8).

[40]Keen P.G.W. & Scott-Morton M.S. (1978) *Decision Support Systems: An Organizational Perspective*, Addison-Wesley, Reading, MA.

[41]Moore J.H. & Chang M.G. (1980) *Data base*, 12(1& 2).

[42]Keen P.G.W. (1980) *Decision support systems: a research perspective*, Decision Support Systems: Issues and Challenges, Pergamon Press, Oxford; New York.

[43]Luconi F.L., Malone T.W. & Morton M.S.S. (1993) *Expert systems: The next challenge for managers*, Decision support systems, 3rd ed., Prentice-Hall, Inc., 365-379.

[44]Nelson R.R., Todd P.A. & Wixom B.H. (2005) *Journal of Management Information Systems*, 21(4), 199-235.

[45]Gachet A. (2004) *Building Model-Driven Decision Support Systems with Dicodess Zurich VDF.*

[46]Stanhope P. (2002) *Get in the Groove: Building Tools and Peer-to-Peer Solutions with the Groove Platform*, Hungry Minds, New York

[47]Codd E.F., Codd S.B. & Salley C.T. (1993) *Providing OLAP (on-line analytical processing) to user-analysts: An IT mandate*, Codd and Date, 32.

[48]Buchanan B.G. & Shortliffe E.H. (1984) *The MYCIN Experiments of the Stanford Heuristic Programming Project*.

[49]Scott-Morton M.S. (1971) *Management Decision Systems; Computer-based support for decision making*, Boston, Boston, Division of Research, Graduate School of Business Administration, Harvard University.

[50]DeSanctis G. & Gallupe R.B. (1987) *Management Science*, 33, 589-609.

[51]Huntington D. (1985) *EXSYS*, version3.2. Expert System Development, Albuquerque, N.M.

[52]Power D.J. (1998) *The On-Line Executive Journal for Data-Intensive Decision Support*, 2.

[53]*Exsys Corvid Tech Overview*, www.exsys.com

[54]Swanson E.B. & Culnan M.J. (1978) *MIS Quarterly*, 31-46.

[55]Little J.D.C. (1975) *Operations Research*, 23(4), 656-673.

[56]Osmond D.L., Smyth T.J., Yost R.S., Hoag D.L., Reid W.S., Branch W., Wang X. & Li H. (2002) *Nutrient Management Support System (NuMaSS) v.2. Soil Management Collaborative research Support Program*, Technical bulletin No. 2002-02, North Carolina State University, Raleigh, NC.

[57]Venkatachalam P. & Mohan B.K. (2003) *IEEE Computer Society 29th International Conference on Data Engineering*, 869-869.

[58]Power D.J. (2000) *Web-based and model-driven decision support systems: concepts and issues*, AMCIS 2000, 387.

[59]Haettenschwiler P. (2001) *Neues anwenderfreundliches konzept der entscheidungsunterstützung*, Gutes entscheiden in wirtschaft, politik und gesellschaft, 189-208.

[60]Holsapple C.W. & Whinston A.B. (2001) *Studies in Informatics and Control*, 10(1), 73-76.

[61]Hackathorn R.D. & Keen P.G.W. (1981) *MIS Quarterly*, 5, 21-26.

[62]Bhargava H. & Power D. (2001) *Decision support systems and web technologies: a status report*, AMCIS 2001, 46.

[63]Armstrong D., Gibbs I. & Johnson F. (2003) *Decision support – More about learning than software packages?*, Australian Farming Systems Conference, Toowoomba, Australia.

[64]Haag S., Cummings M., McCubbrey D.J., Pinsonneault A. & Donovan R. (2000) *Management Information Systems: For The Information age*. McGraw-Hill Ryerson Limited, 136-140.

[65]Marakas G.M. (1999) *Decision support systems in the twenty-first century*, Prentice Hall, Upper Saddle River, NJ.

[66]Geng S., Hess C.E. & Auburn J. (1990) *Journal of Agronomy and Crop Science*, 165(2-3), 73-85.

[67]Jordan V.W.L. & Hutcheon J.A. (1996) *Aspects of Applied Biology*, 47, 301-308.

[68]Boller E.F., Avilla J., Joerg E., Malavolta C., Wijnands F.G. & Esbjerg P. (2004) *OBC/wprs Bulletin*, 27(2),163-205.

[69]EISA (2001) *European Initiative for Sustainable Development in Agriculture*, www.fao.org.

[70]Attachai J., Chanchai S., Thaworn O., Methi E., Phrek G., Benchaphun E., Chada N., Daroonwan K., Honda K., Vinai S., Sukit R., Panjai T., Surat L., Ang C., IM S., Pongdhan S., Suwit L., Krirk P., Roengsak K., Wasu A., Narinthon B. & Thotsaporn S. (2012) *Southeast Asian Studies*, 1(1), 141-162.

[71]Hanley N., Salt C.A., Wilson M. & Culligan-Dunsmore M. (2001) *Journal of Agricultural Economics*, 52(2), 92-109.

[72]Ascough J.C., Dunn G.H., McMaster G.S. & Andales A.A. (2010) *The GPFARM DSS for Agroecosystem Sustainability: Past, Future, and Lessons Learned. International Environmental Modelling and Software Society (iEMSs)*. International Congress on Environmental Modelling and Software Modelling for Environment's Sake, Fifth Biennial Meeting, Ottawa, Canada.

[73]Agatsiva J. & Oroda A. (2002) *International Archives of Photogrammetry And Remote Sensing*, 34(6/W6), 42-49.

[74]Hayman P. (2003) *Decision support systems in Australian dryland farming: A promising past, a disappointing present and uncertain future*. 4th International Crop Science Congress.

[75]Deutscher S. & Bange M.P. (2003) *Advancements in Computerized Decision Support for Australian Cotton Systems*, Proceedings of 3rd World Cotton Conference, Capetown South Africa.

[76]Fu B., Pollino C.A., Cuddy S.M. & Andrews F. (2015) *Journal of Environmental Management*, 157, 127-138.

[77]Quemada M. & Cabreva M.L. (1995) *Soil Science Society of America*, 54, 1059-1065.

[78]Lemberg B., MCSweeney W.T. & Lanyon L.E. (1992) *Journal of Environmental Quality*, 21, 574-578.

[79]Achilea O., Ronen E. & Elharrar G. (2005) *Haifa Nurti-Net–a comprehensive crop Nutrition software, operated over the web*. EFITA/WCCA.

[80]Jorgensen M.S., Detlefsen N.K. & Hutchings N.J. (2005) *FarmN: A decision support tool for managing Nitrogen flow at the farm level*. EFITA/WCCA.

[81]Licklider (2007) *Computers and Electronics in Agriculture*, 57, 168-169.

[82]Gibbons M.M., Fawcett C.P., Warings R.J., Dearne K., Dampney P.M.R.D. & Richardson S.J. (2005) *PLANET Nutrient Management Decision Support System–A standard approach to fertilizer recommendations*. EFITA/WCCA.

[83]Pal S., Sethi I.C. & Arora A. (2009) *J. Ind. Soc. Agril. Statist.*, 63 (1), 91-96.

[84]Patil A.N. (2002) *Decision Support System for Nutrient Management in Wheat, Mustard and Bajra,* Unpublished M.SC thesis IARI, New Delhi, India.

[85]Pal S. (2007) *Decision Support System for nutrient management in crops*, Unpublished M.Sc. thesis IARI, New Delhi, India.

[86]Kumar A. (1992) *Decision Support System for nutrient management in crops*, Unpublished M.SC thesis IARI, New Delhi, India.

[87]Bouma E. (2007) *EPPO Bulletin*, 37(2), 247-254.

[88]Edwards-Jones G. (1993) *Crop Protection*, 12, 565-577.

[89]Arjoon D., Kok R. & Prasher S. (1993) *A decision Support system for pesticide use management,* American Society of Agricultural Engineers. Meeting (USA).

[90]Al Khatib K., Mank G.I., Reisenauer G., Parker R., Westberg H. & Lamb B. (1993) *Weed-Technology*, 7(1), 404-410.

[91]McCown R. L. (2002) *Agricultural Systems*, 74, 11-25.

[92]Samietz J., Graf B., Höhn H., Schaub L. & Höpli H.U. (2007) *EPPO Bulletin*, 37(2), 255-260.

[93]Dowley L.J. & Leonard R. (2000) Decision support system for late blight control. National Potato Conference.

[94]Jeffrey G., Gibson R. & Faminow G. (1992) *Agricultural Economics*, 8, 1-19.

[95]Loi N.K. & Tangtham N. (2004) *Decision Support System for sustainable watershed management in Dong Nai Watershed-Vietnam: Conceptual framework a proposed research techniques. Paper presented in Forest and Water in Warm Humid Asiz*, IUFRO Workshop, 10-12.

[96]Manos B. & Gavezos E. (1995) *Quarterly Journal of International Agriculture*, 34.

[97]Sandstorm P.L. (1996) *Identification of Potential Linkage Zones for Grizzly Bears in the Swan Clearwater Valley Using GIS*, MS Thesis, University of Nontana.

[98]Klik A. (2006) *Agricultural Decision Support System for Soil and Water Conservation Measures*, 4th World congress conference.

[99]Adinarayana J., Maitra S. & Dent D. (2000) *The Land*, 4(2), 111 -130.

[100]Campbell J.C., Radke J., Gkless J.T. & Wirtshafter R.M. (1992) *Environment and Planning*, 24, 535-549.

[101]Chuvieco E. (1993) *International Journal of Geographical Information Systems*, 7, 71-83.

[102]Giupponi C. & Rosato P. (1998) *A farm multicriteria analysis model for the economic and environmental evaluation of agricultural land use*, Multicriteria Analysis for Land-Use Management, Springer Netherlands, 115-136.

[103]Barnard C.S. & Nix J.S. (1993) *Farm Planning and Control*, Cambridge University Press.

[104]Bernardo D.J., Engle D.M., Lochwiller R.L. & McCollum F.I. (1992) *Journal of Range of Management*, 45, 462-469.

[105]Hazell P.B.R. & Norton R.D. (1986) *Mathematical Programming for Economic Analysis in Agriculture*, University of California, Berkeley.

[106]Lee D.J., Tipton T. & Leung P. (1995) *Agricultural Systems*, 49 (2), 101-111.

[107]De Wit C.T., Van Keulen H., Seligman N.G. & Spharim I. (1988) *Agricultural Systems*, 26, 211-230.

[108]Fischer G., Makowski M. & Granat J. (1999) *AEZWIN: An interactive multiple-criteria analysis tool for land resources appraisal*, World Soil Resources Reports (FAO).

[109]GECAFS (2005) *Science Plan and implementation strategy. Earth systems science partnership* (IGBP, IHDP, WCRP, DIVERSITAS) Report No 2, 36.

[110]Vitousek P.M., Mooney H.A., Lubchenco J. & Mellillo J. (1997) *Science*, 277, 494-499.

[111]FOA (2002) *World Agriculture: Towards 2015/2030.*

[112]Steffen W., Sanderson A., Tyson P.D., Jager J., Matson P.A., Moore III.B., Oldfield F., Richardson K., Schellnhuber H.J., Turner II.B.L. & Wasson R.J. (2004) *Global Change and the Earth System: A Planet Under Pressure*, Springer Verlag, Heidelberg, Germany.

[113]Lvovsky L. (1999) *Better Environment Decision making*, The DSS/IPC, pollution management in focus (3).

[114]Zuo H. (1996) *Agroclimatic analysis for mainland East Asia by a GIS approach*, Ph.D. thesis, Australian National University, 264.

[115]White D.H., Howden S.M. & Nix H.A. (1993) *Modelling agricultural and pastoral systems*, Modelling Change in Environmental Systems. John Wiley & Sons Ltd, Chichester, UK, 263-288.

[116]Hunt L.A., Jones J.W., Jhornton P.K., Hoogenboom G., Imamura D.T., Tsuji G.Y. & Singh U. (1994) Accessing data, models and application programs, DSSAT V3. 1-3, University of Hawaii, Honolulu, Hawaii.

[117]Seghi L., Orlandini S. & Gozzini B. (2000) *Survey of Models of Plant Diseases*, Report of the RA VI Working Group on Agricultural Meteorology. CAgM No. 82. WMO/TD No. 2. WMO, Geneva, Switzerland.

[118]Stuth J.W., Hamilton W.T., Cornner J.C. & Sheehy D.P. (1993) *Decision support systems in the transfer of grassland technology*, Grasslands for our World, SIR Publishing, Wellington, New Zealand, 234-242.

[119]Falkenmar M. & Rockstorm J. (2005) *Rain: the neglected resource*, Swedish Water House, Policy Brief, No. 2.

[120]McVicar T.R., Jupp D.L.B., Yang X. & Tian G. (1992) *Linking regional water balance models with remote sensing*, 13th Asian Conf. on Remote Sensing, Ulaanbaatar, Mongolia, B.6.1-B.6.6.

[121]Watkins D.W. & McKinney D.C. (1995) *Reviews of Geophysics*, 33(S2), 941-948.

[122]Sojda R. (1994) *AI Applications*, 8(2).

[123]Bell M.A. & Chung C.L. (2000) *TropRice–a decision support tool for filling the researcherfarmer knowledge gap*, Systems Research for Optimizing Future Land Use in South and Southeast Asia. SysNet REs. Pap. Ser, (2), 217-266.

[124]Lemmon H. (1986) *Science*, 233, 29-33.

[125]Warkentin M., Nair P., Ruth S. & Sprague K. (1990) *Agroforestry Systems*, 11, 71-83.

[126]Hoogenboom G., Jones J.W., Wilkens P.W., Batchelor W.D., Bowen W.T., Hunt L.A., Pickering N.B., Singh U., Godwin D.C., Baer B., Boote K.J., Ritehie J.T. & White J.W. (1994) *Crop models*, DSSAT, University of Hawaii Honolulu, Hawaii, 3, 2-2.

[127]Caldwell R. M. (2003) *Journal of Crop Production*, 9, 383-407

[128]Marrison D.L., Tao B.W. & Frick M.J. (1993) *Computer multimedia instruction versus traditional instruction in agriculture*, International Winter Meeting sponsored by the American Society of Agricultural Engineers, Chicago, IL.

[129]Lynch T., Gregor S. & Midmore D. (2000) *Australian Journal of Experimental Agriculture*, 40, 609-620.

[130]Parker C.G. & Campion S. (1997) *Improving the uptake of decision support systems in agriculture*, First European Conference for Information Technology in Agriculture, Copenhagen.

[131]Hammer G.L., Kropff M.J., Sinclair T.R. & Porter J.R. (2002) *European Journal of Agronomy*, 18, 15-31.

[132]Arnott D. & Pervan G. (2005) *Journal of Information Technology*, 20(1), 67-87.

[133]Mir S.A., Mir I.A., Quadri S.M.K., Ahmad N. & Angchuk S. (2013) *African Journal of Agricultural Research*, 8(48), 6275-6282.

[134]Cain J.D., Jinapala K., Makin I.W., Somaratna P.G., Ariyaratna B.R. & Perera L.R. (2003) *Agricultural Systems*, 76, 457-482.

[135]Jakku E. & Thorburn P.J. (2010) *Agricultural Systems*, 103, 675-682.

[136]Berners-Lee T. (1996) *Computer*, 29(10), 69-77.

[137]King D. (1993) *Intelligent support systems: art, augmentation, and agents*, Decision support systems, 3rd ed., Prentice-Hall, Inc., 137-159.

THE DEFENSE ACTIVATOR FROM YEAST FOR RAPID INDUCTION OF RESISTANCE IN SUSCEPTIBLE PEARL MILLET HYBRID AGAINST DOWNY MILDEW DISEASE

HINDUMATHY C.K.

Department of Biotechnology, Vinayaka Missions University, Salem 636 308 TN, India.
*Corresponding Author: Email- hindumathyck@rediffmail.com

Abstract- Elicitors derived from yeast are shown to be active in eliciting resistance against disease in plants for wide range of pathogens. In the present study elicitor derived from yeast used in the form of seed treatment in pearl millet crop against an oomycetous downy mildew pathogen *Sclerospora graminicola*. The suspension at various concentrations of 0.25, 0.5, 1.0, 1.5 and 2.0mg mL^{-1} for different time intervals of 3, 6, 9 and 12 h were tested. Among the concentrations tested 0.5mg mL^{-1} for a period of 6 h soaking offered 98% seed germination and seedling vigor index 1652. The seed germination and the seedling vigor were significantly greater than the untreated check. The optimum time interval required by the concentration of 0.5mg mL^{-1} was 3 days to offer 78% protection against downy mildew disease. Biochemical studies carried out showed that treated seedlings showed increased level of tyrosine ammonia lysase and poly phenol oxidase and reached a peak at 12 h after inoculation (h.a.i). which was 45% more than their respective uninoculated control. In non-denaturing gel stained for poly phenol oxidase revealed the presence of 3 isoform in elicitor treated seedlings .
Key words- Downy mildew, elicitor, induced resistance and pearl millet.

Introduction

Pearl millet (*Pennisetum glaucum* L.) is one of the major Cereal crop in tropics and semi arid tropics and it is the staple food and fodder in this zone. Downy mildew is one of the major constrain for crop production which is caused by the oomycete, biotrophic pathogen *Scleropora graminicola* (Sacc.) Schroet. The loss due to this disease in pearl millet production worldwide reaches 40% which accounts for a loss of US $ 270 million annually. Hence, effective control strategies for this disease are obviously needed. Though management strategies like resistance breeding and chemical control are practiced, they have their own limitations. Therefore, approaches like inducing systemic resistance are being explored.

In many plant species, resistance may be induced against several pathogens by means of pretreatment with chemical products or biological compounds, from inorganic salt to fungus and bacterial cell wall fractions. Preinoculation with pathogen or nonpathogen can also induce resistance and it is thought pathogen stimulate the same signal transduction (Sequeria 1983). Once resistance is induced, the plant expresses a number of inducible defense response or mechanism that usually coincided with the accumulation of pathogenesis related proteins (Rylas 1996). Induction of resistance mechanism by application of elicitor has been as alternate approach for crop disease control.

Several studies in host pathogen interactions revealed that elicitors derived from microbes were shown to be effective inducer like sulfated oligosaccharides elicits defense response in tobacco (Klazyymsl *et al.* 2003). Cell wall proteins from *Pythium* in sugar beet and wheat (Takenaka *et al.* 2003). Syngrolin, a peptide from Pseudomonas syringe in rice (Waspi *et al.* 1998). Oligandrin a protein elicitor from *P. oligandrum* in tomato (Benhamou *et al.* 2001). Cell wall derived carbohydrate from Phytopthora spp in tobacco (Zhang *et al.* 1998) β-crytogein a protanaceous elicitor from *Phytopthora cryotogea.*

The present study investigates the use of elicitor derived from commercially available baker's yeast as inducer of resistance in pearl millet against downy mildew in susceptible pearl millet cultivar. It also investigates that the induction of resistance is also associated with enhanced production polyphenol oxidase and tyrosine ammonia lyase as biochemical marker.

Materials and methods

Seeds of pearl millet cultivar

Seeds of pearl millet 'HB3' highly susceptible to *S. graminicola* obtained from Project Co-ordinator, ICAR-All India Co-ordinat Pearl Millet Improvement Project, Mandor, Rajasthan, India were used.

A virulent pathotype of downy mildew pathogen *S. graminicola* isolated from pearl millet 'HB3' and maintained on the same cultivar under greenhouse conditions was used for all inoculation experiments.

Elicitor preparation from cell wall of yeast

Glucan preparation

Glucan from cell wall of bakers yeast were obtained according to the method of Sharp *et al.* 1984. Two grams of lyophilized fungal cell wall were suspended in 200mL of 2N trifluro accetic acid (TFA) for 3 h at 80^{0}C under shaking in a water bath and centrifuged at 10,000xg for 15 minutes to remove insoluble residue. The supernatant was filtered through a GF/C filter and residual TFA was eliminated by retroevaporation at 30^{0}C. The evaporated compound was suspended in 20mL ice cold double distilled water, followed by neutralization with 1 N NaOH, and the resultant cell wall was used as elicitor to induce resistance in pearl millet against downy mildew.

Elicitor treatment

Pearl millet seeds 'HB3', disinfected with sodium hypochlorite were soaked in different concentration of aqueous suspension of elicitor prepared from the spores for induction of esistance. Elicitor concentration of 0.25, 0.5, 1.0, 1.5, and 2.0mg mL^{-1} at different time intervals like 3, 6, 9 and 12 h were used. Seeds treated with distilled water served as control.

Determination of seedling vigor analysis

The elicitor obtained from yeast was applied to seeds for their effect on germination and seedling vigor. This study was carried out by the paper towel method. After treatments with different concentrations of elicitor, the seeds were placed on moist germination paper and maintained at 27$\pm$2^{0}C. Seeds treated with sterile distilled water for the same time interval served as control. After 7 days germination percentage and vigor index was calculated. Hundred seedlings were used for all treatments and the experiments were repeated four times.

Study on effect of elicitor

Two-day-old seedlings raised after elicitor treated seeds along with the seedlings raised from distilled water treated seeds were inoculated with 4x10^4 zoospores mL^{-1} of *S. graminicola* by whorl inoculation (Safeeulla 1976).

Disease assessment under greenhouse conditions

Inoculated pearl millet seedlings were transplanted to earthen pots containing soil, sand and manure (1:1:1) and maintained under greenhouse conditions. Plants were observed daily and the progression of downy mildew disease was recorded. Plants were rated as diseased when they had any of the typical symptoms of downy mildew i.e. chlorosis, stunted growth and sporulation. At the end of 30 days disease incidence was recorded as the percent of plants showing symptoms of downy mildew disease. Percentage protection offered by elicitor treatment against *S. graminicola* infection was calculated for seedlings raised from elicitor treated seeds and compared to the control (seedlings raised from water-treated pearl millet seeds).

Studies on optimization of time required for induction of resistance

In this experiment, seeds treated with 0.5mg mL^{-1} yeast elicitor for 3 h were sown in earthen

pots filled with autoclaved soil, sand and manure (1:1:1). The emerging seedlings were inoculated with the zoospore suspension of *S. graminicola* (4x10^4 zoospores mL^{-1}) following the whorl inoculation procedure with a time gap of 1, 2, 3, 4, and 5 days of emergence in different sets of plants. The plants from above experiments were maintained under greenhouse conditions and downy mildew disease incidence recorded as described earlier.

Biochemical studies

Harvesting of samples

Seedlings raised from seeds, treated with elicitor and distilled water were inoculated 2-days after treatment and harvested at 0, 12, 24, 36, 48 and 72 h.a.i. used for the study of tyrosine ammonia lyase (TAL) and polyphenol oxidase (PPO) activity. Inoculated and uninoculated seedling of elicitor treated and distilled water was also harvested at the same time intervals and was stored at -35^{0}C for biochemical studies.

Studies on tyrosine ammonia lyase and poly phenol oxidase were carried out with the samples harvested at 3-day time gap between elicitor treatment and *S. graminicola* inoculation, as this time interval of 3-day gap offered maximum protection.

Tyrosine ammonia lyase

Seedlings of both inoculated and uninoculated were harvested after 0, 12, 24, 36, 48 and 72. For TAL activity seedlings was homogenized in ice cold 0.25M borate buffer (pH 8.7) in an ice bath. The homogenate was centrifuged at 15,000rpm for 15 min at 4^{0}C. The supernatant was used as crude extract. The reaction mixture was prepared with 1 ml of the enzyme extract, 0.5mL of borate buffer pH (8.7), 1.3mL of distilled water and 0.2 mL of 1 M L- tyrosine. Changes in absorbance at 290nm were observed at 30 s intervals for 30 min on a UV- spectrometer. Reaction mixture without substrate served as blank. One unit of enzyme extract produced 3.37nm of cinnamic acid per h. Results were expressed in unit of activity per g of fresh weight.

Assay of polyphenol oxidase activity

Pearl millet seedlings (1 g fresh weight) were ground to a fine powder in liquid nitrogen and extracted with 1mL of extraction buffer (0.1 M Tris-HCl (pH 7.0), 0.1M KCl, 1% (v/v) Triton X-100, 1 mM EDTA and 5% (w/v) PVPP). The extract was centrifuged at 15,000xg for 15 min at 4^{0}C and the supernatant, transferred to a fresh tube served as the enzyme extract. The protein content in the extract was estimated by the dye binding method (Bradford 1976) and PPO activity was assayed spectrophotometrically (Siddiq *et al.* 1992). The standard reaction mixture was prepared with 250µl of 0.2 M sodium phosphate buffer (pH 6.0), 50 µl of 1.0 M catechol and 50µl of enzyme solution. The reaction was carried out at 30°C for 3 min, and PPO activity was measured by monitoring the increase in absorbance at 420nm. One unit of PPO activity was defined as the amount of enzyme that increases absorbance of 0.001 per minute.

Electrophoresis

Non-denaturing gel electrophoresis of soluble proteins in pearl millet seedlings was carried out according to Laemmli 1970 using a vertical mini-gel electrophoresis unit (Biometra, Germany). The resolving gel was 10% and the stacking gel 5% (w/v). The electrode buffer was Tris-base (3.0g Tris-base, 14.4g glycine per liter of distilled water, pH 8.3). Each gel slot was loaded with 40µg protein sample. Electrophoresis was performed at constant voltage of 100V.

Activity staining

Activity staining for PPO in native polyacrylamide gel was done by rinsing the gels in de-ionized water several times and then placing it in 100mM sodium phosphate buffer pH 7.0 containing 10mM 1, 3-dihydroxyphenylalanine (DOPA) on a rotary shaker. After 30 min of incubation, dark bands indicative of PPO isozymes appeared in the gel.

Results

Seed germination and seedling vigour analysis

Table 1- Effect of seed treatment with different concentration of elicitor from yeast call wall on pearl millet 'HB3'. Vigour Index calculation includes product mean root length (MRL) + mean shoot length (MSL) and percent germination (%G)

Concentration of elicitor	Duration of treatment	Germination (%)	Vigor index (MRL + MSL) % G
Control	3	90^{g}	1389^{i}
	6	70^{f}	1373^{e}
	9	63^{de}	1021^{d}
	12	51^{c}	981^{a}
	3	70^{f}	1390^{i}
0.25	6	92^{g}	1490^{j}
	9	72^{f}	1274^{h}
	12	63^{de}	1179^{g}
0.5	3	92^{g}	1491^{j}
	6	94^{g}	1526^{k}
	9	71^{f}	1276^{h}
	12	63^{de}	1137^{de}
1.0	3	70^{f}	1179^{g}
	6	63^{de}	1083^{f}
	9	60^{e}	1065^{e}
	12	50^{d}	1021^{d}
1.5	3	68^{ef}	1027^{d}
	6	56^{d}	987^{c}
	9	51^{c}	945^{bc}
	12	45^{b}	921^{b}
2.0	3	60^{e}	987^{c}
	6	53^{bc}	947^{bc}
	9	47^{b}	849^{b}
	12	38^{a}	673^{a}

- *Results were taken 7 days after inducer treatment (seed treatment) with different concentrations of elicitor.*
- *Results are based on 4 experiments with 100 seedlings per treatment per experiment.*
- *For each experiment and type of data, means in the same column followed by the same letter are not significantly different according to Tukey's HSD Test P= 0.005*

Seeds were treated with aqueous suspension of elicitor at increasing concentrations of 0.25, 0.5, 1.0, 1.5 and 2.0mg mL^{-1} for a period of 3, 6, 9 and 12 h and analyzed for seed germination percentage and seedling vigour index after 7 days. Seeds treated with sterile distilled water served as control. Treatment with 0.5mg mL^{-1} of elicitor for duration of 3h did not affect the seed germination percentage and seedling vigour relative to the control. It even improved germination to 94% and vigour index to 1526 (Table 1), in comparison to the control with 70% germination and a vigour index of 1373 at this time interval. However, lower concentration of elicitor did not enhance seed germination and seedling vigour and higher concentrations of elicitor significantly reduced germination percentage and vigour index when compared to control hence a concentration of 0.5mg mL^{-1} was used for protection studies.

Disease assessment under greenhouse conditions

The elicitor treatment for a period of 3h did not affect germination percent and seedling vigour hence this time period was used for disease assessment studies. Different concentrations of elicitor aqueous suspension at 0.25, 0.5, 1.0, 1.5 and 2.0mg mL^{-1} were applied to seeds of 'HB3' for a period of 3h and seedlings were inoculated with *S. graminicola* after 2-days of seed treatment. Various level of protection over the control was obtained. Elicitor at concentrations of 0.25 and 0.5mg mL^{-1} offered 48.7% and 71% protection, respectively (Figure 1). The percentage of protection decreased with increase of concentration of elicitor i.e., a concentration of 1mg mL^{-1} offered 58.1% protection, 1.5mg mL^{-1} offered 54% protection and a concentration of 2mg mL^{-1} offered 51.8% protection. Control plants recorded 96% downy mildew disease incidence (DMDI). Since maximum protection was observed at 0.5mg mL^{-1} of elicitor, further experiments were carried out using this concentration only.

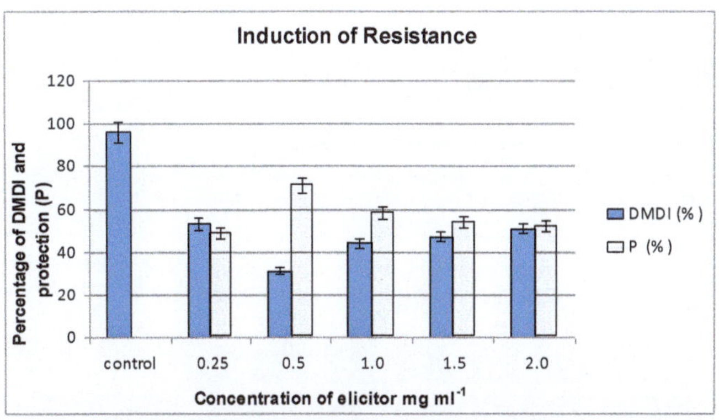

Fig. 1- Effect of seed treatment of pearl millet cultivar HB 3 with different concentrations of elicitor from yeast cell wall on downy mildew disease incidence (DMDI) due to inoculation by *Sclerospora graminicola* and protection (P) % offered against unprotected control

- Seeds of pearl millet were treated with different concentrations of elicitor 0.25, 0.5, 1.0, 1.5 and 2.0 mg ml^{-1} for 6 h. Inoculations were performed 1, 2, 3, 4 and 5 days after treatment.
- Results were taken 30 days after inoculation with 4 x 10^4 zoospores of *S. graminicola.*
- Results are based on 4 experiments with 100 plants per treatment per experiment.

For each experiment and type of data, means followed by the same letter are not significantly different according to Tukey's HSD Test P=0.005.

Studies on optimum time required for induction of resistance

The elicitor when given as seed treatment offered different degrees of protection against downy mildew, ranging from 58 to 71%, depending on the temporal separation between the inducing treatment and the inoculation (Figure 2) A time gap of 1 day after elicitor treatment offered 54% protection, which increased to 60% in the case of 2 days. Maximum protection of 71% was observed for a time gap of 3 days, and did not increase further for 4 and 5 day time intervals. Therefore a 3-day time gap was required for building up of the elicitor induced resistance in the host system for effective control of *S. graminicola* infection.

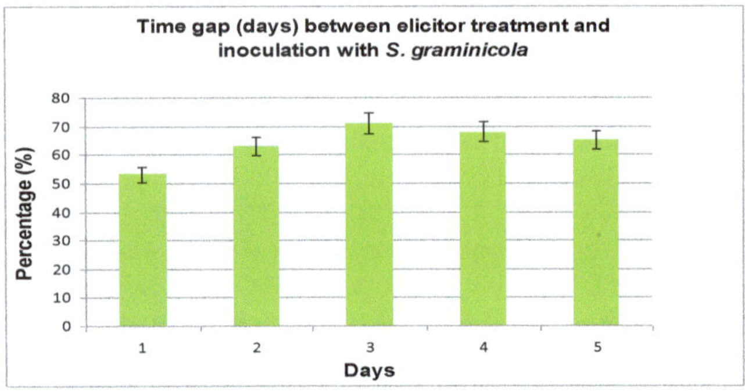

Fig. 2- Effect of time gap (in days) between inducer treatment with elicitor from yeast cell wall (0.5 mg ml^{-1}) and whorl inoculation with Sclerospora graminicola on downy mildew disease incidence in pearl millet. Bars indicate + SE.

For each experiment and type of data, means followed by the same letter are not significantly different according to Tukey's HSD Test P=0.005

Elicitation of TAL activity

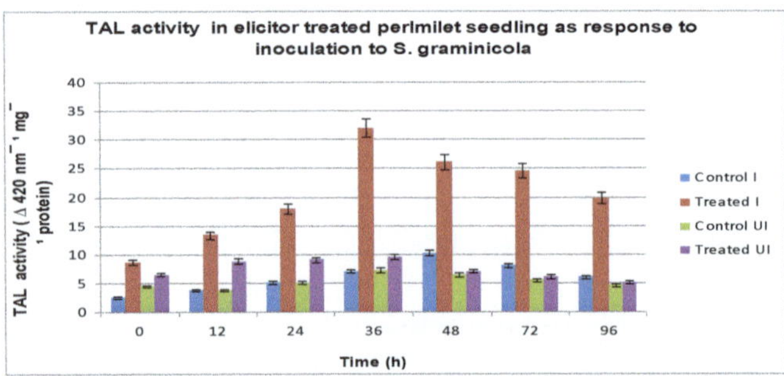

Fig. 3- Tyrosine ammonia lyase specific activity in pearl millet seedlings of elicitor treated pearl millet seedlings as a response to inoculation with Sclerospora graminicola. The data are means of three independent experiments. Bars indicate + SE

TAL activity increased four-fold over time in response to elicitor treatment. TAL activity increased prominently in induced resistant seedlings by 12 h.a.i. reaching a level four-fold greater than the corresponding controls. The rise in TAL activity was transient it showed maximum activity by 36 h.a.i. and decreased by 72 h.a.i (Fig.3) TAL activity in distilled water treated seedlings increased marginally by 24 h.a.i.

Spectrophotometric analysis of polyphenoloxidase

The spectrophotometric determination of PPO activity correlated with the PPO isozyme pattern observed following elicitation. Separation of PPO isoforms was carried out on native anodic PAGE, followed by staining with 10 mM catechol and *o*-phenylenediamine (0.05%, w/v). Elicitor treatment of pearl millet greatly induced the number and intensity of PPO isoforms; induced at 36 h and the highest induction being reached after 48 h.a.i. (Figure 4)

(PPO) activity was estimated in inoculated and uninoculated seedlings of pearl millet. The polyphenol oxidase profile at 0, 12, 24, 36, 48, 72, and 96 h.a.i is depicted in (Figure 4). In the seedlings raised from elicitor treated seeds, a higher level of PPO activity (27 units) was seen at the time of inoculation compared to its distilled water treated control (11 units) at 24 h.a.i. Thus, protected seedlings showed 45% increase in enzyme activity at 48 h after challenge inoculation over the control. The increased enzyme activity was maintained till 72 h.a.i. In distilled water treated seedlings PPO activity did not significantly increase upon inoculation (Figure 4).

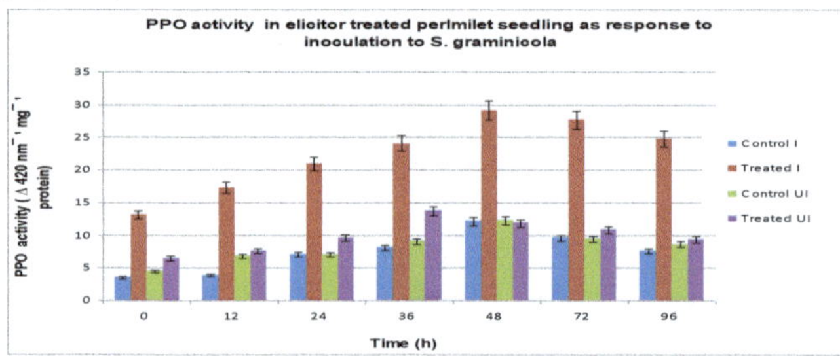

Fig. 4- Polyphenol oxidase specific activity in pearl millet seedlings of elicitor treated pearl millet seedlings as a response to inoculation with Sclerospora graminicola. The data are means of three independent experiments. Bars indicate + SE

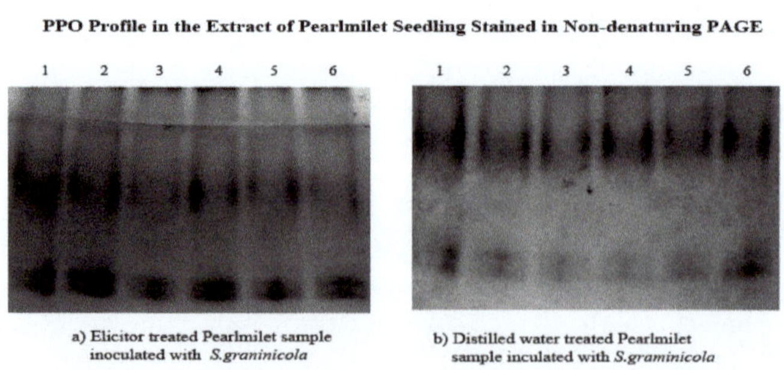

Fig. 5- Poly phenol oxidase profile in extracts of pearl millet seedlings stained in non-denaturing PAGE. 50▯g of protein was loaded for (A) elicitor-treated pearl millet samples inoculated with Sclerospora graminicola; (B) distilled water-treated samples inoculated with Sclerospora graminicola. Lane 1 – 0 h; Lane12 - 24 h.a.i; Lane 3 6 h.a.i.; lane 48 h.a.i, lane 72 h.a.i;

These activity differences were visualized in PPO zymogram upon native PAGE as well (Figure 5). The PPO zymogram pattern showed a high molecular weight isoform and 3 low molecular weight isoforms upon staining the gel with Substrate. Out of these 4 isoforms of polyphenol oxidase detected in the gel, the low molecular weight isoforms were induced upon inoculation in elicitor treated seedlings (Figure 5a). However no such difference in activity staining was observed in distilled water control after *S. graminicola* infection (Figure 5b).

Discussion

Plants in nature are constantly challenged by a diverse array of pathogenic microorganisms. In many cases, their protective mechanisms involved inducible defense mechanisms. The ability of plants to invoke such defense reactions is presumed to be mediated by an initial recognition process between plants and pathogens that involves detection of certain unique signal molecules of incompatible pathogens by receptor-like molecules in plants, with a subsequent resultant cascade of biochemical events that leads to the expression of resistance. This hypothesis is supported by the finding that compounds known as elicitors and isolated from pathogenic microorganisms induce biochemical events characteristic of the resistance responses in plants.

The biocontrol activity of antagonistic yeast has been demonstrated on a variety of commodities. The mode of action of these microbial agents has not been fully elucidated. It has been suggested that their biocontrol activity may be partly associated with production of antibiotics. It has been reported that yeast induces several biochemical defense response in surface wounds (Rodov *et al.* 1994). Induction of defense response like increase activity of chitinase and deposition of papillae on the host cell in Apple fruit by yeast was also reported (Ghaouth *et al.* 1997). Fajardo *et al.* 1991 demonstrated induction of resistance to *Penicillium digitatuminor* in case of orange. This increased resistance was associated with temporal differential induction of chitinase, glucanase and peroxidase activities. Reglinski *et.al* 1990 showed that yeast cell wall extract induce resistance to barley powdery mildew. This elicitor treatment induced PAL activity and enhanced papillae formation in the highly susceptible cultivar leading to a 90% reduction in mildew infection (Reglinski *et al* 1994).

Yeast derived elicitor reduced the severity of infections caused by *Botryis cineria* and *Rhizoctonia solani* on lettuce ((Reglinski *et al.* 1993). It reduced the infection up to 90% in the glass house condition and *Rhizoctonia solani* infections were reduced up to 50-70%. Droby *et al.* 2001 induced resistance against *Penicillium digitatmu* in grape by *Candida oleophola*. When yeast-cellulose formulations applied to the leaf of kiwifruit significantly suppressed the liberations of conodia of *Botrytis cinerea* (Darry 2002).

Elicitor - receptor interactions are presumed to generate signals that activate plant defense responses leading to the induction of stress-related enzymes such as phenylalanine ammonia lyase and the associated accumulation of high levels of phenolic compounds phenol-oxidizing enzymes such as peroxidase polyphenol oxidase, phytoalexins, synthesis of antifungal hydrolytic enzymes such as chitinases, are associated in defense response. TAL and PAL are the key enzymes in phenol biosynthesis. Kuc and Richmond 1977 reported that protection of cucumber and tobacco plants against powdery mildew were accompanied by the increased activity of PAL and PPO similarly activity of PAL was higher in rice resistance *to Xanthomonas oryzae* than the susceptible. Wheeler reported that sweet potato showed increased activity of PAL and PPO when exposed to ethylene. Pre-treatment of susceptible wheat heads with heat killed mycellial wall preparation resulted in the elevation of both POX and PPO activities and induced resistant with *Fusarium graminarium.* Similar results were also reported in Cucumber (Hammerschmidt 1995).

PPO activity is ubiquitous in higher plants, and functions attributed to the enzyme include phenol metabolism and a defense mechanism against pathogens. Several observations have identified a role for PPO in the polymerization of monolignols into olignols, precursor molecules of lignin. The expression of resistance is often accompanied by the activation of phenol-oxidizing enzymes such as POX and PPO. Increase in PPO activity may contribute to defense through the production of oxidized forms of quinones, which can inactivate pectinolytic enzymes produced by pathogens. In this work, PPO activity was enhanced by inducer treatment. Untreated plants did not show any significant fluctuations in PPO activity in the study period. Induced resistant seedlings maintained the PPO activity in a significantly higher level till 12 h.a.i. there was an early hike in the PPO level in inducer treated seedlings. PPO, the peak appeared after 48 h.a.i. of challenge and remained almost static till 72 h.a.i. and diminished gradually, but in the control plants the fall in enzyme activity was sudden. It is because polyphenol oxidase is a nuclear-encoded copper-containing enzyme widely distributed in the plant species and catalyzes the oxidation of phenols to *o*-quinones. The *o*-quinones are highly reactive in plant cells and subsequently involved in defense role. Such increase may provide the plants with enhanced resistance to pathogen invasion by providing increased contents of oxidized quinone derivatives which impede pathogen growth. Similar results were observed in tomato with overexpression of PPO in tomato resulted in enhanced resistance to bacterial disease (Li and Steffens 2002). These results strengthen the assertion of a defense role for the enzyme in plants.

Enhancement of peroxidase, deposition of lignin and callose following the application of Cell wall carbohydrate component of *Aspergillus niger* was reported previously and it was assumed that these enzymes were involved in the activation of resistance mechanism because treatment of *Aspergillus niger* induced resistance to *Sclerospora graminicola* infection up to 67% protection over the control.

In the present study PPO and TAL were induced upon infection in other plant system. The enhancement of these enzymes in the present study indicate that increase in PPO and TAL elicits resistance in pearl millet against downy mildew.

References

[1] Bardford M. (1976) *Annals of Biochemistry* 72, 248-254.

[2] Benhamou N., Richard R.R., Rey P. Tirilly (2001) *Plant Physiology and Biochemistry*, 39, 681-696.

[3] Benhamou N., Richard R.R., Rey P. Tirilly (2001) *Plant Physiology* 39, 681-696.

[4] Dixon R.A. and Harrison M.J. (1990) *Advanced Genetics*, 28, 165-234

[5] Dixon R.A., Harrison M.J. and Lamb C.J. (1994) *Annual Review of Phytopathology*, 32, 479-501.

[6] Droby S., Portat R., Vinocur V., Cohen L., Weiss B. and Dauss A. (2001) *Phytopathology*, 90, 20.
[7] Edreva A., Blancard D., Delon R., Bonnet P. and Ricci P. (2002) *Beitragezur Tabakforschung International*, 20, 53-59.
[8] Enkerli K., Hahn M.G., Mims C.W. (1997) *Canadian Journal of Botany*, 75, 1509-1517.
[9] Fajardo J.E., McCollum T.G., McDonald R.E., Mayer R.T. (1998) *Biological Control*, 13, 143-145.
[10]Ghauth E., Wilson C.L. and Wisneiwski M. (1998) *Phytopathology*, 88, 282-291.
[11]Hammerschmidt R., Kuc J. (1995) *Dordrecht*, Kluwer, 182.
[12]Harllen SAS., Reginaldo S.R., Dirceu M., Bernardo A.H., Maria C.B.P., Ann M. (2004) *Biological Control*, 29, 288-295.
[13]ISTA (1993) *Seed Science and Technology*, 21, 25-30.
[14]Jensen W.A. (1962) *Botanical Histochemistry*.
[15]Keen N.T. (1992) *Plant Molecular Biology*, 19, 109-122.
[16]Kuc J. and Richmond S. (1977) *Phytopathology*, 67, 533-36.
[17]Lax A.R. and Cary J.W. (1995) *Am. Chem. Soc.*, 9, 121-128.
[18]Liang YC Sun, W.C. and Römheld V. (2005) *Plant Pathology*, 54, 678-685.
[19]Lyngkaer M.F., Carver T.L.W. (1999) *Physiological Molecular Plant Pathology*, 55, 151-162.
[20]Mace M.E. and Wilson E.M. (1964) *Phytopathology*, 54, 840-842.
[21]Reglinski T. et al. (1994a) *Annals of Applied Biolo1gy*, 124, 509-517.
[22]Reglinski T. et al. (1994b) *Journal of Plant Disease and Protection*, 101, 1-10.
[23]Rodov V., Yeshos S., D'hallewin G. and Castia T. (1994) *Acta Horticulture*, 241, 301-306.
[24]Ryals J.A., Neuenschwander U.H., Willitis M.G., Molina A., Streiner H.Y. and Hunt M.D. (1996) *Plant Cell*, 8, 1809-1819.
[25]Ryals J., Uknes S., and Ward E. (1994) *Plant Physiology*, 104, 1109-1112.
[26]Safeeulla K.M. (1976) *Mysore*, Wesley press, 304.
[27]Sharp J.K., Valent B., Albeeersheim P. (1984) *Journal of Biological Chemistry*, 259, 111312-11320.
[28]Takenaka S., Nishio Z., Nakamura Y. (2003) *Biological Control*, 93, 1228-1232.
[29]Waspi U., Blank D., Winkler T., Ruedi P. and Dudler R. (1998) *The American Phytopathological Society*, 11, 727-733.

33

APPLICATION OF WIRELESS SENSOR NETWORKS FOR AGRICULTURE PARAMETERS

AWATI J.S.[1], PATIL V.S.[2] AND AWATI S.B.[3]

[1]ETC department, RIT, Sakharale, Sangli, MS, India.
[2]SCOE, Panhala, Kolhapur, MS, India.
[3]DKTE, Ichalkarangi, Kolhapur, MS, India.
*Corresponding Author: Email- jsa.awati@gmail.com

Abstract- In Today's world due to global warming and climate changes there is challenging situation in field of agriculture. Number of advanced methods and technologies are coming in agriculture to reduce cost and improve total productivity. Water is an important resource in agriculture. There is new concept of agriculture in controlled environment. To reduce other cost in agriculture i.e. soil testing. This paper concentrate on Measuring temperature, humidity, and controlling water supply depend on moisture content of soil.
Key words- humidity, WSN, soil, soil moisture, etc.

Introduction

Agriculture is the backbone of the Indian economy. The development of agriculture in terms of area of land under cultivation, use of modern equipment and financial assistance to the farmers is absolutely essential. India is to progress economically with all sectors of the population enjoying its agricultural effect. One of the major problems present today is the less knowledge of the soil content & types, less knowledge of the type of fertilizers to be added, the irrigation amount and pattern depending on the soil porosity and its water retention capacity. In the current Indian scenario analysis of soil to increase crop yields is not being used to a large extent primarily due to the cost involved and the inaccessibility of labs offering such testing facilities. Moreover due to small size of land holdings the procedure of sending soil samples to a far off lab and then taking decision does not seem economically viable.

The technological development in Wireless Sensor Networks made it possible to use in monitoring and control of Agriculture parameters in rural area. Due to uneven natural distribution of rain water it is very crucial for farmers to monitor and control the equal distribution of water to all crops in the whole farm or as per the requirement of the crop. There is no ideal irrigation method available which may be suitable for all weather conditions, soil structure and variety of crops cultures. It is observed that farmers have to bear huge financial loss because of wrong prediction of weather and incorrect irrigation method to crops. In this paper with the evolution in wireless sensor devices, it is possible to uses them for automatic environment monitoring and controlling the parameters of agriculture.

Objective

1. Study some of the agricultural parameters such as soil, Water, crops, Irrigation, chemical and fertilizers issues.
2. Study environmental issues of specific region.
3. Adopt technique for controlling the agricultural parameters:

- Use different sensors to sense the agricultural parameters.
- Collect and transmit the information using Embedded System.
- Using wireless receiver receives the information and does the needful controlling action.
- Design the module which useful in rural development.
- Obtain the readings, observations.
- Measure the performance and benefits to the farmers.

Scope

This paper would take the opportunity to design an instrument that is able to monitor the temperature soil and humidity of an agriculture environment.

In this paper the potential of wireless sensor networks (WSNs) are explored in an original context, the small agriculture of rural area. The goal is to confront an emerging technology with a concrete problem of world-wide dimensions, the sustainability of farming for land-holders living in rural area.
This paper aims to demonstrate the technology of both wireless sensor network and data visualization with the basis for the critical problems, challenges and future goals of development and applications. And we propose a model integrated the two above technologies, emphasis on the processing.

Wireless Sensor Networks
A wireless sensor is a self-powered computing unit usually containing a processing unit, a transceiver and both analog and digital interfaces, to which a variety of sensing units - typically sampling physical data, such as temperature, humidity etc. can be adapted [2]. These sensors automatically organize themselves into an ad-hoc network, which means they do not need any preexisting infrastructure, as do cellular networks such as Zigbee is referred to such a network as an ad-hoc Wireless Sensor Network. Recently, WSNs have raised considerable interest in the computing and communication systems' research community. They have decisive advantages, compared with the technologies previously used to monitor environments via the collection of physical data. Whenever physical conditions change rapidly over space and time, WSNs allow for real-time processing at a minimal cost [3]. Their capacity to organize spontaneously in a network makes them easy to deploy, expand and maintain, as well as resilient to the failure of individual measurement points [4].

Relative humidity
Air, in our normal environment, always holds humidity. The number of water molecules in the air can vary substantially, e.g. it can be as dry as in a desert or as humid as in the tropics. There is an upper limit for the amount of humidity which air can hold at a given temperature. Beyond this limit saturation occurs. If for some reason the humidity level is pushed up to this limit, condensation occurs and fog or water droplets form. Relative humidity tells what percentage of this maximum amount of humidity is present in the air. In contrast to relative humidity, absolute humidity denotes the absolute amount of humidity in the air regardless of the saturation level expressed as the total mass of water molecules per air volume. The maximum possible amount of humidity as well as the actual present amount of humidity in the air is defined by so called water vapor pressures. According to Dalton's law, total air pressure is the sum of the partial vapor pressures of its components and water vapor pressure is one of them. The maximum amount of humidity, which air can hold, is defined by the so-called saturation water vapor pressure. This is a function of temperature.

$P_{water\ vapor}$ = Water vapor pressure, P_{oxygen} = Oxygen pressure,
$P_{nitrogen}$ = Nitrogen pressure, P_{total} = Total air pressure,
P_{others} = Other pressure

$$P_{total} = P_{water\ vapor} + P_{oxygen} + P_{nitrogen} + P_{others} \quad (1)$$

$$\text{Relative Humidity (\%)} = \text{RH (\%)} = [P_{water\ vapor} / P_{saturation}]*100 \quad (2)$$

Soil- Soil is a natural body consisting of layers (soil horizon) of mineral constituents of variable thicknesses, which differ from the parent materials in their morphological, physical, chemical, and mineralogical characteristics. Soil is composed of particles of broken rock that have been altered by chemical and environmental processes that include weathering and erosion. Soil differs from its parent rock due to interactions between the lithosphere, hydrosphere, atmosphere, and the biosphere. It is a mixture of mineral and organic constituents that are in solid, gaseous and aqueous states. In Engineering, soil is referred to as regolith, or loose rock [5].

Soil pH- Soil pH is a measure of the acidity in soils. pH is defined as the negative logarithm (base 10) of the molar concentration of dissolved hydronium_ions (H_3O^+). It ranges from 0 to 14, with 0 being most acidic, 14 being highly basic, and 7 being neutral. Soil pH affects plant nutrient availability by controlling the chemical form of the nutrient. The preferred soil pH varies between plants and can be modified via various soil amendments.

Soil moisture- Soil moisture is the amount of water present in the soil. Gaps between soil particles are called pore spaces or voids. These voids contain various amounts of either water or air. Soil moisture content can be expressed in different basis:

- **Gravimetric:** the mass of water/mass of solid material
- **Volumetric:** the volume of soil/total porosity

The amount of void space within a soil depends on the distribution of particle sizes, and is quantified by soil porosity. Soil moisture is more generally considered within the context of hydrology, where it represents the immediate store of infiltrating rainfall, before it contributes to groundwater recharge [6].

Soil Condition Control- Soil water also affects the crop growth. Therefore, the monitor & control of soil condition has a specific interest, because good condition of a soil may produce the proper yield. The proper irrigations and fertilizations of the crops are varies as per the type, age, phase and climate. The pH value, moisture contains, electric conductivity and the temp of a soil are some key parameters. The pH valves and other parameters will help to monitor the soil condition. The temperature and the moisture can be controlled by the irrigation techniques like drift and sprinkles system.

Humidity sensor- A number of different methods are used for measuring humidity in air. The choice of the most suitable method is usually made by the user based on the local situation. The use of a simple but correctly applied humidity measuring device often permits achieving a better accuracy or meeting the particular requirements.

Main Interface Circuit

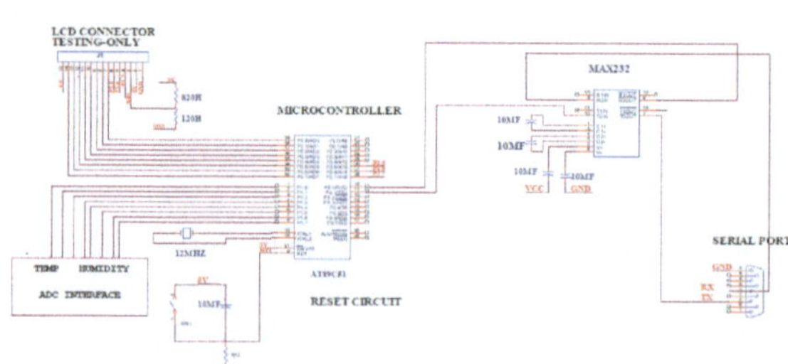

Testing and Results

Tests were done to verify the reliability and accuracy of the Temperature and Humidity monitoring system. Two different set of tests were conducted.

Test conducted in closed room- The location of the test was at our room. The windows and doors were all shut so as to have minimum air-flow from the outside environment. The purpose of the test was to find out how accurate is the Temperature and Humidity monitoring system.

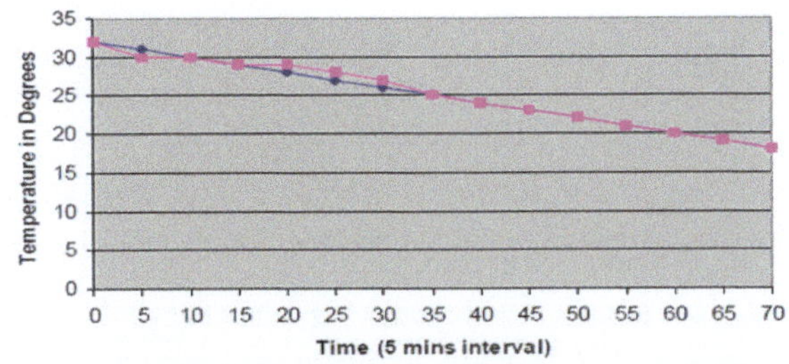

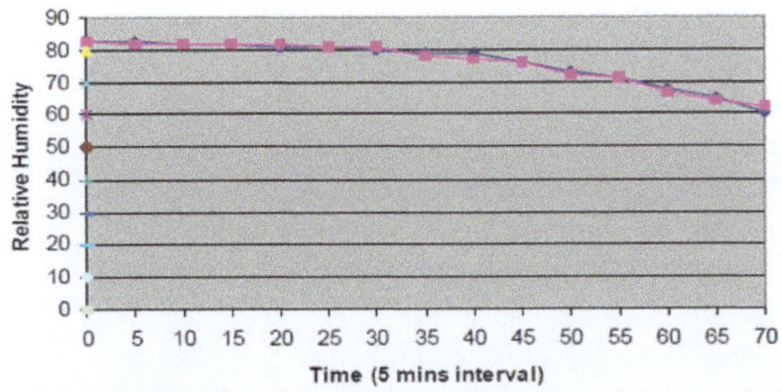

Test conducted with an open environment: The location of the test is at garden.

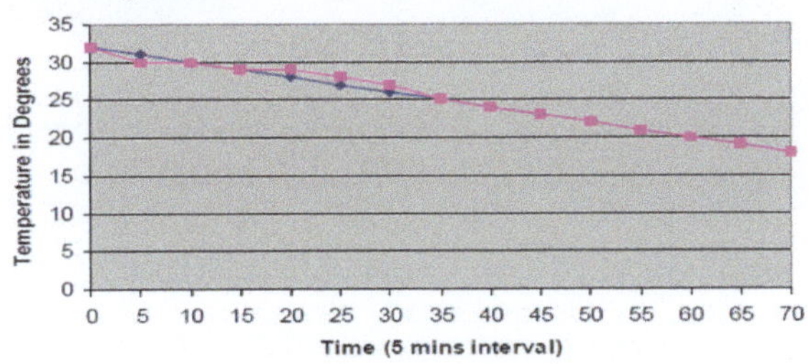

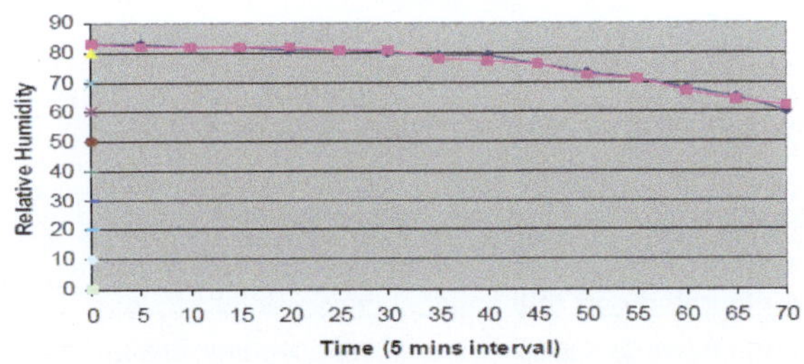

The blue line indicates- The open environment, The pink line indicates- The project prototype.
The second indicates that the results of the first test are quite accurate as the results are almost the same.

Test results-The test results show that both the two test conducted has similar test patterns. The project prototype has similar values after a prolonged time. This could be because it takes a longer period for the sensor to settle down.

Conclusion

The end result is a fully functional Temperature and Humidity Monitoring System, ready for use, from a personal point of view; the paper has been a fulfilling and rewarding experience. This paper deal it both technically and also in terms of how to manage and conduct agriculture parameters. This paper has put into practice a wide range of engineering skills to solve agriculture problems.

References

[1] Eiman Kanjo and Peter Landshoff (2007) IEEE Distributed Systems Online, Urban Computing and Mobile Devices, 8(7).

[2] Burri N., von Rickenbach P. and Wattenhofer R. Dozer (2007) *ultra-low power data gathering in sensor networks,* ACM, 450-459.

[3] Barrenetxea G.,Ingelrest F.,Schaefer G.,Vetterli M.,Couach O.,Parlange M. (2008) *SensorScope: Out-of-the-Box Environmental Monitoring, In The 7th International Conference on Information Processing in Sensor Networks (IPSN).*

[4] Stoianov L., Nachman S., Madden Pipenet (2007) *IEEE International Conference on Information Processing in Sensor Networks.*

[5] Lemos S.G., Noguera A.R., Torre-Neto A., Parra A., Alonso J. (2007) *Journal of Agricultural and Food Chemistry,* 55 4658-4663.

[6] Chaudhary D.D., Nayse S.P., Waghmare L.M. (2011) *International Journal of Wireless & Mobile Networks,* 3, 1.

[7] Zhi Sun and Ian F. Akyildiz (2010) *IEEE Secon,* 978-1-4244-7151-5/10.

GLOBAL ECONOMIC CRISIS: CAUSES, IMPACT ON INDIAN ECONOMY, AGRICULTURE AND FISHERIES

APU DAS[1*], KUMAR N.R.[1], BISWAJIT DEBNATH[1] AND MANDAL S.C.[2]

[1]Central Institute of Fisheries Education (CIFE), Mumbai-400061, Maharashtra, India.
[2]College of Fisheries, CAU, Lembucherra, Agartala, 799210, Tripura, India.
*Corresponding Author: Email- apudascifefec04@gmail.com

Abstract- The global economic crisis started in United States of America mainly due to 'sub-prime mortgages' where interest rate was slower down and there was a great demand for housing loans. Later, American banks repackaged this debt to worldwide financial instruments called 'Collateralized debt obligations' and sold them worldwide, which resulted in unaffordable mortgage payments and many people defaulted or undertook foreclosure. Then this mortgages crisis affected worldwide. Different views on the reasons of the crisis include subprime mortgage, securitization and repackaging of loans, excessive leverage, mismatch between financial innovation and regulation, fair value accounting rules, typical characteristics of US financial system, failure of global corporate governance & complex interplay of multiple factors. Developed countries have so far been the most affected, with a decline in FDI inflows in 2008, mainly due to sluggish market prospects. Flows into developing economies continued to grow in 2008, but at a much lower rate than the year before. Indian economy and agriculture cannot be completely insulated from the global and domestic economic recessions. The impact of economic crisis on Indian agriculture and fisheries were transmitted through three distinct channels, viz., financial sector, exports and exchange rates, and the impact manifests itself in several direct and indirect ways. Some of the impacts were decreased GDP growth rate, high inflation, FDI inflows and international trade.
Key words- Global economic crisis, Sub-prime mortgages, Gross Domestic Product (GDP), Foreign Direct Investment (FDI).

Introduction

Banking and economic crises have been a common phenomenon throughout the modern economic history of mankind. Since the great depression of 1929, the world has witnessed hundreds of such crises and the frequency of the crises has increased over time. According to a UNCTAD (2008a), there were as many as 112 systemic banking crises from the late 1970s until 2001. Most of them, including the current one, have shared some common features: they each started with a hasty process of financial sector reforms, which not only created a vacuum in terms of regulations but also deteriorated the basic economic fundamentals though massive inflows of foreign capital and finally ended up with a change in investor expectations and a consequent mess in the financial markets. Seshan (2009) reported that the financial sector crisis that arose in the latter half of 2007 and was precipitated by the collapse of Lehman Brothers on 23 September 2008 shared most of these features. However, what makes the current crisis exceptional is that it broke out at the very epicentre of global capitalism and its contagion spread very quickly to the entire globe. India, being an integrated part of the global economic order, was also exposed to the adverse impact of the global economic crisis.

The Indian economy looked to be relatively insulated from the global financial crisis that started in August 2007 when the 'sub-prime mortgage' crisis first surfaced in the US. But as the financial meltdown, morphed in to a global economic downturn with the collapse of Lehman Brothers on 23 September 2008, the impact on the Indian economy was almost immediate. Credit flows suddenly dried-up and, overnight, money market interest rate spiked to above 20 percent and remained high for the next month. It is, perhaps, judicious to assume that the impacts of the global economic downturn, the first in the centre of global capitalism since the Great Depression, on the Indian economy are still unfolding. For the first time in 60 years, the IMF (2011) is now reported a global recession with negative growth for world GDP in 2009-10. The Horn (2010) has reported that GDP virtually collapsed in the second half of 2008 and further it declined by as much as nine percent in 2009-10.

Methodology

The data for the study has been collected from secondary sources. The present study focused on origin of the global economic crisis and the impact on Indian economy, agriculture and fisheries.

Result and Discussion

Causes of Economic crisis

i. Sub-prime mortgage

G-20 summit (2008) viewed that the current global economic crisis has originated due to sub-prime mortgage in USA in 2007. With easy availability of credit at low interest rates, real estate prices in US had been rising rapidly since the late 1990s and investment in housing had assured financial return. Kunt et al., (2002) reported that US home-ownership rates rose over the period 1997-2005 for all regions, all age groups, all racial groups, and all income groups. Banks went out of their way to lend to sub-prime borrowers who had no collateral assets. Low income individuals who took out risky sub-prime mortgages were often unaware of the known risks inherent in such mortgages. While on the one hand, they were ever keen to become house-owners, on the other, they were offered easy loans without having any regard to the fact that they were not in a position to refinance their mortgages in the event of the crisis. All this was fine as long as housing prices were rising. But the housing bubble burst in 2007. Home prices fell between 20 per cent and 35 per cent from their peak and in some areas more than 40 per cent; mortgage rates also rose. Sub-prime borrowers started defaulting in large numbers. The banks had to report huge losses. Acharya (2009), Green, King & Dawkins (2010) also explain the main reason of global economic crisis is excessive sub-prime mortgages.

ii. Securitization and Repackaging of Loans

The mortgage market crisis that originated in the US was a complex matter involving a whole range of instruments of the financial market that transcended the boundaries of sub-prime mortgage. An interesting aspect of the crisis emanated from the fact that the banks/ lenders or the mortgage originators that sold sub-prime housing loans did not hold onto them. They sold them to other banks and investors through a process called securitization. Securitization, as a financial process, has gained wide currency in the US in the last couple of decades. Indeed, as recently as 1980 only 10 per cent of US mortgages were securitized compared to 56 per cent in 2006. In the context of the boom in the housing sector, the lenders enticed the naive, with poor credit histories, to borrow in the swelling sub-prime mortgage market. They originated and sold poorly underwritten loans without demanding appropriate documentation or performing adequate due diligence and passed the risks along to investors and securitizes without accepting responsibility for subsequent defaults. These sub-prime mortgages were securitized and re-packaged, sold and resold to investors around the world, as products that were rated as profitable investments. They had a strong incentive to lend to risky borrowers as investors, seeking high returns and were eager to purchase securities backed by sub-prime mortgages. The booming housing sector brought to the fore a system of repackaging of loans. It thrived on the back of flourishing mortgage credit market. The system was such that big investment banks such as Merrill Lynch, Morgan Stanley, Goldman Sachs, Lehman Brothers or Bears Stearns would encourage the mortgage banks countrywide to make home loans, often providing the capital and then the Huge Investment Banks (HIBs), would purchase these loans and package them into large securities called the Residential Mortgage Backed Securities (RMBS).

iii. Excessive Leverage

Global economic outlook (2008) reported that the final problem came from excessive leverage. Investors bought mortgage-backed securities by borrowing. Some Wall Street Banks had borrowed 40 times more than they were worth. In 1975, the Securities Exchange Commission (SEC) established a net capital rule that required the investment banks who traded securities for customers as well as their own account, to limit their leverage to 12 times. However, in 2004 the Securities and Exchange Commission (SEC) allowed the five largest investment banks – Merrill Lynch, Bear Stearns, Lehman Brothers, Goldman Sachs and Morgan Stanley – to more than double the leverage they were allowed to keep on their balance sheets, i.e. to lower their capital adequacy requirements. The institutions that have reported huge losses are those which are highly leveraged. Leveraged investors have had to return the money they borrowed to buy everything from shares to complex derivatives. That sends financial prices even lower. All this led to massive bailout packages in USA, as the government stepped in to buy and lend in a financial market. The role of leverage and credit is, therefore, central to growth. Yet, excessive leverage is fraught with dangerous consequences.

iv. Mismatch between Financial Innovation and Regulation

It is not surprising that governments everywhere seek to regulate financial institutions to avoid crisis and to make sure a country's financial system efficiently promotes economic growth and opportunity. Striking a balance between freedom and restraint is imperative. Conway (2009) reported that the financial innovation inevitably exacerbates risks, while a tightly regulated financial system hampers growth. When regulation is either too aggressive or too lax, it damages the very institutions it is meant to protect.

v. Fair value accounting rules

Fair value accounting rules require banks and others to value their assets at current market prices. The broad aim of fair value accounting is to enable investors, financial system participants, and regulators to better understand the risk profile of securities in order to better assess their position. In order to achieve this, financial statements must, in the case of instruments for which it is economically relevant, be sensitive to price signals from markets, which reflect transaction values. Investors and regulators hold that the fair value accounting standard should not be weakened because it is a key component of accurate and fully transparent financial statements, which in turn are the bedrock of financial activity. But the asset holders maintain that accounting standard should be reformed to fully reflect the reality of financial activities. They have argued that in times of illiquid and falling markets, it has been difficult or impossible to value assets accurately. Singh (2009) reported that the fair-value accounting has resulted in assets being valued at distressed sale prices, rather than at their fundamental value, creating a downward spiral. The requirements of fair value accounting ensured that what began initially as a subprime crisis morphed into a general credit deterioration touching prime mortgages and causing their credit downgrades and system -wide mark downs.

vi. Failure of Global Corporate Governance

The financial system of USA has changed dramatically since the 1930s. Many of America's big banks moved out of the "lending" business and into the "moving business". They focused on buying assets, repackaging them, and selling them, while establishing a record of incompetence in assessing risk and screening for creditworthiness. Hundreds of billions have been spent to preserve these dysfunctional institutions. Singh (2008) told that nothing has been done even to address their perverse incentive structures, which encourage short-sighted behaviour and excessive risk taking. Prudential oversight was lax, allowing poor lending standards, the proliferation of non-transparent securitization structures, poor risk management throughout the securitization chain, and the build-up of excessive leverage by financial institutions. The weaknesses in prudential oversight were partly due to particular characteristics of the US financial system, such as the existence of different regulatory regimes for investment banks, commercial banks and government-sponsored enterprises (Fannie Mae and Freddie Mac), as well as the complex and fragmented supervisory architecture, comprising several federal and state agencies with competing and overlapping mandates.

vii. Typical characteristics of US financial system Failure of Global Corporate Governance

One of the reasons for current crisis in the advanced industrial countries related to the failures in corporate governance that led to non-transparent incentive schemes that encouraged bad accounting practices. Rajadhyaksha (2008) studied that there is inadequate representation and in some cases no representation of emerging markets and less developed countries in the governance of the international economic institutions and standard setting bodies, like the Basle Committee on Banking Regulation. The IMF has observed and stated in The Hindu, (March 11, 2009) that market discipline still works and that the focus of new regulations should not be on eliminating risk but on improving market discipline and addressing the tendency of market participants to underestimate the systemic effects of their collective actions. On the contrary, it has often put pressure on the developing countries to pursue such macro-economic policies that are not only disadvantageous to the developing countries, but also contribute to greater global financial instability.

viii. Complex Interplay of multiple factors

It may be said with a measure of certainty that the global economic crisis is not alone due to subprime mortgage. There are a host of factors that led to a crisis of such an enormous magnitude. The declaration made by the G-20 member states at a special summit on the global economic crisis held on 15th November 2008 in Washington, D.C. identified the root causes of the current crisis and put these in a perspective. During a period of strong global growth, growing capital flows, and prolonged stability earlier this decade, market participants sought higher yields without an adequate appreciation of the risks and failed to exercise proper due diligence. At the same time, weak underwriting standards, unsound risk management practices, increasingly complex and opaque financial products, and consequent excessive leverage combined to create vulnerabilities in the system. Policy-makers, regulators and supervisors, in some advanced countries, did not adequately appreciate and address the risks building up in financial markets, keep pace with financial innovation, or take into account the systemic ramifications of domestic regulatory actions. Major underlying factors to the current situation were, among others, inconsistent and insufficiently coordinated macroeconomic policies, inadequate structural reforms, which led to unsustainable global macroeconomic outcomes. Pan (2009) reported these developments, together, contributed to excesses and ultimately resulted in severe market disruption.

Impact on Indian Economy

Until the emergence of global crisis, the Indian economy was going through a phase of high growth driven by domestic demand – growing domestic investment financed mostly by domestic savings and sustained consumption demand. In fact, consumption and saving were well-balanced. Services sector, led by domestic demand, contributed to the stability in growth. Concomitantly, inflation was also generally low and stable. This overall improvement in macroeconomic performance in India was attributed to calibrated financial sector reforms that resulted in an efficient system of financial intermediation, albeit bank-based; the rule based fiscal policy that reduced the drag on private savings; and forward-looking monetary policy that balanced the short term trade-off between growth and inflation on a continuous basis. Additionally, the phased liberalisation of the economy to trade and capital flows along with a broadly market-driven exchange rate regime enhanced the role of external demand in supporting the growth process, simultaneously exposing the economy to the forces of globalisation. India, though initially somewhat insulated to the global developments, eventually was impacted significantly by the global shocks through all the channels – trade, finance and expectations channels. This raised the issue that whether India is more globalised than what is perceived in terms of conventional trade openness indicators. At the same time, India was also among the first to exhibit strong rebound from the global downturn as compared to many advanced economies.

Impact on Indian GDP growth rate

Below the Indian's GDP growth rate chart is given to show the impact of economic crisis on GDP growth rate for the country. Economic growth is the increase in value of the goods and services produced by an economy. It is conventionally measured as the percent rate of increase in real gross domestic product or GDP. Growth is usually calculated in real terms, i.e. inflation-adjusted terms, in order to net out the effect of inflation on the price of the goods and services produced. Indian GDP growth rate for last five years and for each year quarter growth rate of GDP are compared (Fig 1). It shows that global economic crisis of 2007 had impact on Indian GDP growth rate in the financial year of 2009, where GDP growth rate were lowest in first three quarter of the financial year.

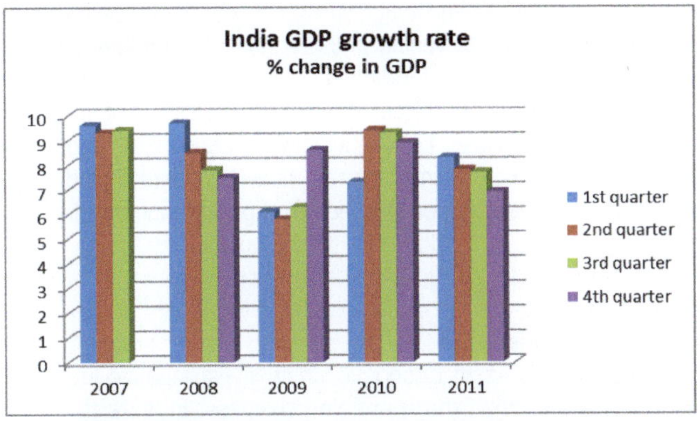

Fig.1- GDP growth rate for last five financial year

Impact on Inflation rate

By January 2010, the domestic growth signals were pointing towards a consolidation of the recovery process. However, sustained increase in food prices was beginning to spill over to manufactured products. Inflation in primary commodities moved up 8.2 in August 2009 to 22.2 per cent by March 2010. An important concern from the point of view of inflation management is the downward rigidity in the primary food articles prices even after a good monsoon. Moreover, the consumption basket is getting diversified more in favour of non-cereals items such as milk, meat, poultry, fish, vegetables and fruits, which are important from the nutritional angle. The decomposition of food inflation indicates that during the recent period the key drivers of food inflation are non-cereals. All together Inflation rate is increasing at higher rate.

Impact on Indian FDI inflows

Foreign Direct Investment (FDI) inflow into the core sectors plays a significant role as a source of capital, management, and technology in transitional economies. It implies that FDI can have positive effects on the host economy's developmental efforts. As mentioned earlier, India has opened its economy and has allowed the entry of multinational corporations (MNCs) as a part of the reform process started in the beginning of 1990s. Like many other countries, India has offered greater incentives to encourage FDI inflows into its economy. The presence of FDI inflow in India was negligible till 1991, but there has been a steady build-up in the actual FDI inflows in the post-liberalisation period. The share of FDI in GDP was merely 0.03 per cent in 1991, which rose to about 3 per cent in 2009-10. Its annual growth during this period was phenomenal. The FDI inflow has been growing rapidly since then with a quantum jump after 2004-05. From US $3250 million in 2004-05, the FDI has leaped to over US $247329 million in 2008-09. However, since February 2008, a reversal in the trend has been observed. A perusal of the monthly inflow of FDI between January 2008 and January 2010 suggests a clear decline over a period of 24 months.

The share of agriculture in the total FDI in India is negligible. The recent data show that agriculture accounted for only about 1 .5 per cent of the total FDI inflows into India. In the agriculture sector, die entry of FDI was confined to plantation crops, food processing industries, agricultural services and agricultural machinery. FDI has been allowed in fertiliser manufacturing also, which have a direct bearing on agriculture but was not allowed in the cultivation of crops or rearing of livestock. However, its entry into the food processing sector can have ramifications on the agriculture sector (though it may be limited). Therefore, though the FDI inflow has slowed down over the past one year, its impact would not be visible on agriculture, as the dependence of agriculture on FDI is minimal (Fig 2).

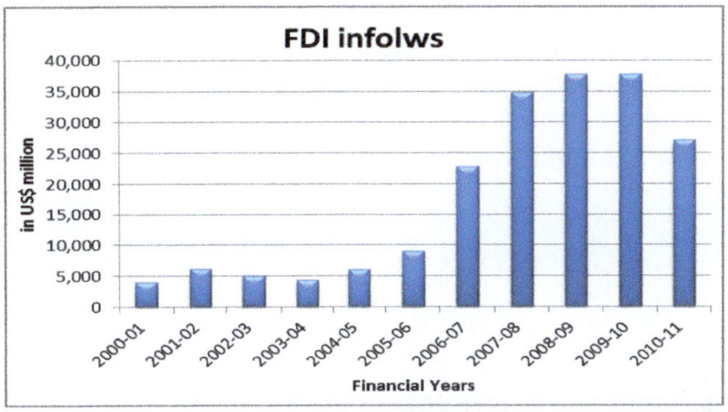

Fig 2- Trend of FDI inflows in India

Impact on Export and Import

Global crisis also affect the Indian export-import market. In 2008-09, export-import both had been reduced due to crisis of 2007. In fact in 2009-10 export and import of India reached in negative. Demand for Indian product has reduced in international market as a result export gone under negative in 2009-10.

Impact on Indian Agriculture

India has opened its market since the beginning of the past decade (more precisely since July 1991) by lowering tariff and non-tariff barriers, as well as liberalising investment policies. Still Indian agriculture is far less vulnerable to the external economic shocks than agriculture in many developing countries. Agricultural trade still accounts for less than 10 per cent of agricultural gross domestic product (Ag. GDP). However, Indian agriculture cannot be completely insulated from the global and domestic economic recessions. The impact of economic crisis is transmitted through three distinct channels, viz., financial sector, exports and exchange rates, and the impact manifests itself in several direct and indirect ways. Some of the implications of the economic crisis are discernible in the short-run, while others may be visible only in the long run. It is difficult to gauge the impact of economic crisis on Indian agriculture in the short run. However, the trends in some broad parameters may indicate its implications and the possible options can be worked out to mitigate its adverse impact. Chandrasekhar (2009) identified the broad indicators for assessing the impact of economic recession on Indian agriculture could be the trends agricultural exports, Ag. GDP, agriculture export & Foreign Direct Investment (FDI) etc. The trends in these indicators have been assessed and briefly discussed in the following sections.

Agricultural Exports

Two remarkable developments have taken place in India's agricultural exports during the post-liberalisation period. one, the agricultural exports have grown at a much faster rate since the initiation of liberal economic policies where agricultural exports in value terms have grown annually from 18.9 per cent during 1990s to 15.2 per cent during 2000s but after 2000 it reduced gradually (Table 1).

It would be interesting to see whether mere has been any divergence from the long-term trend in the export of important agricultural commodities due to economic recession. For this, share of export of agricultural products (including livestock products) in total national export during 2007-08 and 2009-10 can be compared. However, the decline or slowdown in exports cannot be entirely attributed to the economic recession.

Table 1- Trend of agricultural export, 2000 to 2010

Year	Agriculture Exports	Total National Exports	% Agriculture Exports to Total National Exports
2000-01	28657.37	201356.45	14.23
2001-02	29728.61	209017.97	14.22
2002-03	34653.94	255137.28	13.58
2003-04	37266.52	293366.75	12.70
2004-05	41602.65	375339.53	11.08
2005-06	49216.96	456417.86	10.78
2006-07	62411.42	571779.28	10.92
2007-08	79039.72	655863.52	12.05
2008-09	85951.67	840755.06	10.22
2009-10	89522.59	845125.21	10.59

Agricultural GDP

The trend in agricultural GDP during the past two decades suggests that the sector has been growing slowly and steadily, but with occasional slumps. The reasons for slow growth during the 1990s and early 2000s are many, ranging from poor monsoons to depressed agricultural commodity prices in the world market.

The current crisis is expected to have a modest effect on the GDP of agricultural and allied products. Recent trends indicate that the sector is not witnessing similar growth achieved during the previous year. Agricultural GDP is declined by -0.1 per cent in 2008-09 as compared to 5.8 per cent in 2007-08. In 2009-10 and 2010-11 GDP growth rate for agriculture sector was 0.4 percent and 6.6 percent respectively, Economic survey of India (2011). The trends in Ag.GDP seem to have weak links with the present recession (Fig 3).

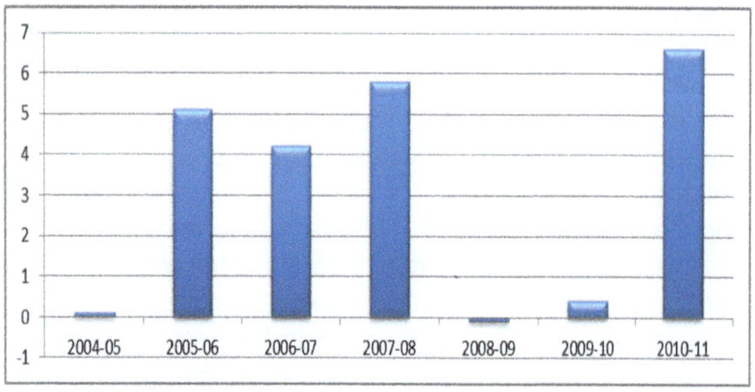

Fig. 3- GDP growth rate of agri-allied sectors

Agricultural growth has been accorded priority to improve the distributional aspects by which several schemes like National Food Security Mission, Rashtriya Krishi Vikas Yojana (RKVY), substantial increase in the flow of agricultural credit, waiving off agricultural loans, etc. have been launched to foster growth in this sector. These schemes are likely to taper off the adverse impact of the economic recession on agriculture to a large extent. However, the economic crisis may put downward pressure on farm production in the short-run. Even though the government provides a shield to the farmers by intervening in the agricultural markets to realise stabilised income, its intervention is limited to a few commodities in some states. Therefore, in spite of government's efforts, farm income is expected to have slightly adverse impact due to economic recession. It is important to note that rainfall and other weather parameters influence agricultural growth significantly.

Impacts on Fisheries

Economic Crisis in Fisheries

Roughly speaking, the Asian Economic Crisis gives negative impacts to fisheries and any fisheries-related business. In domestic fish markets, demand for fisheries products is on sharp decline. Wholesale and retail prices sharply fall down. Export of Indian fisheries products mainly for Asian markets is in a severe slump. Exporters, processing companies, and any type of fish dealers suffer from extreme market slump. Their financial positions get worse, which causes a sharp decline in fish prices in production sites. Moreover, fishers and fish farmers find it very hard to raise capital for investment and operation, by depending on fish traders. Since prices of productive materials rise, the rate of profit decreases. Before the Crisis, exporters were prevented from further expansion of trading products with lower additional value. In a much contrast, domestic-oriented production remained in depression.

Impact on Fisheries Export

However, the decline or slowdown in exports cannot be entirely attributed to the economic recession. Quantity exported in the year of 2006-07 was 612 thousand tonnes and it has been reduced in the year 2007-08 to 541 thousand tonnes (Table 2), it shows the negative impact of economic crisis on fisheries export.

However, after the outbreak of global Economic Crisis, demand for Indian fisheries has suddenly declined in foreign markets. It is reported that, in European Union, wholesale prices of Indian fishes imported sharply fall down. Collectors have to reduce their scale of transaction and reduce purchase prices of Indian fishes/shrimp. Farm gate price is almost half of the highest at peak. This causes damage to small-scale fishers who are engaged in catching fry and young fish.

Table 2- Trend of fisheries export

Year	Export of marine products	
	Qty ('000 tonnes)	Value (Rs in crore)
2006-07	612	8363
2007-08	541	7620
2008-09	602	8608
2009-10	664	9921
Source: Economic survey, 2011		

Conclusion

While the developed world, including the U.S, the Euro Zone and

Japan, has plunged into recession, the Indian Economy is being affected by the spill-over effects of the global financial crisis, Chidambaram (2008) & Seshan (2008). Great savings habit among people, strong fundamentals, strong conservative and regulatory regime have saved Indian economy from going out of gear, though significant parts of the economy have slowed down and there is a wide variance of opinion about how long it will continue. It is expected that growth will be moderate in India. The most important lesson that we must learn from the crisis is that we must be self-reliant. Though World Trade Organization (WTO) propagates free trade, we must adopt protectionist measures in certain sectors of the economy so that recession in any part of the globe does not affect our country.

Acknowledgement

Authors are highly grateful to Dr. W.S. Lakhra, Director, CIFE, Mumbai for proving us need full facilities for preparation of this manuscript and also thanks to Dr. R.S. Biradhar, Joint Director, CIFE, Mumbai for his support and valuable suggestions.

References

[1] Acharya S. (2009) *Gloabal crisis, India and Agriculture, 9th convocation*, Rajasthan Agriculture University (RAU).
[2] Chandrasekhar C.P. and Ghosh Jayanthi (2008) India and the Global Financial Crisis. From *http://www.macroscan.org*.
[3] Chidambaram P. (2008) Spill - over effects of global crisis will be tackled. *The Hindu, Daily*, 15.
[4] Conway E. (2009) *WEF, Global crisis has destroyed 40pc of world wealth*, Internet edition.
[5] *Declaration of the G-20 Summit on Financial Markets and the World Economy* 2008, Washington DC.
[6] *Economic survey of India*, 2011.
[7] *Global Economic Outlook*, 2008.
[8] Green D., King R. and Dawkins M.M. (2010) *Impact and Response*, Oxfam Research Report.
[9] Horn Z.E. (2010) *Women in Informal Employment, Globalizing and Organizing* (WIEGO).
[10] IMF (2011) *World economic outlook update, rapidly weakening prospects call for new policy stimulus*. Washington, DC.
[11] Kunt D., Asli and Enrica Detragiache (2002) *J. of Mone. Econ.*, 49 (7), 122-126.
[12] Pan P.P. (2009) *Economic Crisis Fuels Unrest in East Europe*, The Washington Post.
[13] Rajadhyaksha N. (2008) *The Hindustan Times*.
[14] RBI's Bulletin (2011) *Foreign Investment Inflows, Table No.* 44.
[15] Seshan P.A. (2008) *Concern over impact of U.S. crisis*, The Hindu, 44.
[16] Seshan P.A. (2009) *The Hindu (Editorial)*, 35.
[17] Singh N.K.. (2008) *The Hindustan Times*.
[18] Singh N.K. (2009) *Mail Today*.
[19] UNCTAD (2008a) *World Investment Prospects Survey*. New York and Geneva.

EVALUATION OF SPROUTING BROCCOLI (*Brassicae oleraceae* var. *italica*) GENOTYPES FOR GROWTH, YIELD AND QUALITY

THAPA U.* AND RAI R.

Department of Vegetable Crops Faculty of Horticulture, Bidhan Chandra Krishi Viswavidyalaya, Mohanpur, Nadia -741252, WB, India.
*Corresponding Author: Email- drumesh.thapa@gmail.com

Abstract- Vegetables play a very important role in our daily diet. As an unconventional vegetable "Broccoli" is yet to gain the desired popularity in our country. It is very rich source of various anti-cancer agents as well as Vitamin C and dietary fibre. However, considerable attention is being given on the production technology of Broccoli which are rich in nutrient content and greater yield potential. But yet, No systematic work has been done on evaluation and commercialization of high value nutrient rich this cole crops. Therefore, the present study were carried out at Horticultural Research Station, Mondouri, Bidhan Chandra Krishi Viswavidyalaya, Mohanpur, Nadia in West Bengal during [2010-2012] rabi season with objectives to standardize the production technology of sprouting broccoli. Cultivation of these value added vegetables can boost the income of farmers due to very high market price and export demand. The investigations were followed in Randomized Block Design with three replications. Twelve varieties of Broccoli [Viz. Fiesta , Princess , Sultan , Nokguk , Early You, KE-180 , Priya , Puspa , Grandsino , Rapido , Prema and Packman] were evaluated. Observations recorded under the physical and chemical parameters of different varieties of genotype. In general, the performances of this crop with different hybrid varieties proved that there is ample scope to grow broccoli vegetable due to prevailing suitable agro - climatic condition as well as the gaining importance as potential vegetable for export. Among all the varieties of Broccoli Nokguk was found superior, which gave higher yield [145.47q/ha] in combination with best head formation. Whereas, in chemical analysis KE-180 variety was found outstanding position in chlorophyll b, reducing sugar, total sugar and carotene content among all the other genotypes taken under observation. So, proper management practices are highly required for the cultivation of this crop.

Keywords- Evaluation, broccoli, varieties, quality and yield

Introduction

India is endowed with a wide range of tropical, sub-tropical and temperate vegetable crops. But still there are some vegetables which are lesser known or rare to most of our growers and consumers. Our farmers can earn a lot of profit by growing this rare or unusual high value Cole vegetables nearby big cities (peri-urban areas) and towns as they attract very high prices in cosmopolitan markets, star hotels and places of tourists' interest. They can also be exported to foreign especially European countries where their cultivation is not possible throughout the year in open field conditions. But due to lack of information about their cultural practices for our conditions the production or availability of these vegetables is still meager. Chinese cabbage, Sprouting broccoli , Red cabbage and Brussels sprouts, etc. have opened up new opportunities for vegetable growers of our country for diversification and off-season production for high market in metropolis. But due to lack of preference in food among Indians some of the introduced vegetables could not get popularity though they are rich in protein, carbohydrates, minerals, vitamins and fibers etc. However, with the growing tourist industry and nutritional awareness among people, these vegetables are gaining popular. Among the Cole crops broccoli is more nutritious than other Cole crops, such as cabbage, cauliflower and kohlrabi. It is fairly rich in carotene and ascorbic acid and contains appreciate quantities of thiamin, riboflavin, niacin and iron. Realizing the tremendous potential of sprouting broccoli in domestic and foreign market, the cauliflowers growers of terai zone of West Bengal (India) are gradually adopting the broccoli cultivation [3]. To popularize this high value Cole crops and its variety among the marginal and small farmers, proper demonstration should be adopted through personal contact approach, monitoring, motivation and awareness creation about benefits. Efforts are being made on long way in making farmers and consumers of West Bengal, India in adopting new exotic vegetables and varieties, which created an impact on ground level and helped the state in a big way in addressing/lessening the problems of hunger, under nutrition and malnutrition not only in West Bengal but elsewhere in the country as well. But no systematic work has been done on evaluation of these vegetables in our region. However, State is facilitated with good and congenial agro-climatic condition for cultivation of these crops. Therefore, present studies were aimed at promotion of high value Cole vegetables by identifying new promising varieties with high productivity under wide range of environmental conditions, better horticultural characteristics and market opportunities.

Materials and Methods

The present studies were carried out at Horticultural Research Station, Mondouri, Bidhan Chandra Krishi Viswavidyalaya, Mohanpur, Nadia in West Bengal during rabi season [2010-2012] with the principle objective to standardize the production technology of sprouting broccoli. The investigation details are as follows:

Broccoli [*Brassica oleraceae* var. *italica*] were planted in the rabi season. The design of experimental site was Randomized Block Design replicated thrice utilizing twelve genotypes showing diverse features. Genotypes taken under observations were Fiesta, Princess, Sultan, Nokguk, Early you, Priya, Puspa, Grandsino, Rapido, Prema, and Packman. The transplanting of seedlings were accomplished on first week of November with the spacing of [60cm x 45cm]. Applied fertilizer doses are in NPK ratio of [100:80:100] kg per hectare. Regular cultural practices, crop protection measures were adopted as per the requirements of crop. Observations were taken under physical, yield and quality attributing parameters. Mean value of randomized data were analysed by following standard statistical technique [2].

Results and Discussion

The twelve different varieties of Broccoli were varied significantly. The height of the plants varied from [30.78 to 41.69 cm]. From the data it revealed that the variety, Nokguk recorded significantly maximum plant height [41.69 cm] while Early You variety recorded the minimum [30.78 cm]. The lowest plant height observed in some other varieties might be due to its inherent genotypic characteristics or for the variations in agro-climatic condition. The number of leaves per plant is an important character that might influence the yield. The cultivars included in the study produced an average variation of [11. 33 to 20.20] leaves per plant. The maximum number of leaves per plant was recorded as [20.20] in variety Fiesta, followed by Nokguk [19.00], Grand Sino [18.077], and Priya [15.30]. The lowest number of leaves was noticed in the variety Puspa [11.33] Sultan [11.93], Early You [12.06] and Princess [13.07]. Lower number of leaves in some cultivars was probably due to slow rate in leaf initiation which would be an inherent character of the cultivars. This is in agreement with previous investigation in which some of the cultivars were included [1-3, 5]. In determining the photosynthetic efficiency of the leaves, the surface area of the leaves is an important factor. In this investigation, plant spread in each cultivar were recorded and found significant differences. The range of plant spread was [4515.87 to 6641.81cm^2]. It has been found from the experiment results, the maximum plant spread [6641.81cm^2] was obtained with variety Princes. The minimum plant spread of [4515.87 cm^2], with variety Puspa was recorded.

It has been found from the experimental results that the highest stem diameter was measured in variety Prema [3.77 cm] followed by Grandsino [3.70 cm]. Similarly the higher site in diameter of stem was observed with variety Princes [3.67 cm], Nokguk [3.67 cm], and Early You [3.66 cm]. From the [Table-1] it is clear that among the above mentioned varieties there were significant differences among themselves. Rest of the other varieties different significantly from the above mentioned one. However, the lowest diameter of stem was obtained with variety Fiesta [3.32cm]. This similarity and dissimilarity among the varieties in stem diameter may be attributed to the variability in their genetic configuration. In respect of the stem length, statistically parity was observed. Among the twelve varieties the minimum stem length [20.42 cm] was observed in variety Early You. This showed that the cultivars represent a good range of genetic diversity in response of stem length.

Yield and yield attributing characters due to different varieties showed a significant differences effect. In respect of earliness of head initiation and days required to harvesting, the cultivars under study were found significant. The average number of days to head initiation varied from [59.00 to 70.00]. The cultivar Early You, Sultan, Puspa, Rapido, Prema and Packman found earlier and Grandsino found very late in respect of head initiation. The average period required days to harvesting varied from [72.00 to 85.67]. The highest head diameter was recorded in Nokguk [23.63cm] followed by Early You [21.70 cm].

Table 1- Varietal evaluation of different varieties of Broccoli Significant increase in head weight was observed in response to different varieties.

Variety	Days to Head initiation (Days)	Days to harvest (Days)	Plant height (cm)	Plant spread (cm^2)	Number of leaves	Leaf area (cm^2)	Head diameter (cm)	Stem diameter (cm)	Stem length (cm)	Number of sprout	Sprout weight (gm.)	Sprout yield/plot (kg)	Head weight (gm.)	Yield (q/ha)
Fiesta	67.67	80.67	40.74	5009.58	20.2	4961.14	21.64	3.32	30.29	3.67	25.72	1.5	233.83	93.53
Princies	61.33	72	34.53	6641.81	13.07	1041.11	20.27	3.67	22	4.76	45.3	3.4	320.28	119.89
Sultan	59	72	32.47	4987.08	11.93	812.32	19.99	3.39	21.31	3.92	64.68	4	259.56	103.83
Nokguk	63.67	78.33	41.69	5543.19	19	464.39	23.63	3.67	28.92	1.35	57	1.24	375	145.47
Early you	59	72	30.78	5775.74	12.06	887.18	21.7	3.66	20.42	5.16	35.79	2.94	333.67	128.11
KE-180	63.67	76	38.47	5607.93	13.47	852.52	20.67	3.7	25.07	4.46	54.79	3.84	251.67	100.67
Priya	63.67	73	36.61	5913.78	15.3	1022.68	20.14	3.63	26.67	6.1	33.43	3.26	255.83	102.33
Puspa	59	73	33.07	4515.87	11.33	896.48	20.17	3.49	21.67	4.77	31	2.36	289	115.6
Grandsino	70	78.33	34.7	5708.6	18.07	720.55	21.1	3.72	26.4	2.4	76.55	2.95	281.67	112.67
Rapido	59	72	40	5731.59	15.2	945.95	20.65	3.55	28.53	3.81	30.25	1.84	253.33	114.83
Prema	59	74.4	35.2	5367.55	13.3	1112.09	21.57	3.77	23.6	4.06	27.67	1.8	263.89	108
Packman	59	77.33	36.2	4785.67	13.4	1108.49	19.37	3.65	24.3	4.69	40.93	3.05	279.6	111.6
SEm (±)	3.0965	2.6328	1.403	520.038	1.0502	1045.85	0.9925	0.1782	1.8655	0.4477	4.493	0.345	34.081	13.004
CD at 5%	6.3651	5.4118	2.884	1068.96	2.1587	2149.79	2.0402	0.3662	3.8846	0.9203	9.236	0.7104	70.055	26.731

The maximum head weight of [375.00gm] was found with Nokguk, variety. The varieties which produced comparatively more head weight are namely Early You, Prema, Fiesta, Princes, Packman and Puspa. The highest head weight might be due to resulted from the highest head diameter of the respective varieties. The minimum head weight of [233.83.gm] was obtained with Grandsino variety. There was a significant and positive effect of different varieties on head yield [q/ha] Nokguk performed the highest results in head yield [149.74q/ha] and the other two varieties showed statistically similar results [120.86 q/ha, 119.72 q/ha]. It indicates that next to Nokguk, there two varieties, Early You and Princes have ability to produced good head yield. The other remaining varieties has produced similar head yield. The tabulated data [Table-1.] showed clearly that the best quality of more number of sprout (spears) was recorded from the variety Priya followed by Early You. The lowest numbers of sprout were observed from Nokguk variety. The differences in number of sprout among these varieties may be due to their own genetic characters. Results obtained in [Table-1] reflect significant differences in the sprout weight of the different varieties. The highest sprout weight was obtained from Grandsino followed by Nokguk.

Quality parameters were also analysed in [Table-2] such as total soluble solids, reducing sugar, non-reducing sugar, total sugar, ascorbic acid, chlorophyll 'a', chlorophyll 'b', total chlorophyll and carotene. Among all the varieties of Broccoli, variety KE-180 reported maximum chlorophyll 'b' [1.65 mg/lit], carotene [3.13ug/100g], total sugar [3.22%] and reducing sugar [2.91%]. Total chlorophyll was found maximum [1.46 mg/lit] in variety Priya. Variety Prema gave maximum TSS of [11.25*Brix] in compare to other varieties and variety Sultan have relatively high ascorbic acid of [98.5 mg/100gm].

Table 2- Varietal evaluation on Quality attributes of broccoli

Variety	Chlorophyll 'a'(mg/lit)	Chlorophyll 'b'(mg/lit)	Total Chlorophyll (mg/lit)	T.S.S (*Brix)	Carotene (ug/100g)	Ascorbic acid (mg/100 gm.)	Total sugar (%)	Reducing sugar (%)	Non-Reducing sugar(%)
Fiesta	0.0926	0.709	0.514	11.08	2.02	89.7	3.17	2.75	0.42
Princies	0.747	1.333	2.023	8.58	2	82	2.23	1.98	0.25
Sultan	0.334	0.798	0.997	8.42	0.24	98.5	2.34	2.01	0.33
Nokguk	0.158	0.488	0.526	9.17	2.23	89.1	2.12	1.86	0.25
Early you	0.453	1.285	1.452	10	2.19	88.7	2.45	2.08	0.36
KE-180	0.223	1.65	1.218	10.233	3.12	95.8	3.22	2.91	0.32
Priya	0.71	0.349	1.468	11.17	2.38	87	1.93	1.64	0.29
Puspa	0.351	0.808	1.084	8.42	1.71	94.5	2.18	1.91	0.26
Grandsino	0.287	0.312	0.676	10.37	1.99	84.8	2.31	2.02	0.29
Rapido	0.155	0.697	0.623	8.33	1.07	91.5	2.47	2.12	0.35
Prema	0.09	0.246	0.285	11.25	1.96	92	1.83	1.52	0.32
Packman	0.244	1.054	0.959	10.33	2.39	88.1	3.18	2.82	0.36
SEm (±)	0.173	0.2655	0.3564	0.2906	0.2882	1.894	0.0148	0.0223	0.0302
CD at 5%	0.3556	0.5458	0.7326	0.5973	0.5925	2.588	0.0304	0.0457	0.0621

The present experiment revealed that the yield and yield attributing characters significantly differed within the different varieties. On the basis of performance of varieties related to head yield and concerning yield attributing characters, Nokguk performed the highest head yield and other two varieties Early You, and Princes are also considered suitable for positive response for boosting higher yield.

References

[1] McCall D. Sorensen L. and Jensen B.D. (1996) *S.P. Rapport Statens Planteavlsforsog*, 8, 32.

[2] Panse V.G. and Sukhatme P.V. (1985) *Statistical Methods for Agricultural Workers, 4th ed., ICAR, New Delhi*, 347.

[3] Rooster L.de., Callens D., de-Rooster L. (1998) *Proeftuinnieuws*, 8(22), 31-32.

[4] Saha P., Chatterjee R. and Mukhopadhyay D. (2006) *Crop Research Hisar,* 32(3), 396-400.

[5] Vanparys L. (1998) *Proeftuinnieuws*, 8(14), 29-31.

AN INTERACTIVE SPREADSHEET FOR DRIP IRRIGATION SYSTEM UNIFORMITY PARAMETERS EVALUATION

EL-NEMR M.K.

Department of Agricultural Engineering, Faculty of Agriculture, Kafrelsheikh University, Egypt.
*Corresponding Author: Email- moknemr@agr.kfs.edu.eg

Abstract- It is very important to have a fast utility for the irrigation system evaluation to help the system designer and operator to evaluate the system performance and make the reactions towards any system operation problem if exists. Microsoft excel 2010 spreadsheet was successfully created in order to evaluate the drip irrigation system hydraulic performance. The spreadsheet was designed to provide the user with Friendly user interface, easy data input supported with input values check, giving more clarification for emitters' flow distribution supported with charts, and Easy printable outputs Supported with user help work sheet. The uniformity parameters Included Uniformity coefficient, statistical uniformity, manufacturer's coefficient of variation, emitters' flow rate variation, distribution uniformity, and emission uniformity. Emitter exponent also can be calculated by this spreadsheet. The spreadsheet was evaluated by researchers involved with drip irrigation field as excellent rate, beside there was an agreement between spreadsheet calculations and evaluators' in addition to saving the calculation time .
Key words- Drip, Evaluation, Irrigation, Spreadsheet, Uniformity.

Introduction

Drip irrigation system is widely spread as it can be used safely for most soil types; beside it has high theoretical application efficiency. The use of drip irrigation has been increase in most crop commodities, especially vegetable and fruit crops, to improve water use efficiency [1] and water supply [2]. Nowadays how uniform is the distribution of water on land surface is accepted as one of the key criteria for evaluating irrigation system performance [3]. Poor irrigation-water-application uniformity can be a cause of low crop-yields. Improved yields could result from maintaining the soil moisture at an optimum level through more frequent water and nutrient application [4]. This can result from a uniform water application and effective scheduling. The uniform application of water will cause more uniform distribution of water in the root zone as the behavior of applied water affects the soil moisture pattern under point source trickle emitter [5]. Final water application variability in drip system or any other irrigation system will depend on the manufacturing variation and the hydraulic variability [6]. In order to conserve water resources, close attention has to be paid to the performance of irrigation systems [7]. They added that, the regular irrigation system evaluation may assure how maintained is the system and how is its performing according to design. The irrigation system must also be managed correctly and effectively. The drip irrigation system offers the highest irrigation uniformity compared with other irrigation systems [8]. Evaluation of drip irrigation system is important for several reasons .a- the design engineer in establishing whether the desired emitter discharge uniformity specification are being met , b- the irrigation in deciding whether the system can be operated efficiently , c- determining the proper operation of the system and its components to take remedial action as required [9]. Several methods have been developed to aid in design of such a system [9]. These include the Emission uniformity (EU) concept [10], Emitter flow variation concept [11], and statistical uniformity concept [12]. A successful uniform drip irrigation system application depends on the physical and hydraulic characteristics of the drip tubing [13]. Drip irrigation system efficiency depends on application uniformity. In surface drip irrigation systems, uniformity can be evaluated by direct measurements of emitter flow rates. In most cases, irrigation non-uniformity is the major source of reduced crop yields [14, 15]. It is very important to have a fast utility for the irrigation system uniformity evaluation to help the system designer and operator to evaluate the system performance and make the reactions towards any system operation problem if exists. Spreadsheets are easy, widely spread way to make needed fast calculations. One of the beauties of using spreadsheet is that it is possible to setup calculations, change some cell values and look at the effect on the result immediately [16]. Microsoft Excel is used in solving optimization problems of one and several variables without constraints [17]. Excel spreadsheet gives accurate calculations, easy user interface, and Compatible with most computer operating systems. This work introduces a Microsoft Excel spreadsheet that will calculate uniformity measures and emitter exponent to evaluate the drip irrigation system performance. There is numerous models capable of simulating crop behavior under water stress conditions but, none of them takes into consider the effect of irrigation water uniformity on yield [18]. They developed a model to simulate the effect of uniformity on yield and the repercussion on growth margin. University of California introduced an Excel worksheet for the emission uniformity calculation and evaluation [19]. [20] used spreadsheet as a simple conceptual model that the water losses in rice field has been developed and quantified in. [21] developed a simple design procedure for hydraulic analyses and uniformity evaluation and computer simulation model called Back Step Method to verify and evaluate the design procedure. This study indicated that emitter discharge is considered to be a spatial variable, which overcomes the uniform emitter discharge assumption used in the methods presented by [22, 23]. The spreadsheet was designed in order to give the following features:

1. Easy calculation for all uniformity parameters with friendly user interface.
2. Possibility for using wide range of emitters' flow rates samples, in addition to the clarification for emitters' flow distribution supported with charts.
3. Improve the spreadsheet user with values checking ability in order to notice any mistakes before applying his results.
4. Supported with user help beside easy printable outputs.

Methodology

A. Spreadsheet Design

A spreadsheet was created using Microsoft Excel 2010 to evaluate the drip irrigation laterals performance under certain operating pressure head. The spreadsheet contains three worksheets. The first one called inputs and evaluation, on which the required input data, outputs and a chart for emitters' flow rate distribution along lateral. The second one contains a chart to conclude the statistical relationship between emitter flow rate and operating pressure head. The third worksheet is a user guide help worksheet to explain how to use the spreadsheet step by step and clarifying the listed abbreviations and concepts which may face the user of the spreadsheet. Figure.1 a, b, and c shows the three worksheets.

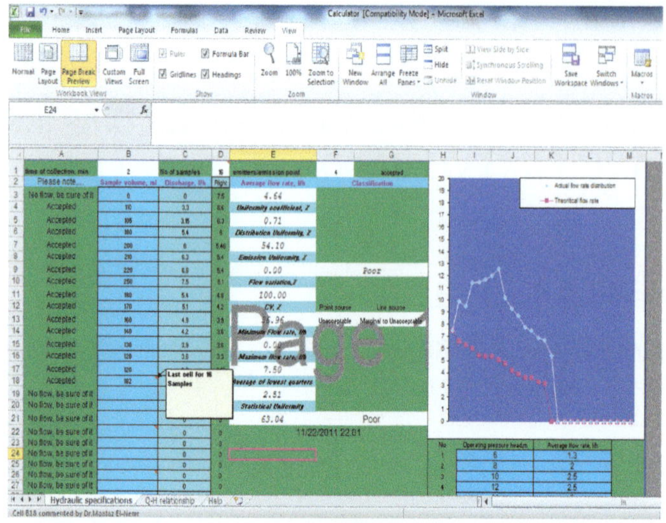

Fig. 1a-

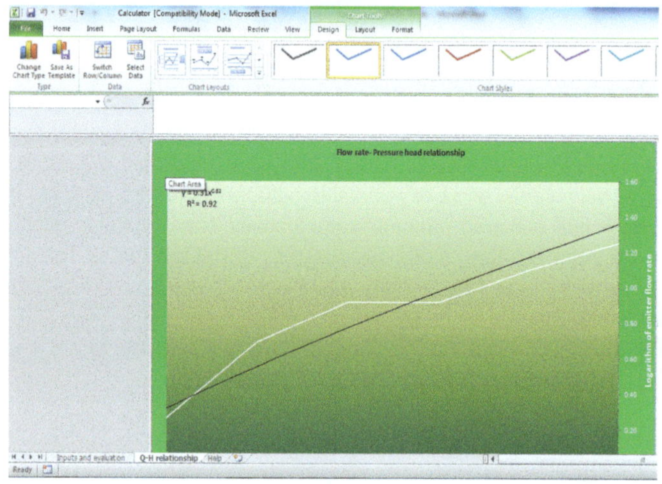

Fig. 1b-

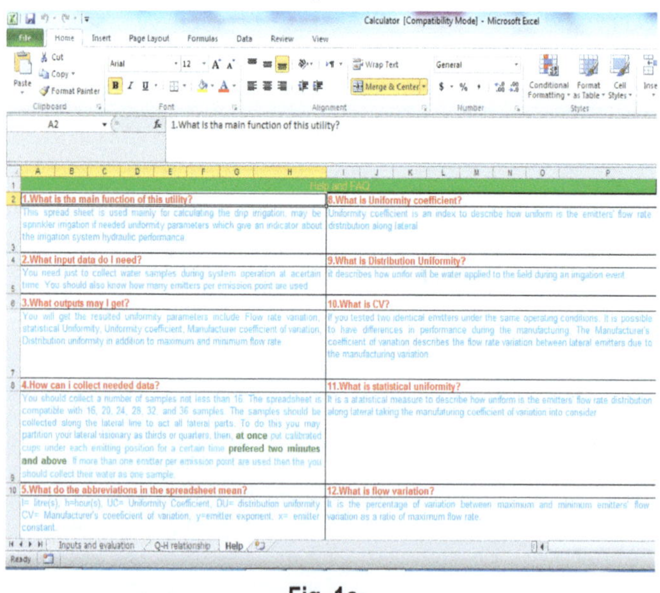

Fig. 1c-

Fig.1- Worksheets used in the evaluation spreadsheet- a- Input data and uniformity parameters calculation worksheet b-Flow rate-pressure head diagram. c- Help worksheet.

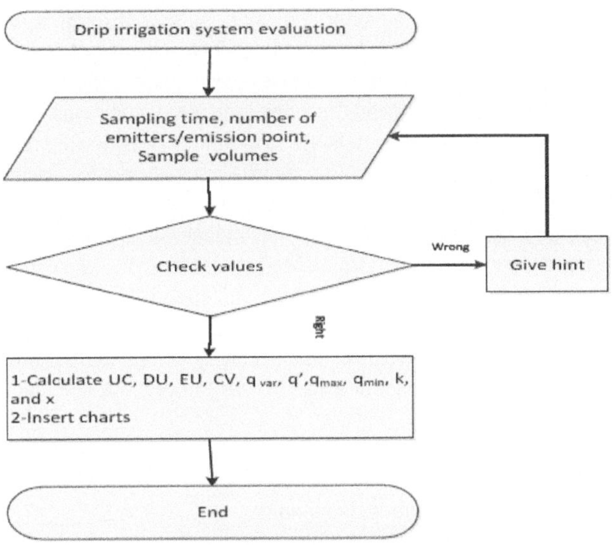

Fig.2- Data flow chart

B. Input data and values checking

The required input data includes number of decided emitters' samples, sampling time, number of emitters per emission point and the collected water samples' volumes. The water samples' volume should be put in ml. The spreadsheet deals with 16, 20, 24, and 28 sample of collected water. If the user reached the last cell he should use for his samples, he will be notified by a comment to stop. At the number of samples input cell, the user will be notified to choose a number from the previously mentioned compatible numbers of samples. If the user tried to input a different number instead of the mentioned numbers, he will be notified by a message that his input is wrong. The spreadsheet is designed to deal with number of emission point greater than zero otherwise the user will be notified to change this value. The sampling time should be entered in minutes or fractions of minutes. If the entered time value is less or equals zero, the cell will turn into red color and strikethrough to notify the user for the error. Depending on samples' volume and sampling time, the flow rate for each emitter will be calculated in lh^{-1} unit and show the flow rates in descending way in dependent cells. Figure.3 shows the input data worksheet.

C. Outputs

The spreadsheet user will calculate and show the Uniformity coefficient, Manufacturers' coefficient of variation. Distribution uniformity, emission uniformity, emitters' flow rate variation, average emitters' flow rate lh^{-1}, descending flow rate distribution along lateral, and average of low quarter emitters' flow rates. Theoretically the flow rate distribution along lateral is descending. A chart describing the flow rate distribution depending on the order of emitters' flow rates along lateral which is put by the user compared to theoretical distribution (Flow rates reduces towards the end of lateral) will be shown to the user in order to help him if he has a problem in a certain emission point like partial clogging or landscape,..etc (Figure 1a).

D. Uniformity Parameters

The uniformity parameters including Uniformity coefficient (UC), Statistical uniformity U_s, Distribution uniformity (DU), Emission uniformity (EU), Manufacturing coefficient of variation (CV), and emitter flow rate variation (q_{var}) were used to evaluate the drip irrigation system performance. The degree of emitter flow variation is expressed by the uniformity coefficient as defined by the following equation [24]:-

$$UC = 1 - \left(\frac{\sum_{i=1}^{i=n} |q_i - q'|}{q' \times n} \right) \times 100 \ldots\ldots\ldots\ldots 1$$

Where n = number of observed emitter or cans, q_i = emitter flow rate, lh^{-1}, q` = average of emitters flow rates, lh^{-1}. Statistical uniformity was calculated using equation 2 [25]. By using the statistical treatment, all of the various factors such as emitter manufacturing variation, lateral line friction, elevation difference, and emitter plugging are included [12].

$$U_s = 100(1 - U_s) \ldots\ldots\ldots\ldots\ldots\ldots 2$$

The emitter manufacturer's coefficient of variation was calculated as follows [10]:-

$$CV = \frac{S_q}{q'} \ldots\ldots\ldots\ldots\ldots\ldots\ldots\ldots 3$$

Where:-

S_q = standard deviation of emitters flow rate.

The emitter manufacturer's coefficient of variation is a measure of the variability of discharge of a random sample of a given make, model and size of emitter, as produced by the manufacturer and before any field operation or aging has taken place [26]. The emitters' flow rate variation was calculated using the following equation [25].

$$q_{var} = \frac{q_{max} - q_{min}}{q_{max}} x\ 100 \ldots\ldots\ldots\ldots 4$$

Where q_{max}= maximum emitter flow rate lh^{-1}, and q_{min}= minimum emitter flow rate , lh^{-1}. Distribution uniformity was calculated using the following equation [27]: -

$$DU = 100x \frac{q'_{lq}}{q'} \ldots\ldots\ldots\ldots\ldots\ldots\ldots\ldots\ldots\ldots 5$$

Where: -

q'_{lq} =mean of lowest one-fourth of emitter flow rates, lh^{-1}.

Emission uniformity expresses the uniformity of emitters under constant pressure [28]. The emission uniformity in percent was calculated by the following formula [29]:

$$Eu = 100(1 - \frac{1.27CV}{N_p^{0.5}}) \frac{q_{min}}{q'} \ldots\ldots\ldots\ldots\ldots\ldots 6$$

Where:-N_p= the number of emitters per plant.

2.3 Flow rate- Pressure head relationship

Flow rate – pressure relationship was calculated by using equation 7[29]:

$$q = kh^x \ldots\ldots\ldots\ldots\ldots\ldots\ldots\ldots\ldots\ldots\ldots\ldots\ldots\ldots 7$$

Where: -

h = operating pressure head in meters, x = emitter discharge exponent, and k = emitter discharge coefficient.

The emitter exponent shows how the emitter flow rate will be affected by pressure variation. The exponent of discharge (x) is affected by the type of emitter, the lower the value of x; the less the discharge will be affected by pressure variation. Emitter with exponent less than 0.5 is considered pressure compensating emitter. When x approaches 1 the emitter is considered laminar flow type or non-compensating emitter [30]. Six operating pressure head and the facing values of emitters' flow rates average will be put in a table in the input data worksheet. A chart describing the relationship between the logarithm of operating pressure head on the horizontal axis and the logarithm of emitter flow rate on the vertical axis will be shown on q-h relationship worksheet (Figure 1b). An exponential trend line will describe the relationship as shown in equation 7. The equation and R squared will be shown on the chart.

E. Uniformity parameters evaluation

U_s, EU, and CV were evaluated in order to evaluate the hydraulic performance of the system at certain operating pressure head. American Society of Agricultural Engineers [31] recommended a classification of manufacturer's coefficient of variation. Table 1 lists the recommendations for classifying the manufacturer's coefficient of variation.

Table 1- Classification of CV values range

Emitter type	Cv range	Classification
Point source	<0.05	Good
	0.05 to 0.10	Average
	0.10 to 0.15	Marginal
	>0.15	Unacceptable
Line source	<0.10	Good
	0.10 to 0.120	Average
	>0.2	Marginal to Unacceptable

The general criteria for EU values are shown in table 2 [32].

Table 2- General criteria for EU

EU, %	Evaluation
90% or greater	Excellent
80-90%	Good
70-80%	Fair
Less than 70%	Poor

The general criteria for an acceptable statistical uniformity coefficient are shown in table 3 [9].

Table 3- Values of statistical uniformity and its evaluation

Us	Evaluation
90% or greater	Excellent
80-90%	Very good
70-80%	Fair
60-70%	Poor
Less than 60%	Unacceptable

F. System Development

A questionnaire was prepared to ask possible interested users to evaluate the spreadsheet (Table 5, Annex A). A Sample of 10 interested persons in drip irrigation study varied between master and Ph.D. degrees holders. This questionnaire also was designed to get a sight for the suggested future work. The spreadsheet purpose and use method was explained before applying the questionnaire. It was asked from the specialists to evaluate how helpful do they expect the spreadsheet will be for farmers and farming objectives. They were also requested to evaluate the interface, how easy is the data input, how are their satisfied to the outputs, outputs printout, and the help worksheet. They were asked to evaluate the previously mentioned indicators as spreadsheet features indicators with a value from 0 to 10. The average of each indicator was used to describe the final user opinion about each system feature indicator. A rate for each indicator values was put and explained to the sample persons to take into consider during evaluation. Table 4 shows the rate which is equivalent for each range of degrees. A case study of 20 water samples was given to the researchers to calculate the uniformity measures and compare the results they got with the spreadsheet results in order to evaluate the spreadsheet accuracy and calculation speed. The average of difference percentage between the spreadsheet and their calculations was used to evaluate the accuracy. They evaluation samples used the spreadsheet also in order to compare calculation time.

Table 4- Rates equivalent to evaluation indicators degrees in users questionnaire

Value range	0-4.9	0.5-6.4	6.5-7.4	7.5-8.5	8.5-100
Rate	fail	Poor	Good	Very good	Excellent

Results

Figure 3 shows the degree of each features indicators. The help worksheet had the highest degree (9.5) while checking values help had the lowest one (7.5). The questionnaire analysis showed that, the researcher found the spreadsheet Excellent for farming objects. The average of all indicators was 8.7 which will give excellent rate for this spreadsheet application. The calculation of the spreadsheet agreed with the calculations of the researchers. The uniformity parameters and evaluation took 82 minutes in average while the calculation with the spreadsheet took 3 minutes in average including printout.

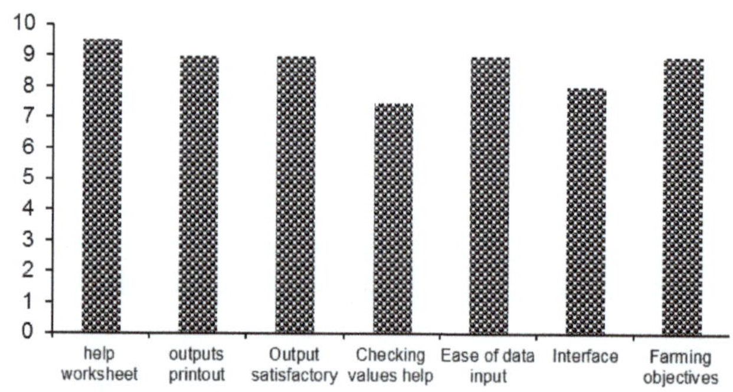

Fig. 3- Spreadsheet evaluation indicators comparison

Discussion

This spreadsheet gave a useful utility to calculate and evaluate the drip irrigation system uniformity parameters. The evaluation of the spreadsheet showed that all the indicators used showed a rate between very good and excellent. The major advantages of using Excel spreadsheet like the possibility of calculations setup, changing cell values and watching the effect [16,17]_was successfully used in this spreadsheet design and achievement. The University of California 2011 introduced a worksheet for emission uniformity calculation and evaluation [19]. In this work the CV, EU, DU, UC, U_s, and q_{var} were calculated in addition to the evaluation of EU, Us, and CV. The emitter exponent calculation and charts support were also provided. Help worksheet which had the highest degree is considered one of the main features of this spreadsheet. This spreadsheet may also help if combined with a crop productivity simulation model in order to put the effect of drip irrigation system uniformity parameters in consider as one of the most important factors affecting the crop yield although it is neglected in most simulation models [18]. The checking values support of this spreadsheet should be improved as it had the lowest degree with using some advanced methods like macros which is supported by the Excel program in order to give the user higher possibility to check his values without making effect on calculations if wrong. The evaluators gave a rate excellent for the farming object which may mean it is useful to ensure the good performance of the drip irrigation system to avoid lack of crops [14,15] and to take reaction towards any design or operating problems if exists [9].

Conclusion

An interactive Microsoft excel 2010 spreadsheet was successfully created in order to evaluate the drip irrigation system hydraulic performance. The spreadsheet was designed to provide the user with Friendly user interface, easy data input supported with input values check, giving more clarification for emitters' flow distribution supported with charts, and Easy printable outputs. The uniformity parameters Included Uniformity coefficient, statistical uniformity, manufacturer's coefficient of variation, emitters' flow rate variation, distribution uniformity, and emission uniformity. Emitter exponent also can be calculated by this spreadsheet. A user guide help sheet is also provided. The spreadsheet was evaluated by drip irrigation interested researchers in the drip irrigation field. It was observed that the checking values feature had the lowest degree which made a very good rate while the help spread sheet had the maximum degree with an excellent rate. The spreadsheet was evaluated in general as excellent. The spreadsheet calculations agreed with evaluators calculations and saved time of calculation time. There should be some improvements on checking values like the spreadsheet should stop calculations with wrong input values. A wider range of emitter samples is also required.

Table 5- Spreadsheet evaluation questionnaire

Name	
Degree	(M.Sc.- Ph.D.)
Do you know how to evaluate drip irrigation system?	
Were you interested in system hydraulic performance during your study?	
If yes how did you calculate them?	Manual – calculator- computer model (Other, please specify)-
Do you find this spreadsheet model helpful for your research field?	
Please give a degree from 0 to 10 for the following items:	
Farming objectives	
Interface	
Ease of data input	
Output satisfaction	
Outputs print out	
Help worksheet	
Accuracy	
Speed	
Suggested future work and modifications.	

References

[1] Jayapiratha V., Thushyanthy M. and Sivakumar S. (2010) *Asian journal of agricultural research* 4(3), 139-147.

[2] Shiri-e-Janagrad M., Tobeh A., Hokmalipour S., Jamaati-e-Somarin S., Abbasi A. and Shahbazi K. (2009) *Asian journal of plant sciences* 8(6),390-399.

[3] Yavuz M.Y., Demırel K., Erken O., Bahar E. and Devecıler M. (2010) *Afr. J. Agric. Res.*, 5: 532-538

[4] Phene C.J. (1991) *Southern African Irrigation Symposium,* 34-41.

[5] Ainechee G., Boromand-Nasb S. and Behazd M. (2009) *Journal of applied science* 9(6), 1170-1174.

[6] Gil M., Rodriguez-Sinobas L., Juana L., Sanchez R. and Losada A. (2008) *Irrig. Sci.*, 26, 451-458.

[7] Ascough G.W. and Kiker G.A. (2002) *Water SA*, 28: 235-242.

[8] Kirnak H., Dogan E., Demir S. and Yalcin S. (2004) *Turk. J. Agric.* For., 28, 223-230.

[9] Bralts V.F. (1986) *Operational Principles. In: Trickle Irrigation for Crop Production Design: Operation and Management,* 216-240.

[10]Keller J. and Karmeli D. (1974) *Trans. ASABE.*, 17(4), 678-684.

[11]Wu I.P. and Gitlin H.M. (1974) *Hawaii Agricultural Experiment Station Technical Bulletin*, 96, University of Hawaii, 29.

[12]Bralts V.F., Wu I.P. and Gitlin H.M. (1981) *Amer. Soc. Agric. Eng.* 24(5),1234-1240.

[13]Al-Amound A.I. (1995) *J. Agric. Eng. Res., 60*: 1-5.

[14]Wu I.P. (1987) *Agric. Water Manage.*, 32: 275–284.

[15]Bhatnagar P. and Srivastava R. (2003) *Irrig. Sci.*, 21: 151-157.

[16]Peker M. and Halat E. (2009) *Journal of applied science.* 9 (7),1301-1308.

[17]Radwan F.A.A. (2002) *Journal of applied science.* 2(1), 92-96.

[18]Lopez-Mata E., Tarjuelo J.M., de Juan J.A., Ballesteros R. and Dominguez A. (2010) *Agric. Water Manage.*, 98: 190-198.

[19]UCCE (2011) *http://cesonoma.ucdavis.edu/.*

[20]Islam M.J., Mowla G., Islam M.S. and Leeds-Harrison P.B. (2003) *Pak. journal of biological science* 6(18), 1600-1607.

[21]Sharaf G.A. (2004) *J. Agric. Eng.*, 21: 103-124.

[22]Keller J. and Bliesner R.D. (1990) *Sprinkler and Trickle irrigation.* Van Nostrant Reinhold, New York.

[23]Wu I.P. and Gitlin H.M. (1975) *J. Irrig. Drainage Div.,* 101: 323 -326.

[24]Christiansen J.E. (1942) *Trans. Am. Soc. Civ. Eng.*, 107: 221-239.

[25]Bralts V.F. and Kesner C.D. (1983) *Drip irrigation field uniformity estimation. Trans. ASABE, 26: 1369-1374.*

[26]ASAE (1996) *American Society of Agricultural Engineer*, 864.

[27]Kruse E.G. (1978) *J. Irrig. Drainage Div.*, 104, 35-41.

[28]ASAE. (1994) *American Society of Agricultural Engineers.*, 724-727.

[29]Keller J. and Karmeli D. (1975) *Rain Bird Sprinkler Manufacturing Corporation*, 133.

[30]James L.G. (1988) *Principles of farm irrigation system design,* John willey & sons, inc.

[31]ASAE (1985) *American Society of Agricultural Engineers.*

[32]Merriam J.L. and Keller J. (1978) *Farm Irrigation System Evaluation: A Guide for Management.*, 271.

GENETIC CONTROL OF SEED SIZE IN COWPEA (*Vigna unguiculata* (L.) Walp)

EGBADZOR K.F.[1,2*], DADOZA M.[3], DANQUAH E.Y.[1], YEBOAH M.[1], OFFEI S.K.[5] AND OFORI K.[4]

[1]West Africa Centre for Crop Improvement, University of Ghana, Legon, Accra, Ghana.
[2]CSIR - Plant Genetic Resources Research Institute, Bunso, E.R. Ghana.
[3]Ministry of Food and Agriculture, Ketu North District, V.R. Ghana.
[4]Department of Crop Science, University of Ghana, Legon, Accra, Ghana.
[5]The Biotech Centre, University of Ghana, Legon, Accra, Ghana.
*Corresponding Author: Email- kenfafa@yahoo.com

Abstract- Seed size contributes to yield and it is also a consumer trait of cowpea in Ghana and other parts of the world. Ghanaian consumers tend to prefer large seeded cowpea to small sizes although, Ghanaian varieties have smaller seeds compared to the imported types. This might be a contributing factor to high importation of cowpea to Ghana. To breed cowpeas with large grains, it is paramount to understand the inheritance of the trait. Generation mean analysis was performed to estimate gene actions on seed size of cowpea using CB27 and Gh3710 as the large and small seeded parents respectively. It was observed that eight genes control seed size and that small seed is partially dominant over large seed. The estimated broad sense heritability was 84% while the narrow sense was 35%. Both additive and non-additive genes with their interactions in the exception of additive - additive had significant effect on seed size inheritance in cowpea. The inheritance of seed size was observed to be complex, it was suggested that combinations of methods such as single seed descent and backcross methods with selection from early generations through advanced level would help improve seed size of cowpea.

Keywords- Additive Gene, Cowpea, Dominance, Generation Mean, Heritability, Seed size

Introduction

Seed size is of interest to many people for the reason that 70-80% of the world's food calories and 60-70% protein consumed by man come from seeds [1]. The effort to increase the number of seeds or their sizes relates to increasing food production [2]. The genetic control of the final seed size is complex as a number of factors, although, discovered (mainly in *Arabidopsis thaliana*), the control of the trait is not well understood. According to Sundaresan [3] three different genes are involved in endosperm size control with *IKU1* acting upstream of *MINI3* and *IKU2*. A factor named GW2 was discovered in rice to induce cell proliferation in grains thereby increasing its size [4]. Li, et al [5] reported that "*DA1* plays a major role in setting final sizes of both seeds and organs by restricting the duration of proliferative growth". In addition to *AUXIN RESPONSE FACTOR2*, *APETALA2*, and *DA1*, Adamski, et al [6] reported *KLU* which acts independently of other maternal factors to influence seed coat cell proliferation. EOD3 and CYP78A9 gene families were reported to control seed size in Arabidopsis in both overlapping and distinct manner [7].

Each of the three major components of a seed namely; the seed coat, the endosperm (cotyledon) and the embryo contribute to its final size. The integument (seed coat) determines the maximum volume the seed can grow [7]. The size of the integument itself is specified by its cell number and size [2]. In most plants, the endosperm is the major contributor to the volume of the mature seed; however, in *Arabidopsis* and some other dicots, the endosperm is consumed, being replaced by the embryo, which then constitutes most towards the mature seed size [3].

Cowpea seed size which is an important component of seed yield is also an essential market trait as consumers in Ghana tend to prefer larger seeds to smaller ones [8]. Survey in some markets in Ghana has revealed that the varieties in cultivation in the country generally smaller seeds than imported ones. It will therefore be imperative to breed larger seeded cowpea. However, the mode of inheritance of the trait must first be understood before setting it as a breeding goal.

Inheritance of seed size of cowpea has been of interest to plant breeders for very long time. Drabo, et al [9] indicated that contradictory reports have been given on inheritance of seed size in cowpea since 1950. They also reported that cowpea seed weight (size) is governed by many genes acting mainly additively with small size partially dominant over large. Additive gene effect is desirable as progress can be made through selection. Yilwa [10] reported large seed to be dominant over small seed, continuing the contradiction and therefore warranting the study of the trait among genotypes available to the individual for cowpea improvement purposes.

Cowpea seed is formed through the process of double fertilization as in angiosperms. The size of the seed formed following double fertilization is affected by maternal, paternal and positional effects jointly [11]. The positional effect is the contribution of the zygote to the seed size.

Relationship between genetic effects and the way traits are transmitted from parents to their progenies is important in crop improvement. Several genetic designs have been developed to measure the relative importance of additive, dominance and epistasis which are the main gene actions in the expression of quantitative traits like seed size. Generation mean analysis proposed by Mather and Jinks is a design favored by breeders and geneticists working with self-pollinated species [12].

Materials and Methods

Cowpea genotypes, CB27 and Gh3710 with respective seed weights of - and - were used for the experiment. CB27 was obtained from the University of California Riverside while Gh3710 was obtained from CSIR-PGRRI. Both genotypes flowered about the same number of days after planting (32 days). The parents were crossed and subsequently the generations, F_1, F_2, BC_1, BC_2 together with the parents used for generation mean analysis for seed size. The experiment was conducted at Ohawu [Latitude = 6*03'50.74812"N; Longitude = 0*54'49.85784"E] in the coastal savanna of Ghana in 2013.

The experiment was conducted on field under rain-fed condition. The experiment was laid in a randomized complete block design with 10 replications for the parents and F_1 populations while the Backcrosses and F_2 were replicated 30 and 60 times respectively. The between row spacing was 100 cm and the within row 50 cm, while between block was 200 cm.

One hundred seed mass per plant was weighed with a digital scale in grams.

Results

Differences in Seed Weight of Various Generations

Analysis of variance showed significant differences between 100 seed weight of the various generations of plants used as treatments. Boxplot of the seed weight is shown in [Fig-1].

The means and standard errors (SE) of the various generations are shown in [Table-1].

Table 1- Average of 100 seed weight (g) and standard errors for various cowpea generations with a matrix showing LSDs (5%) with non-significant pairs indicated with asterisks (*)

Generation	Mean ± SE	LSD					
1 (BCP_1)	16.79 ± 0.2340	-					
2 (BCP_2)	13.52 ± 0.2340	0.63	-				
3 (F_1)	15.92 ± 0.4243	0.97*	0.97	-			
4 (F_2)	15.98 ± 0.1892	0.63	0.63	0.97*	-		
5 [CB27 (P_1)]	22.58 ± 0.4243	0.97	0.97	1.1	0.97	-	
6 [Gh3710 (P_2)]	11.20 ± 0.4243	0.97	0.97	1.1	0.97	1.1	-
	Generation	1	2	3	4	5	6

Note: The number of observations for P_1 P_2 and F_1 = 10; BCP1 and BCP2 = 30 and F_2 = 60

Matrix pairing of the generations is also shown in the Table and the generations with significant differences in 100 seed weight indicated with asterisks. There were only two non-significant pairs of generations; F_1 against F_2 and BCP_1.

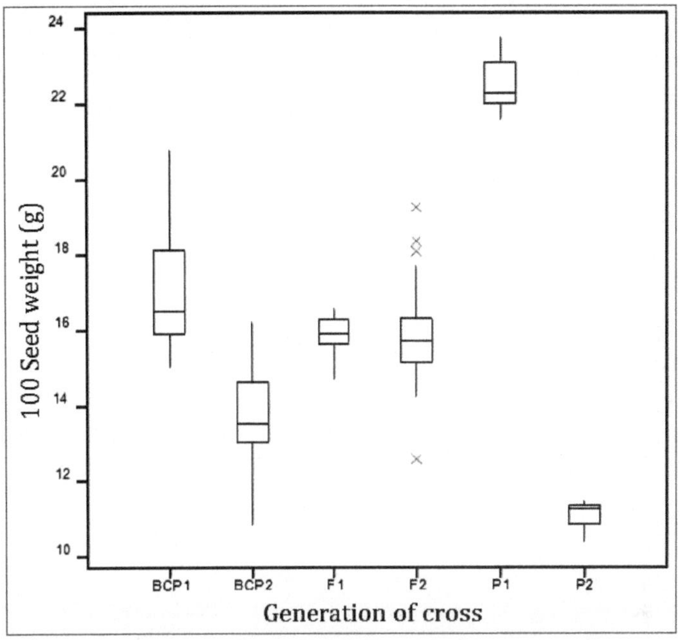

Fig. 1- Boxplot of 100 seed weight (g) of different generation of cowpea

Heritability of 100 Seed Weight in the Broad Sense

Broad sense heritability (H_b^2), of 100 seed weight of cowpea was estimated by the formula of Ito & Brewbaker [13] as shown in [Eq-1].

$$H_b^2 = (VF_2 - VE) / VF_2: \quad (1)$$

where,

H_b^2 = Broad sense heritability

VE = Error variance = 1/4 ($VP_1 + VP_2 + 2VF_1$) = 0.39255

VF_2 = Variance of F_2 family = 2.4722

VP_1 = Variance of parent 1 = 0.3686

VP_2 = Variance of parent 2 = 0.2084

VF_1 = Variance of F_1 family = 0.4966

H_b^2 = 0.84

The estimated broad sense heritability for 100 seed weight is therefore 84%.

Narrow Sense Heritability

The method of Warner [14] was used to estimate narrow-sense heritability (h^2ns) [Eq-2]

$$h^2ns = [2VF_2 - (VBCP_1+VBCP_2)] / VF_2 \quad (2)$$

where,

VF_2, $VBCP_1$ and $VBCP_2$ are the variances of the F_2, CB27/F_1 and Gh3710/F_1 respectively.

VF_2 = 2.472

$VBCP_1$ = 2.54

$VBCP_2$ = 1.539

h^2ns = 0.35

The estimated narrow sense heritability for 100 seed weight is therefore, 35%.

Number of Effective Factors (Genes)

The number of effective factors controlling seed weight was estimated using the method of Burton [15], shown in [Eq-3]

$$k = [0.25(0.75 - h + h^2)\, D^2] / (VF_2 - VF_1) \qquad (3)$$

where,

$D = P_1 - P_2: = 11.38$

$h = (F_1 - P_2) / P_1 - P_2 = 0.4148$

k = minimum number of effective factors\

VF_2 = Variance of F_2 population = 2.47222

V^2F_1 = Variance of F_1 population = 0.4966

P_1 = mean of parent 1 = 22.58

P_2 = mean of parent 2. = 22.58

Number of effective factors controlling seed mass = 8

The Degree of Dominance

The degree of dominance is the deviation from the mid-parent value towards one of them as illustrated [Fig-2].

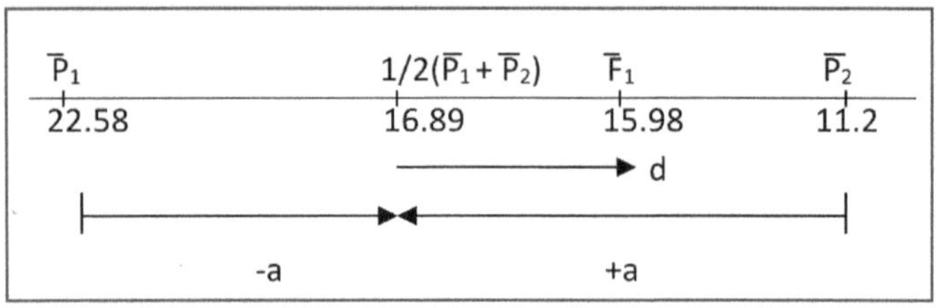

Fig. 2- Diagrammatic representation of the genotypic values based on Falconer [16]

d = F1 – ½ (P1 +P2)

= 15.98 – ½ (22.58 + 11.2)

= - 0.91 P2 is dominant over P1 making d negative

a = ½ (P1 – P2)

= 5.69

Degree of dominance = d/a = 0.1599 = 16%

The mean value, that is, the F1 value is closer to the smaller seeded variety than to the larger which means that the former is the dominant variety.

Gene Action Controlling 100 Seed Weight

Generation mean analysis was used to estimate genetic control of seed weight in cowpea

The GMA model is stated below:

$$\mathbf{Y = m + \alpha a + \beta d + \alpha^2 aa + 2\alpha\beta ad + \beta^2 dd}$$

α and β are the coefficients for a and d, respectively.

Y = the observed mean

m = mean = mean of the F2

a = pooled additive effects

d = pooled dominance effects

aa = additive x additive gene interaction effects

ad = additive x dominance gene interaction effects

dd = dominance x dominance gene interaction effects

The mode of inheritance of 100 seed weight in cowpea was estimated by generation mean analysis with six generations (P_1, P_2, F_1, F_2, BCP_1 and BCP_2) of CB27 and Gh3710. Following significant differences in 100 seed weight of the various generations, their means and variances were used to perform generation mean. Vital results from regression analysis from SAS are presented in [Table-2]. The generation mean analysis indicated that there were additive and dominance gene effects and their interactions in the inheritance of the trait.

Table 2- Analysis of variance and parameter estimates for genetic control of 100 seed weight of cowpea obtained through generation mean analysis with CB27 and Gh3710.

Variable	DF	Parameter Estimate	Standard Error	t Value	Pr > |t|	Type I SS	Type II SS
Intercept	1	18.747	1.722	10.89	<.0001	100712	487.836
Rep	1	0.008	0.012	0.65	0.5151	0.743	1.753
a	1	-5.678	0.217	-26.21	<.0001	3196.69	2828.02
d	1	-9.706	3.469	-2.8	0.0058	39.062	32.234
aa	1	-1.99	1.682	-1.18	0.2387	36.671	5.763
ad	1	4.775	0.674	7.09	<.0001	200.316	206.906
dd	1	6.61	1.917	3.45	0.0007	48.964	48.964

Note: (a) additive; (d) dominance; (aa) additive-additive; (ad) additive-dominance; (dd) dominance-dominance gene effects.

Discussion

Significant differences were observed between the various generations of plants from the cross of CB27 and Gh3710 in terms of 100 seed weight. The means of F_1, F_2 and the backcrosses fell between the two parental means of 11.2g and 22.58g, showing no transgressive segregates. This result is similar to that of Lopez, et al [17], although, comparably each of their generations had smaller weight with their small and large seeds having 100 seed weight of 4.82g and 15.14g respectively. Lucas, et al [18] on the other hand used slightly heavier genotypes (11.6g and 26.41g) as the parents in QTL studies for seed size.

The broad sense heritability estimated for 100 seed weight in this research was 84%. This value is very close to 83.3% reported [19]. Drabo, et al [9] also reported heritability estimate of 85% while Manggoel, et al [20] reported 86.84%. Omoigui, et al [21] and Tchiagam, et al [22], however, reported broad sense heritability as high as 98.9% and 90% respectively for the trait. Egbadzor, et al [23], however, used association mapping with SNP markers and reported 96% broad sense heritability for 100 seed weight in cowpea.

The low narrow sense heritability observed may have been caused by the significant additive dominance and dominance dominance epistatic effects. However, such low narrow sense heritability has been reported: Adeyanju, et al [24] for example reported 15 and 25% from two different crosses. Relatively higher narrow sense heritability for 100 seed weight of 67% was reported for cowpea by Tchiagam, et al [22]. These reports suggest that breeding methodology to improve seed size in cowpea may be influenced by the genotype and environment in which the breeding is to take place.

Findings in this experiment confirm that of Drabo, et al [9] that small seed size is partially dominant over larger seed. Partial dominance genetic variance was evident in that the F_1 of CB27/Gh3710 resembled Gh3710 more than CB27 in terms of seed size. This means that small seed size is partially dominant over large seed. Dominance gene effect was observed to be important in seed size inheritance in lentil as well [25]. However, the finding in this experiment was contrary to the observation of Lopez, et al [17] who stated that large seed in cowpea is rather partially dominant over the small. These contradictory findings may be due to the genotypes used in the various studies. It was also observed that seed from none of the generations namely: P_2, F_1, F_2, BCP_1 or BCP_2 equaled the weight of P_1 (CB27). Similarly, Fatokun, et al [19] had no F_2 seeds as heavy as the heavy parent used in their experiment.

The number of gene pairs estimated to control seed weight was eight, the exact number reported by Drabo, et al [9] in spite of different genotypes used in the separate trials. Lopez, et al [17], however, estimated five factors. The genetic effect of these genes, being it additive, dominant or their interactions would determine progress that could be made in breeding.

Additive and non-additive gene effects with their interactions were shown to significantly control seed size in cowpea in the exception of additive - additive (aa) in the current experiment. This means that (a), (d), (ad) as well as (dd) were significant. However, the magnitude of the significant fixable gene action (a) has smaller magnitude compared with the non-fixable (d), (ad) and (dd)) [26]. Gene interaction observed was "duplicate" because (d) was negative while (dd) was positive [25]. The negative (d) also shows that small seed is dominant over large seed. The presence of two or more of the eight factors estimated to control seed size in this experiment would act in duplicate manner [27] to affect the trait. Although, (aa) gene effect was not significant, the associated negative sign indicates the dispersion of seed weight genes in both parents (CB27 and Gh3710).

Significant (a) gene effect is an assurance that gains can be made with selection in the long term [28] and in this case for seed size in cowpea. Seeds of the backcross generations in this experiment are indication of the practicality of this principle by being significantly larger or smaller than the F_1s. The backcross to the larger seeded parent (CB27) resulted in larger seeds while the F_1 crossed to the small seeded parent (Gh3710) were also smaller than the F_1.

Both fixable and non-fixable gene effects are important and therefore make breeding for cowpea seed size complex. The presence of significant (a) gene effect would make selection from early generation effective while selection at advanced generations would also be effective because of significant (ad) and (dd). The implication is that selection for seed size for cowpea should commenced from the segregation population through advanced level where varieties are at evaluation stage.

Conclusion

It was observed in the experiment that eight genes control 100 seed weight in cowpea using the genotypes CB27 and Gh3710 as parents in generation mean analysis. Small seed size in partially dominant over large seed. Additive, dominant, additive - dominant and dominant - dominant gene actions were found to be significant. The inheritance of seed weight is complex and suggesting that a combination of different approaches to improve the trait is a panacea to succeed contrary to the use of a single approach. Such methods as combination of single seed descent and backcrosses with selection from early generation through to advanced state may be used to make progress.

Acknowledgement

The authors thank Alliance for a Green Revolution in Africa (AGRA) for providing the financial support for this research. They also expressed appreciation for the support given by staff of Ketu - North District of the Ministry of Food and Agriculture, Ghana, especially their director, Mr. Samuel Laryea - Tetteh. Last but not the least, staff of CSIR - PGRRI, notably, Messrs.' Daniel Ashie Kotey and Dickson Gamedoagbao are commended for their support.

Conflict of Interest: None declared.

References

[1] Lu J., Zhanga C., Baulcombe D.C. and Chen Z.J. (2012) *Proceedings of the National Academy of Sciences*, 109(14), 5529-5534.

[2] Herridge R.P., Day R.C., Baldwin S. and Macknight R.C. (2011) *Plant Methods*, 7(1), 3.

[3] Sundaresan V. (2005) *Proceedings of the National Academy of Sciences,* 102(50), 17887-17888.

[4] Song X., Huang W., Shi M., Zhu M. and Lin H. (2007) *Nature Genetics*, 39(5), 623-630.

[5] Li Y., Zheng L., Corke F., Smith C. and Bevan M.W. (2008) *Genes Dev.*, 22(10), 1331-1336.

[6] Adamski N.M., Anastasiou E., Eriksson S., O'Neill C.M., Lenhard M. (2009) *Proceedings of the National Academy of Sciences*, 106(47), 20115-20120.

[7] Fang W., Wang Z., Cui R., Li J., Li Y. (2012) *The Plant Journal*, 70(6), 929-939.

[8] Langyintuo A.S., Lowenberg-DeBoer J., Faye M., Lambert D., Ibrod G., Moussa B., Kergna A., Kushwaha S., Musa S. and Ntoukam G. (2003) *Field Crops Research*, 82(2) 215-231.

[9] Drabo I., Redden R., Smithson J.B. and Aggarwal V.D. (1984) *Euphytica*, 33, 929-934.

[10]Yilwa V. (2012 *Genetics of Cowpea - Inheritance of Qualitative and Quantitative Traits*, Lambert Academic Publishing, 175-178.

[11]House C., Roth C., Hunt J. and Kover P.X. (2010) *Proceedings of the Royal Society B: Biological Sciences*, 277(1695), 2885-2893.

[12]Piepho H.P and Mohring J. (2010) *Crop Science*, 50, 1674-1680.

[13].Ito G.M. and Brewbaker J.L. (1991) *J. Amer. Soc. Hort. Sci.*, 116, 1072-1077.

[14]Warner J.N. (1952) *Agron. J.*, 44, 427-430.

[15]Burton W.G. (1951) Agronomy Journal, 43, 409-417.

[16]Falconer D.S. and Mackay T.F.C. (1996) *Introduction to Quantitative Genetics*, Pearson Education ltd, England.

[17]Lopes F.C.C., Gomes R.L.F. and Filho F.R.F. (2003) *Scientia Agricola*, 60(2), 315-318.

[18]Lucas M.R., Huynh B.L., Vinholes P.S., Cisse N., Drabo I., Ehlers J.D., Roberts P.A., Close T.J. (2013) *Frontiers in Plant Sciences*, 4(95), doi: 10.3389/fpls.2013.00095.

[19]Fatokun C.A., Menancio-Hautea D.I., Danesh D. and Young N.D. (1992) *Genetics*, 132, 841-846.

[20]Manggoel W., Uguru M.I., Ndam O.N. and Dasbak M.A. (2012) *Journal of Plant Breeding and Crop Science*, 4(5), 80-86.

[21]Omoigui L.O., Ishiyaku M.F., Kamara A.Y., Alabi S.O., Mohammed S.G. (2006) *African Journal of Biotechnology*, 5(13), 1191-1195.

[22]Tchiagam J.B.N., Bell J.M., Ngakeu D.F., Njintang N.Y. and Youmbi E. (2011) *Agric. Biol. J. N. Am.*, 2(4), 698-707.

[23]Egbadzor K.F., Yeboah M., Danquah E.R., Ofori K and Offei S.K. (2013) *Int. J. Plant Breed. Genet.*, 7(2), 115-123.

[24]Adeyanju A.O., Ishiyaku M.F., Echekwu C.A. and Olarewaju J.D. (2012) *African Journal of Biotechnology*, 11(46), 10473-10483.

[25]Khodambashi M., Bitaraf N. and Hoshmand S. (2012) *J. Agr. Sci. Tech.*, 14, 609-616.

[26]Hakizimana F., Ibrahim A.M.H., Langham M.A.C., Rudd J.C. and Scott D. (2004) *Euphytica*, 139, 133-139.

[27]Suzuki D.T., Griffiths A.J.F., Miller J.H. and Lewontin R.C. (1989) *An Introduction to Genetic Analysis*, W.H. Freeman and Company, New York. P83.

[28]Acquaah G. (2007) *Principles of Plants Genetics and Breeding*, Blackwell Publishing.

[29]Fatokun C.A. (2002) *Proceedings of the World Cowpea Conference III*, IITA, Ibadan, Nigeria.

[30]Park S.O., Coyne D.P., Jung G., Skroch P.W. Arnaud-Santana E., Steadman J.R., Ariyarathne H.M. and Nienhuis J. (2000) *J. Amer. Soc. Hort. Sci.*, 125(4), 466-475.

EVALUATION OF AGRONOMIC PERFORMANCE OF DROUGHT - TOLERANT QTL INTROGRESSION HYBRIDS OF MILLET (*Pennisetum glaucum* L R. BR.) IN THE GUINEA SAVANNAH ZONE OF GHANA

DEMUYAKOR B.[1]*, GALYUON I.[3], KYEREH S.[2], AHMED M.[2]

[1]Department of Biotechnology, Faculty of Agriculture, University for Development Studies, Box TL 1882, Tamale, Ghana.
[2]Department of Agronomy, Faculty of Agriculture, University for Development Studies, Box TL 1882, Tamale, Ghana.
[3]Department of Molecular Biology and Biotechnology, University of Cape Coast, Cape Coast, Ghana.
*Corresponding Author: Email- bawainghana2@yahoo.co.jp

Abstract- The development of crops adapted to the semi-arid and arid regions would enhance food security in Africa. As part of research towards ensuring food security, field trials were carried out to evaluate the performance of sixteen hybrids of drought tolerant Quantitative trait loci (QTL) introgression lines of pearl millet in a Guinea-Savannah zone of Ghana, where the crop is one of the cereals popularly cultivated by farmers as a staple. The hybrids were obtained from ICRISAT in India and the validation trial was laid out in a randomized complete block design with (4) replicates. All the sixteen lines (843A × H77/833-2, ICML03048 × H77/833-2, ICML03050 × H77/833-2, ICML03051 × H77/833 -2, ICML03052 × H77/833-2, ICML03056 × H77/833-2, ICML03057 × H77/833-2, 841B × H77/833-2, 863B × H77/833-2, 843A × ICMR01029, 843A × ICMR01031, 843A × ICMR 01040, 843A × ICMR 01046, 843A × ICMR 01042, 843A × ICMR 01044 and 843A × PRLT2/89-33) were able to withstand the environmental conditions during the trial period. Three entries (843A × PRLT 2/89-33, 843A × ICMR 01046 and 843A × ICMR 01029) took fewer days to boot. Two hybrids (843A × H77/833-2 and 843A × PRLT 2/89-33) had the highest values of 100-seed weight. The hybrid 863B × H77/833-2 was outstanding in terms of plant height, fresh and dry stover weights, ear head length and diameter; 100-seed weight, head yield and grain yield. It is, therefore, concluded that marker-assisted selection could be used to breed for high yielding millet cultivars in drought prone savannah agro-ecological zones of northern Ghana to enhance yields and improve food security.

Keywords- Food security, QTL introgression, pearl millet, yield

Introduction

Millet is a collective term referring to a number of small seeded annual grasses that are cultivated as grain crops, primarily, on marginal lands in dry areas in temperate and subtropical regions. The most important species are pearl millet (*Pennisetum spp.*), finger millet (*Eleusinian caracara*), proso millet (*Panicum miliaceum*), and foxtail millet (*Setaria italica*). Pearl millets (*Pennisetum glaucum, P. typhoides, and P. americanum*) are the most widely grown and accounts for 50% of global millet production [1]. *Pennisetum glaucum* (L) R. Brown, is commonly known as pearl, bulrush, cattail, or spiked millet in English; *bajra* in Hindi; *kambu* in Tamil, and *dukhn* in Arabic, and *kyi* in Dagaare. Among the cereals, it is the crop of the driest region. It is drought-resistant and yields well on infertile sandy soils which could be unsuitable for other important cereal groups. It thrives well on light-textured and well-drained soils, and does not tolerate water-logging and flooding [2-3]. Millet is grown in areas with a rainfall of 125-900mm[4]. Hence, the crop is normally reserved for areas where maize and sorghum fail [5], because of low rainfall or adverse soil factors [6]. Millet grain is nutritious and form a major staple food for over 500 million people in some parts of Africa and India [2].

Pearl millet provides cheap source of protein, minerals, and vitamins to the poorest of the poor who need such ingredients the most. The dehusked pearl millet has a protein content of 12.5% and calcium (344 mg/100 g) compared to wheat grain (11.8%) and (41 mg/100 g), respectively. After dehusking, the grain may be cooked like rice or ground into flour to make porridge, or cake. Also, the grain is used to produce malt, or feed livestock and poultry, while the green plants provide a useful fodder. The straw may be fed to livestock, or used for bedding, thatching, fencing, and fuel [7]. In Ghana, two varieties- early and late maturing millets- are grown in the Northern or Guinea/Sudan savanna zones. The early millets, which are usually cultivated to prevent prolonged hunger from the lean season, are susceptible to drought spells. This crop usually withers and dies off when there is low rainfall which widens the hunger gap between the rainy and dry seasons. For this reason, hybrids of drought-tolerant QTL introgression lines of millets were developed through international collaborative research works. However, the performance of these varieties needs to be assessed in the growing regions before introducing them to farmers. The objective of this field trial was to evaluate and identify which of the new drought-tolerant QTL introgression hybrids of millet would best withstand drought, give best growth parameters and grain yields in the Guinea Savannah zone of Ghana, where the crop is popularly grown for food.

Materials and Methods

Experimental Site and Agro-Climatic Details

The study was conducted at the experimental field of the Faculty of Agriculture, University for Development Studies, Nyankpala Campus, Northern Region of Ghana. The field is located on latitude 9°25"N, longitude 0°58"W and at an altitude of 183m above sea level [8]. The area falls within the Guinea Savannah zone. The soil is classified as a Haplic Lixisol and locally referred to as the 'Kpalsawgu' series, which consist of yellow-brown clay and silt, developed from local colluviums [9]. The rainfall pattern is unimodal, which starts from May to October with an average annual value of 1043 mm. The average annual temperature and humidity are 28.3° C and 58%, respectively.

Plant Culture and Treatments (Experimental Crops)

Sixteen (16) hybrids of drought-tolerant QTL introgression lines of pearl millet [Table-1] were obtained from the International Crop Research Institute of Semi-Arid Tropics (ICRISAT) in India.

Table 1- Sixteen drought-tolerant QTL introgression hybrids of pearl millet

Entry Number	Entry Name
1	843A × H77/833-2
2	ICML 03048×H77/833-2
3	ICML 03050×H77/833-2
4	ICML 03051×H77/833-2
5	ICML 03052×H77/833-2
6	ICML 03056×H77/833-2
7	ICML 03057×H77/833-2
8	841B × H77/833-2
9	863B × H77/833-2
10	843A × ICMR 01029
11	843A × ICMR 01031
12	843A × ICMR 01040
13	843A × ICMR 01046
14	843A × ICMR 02042
15	843A × ICMR 02044
16	843A×PRLT 2/89-33

QTL= Quantitative trait loci.

Land Preparation and Plot Layout

The field was ploughed, harrowed twice and ridged with a farm tractor. The experiment was laid out in a randomized complete block design (RCBD) with four replications. In each replication, there were 16 plots with each plot measuring 2m x 1.2m (2.4 m^2) and 1m between adjacent plots and blocks.

Planting and Other Management Practices

The sixteen drought-tolerant QTL introgression hybrids were randomly assigned to the plots in each block. About 6-8 seeds of each hybrid were sown per hill on 16th August, 2008 and later thinned (two weeks after emergence) to three seedlings per hill at a spacing of 75cm between rows x 15cm within rows. The total plant population was 54 plants $plot^{-1}$. Four weeks after planting (4WAP) NPK, 15 -15-15 fertilizer was applied at the rate of 42 g $plot^{-1}$ and 2 weeks later (6WAP) urea was applied as top dressing at the rate of 30 g $plot^{-1}$ to meet the crops' total nitrogen requirement. Weeding was done manually using a hoe at 5 WAP.

Rainfall Distribution

Daily rainfalls were recorded in millimeters in order to ascertain the rainfall distribution during the growing period of the crop.

Measurement of growth Characteristics

Three weeks after planting (3WAP), 5 plants were randomly selected from each plot and tagged for the measurement of growth characteristics at 3, 6 and 9 WAP.

Plant Height (cm)

Plant height was measured from the base to tip at 3, 6 and 9 WAP. The mean height from the 5 randomly selected plants was taken as the score for each plot.

Fresh and Dry Stover Weights per Plant (g)

Five plants were randomly selected from each plot at harvest and separated into leaf and stem materials and the fresh weight taken immediately using a balance (Sartorius, TE 612). The mean of the 5 selected plants was computed and used as the fresh stover weight for each plot. The dry stover weight was determined by drying the plant parts to constant weight in an oven set at 80 °C and the mean dry weight computed for each plot.

Days to 75% flowering

The plants were monitored daily and the number of days taken in each plot for 75% of the millet plants to flower recorded.

Measurement of Yield Traits

The millet plants were harvested at physiological maturity (14 weeks after planting) at the hard dough stage and the pertinent yield data recorded.

Ear Head Length and Diameter (cm)

The ear head lengths of the 5 randomly selected millet plants were measured separately and the mean computed for each plot. Head diameter was determined using a measuring tape wrapped around the head to find the circumference (cm), which was then divided by π (3.14) to get the diameter of each head. The average diameter was computed for each plot.

Hundred-Seed Weight (g)

Seeds were obtained from the five heads randomly selected from each plot. One hundred (100) seeds were then selected randomly from each head and the weight determined. The average weight was determined and used as the hundred seed weight for the plot.

Head Yield (g)

All millet heads in a plot were carefully harvested and weighed. The average head yield for each plot was determined by dividing the total yield (g) by the number of heads for each plot.

Grain Yield per Plot (kg)

All millet heads in each plot were carefully harvested, threshed using a threshing machine and the chaff separated from the grain. The grains collected for each plot were then weighed and recorded as the grain yield for the plot.

Statistical Analysis

Data obtained was subjected to analysis of variance (ANOVA) using GENSTAT Discovery statistical software (Third Edition). Where significant differences were found, means were separated using the least significant difference (LSD) at 5% level of significance.

Result and Discussion

Plant Height

Plant heights at 3, 6 and 9 WAP varied significantly ($p < 0.05$) among the hybrids [Table-2]. The average plant heights at 3, 6 and 9 WAP were 3.8, 44.2 and 85.6cm, respectively. Some hybrids (843A × H77/833-2, ICML 03051×H77/833-2, ICML 03052×H77/833 -2, ICML 03057×H77/833-2, 841B × H77/833-2, 863B × H77/833-2, 843A × ICMR 01031, 843A × ICMR 01046 and 843A × ICMR 02044) were taller than the average height while others (ICML 03048×H77/833-2, ICML 03050×H77/833-2, ICML 03056×H77/833-2, 843A × ICMR 01029, 843A × ICMR 01040, 843A × ICMR 02042 and 843A×PRLT 2/89-33) were shorter than the average height of 85.6cm at 9 WAP. The hybrids 843A × H 77/833-2, ICML 03051×H 77/833-2, ICML 03052×H 77/833-2, 863B × H 77/833-2, 843A × ICMR 01046, 843A × ICMR 02044 were significantly ($p < 0.05$) taller than the others at 9 WAP. The differences among the plant heights could be attributed to variations in genotype.

Table 2- Plant heights (cm) of drought - tolerant QTL introgression millet hybrids at 3, 6 and 9 WAP

Entry Number	Entry name	PH(3WAP)	PH(6WAP)	PH (9WAP)
1	843A × H 77/833-2	3.7^{ab}	65.2^{a}	90.2^{ab}
2	ICML 03048×H 77/833-2	3.2^{b}	28.1^{b}	80.0^{ab}
3	ICML 03050×H 77/833-2	3.4^{ab}	39.1^{b}	82.4^{ab}
4	ICML 03051×H 77/833-2	4.1^{ac}	39.3^{b}	97.3^{ab}
5	ICML 03052×H 77/833-2	3.4^{ab}	47.5^{a}	93.5^{ab}
6	ICML 03056×H 77/833-2	3.6^{ab}	30.7^{b}	84.0^{ab}
7	ICML 03057×H 77/833-2	3.7^{ab}	33.7^{b}	87.6^{ab}
8	841B × H 77/833-2	3.8^{ab}	41.0^{b}	85.9^{ab}
9	863B × H 77/833-2	4.1^{ac}	36.9^{b}	105.8^{a}
10	843A × ICMR 01029	3.9^{ab}	39.1^{b}	71.9^{b}
11	843A × ICMR 01031	4.4^{ac}	59.0^{a}	89.2^{ab}
12	843A × ICMR 01040	3.6^{ab}	55.8^{a}	83.8^{ab}
13	843A × ICMR 01046	4.9^{c}	56.1^{a}	95.7^{ab}
14	843A × ICMR 02042	3.7^{ab}	44.1^{b}	44.1^{c}
15	843A × ICMR 02044	4.2^{ac}	62.6^{a}	96.5^{ab}
16	843A×PRLT 2/89-33	3.8^{ab}	28.5^{b}	81.0^{ab}
SED		0.4248	11.08	13.26
P-value		0.037	0.012	0.024

Means within a column with different superscripts are significantly (P < 0.05) different. PH = Plant height; WAP = Weeks after planting; SED = Standard error of differences of means.

Fresh and Dry Stover Weights

The biomass yields (fresh and dry stover weights) of the hybrids of pearl millets are shown in [Table-3]. The hybrids did not vary significantly ($P > 0.05$) in terms of their fresh stover and dry stover weights. The average fresh and dry weights recorded were 382.8 g and 283.6 g, respectively. Even though there was no significant ($p > 0.05$) difference between the hybrids, eight hybrids (lines ICML 03050×H77/833-2, ICML 03051×H77/833-2, ICML 03056×H77/833-2, 863B × H77/833-2, 843A × ICMR 01029, 843A × ICMR 01046, 843A × ICMR 02044, and 843A×PRLT 2/89-33) had fresh stover weights above the mean score while six of the hybrids (ICML 03050×H77/833-2, ICML 03051×H77/833-2, ICML 03056×H77/833-2, 863B × H77/833-2, 843A × ICMR 02044, and 843A×PRLT 2/89-33) had dry stover weight below the mean. The insignificant difference in fresh stover and dry stover weights among the millet lines is an indication that these traits were not influenced by the QTL introgressed.

Table 3- Fresh and dry Stover weights of drought-tolerant QTL introgression hybrids of pearl millet

Entry Number	Hybrid	Fresh Stover Weight	Dry Stover Weight
1	843A × H77/833-2	368.5	280
2	ICML 03048×H77/833-2	370	261.8
3	ICML 03050×H77/833-2	394.5	296.5
4	ICML 03051×H77/833-2	387.2	288.5
5	ICML 03052×H77/833-2	369.5	272.5
6	ICML 03056×H77/833-2	385.2	287
7	ICML 03057×H77/833-2	369	275.2
8	841B × H77/833-2	382.8	276.8
9	863B × H77/833-2	426.5	304.5
10	843A × ICMR 01029	383.2	270.5
11	843A × ICMR 01031	358.5	268.2
12	843A × ICMR 01040	375.5	283.8
13	843A × ICMR 01046	390.8	283.5
14	843A × ICMR 02042	346	252.5
15	843A × ICMR 02044	397.2	295.5
16	843A×PRLT 2/89-33	420	341
Mean		382.8	283.6
SED		23.16	21.51
P -value		0.11	0.069

SED = Standard error of difference of means.

Days to 75% Flowering

Days to 75% booting/flowering among the sixteen drought-tolerant QTL introgression hybrids of millet was similar ($p > 0.05$) as shown in [Fig-1]. The number of days to 75% booting ranged between 58 and 66 days. This is an indication that the introgression of drought-tolerant QTL did not affect days to 75% flowering among millet lines.

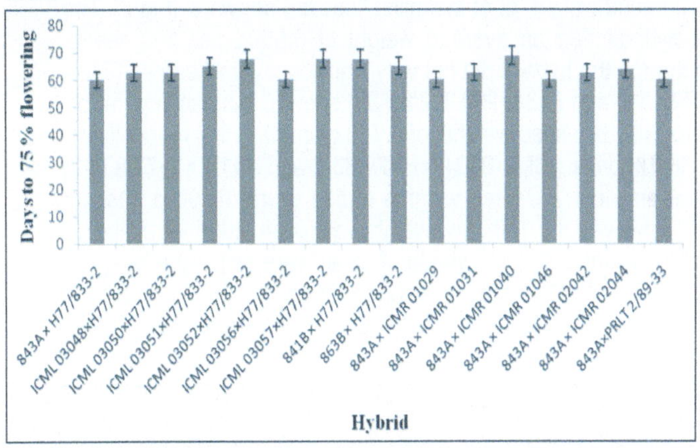

Fig. 1- Number of days to 75% booting of drought-tolerant QTL introgression pearl millet hybrids. The error bars represent Standard error.

QTL= Quantitative trait loci.

Ear Head Length and Diameter

The length and diameter of ear head of the sixteen hybrids of pearl millet are shown in [Table-4]. The average ear head length and ear head diameter recorded were 15.18cm and 2.044cm, respectively. The ear head length and diameters differed significantly ($p < 0.05$) among the hybrids. The hybrids ICML 03048×H77/833-2, ICML 03050×H77/833-2, ICML 03051×H77/833-2, ICML 03052×H77/833-2, ICML 03056×H77/833-2, ICML 03057×H77/833-2, 841B × H77/833-2, 863B × H77/833-2843A × ICMR 01046, and 843A×PRLT 2/89-33 had significantly ($p < 0.05$) greater ear head lengths when compared to the other hybrids, while the longest ear head diameter was observed in 863B × H77/833-2, 843A × ICMR 01029, 843A × ICMR 01046, and 843A×PRLT 2/89-33. Most of the hybrids had ear head length and ear head diameter within the usual range of values (15-45cm and 2-3cm, respectively) reported for pearl millet [3], which suggests that ear yield was not influenced by the introgression QTL for drought tolerance.

Table 4- Ear head length (cm) and ear head diameter (cm) of pearl millet hybrids.

Entry Number	Hybrid	Ear head length (cm)	Ear head diameter (cm)
1	843A × H77/833-2	13.58^{b}	2.062^{b}
2	ICML 03048×H77/833-2	15.60^{a}	1.790^{c}
3	ICML 03050×H77/833-2	15.14^{a}	1.710^{c}
4	ICML 03051×H77/833-2	16.43^{a}	1.838^{c}
5	ICML 03052×H77/833-2	15.92^{a}	1.762^{c}
6	ICML 03056×H77/833-2	15.87^{a}	1.703^{c}
7	ICML 03057×H77/833-2	16.09^{a}	1.718^{c}
8	841B × H77/833-2	15.82^{a}	1.757^{c}
9	863B × H77/833-2	16.37^{a}	2.473^{a}
10	843A × ICMR 01029	14.71^{b}	2.297^{a}
11	843A × ICMR 01031	14.93^{b}	2.268^{b}
12	843A × ICMR 01040	13.76^{b}	2.065^{b}
13	843A × ICMR 01046	15.44^{a}	2.343^{a}
14	843A × ICMR 02042	13.81^{b}	2.180^{b}
15	843A × ICMR 02044	14.06^{b}	2.150^{b}
16	843A×PRLT 2/89-33	15.37^{a}	2.595^{a}
SED		0.739	0.0963
P-value		0.001	0.001

Means within the same column with no superscript in common are significantly different SED= Standard error of differences of means.

Hundred-Seed Weight

The 100-seed weight (g) of the millet lines are shown in [Fig-2]. The millet hybrids had an average weight of 0.650g per 100 seeds. Even though the hybrids did not vary significantly, two entries (843A × H77/833-2 and 843A×PRLT 2/89-33) showed outstanding performances with 100-seed weights of 0.753 g and 0.815 g, respectively, while the hybrids (ICML 03048×H77/833-2 and 841B × H77/833-2) had the smallest 100-seed weights (0.555 g and 0.500 g, respectively). However, the 100-seed weight values recorded for all the 16 drought-tolerant QTL line hybrids of millet were within the range of values (0.3 -0.8 g/100 seeds) reported for pearl millet in India [10]. This indicates that QTL introduced did not adversely affect seed weight among the hybrids.

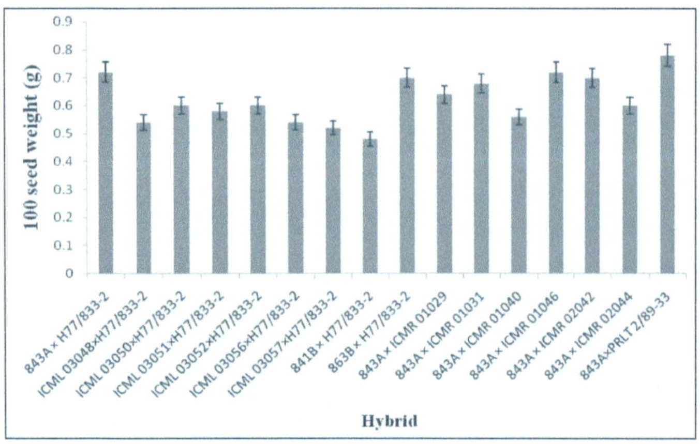

Fig. 2- Average hundred-seed weight (g) of pearl millet hybrids. Error bars represent standard error.

Head Yield Performance

The head yield was significantly ($p > 0.05$) influenced by the genotype of millet [Fig-3]. The average head yield observed was 76.6 g. Six entries, namely 843A × H77/833-2, ICML 03057×H77/833-2, 863B × H77/833-2, 843A × ICMR 01046, 843A × ICMR 02044 and 843A×PRLT2/89-33, performed above average while ten (ICML 03048×H77/833-2, ICML 03050×H77/833-2, ICM 03051×H77/833-2, ICML 03052×H77/833-2, ICML 03056×H77/833-2, 841B × H77/833-2, 843A × ICMR 01029, 843A × ICMR 01031, 843A × ICMR 01040, and 843A × ICMR02042) had values below the mean. The hybrids 863B × H77/833-2, 843A × H77/833-2 and 843A × ICMR 01046 had significantly ($p < 0.05$) higher head yield than the other hybrids.

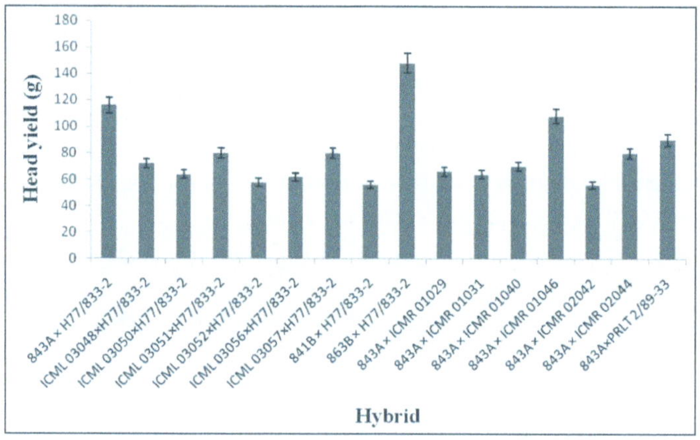

Fig. 3- The head yield (g) of pearl millet hybrids.

Error bars represent standard error.

Grain Yield

Grain production differed significantly ($P < 0.05$) among the drought -tolerant QTL introgression hybrids of millet [Fig-4]. The highest grain yield (kg/ha) was recorded in line 9 (863B × H77/833-2), while the lowest yield was observed in line 14 (843A × ICMR 02042). Indeed line 9, which produced the greatest grain yield, was the tallest with the highest values for ear head length, ear head diameter, 100-seed weight and head yield. Hence entry number 9 (863B × H77/833-2) performed best among the 16 hybrids.

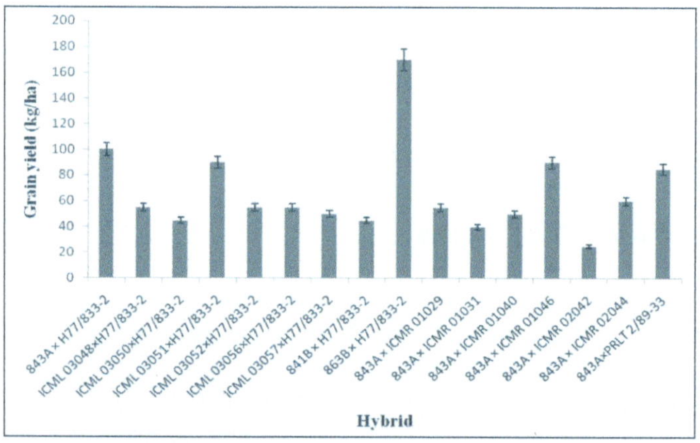

Fig. 4- Mean grain yield of drought-tolerant QTL introgression hybrids of millet. Error bars represent standard error.

QTL= Quantitative trait loci.

Rainfall Distribution

The average daily rainfall distributions (mm) during the field trial are shown in [Fig-5]. The rainfall was higher than 100mm in the 2nd and 8th weeks, but very low during the maturity stage (12th to 14th weeks). Although reports indicated that millet can grow in areas with rainfall ranging between 125 and 900mm[4], the average rainfall recorded from the 4th week after planting to maturity was far below the average requirement for the plant. The crop needs water for growth and development at this critical period. Therefore, insufficient water supply can reduce crop yield. Despite the inadequate rainfall throughout most of the growing season, all the drought-tolerant QTL introgression hybrids of millet were able to grow normal and yield grain.

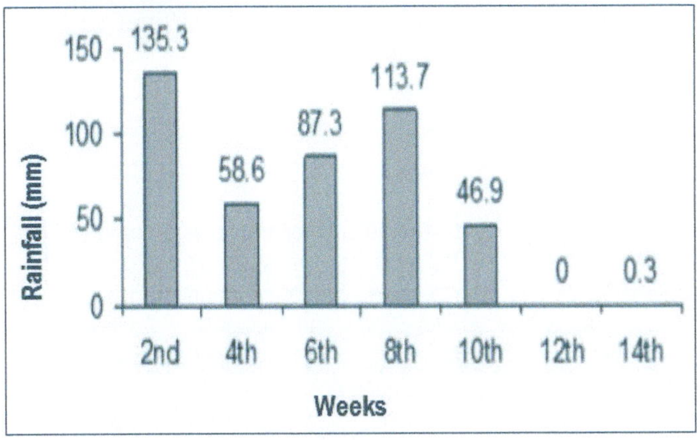

Fig. 5- Rainfall distribution (mm) from sowing to harvesting of drought-tolerant QTL introgression hybrids.

Conclusion and Recommendations

The study conducted revealed that the hybrid 863B × H77/833-2 had the best performance in terms of plant height, fresh and dry stover weights, ear head length and diameter, 100-seed weight, head yield, and grain yield. Also, accessions 843A × PRLT 2/89-33, 843A × ICMR 01046, and 843A × ICMR 01029 took fewer days to boot. Generally, the following accessions 843A × H77/833-2, ICML03048 × H77/833-2, ICML03050 × H77/833-2, ICML03051 × H77/833-2, ICML03052 × H77/833-2, ICML03056 × H77/833-2, ICML03057 × H77/833-2, 841B × H77/833-2, 843A × ICMR01031, 843A × ICMR 01040, 843A × ICMR 01042, and 843A × ICMR 01044 recorded poor grain filling and quality. Even though all the hybrids survived the inadequate rainfall for most of the growing periods, grain yield produced was very varied. Therefore, there will be the need to conduct more locational trials to evaluate the agronomic performance of the hybrids before recommending the best performing hybrids to farmers. Otherwise, these hybrids with the drought tolerance should be crossed with high yielding but drought susceptible cultivars to breed for high grain yield, which is what the farmers and their households and animals rely on for sustenance.

References

[1] FAO and ICRISAT (1996) *The World Sorghum and Millet economics, facts, trends and outlook*, 31-51.

[2] Onwueme T.C. and Sinha T.D. (1991) *Field Crop Production of Tropical Africa, Principles and Practice*, Technology Centre For Agriculture and Rural Co-operation, 190-199.

[3] Panda S.C. (2005) *Agronomy*, AGROBIOS, 429-436.

[4] Jain S.V. and Muthur C.M. (1961) *Indian Journal of Agronomy*, 5(3), 151-156.

[5] Krishnaswamy N. (1962) *Bajra: Pennisetum typhoides S & H*, Indian Council of Agriculture Research Cereal Crop Series II.

[6] Burton G.W. (1951) *Advance. Agron.*, 3, 197-241.

[7] Rachie K.O. and Majumuder J.V. (1980) *Pearl Millet*, Pennsylvania State University Park, 1-265.

[8] Nyankpala Agricultural Experimental Station (1992) *NAES Annual Report,* Nyankpala, Ghana.

[9] FAO/UNESCO (1987) *Soil Map of the World Revised Legend, World Soils Resource Report* 60. FAO, Rome, 41.

[10]Murty B.R., Upadhyay M.K. and Manchanda P.L. (1967) *Indian Journal Genetic Plant Breed*, 27.

MAXIMIZING SUGAR BEET YIELDS WITH DECREASING MINERAL FERTILIZATION POLLUTION

SEADH S.E.*
Agronomy Department, Faculty of Agriculture, Mansoura University, Egypt.
*Corresponding Author: Email- seseadh04@mans.edu.eg

Abstract- In order to increase sugar beet productivity beside decreasing mineral fertilization pollution two field experiments were carried out at El-Garayda Village, Bialla Center, Kafr El-Sheikh Governorate during 2010/2011 and 2011/2012 seasons. This study aimed to determine the effect of nitrogen, phosphorus and potassium (NPK) fertilizers levels (100, 75 and 50% decrease from recommended dose) and foliar fertilization treatments as follows; control treatment, spraying with water, Melagrow, Amino-Power and Melagrow + Amino-Power on sugar beet (*Beta vulgaris* L.) cv. Gloria growth, yield components, yield and quality parameters. A strip plot design with four replications was used. The results showed that increasing NPK levels from 50 up to 100% of the recommended dose significantly affected all studied characters in both seasons. Application 100% of the recommended dose of NPK (80 kg N + 30 kg P_2O_5 + 48 kg K_2O/fed) was the most effective treatment on yield. Foliar fertilization treatments significantly affected all studied characters, except purity % in both seasons. Spraying with Melagrow + Amino-Power solution produced highest values of all studied characters. Maximum values of sugar beet growth, yields and its components could be achieved by fertilizing with 100% of the recommended dose of NPK and spraying with Melagrow + Amino-Power. It can be recommended that mineral fertilization with 75% of the recommended dose of NPK and spraying with Melagrow + Amino-Power, which exceeded yields and root quality upon mineral fertilizing with 100% of the recommended dose of NPK + without spraying, and also possibility to reduce agriculture costs and mineral fertilization pollution.

Keywords- Sugar beet, NPK levels, mineral fertilization pollution, foliar fertilization, yields, quality

Introduction

Mineral synthetic fertilizers are composed mostly of nitrogen, phosphorous and potassium, which produce the greatest impact on plant growth. Although these fertilizers may produce impressively quick results in field where growth improved, the excessive and uncontrolled use of these synthetic compounds can lead to fertilizer pollution. The excessive use of synthetic fertilizer may increase yields in the short term and cause serious water pollution and harm to humans.

Moreover in Egypt, there was a gap between sugar consumption and production due to steady increases in the country population and average consumption of sugar beside limited cultivated area. Increasing sugar crops cultivated area and sugar production per unit area are considered the important national target to minimize the gab between sugar consumption and production. Improvement of sugar beet productivity can be achieved through many factors such as fertilization management and foliar fertilization.

NPK fertilization is among the vital factors affecting growth, yield and quality of sugar beet, thus application the suitable rates of nitrogen, phosphorus and potassium one of the favorable factors for increasing sugar beet productivity and quality. The proper fertilization reference can be given only based on the soil fertility. It must be determining optimum nitrogen rate, which produce the maximum root, sugar yields and quality parameters. In this respect, [10,13,15,17,22] stated that raising nitrogen fertilizer rates up to 90 kg N/fed increased all yield components and root, sugar yields significantly. Whereas, highest quality parameters values were obtained from the lowest nitrogen rate. [1,8,19,20,23,24] concluded that increasing nitrogen fertilizer levels up to 125 kg N/fed under newly reclaimed soil conditions significantly increased yield components, root and sugar yields. While, remarkable reduction were found on quality traits. Phosphorus fertilization is an essential nutrient to sugar beet production [7]. Sims and Smith [25] reported that root yield significantly less in the control treatment (0 kg P_2O_5/ha) compared to apply others (15,30 and 45 kg P_2O_5/ha). Marinkovic B., et al [16] showed that by increasing amounts of phosphorus from 50 to 100 and 150 kg P_2O_5/ha, root and sugar yields were increased. Potassium plays an important roles in photosynthesis, protein synthesis, translocation of assimilates as well as increasing plant growth and yield. In this concern, Amer M.A., et al [3] found that adding potassium up to 90 kg K_2O/fed resulted a significant increase in root and sugar yields/fed, TSS, sucrose and purity percentages. Gobarah Mirvat E., et al [9] stated that application potassium fertilizer at the rate of 114 kg K_2O/ha significantly increased root weight and dimensions, sucrose and purity percentages as well as root, top and sugar yield. Potassium uptake is much affected by nitrogen level and in most cases, potassium is more effective at higher nitrogen level, which is the case especially to modern high yielding varieties. In this regard, Abdel-Motagally and Attia [1] and Seadh S.E., et al [23] concluded that application the highest level of NK (125 kg N + 72 kg K_2O/fed) improved root and sugar yields as well as sugar beet grown in a sandy soil. Marinkovic B., et al [16] concluded that the optimal fertilization variant for the root yield was 100 kg N + 50 kg P_2O_5 + 50 kg K_2O/ha. While, the optimal fertilization variant for the quality and sugar yield was 50 kg N + 50 kg P_2O_5 + 50 kg K_2O/ha.

Foliar fertilization of nutrients is considering one of various techniques to improve fertilizer efficiency in order to increase productivity and improve quality of crop product [21]. This procedure can improve nutrient utilization and lower environmental pollution through reducing the amount of fertilizer added to soil [2]. Natural plant growth takes place through regular photosynthesis. The foliar application with Melagrow as a natural plant growth promoter for different crops was a great importance which promotes growth, increases yield and improves quality due to its contain phosphorus, potassium, boron and brassinolides and combined effects of growth regulators such as oxen, cytokines, gibberellins, ethylene and hydrogen cyanamid [4]. Amino acids is a well known biostimulant which has positive effects on growth, yield and significantly reduces the injuries caused by abiotic stresses [14]. Tugnoli and Bettini [27] concluded that use of amino acids and biostimulants as foliar spraying makes it possible to quickly overcome the nutritional and environmental deficiencies that arise during the crop growth cycle. Diána and Győri [6] stated that foliar fertilization significantly increased root and the sugar yields, while the quality parameters have not changed. Further, Grzebisz W., et al [12] found that root yield had increased by 31.6% and 22.1% for the NPK + Mi (spraying with multi-microelement fertilizer composed of cations) and NPK + MiB (spraying with multi-microelement fertilizer plus boron) treatments, respectively as compared with NPK (plot fertilized only with NPK) treatment. Almost the same degree of increase was noted for yield of recoverable sugar.

Therefore, this experiment aimed to study the response of sugar beet cv. Gloria to decreasing different NPK rates and compensate this reduction in NPK rates with foliar fertilization treatments to achieve maximum productivity and quality of sugar beet in a same time decreasing mineral fertilization pollution.

Materials and Methods

Two field experiments were carried out at El-Garayda Village, Bialla Center, Kafr El-Sheikh Governorate during 2010/2011 and 2011/2012 seasons. The main objective of this study was to determine the effect of nitrogen, phosphorus and potassium (NPK) fertilizers levels and foliar fertilization treatments on sugar beet (*Beta vulgaris* L.) cv. Gloria growth, yield and its components and quality parameters.

The experiments were laid-out in strip plot design with four replications. The vertical plots were assigned with three levels of NPK fertilizers as follows: a- 100% of the recommended level (80 kg N + 30 kg P_2O_5 + 48 kg K_2O/fed). b- 75% of the recommended level (60 kg N + 22.5 kg P_2O_5 + 36 kg K_2O/fed). c- 50% of the recommended level (40 kg N + 15 kg P_2O_5 + 24 kg K_2O/fed). Nitrogen and potassium fertilizers in the forms of urea (46%N) and potassium sulphate (48% K_2O), respectively were applied as a side-dressing in two equal portions, one half after thinning (35 days from sowing) and the other before the third irrigation (70 days from sowing). Calcium superphosphate (15.5% P_2O_5) was applied during soil preparation.

The horizontal plots were occupied with five foliar fertilization treatments as follows: 1- Without spraying (control treatment). 2- Spraying with irrigation water. 3-Spraying with Melagrow at the rate of 10 g Melagrow/200 liter water/fed). 4. Spraying with Amino-Power at the rate of 100 ml/200 liter water/fed. 5- Spraying with Melagrow at the rate of 10 g Melagrow/200 liter water/fed + Amino-Power at the rate of 100 ml/200 liter water/fed. Spraying was conducted by hand sprayer twice at the aforementioned rates after 50 and 70 days from sowing.

Melagrow is produced by Green India Co., and obtained from Gaara Foundation for Import and Export. Melagrow is considering as natural growth promoter extracted from pollen of cabbage flowers. Melagrow is combined effects of many growth regulators such as oxen, cytokines, gibberellins, ethylene and hydrogen cyanamid. The chemical composition of Melagrow is 20% phosphorus, 10% potassium, 3% boron and 0.2% brassinolide. Natural brassinolides (0.2%) is natural plant growth promoter for growth, yield increases and improves quality.

Amino-Power® was obtained from Union For Agriculture Development Co. Amino-Power is consider also as natural growth promoter, which contain free amino acids in a form of Levogire "L-Amino Acids" (Glycin, Seriocin, Glutamic, Valin, Broline, Hydroxy broline, Methionine iso lucin, lucin, Phenyl, Alanine, Serine, Asparatic, Arginine, Hydroxy Lysine and Hestedine) in addition microelements (Fe 1500 ppm, Zn 500 ppm and Mn 500 ppm) and citric acid (3.5% potassium citrate).

The soil was clay texture with an electrical conductivity (EC) of 1.65 dS/m and a pH of 7.88. Each experimental basic unit included 5 ridges, each of 60 cm apart and 3.5 m long, comprising an area of 10.5 m^2 (1/400 fed). The preceding summer crop was rice (*Oryza sativa* L.) in both seasons.

Experiments were sown on September 15th and 18th in the first and second seasons, respectively. Sugar beet was hand sown 3-5 balls/hill using dry sowing method on one side of the ridge in hills 20 cm apart. Plants were thinned to one plant/hill (35000 plants/fed) at the age of 35 days from planting. The recommended agricultural practices for growing sugar beet were followed except the factors under study.

Studied Characters

A. Growth Characters

Five guarded plants from the outer ridges of each plot were chosen randomly at 150 days from sowing to determine the following growth characters:

1. Total chlorophyll: Leaf chlorophyll content was assessed by SPAD-502 (Minolta Co. Ltd., Osaka, Japan).

2. Leaf area/plant (cm^2). It was determined using Field Portable Leaf Area Meter AM-300 (Bio-Scientific, Ltd., Great Amwell, Herforshire, England).

B. Yield, Yield Components and Quality Characters

At harvest, five plants were chosen at random from the outer ridges of each plot to determine yield components and quality characters as follows:

1. Foliage fresh weight (g/ plant).
2. Foliage length (cm).
3. Root fresh weight (g/plant).
4. Root length (cm).
5. Root diameter (cm).
6. Total soluble solids (TSS %) in roots was measured in juice of fresh roots by using Hand Refractometer.
7. Sucrose percentage (%) was determined Polarimetrically on lead acetate extract of fresh macerated roots according to the method of Carruthers and Oldfield [5].
8. Apparent purity percentage (%). It was determined as a ratio between sucrose % and TSS % of roots as the method outlined by Carruthers and Oldfield [5].

Plants that produced from the two inner ridges of each plot at harvesting time were collected and cleaned, and then roots and tops were separated and weighted in kilograms and converted to estimate:

1. Root yield (t/fed).
2. Top yield (t/fed).
3. Sugar yield (t/fad). It was calculated by multiplying root yield by sucrose percentage.

All obtained data were statistically analyzed according to the technique of analysis of variance (ANOVA) for the strip plot design as published by Gomez and Gomez [11], using MSTAT statistical package (MSTAT-C with MGRAPH version 2.10, Crop and Soil Sciences Department, Michigan State University, USA). Least Significant Difference (LSD) method was used to test the differences between treatment means at 5% level of probability [26].

Results and Discussion

A. Effect of NPK Levels

The effect of NPK levels on total chlorophyll, leaf area/plant, foliage fresh weight, foliage length, root fresh weight, root length and diameter [Table-1], TSS, sucrose and apparent purity percentages, root, top and sugar yields [Table-2] was significant in the two growing seasons. It can be stated that growth characters, yields and its components significantly increased as a result of increasing NPK from 50 to 75 and 100% of the recommended rate and the differences among them were significant in both seasons. Application 100% of the recommended level of NPK (80 kg N + 30 kg P_2O_5 + 48 kg K_2O/fed) produced the highest values of growth characters, root, top and sugar yields and its components in both seasons. While, fertilizing beet plants with 75% of the recommended level of NPK (60 kg N + 22.5 kg P_2O_5 + 36 kg K_2O/fed) came in the second rank, then using 50% of the recommended level of NPK (40 kg N + 15 kg P_2O_5 + 24 kg K_2O/fed), which resulted in the lowest values of growth characters and yields and its components in both seasons.

The increases in yields and its components as a result of increasing NPK levels can be ascribed to the role of nitrogen in improving vegetative growth through enhancing leaf initiation, color and vigor of the leaf canopy as well as meristematic activity which contribute to the increases in number of cells in addition to cell enlargement. Moreover, nitrogen encourages plants to uptake other elements, enhancing net assimilation rate and activating accumulation of carbohydrates, which translated from leaves to developing roots which in turn enhanced root length, diameter, root fresh weight and finally root and sugar yields per unit area [18]. Also, the role of phosphorus in energy transfer within the plants and in maintaining the structural integrity of the plant cell membranes as well as carbohydrate assimilation and improving plant growth, which was reflected in increasing root dimensions and weight [7,25]. Beside, the role of potassium in building up metabolites and activating starch synthetase enzymes and carbohydrates accumulation which transferred from leaves to developing roots consequently enhanced root and chemical constituents [18]. Increasing potassium fertilizer level significantly increased root weight and dimensions, sucrose and purity percentages, root, top and sugar yields [3,9]. These results are in coincidence with those stated by [1,16,23].

The optimum means of sucrose and purity percentages were obtained from using 50% of the recommended level of NPK in both seasons. Whereas, the optimum means of TSS % were achieved when application 75% of the recommended level of NPK in the two growing seasons. However, application 100% of the recommended level of NPK produced the lowest values of all quality parameters in both seasons. The decrease in quality parameters due to excessive nitrogen application can be ascribed to its role in increasing root weight and diameter, tissue water content, increasing non-sucrose substances such as proteins and alpha amino acids, and hence decreasing sucrose content in roots. Raising nitrogen fertilizer levels caused a significant decreases in quality parameters of sugar beet [10,15,17]. Furthermore, highest quality parameters values were obtained from the lowest level of nitrogen [16].

B. Foliar Fertilization Treatments Effect

Foliar fertilization treatments significantly affected total chlorophyll, leaf area/plant, foliage fresh weight, foliage length, root fresh weight, root length and diameter [Table-1], TSS %, sucrose %, root, top and sugar yields [Table-2] in both seasons. It could be observed that spraying sugar beet plants twice after 50 and 70 days from sowing with water, Melagrow, Amino-Power and the mixture of Melagrow + Amino-Power caused a gradual increases in all studied characters compared with without spraying (control treatment) in the two growing seasons. Highest values of all studied characters were obtained as a result of spraying sugar beet plants with the mixture of Melagrow + Amino-Power treatment in both seasons under the environmental conditions of studied region. In addition, spraying sugar beet plants with Amino-Power treatment ranked after previously mentioned treatment, followed by spraying sugar beet plants with Melagrow treatment then spraying sugar beet plants with water in the two growing seasons. On the other wise, the control treatment (without spraying sugar beet plants) gave the lowest values of these characters in both seasons.

The desirable effect of spraying sugar beet plants with the mixture of Melagrow + Amino-Power may be ascribed to the role of Melagrow in promotes growth, increases yield and improves quality due to its contains macronutrients (phosphorus and potassium), micronutrients (boron) and natural brassinolides as natural plant growth promoter as well as its combined effects of oxen, cytokines, gibberellins [4]. Furthermore, the role of Amino-Power as natural growth promoter also, which contains free amino acids in a form of Levogire (L-Amino Acids) in addition microelements (Fe, Zn and Mn) and citric acid, which has positive effects on plant growth and yield [14]. Foliar fertilization treatments of sugar beet significantly increased root and sugar yields [6,12,27].

Table 1- Means of total chlorophyll, leaf area/plant, foliage fresh weight, foliage length, root fresh weight, root length and diameter as affected by NPK levels and foliar fertilization treatments and their interaction during 2010/2011 and 2011/2012 seasons.

Treatments	Characters													
	Total chlorophyll (SPAD)		Leaf area/plant (cm^2)		Foliage fresh weight (g)		Foliage length (cm)		Root fresh weight (g)		Root length (cm)		Root diameter (cm)	
Seasons	2010/2011	2011/2012	2010/2011	2011/2012	2010/2011	2011/2012	2010/2011	2011/2012	2010/2011	2011/2012	2010/2011	2011/2012	2010/2011	2011/2012
NPK Levels														
100% NPK	49.13	47.33	3839.3	3773	497	487	41.86	39.4	877.1	855.3	32.43	31.86	14.7	14.06
75% NPK	47.06	45.13	3704.8	3643.6	410.8	395.9	38.93	37.06	746	741	30.96	30.2	13.8	13.4
50% NPK	43.53	41.26	3282	3216.3	311.8	299.2	30.66	28	607.6	593.3	24.56	23.83	12.35	11.66
LSD at 5%	0.8	0.96	47.8	47.9	39.3	39.7	0.68	0.55	30.6	44	0.5	0.4	0.32	0.24
Foliar Fertilization Treatments														
Without	40.77	39.22	3261.1	3191.6	338.5	326.1	31.66	29.22	617.8	605.5	26.83	25.88	12.44	11.77
Water	43	40.33	3346.6	3257.7	377.4	357.1	33.66	31.11	662	641.1	27.61	26.72	12.92	12.33
Melagrow	48.11	46	3720.7	3661.6	414.2	405.8	38.44	35.77	733.3	723.3	29.83	29.33	13.66	13.22
Amino-Power	49.66	47.77	3797.7	3752.2	437.2	425.5	39.88	37.88	820.5	808.8	30.72	30.22	14.33	13.72
Melagrow + Amino-Power	51.33	49.55	3917.2	3858.3	465.4	455.5	42.11	40.11	884.3	870.5	31.61	31	14.72	14.16
LSD at 5%	0.83	0.94	34.2	38.7	15.8	11.2	0.69	0.39	34.9	34.9	0.44	0.36	0.34	0.25
C- Interaction:	NS	NS	*	*	NS	NS	*	*	*	*	NS	NS	*	NS

Table 2- Means of total soluble solids (TSS), sucrose and apparent purity percentages, root, top and sugar yields as affected by NPK levels and foliar fertilization treatments and their interaction during 2010/2011 and 2011/2012 seasons.

Treatments	Characters											
	TSS (%)		Sucrose (%)		Apparent purity (%)		Root yield (t/fed)		Top yield (t/fed)		Sugar yield (t/fed)	
Seasons	2010/2011	2011/2012	2010/2011	2011/2012	2010/2011	2011/2012	2010/2011	2011/2012	2010/2011	2011/2012	2010/2011	2011/2012
NPK Levels												
100% NPK	22.03	21.4	17.42	17.11	79.11	78.92	25.579	23.835	12.009	11.591	4.468	4.09
75% NPK	22.56	22.11	18.04	17.78	79.99	80.45	21.778	21.387	10.645	10.4	3.935	3.809
50% NPK	21.82	21.68	18.76	18.55	85.96	86.67	17.205	16.758	8.467	8.193	3.235	3.115
LSD at 5%	0.28	0.27	0.15	0.2	1.3	1.29	0.787	0.685	0.704	0.695	0.137	0.142
Foliar Fertilization Treatments												
Without	21.63	21.28	17.67	17.37	81.31	81.66	18.111	17.489	8.679	8.478	3.187	3.025
Water	21.9	21.43	17.8	17.56	81.62	81.99	18.991	18.231	9.206	9.022	3.367	3.188
Melagrow	22.2	21.78	18.11	17.87	81.74	82.09	21.737	20.856	10.353	10.05	3.919	3.711
Amino-Power	22.36	21.96	18.31	18.04	81.84	82.15	23.431	22.417	11.349	10.983	4.269	4.025
Melagrow + Amino-Power	22.61	22.2	18.48	18.22	81.93	82.19	25.333	24.308	12.282	11.772	4.657	4.408
LSD at 5%	0.24	0.24	0.1	0.09	NS	NS	0.77	0.778	0.461	0.418	0.156	0.143
C- Interaction:	NS	NS	*	*	NS	NS	*	*	*	*	*	*

C. Interaction effect

The interaction between both studied factors (NPK levels X foliar fertilization treatments) had a significant effect on leaf area/plant, foliage length, root fresh weight, sucrose %, root, top and sugar yields/fed in both seasons and root diameter in the first season only.

The highest values of leaf area/plant [Fig-1], foliage length [Fig-2], root fresh weight [Fig-3], root diameter [Fig-4], root yield/fed [Fig-5], top yield/fed [Fig-6] and sugar yield/fed [Fig-7]were produced from mineral fertilizing sugar beet plants with 100% of the recommended level of NPK (80 kg N + 30 kg P_2O_5 + 48 kg K_2O/fed) beside spraying with the mixture of Melagrow + Amino-Power. While, the lowest values of these characters were resulted from fertilizing sugar beet plants with 50% of the recommended level of NPK (40 kg N + 15 kg P_2O_5 + 24 kg K_2O/fed) without spraying treatment (control treatment) in both seasons. Whereas, the highest values of sucrose % [Fig-8] were obtained when mineral fertilizing sugar beet plants with 50% of the recommended level of NPK in addition spraying with the mixture of Melagrow + Amino-Power, while the lowest values of this character were resulted from fertilizing sugar beet plants with 100% of the recommended level of NPK without spraying treatment (control treatment) in both. The second best interaction treatment that increased sugar beet yields was mineral fertilizing with 100% of the recommended level of NPK + spraying with Amino-Power followed by mineral fertilizing with 75% of the recommended level of NPK + spraying with the mixture of Melagrow + Amino-Power. From obtained results that graphically illustrated in figures 1 through 8, foliar spraying sugar beet plants with the mixture of Melagrow + Amino-Power and fertilizing with 75% of the recommended level of NPK significantly exceeded the common treatment that most farmer are used (mineral fertilizing with 100%

of the recommended level of NPK without spraying treatment) in both seasons.

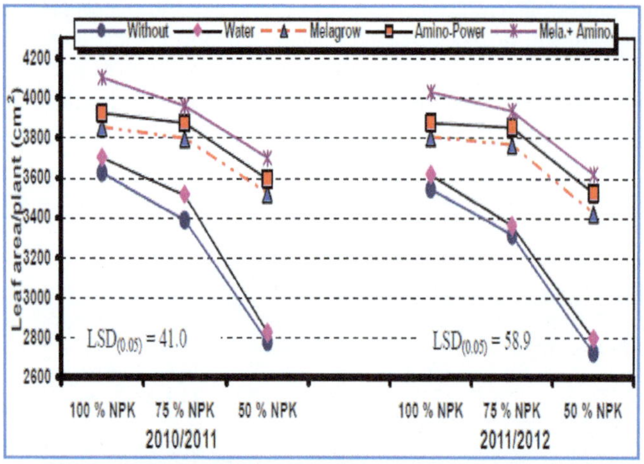

Fig. 1- Leaf area/plant (cm^2) of sugar beet as affected by the interaction between NPK levels and foliar fertilization treatments during 2010/2011 and 2011/2012 seasons

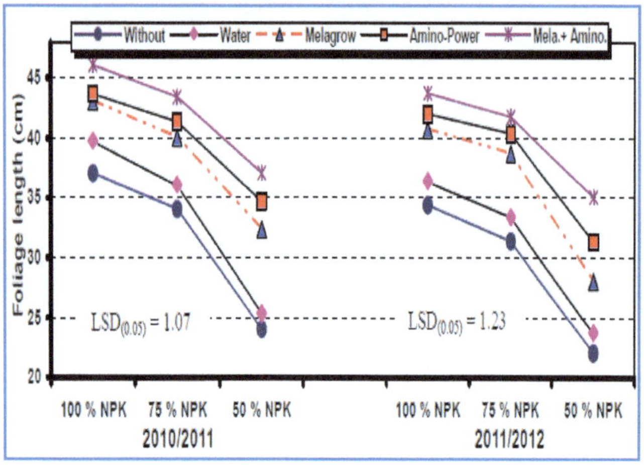

Fig. 2- Foliage length (cm) of sugar beet as affected by the interaction between NPK levels and foliar fertilization treatments during 2010/2011 and 2011/2012 seasons.

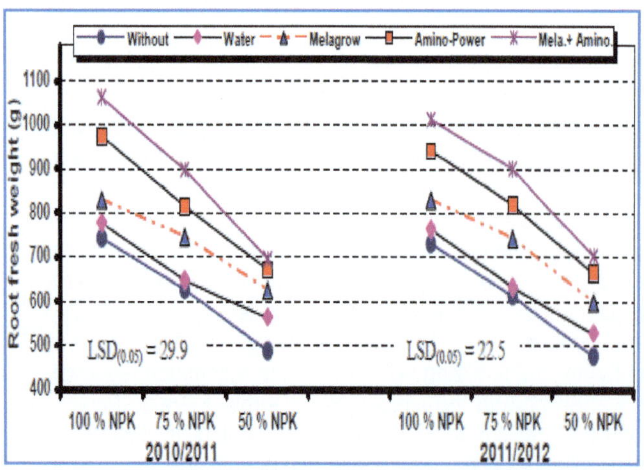

Fig. 3- Root fresh weight (g) of sugar beet as affected by the interaction between NPK levels and foliar fertilization treatments during 2010/2011 and 2011/2012 seasons.

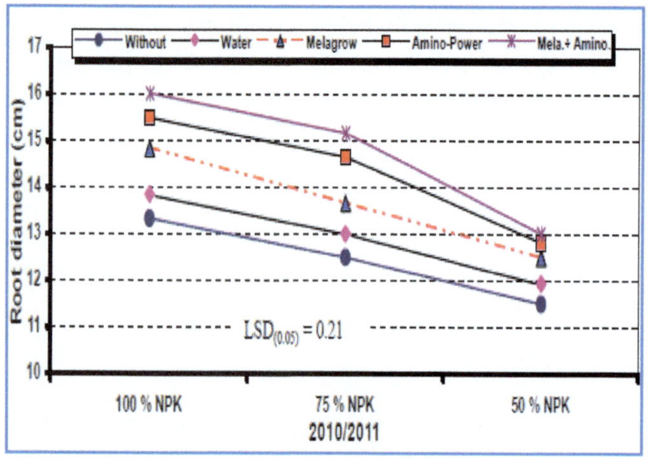

Fig. 4- Root diameter (cm) of sugar beet as affected by the interaction between NPK levels and foliar fertilization treatments during 2010/2011 season

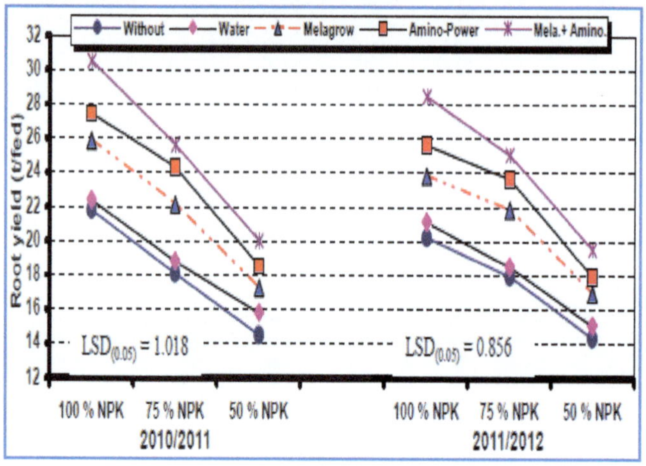

Fig. 5- Root yield (t/fed) of sugar beet as affected by the interaction between NPK levels and foliar fertilization treatments during 2010/2011 and 2011/2012 seasons.

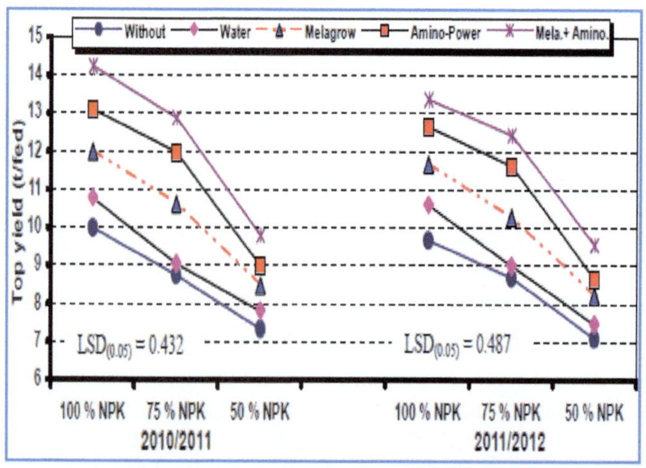

Fig. 6- Top yield (t/fed) of sugar beet as affected by the interaction between NPK levels and foliar fertilization treatments during 2010/2011 and 2011/2012 seasons.

Accordingly, the favorable treatment that increased sugar beet yields and quality parameters in the same time reduces agriculture costs and environmental pollution was mineral fertilizing sugar beet plants with 75% of the recommended level of NPK and foliar spraying with the mixture of Melagrow + Amino-Power.

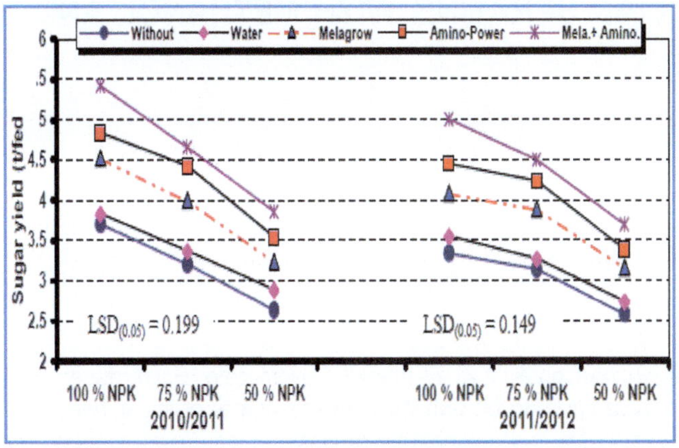

Fig. 7- Sugar yield (t/fed) of sugar beet as affected by the interaction between NPK levels and foliar fertilization treatments during 2010/2011 and 2011/2012 seasons.

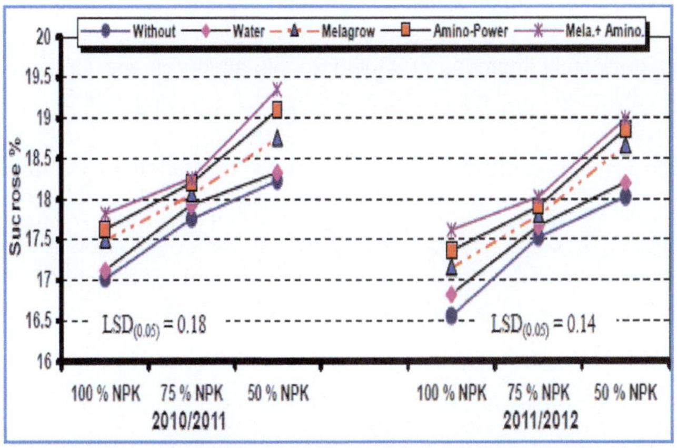

Fig. 8- Sucrose % of sugar beet as affected by the interaction between NPK levels and foliar fertilization treatments during 2010/2011 and 2011/2012 seasons.

Conclusion

Maximum sugar beet growth, yields and its components were achieved by mineral fertilizing with 100% of the recommended level of NPK (80 kg N + 30 kg P_2O_5 + 48 kg K_2O/fed) and foliar spraying with the mixture of Melagrow + Amino-Power twice after 50 and 70 days from sowing. To reduce agriculture inputs and environmental pollution, it could be recommended that mineral fertilizing with 75% of the recommended level of NPK (60 kg N + 22.5 kg P_2O_5 + 36 kg K_2O/fed) and foliar spraying with the mixture of Melagrow + Amino-Power, increased yields and quality upon mineral fertilizing with 100% of the recommended level of NPK + without spraying.

References

[1] Abdel-Motagally F.M.F. and Attia K.K. (2009) *Int. J. Agric. Biol.*, 11(6), 965-700.

[2] Abou El-Nour E.A.A. (2002) *Pak. J. Bio. Sci.*, 5(3), 259-262.

[3] Amer M.A., Attalah M.Z., Ibrahim M.F. and El-Taweel F. (2004) *J. Agric. Sci., Mansoura Univ.*, 29(6), 2963-2978.

[4] Attia A.N.E., Badawi M.A., Seadh S.E. and El-Hety S.M.S. (2011) *J. Plant Production*, 2(12), 1703-1716.

[5] Carruthers A. and Oldfield J.F.T. (1960) *Int. Sugar J.*, 63, 72-74.

[6] Diána U. and Győri Z. (2007) *Analele Universităţii din Oradea, Fascicula: Protecţia Mediului*, XII, 187-192.

[7] Draycott A.P. (1993) *Nutrition, The Sugar Beet Crop.*, 238-277.

[8] El-Geddawy I.H., El-Shafai A.M.A. and Azzazy N.B. (2006) *J. Agric. Sci. Mansoura Univ.*, 31(1), 43-54.

[9] Gobarah Mirvat E., Mekki B.B., Mohamed Magda H. and Eawfik M.M. (2011) *American-Eurasian J. Agric. & Environ. Sci.*, 10(4), 626-632.

[10] Gomaa A.M., Freweez H. and Abo-El-Wafa A.M. (2005) *Assiut J. of Agric. Sci.*, 36(3), 39-56.

[11] Gomez K.N. and Gomez A.A. (1984) *Statistical Procedures for Agricultural Research., 2nd ed.*, 95-109.

[12] Grzebisz W., Przygocka Cyna K., Lukowiak R. and Biber M. (2010) *J. Elementol*, 15(3), 493-507.

[13] Kandil A.A., Badawi M.A., El-Moursy S.A. and Abdou U.M.A. (2002) *J. Agric. Sci. Mansoura Univ.*, 27(11), 7257-7266.

[14] Kowalczyk K. and Zielony T. (2008) *Conf. of Biostimulators in Modern Agriculture*, Warsaw, Poland.

[15] Leilah A.A., et al. (2007) *Sci. J. of King Faisal Uni.*, 8(1), 87-100.

[16] Marinkovic B., Crnobarac J., Jaćimović G., Marinković D., Mircov D.V. and Rajici M. (2008) *J. of Agri. Res.*, 40(2), 99-104.

[17] Monreala J.A., Jiméneza E.T., Remesala E., Morillo-Velardeb R., García-Mauriñoa S. and Echevarríaa C. (2007) *Enviro. and Exper. Botany*, 60(2), 257-267.

[18] Moore A., Stark J., Brown B. and Hopkins B. (1999) *University of Idaho College of Agricultural and life Science*, CIS1174, 1-7.

[19] Nemeat Alla E.A.E., Aboushady K.A. and Yousef N.O. (2007) *J. Agric. Sci. Mansoura Univ.*, 32(10), 8069-8078.

[20] Ramadan B.S.H. (2005) *J. Agric. Sci. Mansoura Uni.*, 30(2), 773-783.

[21] Romheld V. and El-Fouly M.M. (1999) *2nd International Workshop on Foliar Fertilization*, Bangkok, Thailand, 1-34.

[22] Seadh S.E. (2004) *Agricultural Studies on Sugar Beet Crop. Ph.D. Thesis, Fac. of Agric. Mansoura Univ.*, Egypt.

[23] Seadh S.E., Farouk S. and El-Abady M.I. (2007) *African Crop Sci. Conf.*, 8, 147-153.

[24] Seadh S.E. (2008) *Meeting the Challenges of Sugar Crop & Integrated Industries in Developing Countries*, Egypt, 110-115.

[25] Sims A.L. and Smith L.J. (2001) *J. of Sugar Beet Res.*, 38(1), 1-17.

[26] Snedecor G.W. and Cochran W.G. (1980) *Statistical Methods. 7th ed.*, USA.

[27] Tugnoli V. and Bettini G. (2003) *1st Joint IIRB-ASSBT Congress*, San Antonio, USA, 655-660.

SUGAR BEET PRODUCTIVITY AS AFFECTED BY FOLIAR SPRAYING WITH METHANOL AND BORON

ABIDO W.A.E.*
Agronomy Department, Faculty of Agriculture, Mansoura University, Egypt.
*Corresponding Author: Email- madawy78@mans.edu.eg

Abstract- Two field experiments were carried out at Kafr El-Garayda Village, Bialla Center, Kafr El-Sheikh Governorate, Egypt during 2010/2011 and 2011/2012 seasons to study the effect of methanol and boron foliar spraying on growth, yield and quality of sugar beet. This aimed to study four aqueous methanol solutions control without 0, 15, 30 and 45% (v/v) sprayed three times during growth stages which occupied the vertical plots and four boron concentrations control without 0, 40, 80 and 120 ppm as boric acid applied as foliar spraying at two times during growth season and assigned in the horizontal plots.

Results indicated that foliar application of 30% methanol solution significantly increased all studied characters *i.e.* total chlorophyll, leaf area/plant, foliage and root fresh weight, foliage and root length and root diameter, total soluble solids, sucrose and apparent purity percentages, root, top and sugar yields/ha, whereas harvest index was decreased. Moreover, increasing methanol concentration to 45% tended to decrease all above mentioned characters. Application of 80 ppm boron significantly improved root yield and its attributes and root quality, on contrarily harvest index was decreased. Maximum top, root, sugar yields/ha and root quality produced by foliar spraying of 30% methanol and 80 ppm boron.

Keywords- Sugar beet, (*Beta vulgaris*, L.), methanol, Boron, yield and quality

Introduction

In Egypt, there is a gap between sugar consumption and production due to steady increases in population and average consumption of sugar beside limited cultivated area of both sugar beet (*Beta vulgaris*, L) and sugar cane. To increase sugar beet productivity per unit area sugar beet plants subjected to *i.e.* methanol and boron foliar fertilization since production of biomass by plants depends to great extent on environmental factors *i.e.* water supply, air temperature and carbon dioxide concentration in the canopy [24]. Increasing the dioxide carbon content in the air increase yield and plants accumulated more carbohydrates because almost 90% of plant dry matter weight is resulted from CO_2 assimilation during photosynthesis [1,12,15,17]. Photo respiration can be minimized with methanol spray, since 25% of carbon wastes during photorespiration [6]. Similar conclusions were reported by [2,18,23,25]. Moreover, Sadeghi-Shoae M., et al [20] concluded that foliar application of methanol at 14% increased root yield to 25% in comparison with the control.

Foliar spraying of boron increase root yield since roots absorbed boric acid and its uptake depends on soil pH and soil boron content [8] due to chloroplast formation and sink limitations [22] and changes in cell wall which effects of boron deficit and led to secondary effects in plant metabolism, development and growth [16]. Gobarah Mirvat E., et al [10] revealed that application of boron rates from zero up to 1.5 Kg/acre increased root length, diameter and root yield. Moreover, increasing boron fertilizer up to 2.0 Kg/acre resulting highest sugar yield (6.611 ton/acre). Sucrose and juice purity percentages were also increased by adding higher concentration of boron might be attributed to decrease Na and K uptake in root juice. Similar results were recorded by [7,9,13,14]. Armin M. and Asgharipour M.R. [3] indicated that highest root yield and sucrose concentration were obtained by spraying with 12% boric acid.

Therefore, this study was undertaken to study the response of sugar beet cv. Gloria to foliar application with methanol and boron to achieve maximum root productivity and its quality.

Materials and Methods

Two field experiments were conducted at Kafr El-Garayda Village, Bialla Center, Kafr El-Sheikh Governorate, Egypt during 2010/2011 and 2011/2012 seasons. The main objectives of this study were aimed to study the effect of foliar application with methanol and boron concentrations on sugar beet (*Beta vulgaris* L.) cv. Gloria productivity and quality.

Two field experiments were laid-out in strip plot design with four replications. The vertical plots were occupied by four concentrations of methanol *i.e.* control without 0, 15, 30 and 45% (v/v) methanol and each solution contained 0.2% glycine to reduce the probability of methanol toxicity [19]. These solutions were sprayed on sugar beet foliage three times every two week. The first foliar application was applied at 80 days after planting. The horizontal plots were assigned to four foliar application of boron in the form of boric acid *i.e.* control without 0, 40, 80 and 120 ppm these solutions were sprayed on foliage parts of sugar beet two times (50 and 70 days after planting). The foliar solutions volume was to 200 L/ha conducted by hand sprayer. The soil of experimental site was clayey with an electrical conductivity (EC) of 1.65 dS/m and a pH of 7.88.

Each plot consist 5 ridges, each of 60 cm apart and 3.5 m long, comprising an area of 10.5 m^2 (1/400 ha). The preceding summer crop was rice (*Oryza sativa* L.) in both seasons. The experimental soil was fertilized with 71.4 kg P_2O_5/ha in the form of calcium superphosphate (15.5% P_2O_5) during soil preparation, nitrogen in the form of ammonium nitrate (33.0% N) at a rate of 297.5 kg N/ha added at four equal doses before second, third, fourth and fifth irrigations and potassium at the rate of 114.24 kg K_2O/ha in the form of potassium sulphate (48% K_2O) in two equal portions added before second and fourth irrigations. Sowing took place on September 15th and 18th in the first and second seasons, respectively. Sugar beet balls were hand sown (3-5 balls/hill) using dry planting method on one side of the ridge and hills 20 cm apart. Experimental plots were irrigated immediately after planting, then irrigation frequently every 10 days. Plants were thinned after 30 days from planting to one plant/hill to produce 83300 plants/ha. Plants were kept free from weeds, which were manually controlled by hoeing two times before the second and third irrigations. In general, the agricultural practices for growing sugar beet according to Ministry of Agriculture recommendations were followed.

Studied Characters

Growth, Yield Component and Quality Characters

A representative samples were taken during the growth period (150 days from planting), *i.e.* five guarded plants were chosen at random from second and fourth ridges of each plot to determinate the following traits:

1. Total chlorophyll (SPAD): Leaf chlorophyll content was assessed by SPAD-502 (Minolta Co. Ltd., Osaka, Japan).
2. Leaf area/plant (cm^2): It was determined using Field Portable Leaf Area Meter AM-300 (Bio-Scientific, Ltd., Great Amwell, Herforshire, England).

At maturity (190 DAP) five guarded plants were chosen at random from the second and fourth ridges of each plot to determine yield components and quality characters as follows:

3. Foliage fresh weight (gm/ plant).
4. Foliage length (cm).
5. Root fresh weight (gm/plant).
6. Root length (cm).
7. Root diameter (cm).
8. Total soluble solids (TSS %) in roots was measured in juice of fresh roots by using Hand Refractometer
9. Sucrose percentage (%) was determined Polarimetrically on lead acetate extract of fresh macerated roots according to the method of Carruthers A. and Oldfield J.E.T. [5].
10. Apparent purity percentage (%): It was determined as a ratio between sucrose % and TSS % of roots as the method outlined by Carruthers A. and Oldfield J.E.T. [5].

Yield of three inner ridges of each plot were harvested and cleaned. Roots and tops were separated and weighted to estimate:

11. Root yield (ton/ha).
12. Top yield (ton/ha).
13. Sugar yield (ton/ha): was calculated by multiplying root yield (ton/ha) by sucrose %.
14. Harvest index (HI): was calculated by using the following equation:

$$HI = \frac{\text{Root yield (ton/ha)}}{\text{Top yield (ton/ha)} + \text{Root yield (ton/ha)}}$$

All obtained data were statistically analyzed according to the technique of analysis of variance (ANOVA) for the strip plot design as published by Gomez and Gomez [11] by using by using MSTAT-C Statistical package MSTAT-C with (MGRAPH version 2.10). Least Significant Difference (LSD) method was used to test the differences between treatment means at 5 and 1% level of probability [21].

Results and Discussion

A. Effect of Methanol Concentrations

Results showed that all studied characters *i.e.* total chlorophyll, leaf area/plant, foliage fresh weight, foliage length, root fresh weight, root length, root diameter [Table-1], total soluble solids, sucrose, apparent purity percentages, root yield, top yield, sugar yield and harvest index [Table-2] were significantly affected by spraying 30% methanol in both seasons and then decreased due to application of 45% methanol. It could be stated that foliar application of 30% methanol solution significantly increased all studied growth, yield and quality of sugar beet traits and consequently produced the highest values. It could be noticed that applying 30% methanol caused significant increases in total chlorophyll, leaf area/plant, foliage fresh weight, foliage length, root fresh weight, root length, root diameter, total soluble solids, sucrose (%), apparent purity percentages, root yield/ha, top yield/ha and sugar yield/ha by 12.38, 8.95, 40.92, 10.53, 17.20, 18.32, 19.53, 5.17, 10.8, 7.76, 16.31, 30.53 and 25.5%, respectively as an average of both seasons compared with control treatment. Harvest index was decreased by 4.04% as an average of both season compared to control treatment. Generally, foliar application of 45% methanol solution came in the second rank after 30% methanol and then followed by foliar application of 15% methanol solution, whereas control treatment recorded the lowest values for all studied characters, except (HI) was the higher one. The favorable effect of 30% methanol might be due to the treating plants with methanol could be enhance their net photosynthesis, therefore improving the yield. Nonomura and Benson [19] showed that methanol reduces the plants photorespiration and the rapidly oxidized. Also methanol causes to delay senescence in leaves and influences on ethylene production in plant which this causes to increase photo synthesis activity [25]. Moreover, Zbieć I., et al [24] observed that yield of roots increased by 10% using 20 or 30% methanol solutions. These results are in accordance with those reported by [17,18,20].

B. Effect of Boron Concentrations

A significant effect was detected due to boron concentrations application on total chlorophyll, leaf area/plant, foliage fresh weight, foliage length, root fresh weight, root length, root diameter [Table-1], total soluble solids, sucrose, apparent purity percentages, root yield, top yield, sugar yield and harvest index [Table-2] in both seasons. Increasing boron concentrations up to 80 ppm significantly increased all studied traits. While, application boron at 120 ppm came in the second rank with respect to these characters. Foliar spraying of boron at 80 ppm increased total chlorophyll, leaf area/plant, foliage fresh weight, foliage length, root fresh weight, root length, root diameter, total soluble solids, sucrose (%), apparent purity percentages, root yield/ha, top yield/ha and sugar yield/ha by 12.77, 9.53, 31.34, 10.83, 9.72, 16.68, 15.24, 2.48, 9.75, 7.39, 11.27, 19.01 and 20.14%, respectively as an average of two seasons compared with the control treatment. While harvest index decreased by 1.91% as an average of two seasons compared to the control treatment. The positive effect of boron may be due to the boron role in cell elongation where, in case of boron deficiency, plant leaves were smaller, stiff and thick [4]. Gobarah Mirvat E. and Mekki B.B. [10] indicated that root yield, sucrose and juice purity percentage increased by boron addition which may be attributed to decrease Na and K uptake in root juice. These results are in harmony with those obtained by [3,7,9,13].

Table 1- Means of total chlorophyll, leaf area/plant, foliage fresh weight, foliage length, root fresh weight, root length and diameter as affected by methanol, boron concentrations and their interaction during 2010/2011 and 2011/2012 seasons

Treatments	Characters													
	Total chlorophyll (SPAD)		Leaf area/plant (cm^2)		Foliage fresh weight (g)		Foliage length (cm)		Root fresh weight (g)		Root length (cm)		Root diameter (cm)	
Seasons	2010/2011	2011/2012	2010/2011	2011/2012	2010/2011	2011/2012	2010/2011	2011/2012	2010/2011	2011/2012	2010/2011	2011/2012	2010/2011	2011/2012
A- Methanol concentrations (%)														
Control without 0	42.08	42.86	3321.6	3372.9	229.91	238.33	41.02	41.8	843.75	851.08	21.71	22.98	11.6	11.92
Methanol 15%	44.47	45.34	3417	3509.5	291.25	303.75	43.09	44.09	906.25	914.58	24.32	25.87	13.07	13.57
Methanol 30%	47.91	49.04	3647	3707	385	407.91	46.12	47.57	1025	1022.16	26.89	27.82	14.25	14.99
Methanol 45%	45	46.12	3534.8	3559.5	331.66	348.33	44.02	45.02	931.66	927.25	25.71	26.27	13.66	13.92
F. test	**	**	**	**	**	**	**	**	**	**	**	**	**	**
LSD at 5%	0.72	0.39	36.3	36.3	24.7	15.42	0.86	0.85	7.94	23.4	0.54	0.51	0.32	0.28
B- Boron concentration (ppm)														
Control without 0	41.7	42.47	3284.1	3336.2	259.5	265	40.76	41.58	867.08	881	22.65	23.12	11.8	12.41
Boron 40 ppm	44.46	45.56	3433.3	3510.4	285	296.66	43.4	44.52	905.83	915.83	23.6	25.35	12.96	13.34
Boron 80 ppm	47.31	48.26	3641.6	3676.6	374.58	389.58	45.75	46.6	975	961.66	26.81	28.15	14.1	14.46
Boron 120 ppm	45.99	47.06	3561.5	3625.8	318.75	347.08	44.34	45.77	958.75	956.58	25.57	26.32	13.72	14.19
F. test	**	**	**	**	**	**	**	**	**	**	**	**	**	**
LSD at 5%	0.59	0.44	34	27.1	26.49	17.38	0.74	0.31	10.82	10.01	0.46	0.23	0.48	0.46
C- Interaction:	**	**	**	**	**	**	**	**	**	**	**	**	**	**

Table 2- Means of total soluble solids (TSS), sucrose and apparent purity percentages, root, top and sugar yields and harvest index (HI) as affected by methanol, boron concentrations and their interaction during 2010/2011 and 2011/2012 seasons.

Treatments	Characters													
	Total soluble solids (TSS%)		Sucrose (%)		Apparent Purity (%)		Root yield (ton/ha)		Top yield (ton/ha)		Sugar yield (ton/ha)		Harvest index (HI)	
Seasons	2010/2011	2011/2012	2010/2011	2011/2012	2010/2011	2011/2012	2010/2011	2011/2012	2010/2011	2011/2012	2010/2011	2011/2012	2010/2011	2011/2012
A- Methanol concentrations (%)														
Control without 0	21.67	21.86	16.45	16.73	75.86	76.5	57.217	57.75	15.017	15.467	9.429	9.677	0.792	0.789
Methanol 15%	22.18	22.6	17.1	17.4	77.11	76.98	59.081	60.352	16.686	17.283	10.117	10.517	0.78	0.778
Methanol 30%	22.8	23.11	18.45	18.75	80.87	81.12	67.882	69.595	21.284	22.621	12.554	13.097	0.762	0.755
Methanol 45%	22.7	22.73	18.05	18.2	79.49	80.04	62.682	63.912	19.12	19.682	11.335	11.654	0.767	0.765
F. test	**	**	**	**	**	**	**	**	**	**	**	**	**	**
LSD at 5%	0.23	0.24	0.34	0.33	1.39	1.55	0.381	0.376	0.143	0.269	0.121	0.127	0.003	0.007
B- Boron concentration (ppm)														
Control without 0	22.04	22.28	16.42	16.7	74.49	74.93	57.398	58.35	15.527	16.638	9.443	9.7627	0.787	0.779
Boron 40 ppm	22.24	22.5	17.45	17.72	78.41	78.7	61.758	62.696	18.173	18.692	10.819	11.152	0.774	0.772
Boron 80 ppm	22.63	22.82	18.2	18.5	80.36	81	64.359	66.111	19.565	20.141	11.759	12.295	0.768	0.768
Boron 120 ppm	22.45	22.7	17.98	18.17	80.08	80.01	63.346	64.455	18.844	19.585	11.414	11.738	0.771	0.768
F. test	**	**	**	**	**	**	**	**	**	**	**	**	**	**
LSD at 5%	0.19	0.05	0.29	0.2	1.15	0.79	0.354	0.197	0.278	0.243	0.083	0.068	0.007	0.006
C- Interaction:	NS	NS	NS	**	NS	**	**	**	**	**	**	**	**	*

C. Interactions Effect

Regarding the interaction effect between methanol and boron concentrations. The results clearly showed a significant effect on total chlorophyll [Fig-1], leaf area/plant [Fig-2], foliage fresh weight [Fig-3], foliage length [Fig-4], root fresh weight [Fig-5], root length [Fig-6], root diameter [Fig-7], root yield [Fig-8], top yield [Fig-9], sugar yield [Fig-10] and harvest index [Fig-11] in both season as well as sucrose [Fig-12] and apparent purity [Fig-13] in the second season only.

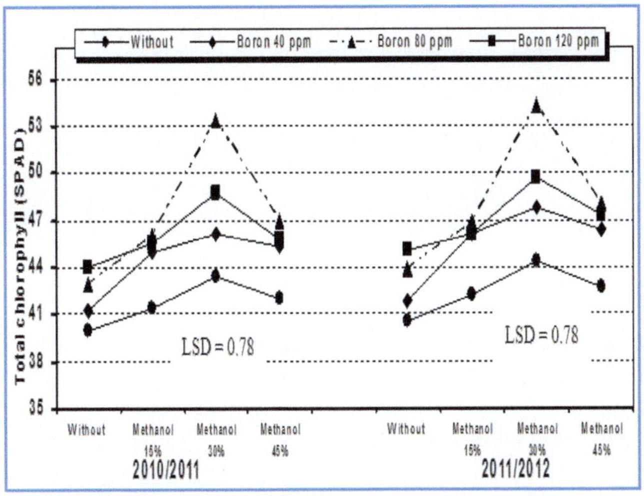

Fig. 1- Total chlorophyll (SPAD) of sugar beet as affected by the interaction between methanol and boron treatments during 2010/2011 and 2011/2012 seasons

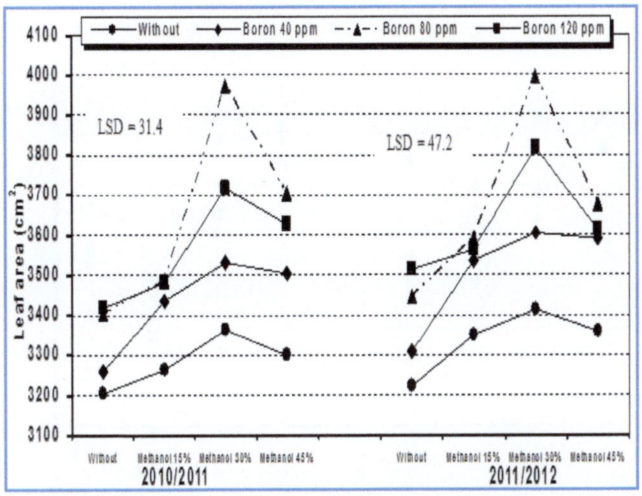

Fig. 2- Leaf area (cm^2) of sugar beet as affected by the interaction between methanol and boron treatments during 2010/2011 and 2011/2012 seasons

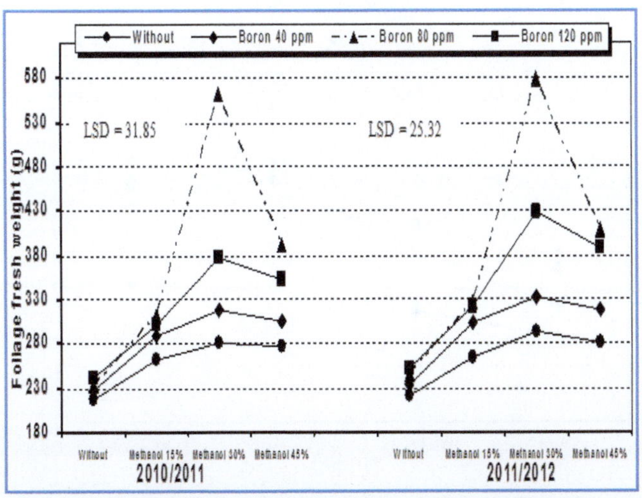

Fig. 3- Foliage fresh weight (g) of sugar beet as affected by the interaction between methanol and boron treatments during 2010/2011 and 2011/2012 seasons

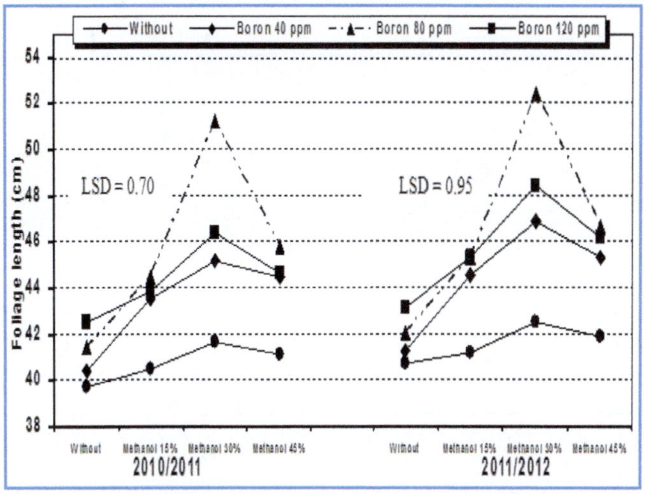

Fig. 4- Foliage length (cm) of sugar beet as affected by the interaction between methanol and boron treatments during 2010/2011 and 2011/2012 seasons

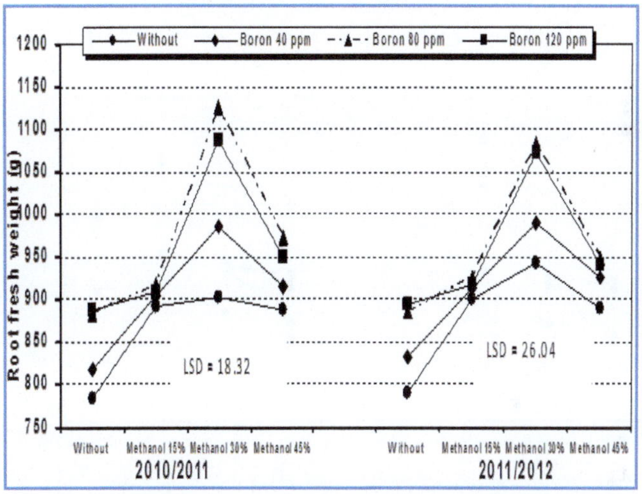

Fig. 5- Root fresh weight (g) of sugar beet as affected by the interaction between methanol and boron treatments during 2010/2011 and 2011/2012 seasons

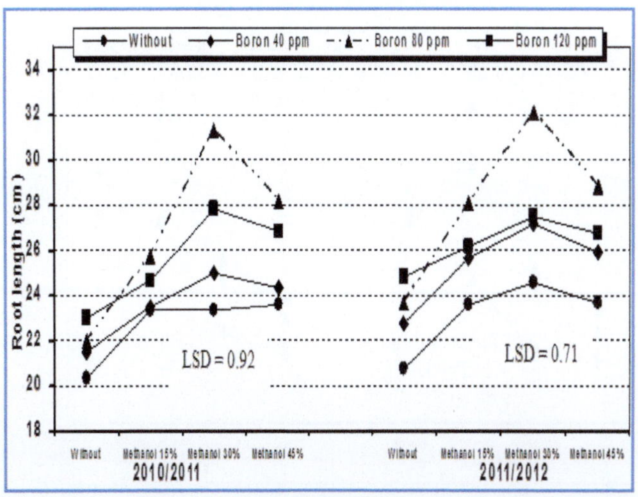

Fig. 6- Root length (cm) of sugar beet as affected by the interaction between methanol and boron treatments during 2010/2011 and 2011/2012 seasons

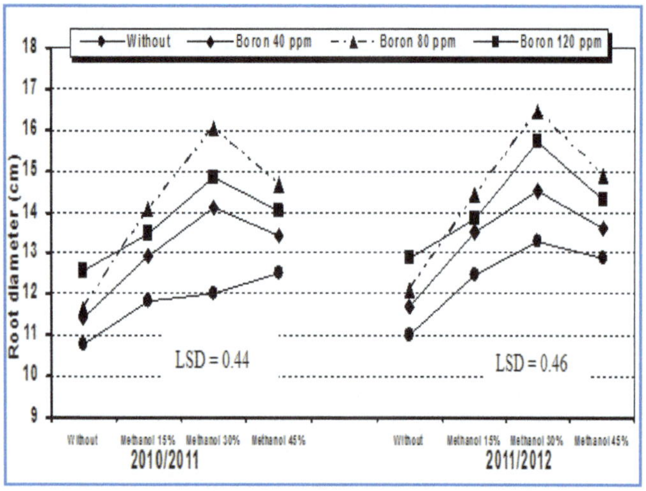

Fig. 7- Root diameter (cm) of sugar beet as affected by the interaction between methanol and boron treatments during 2010/2011 and 2011/2012 seasons.

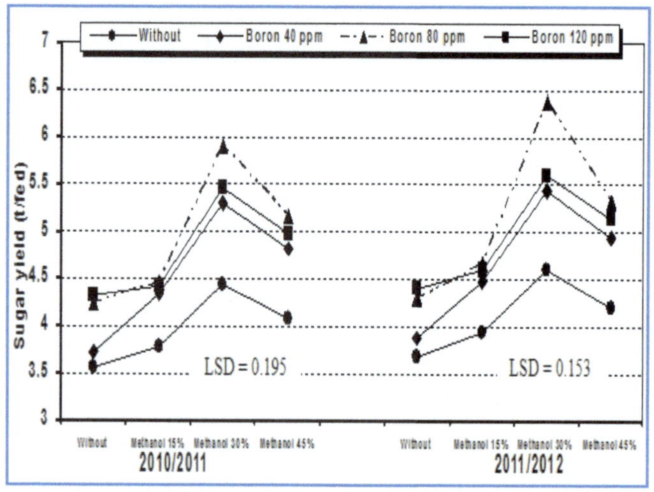

Fig. 10- Sugar yield (t/fed) of sugar beet as affected by the interaction between methanol and boron treatments during 2010/2011 and 2011/2012 seasons.

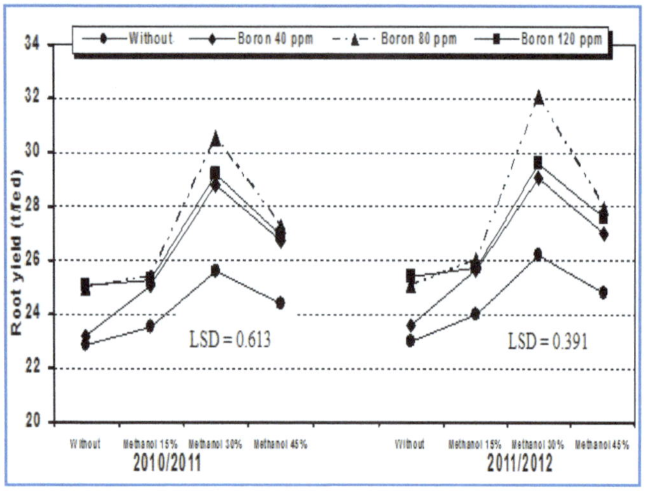

Fig. 8- Root yield (t/fed) of sugar beet as affected by the interaction between methanol and boron treatments during 2010/2011 and 2011/2012 seasons

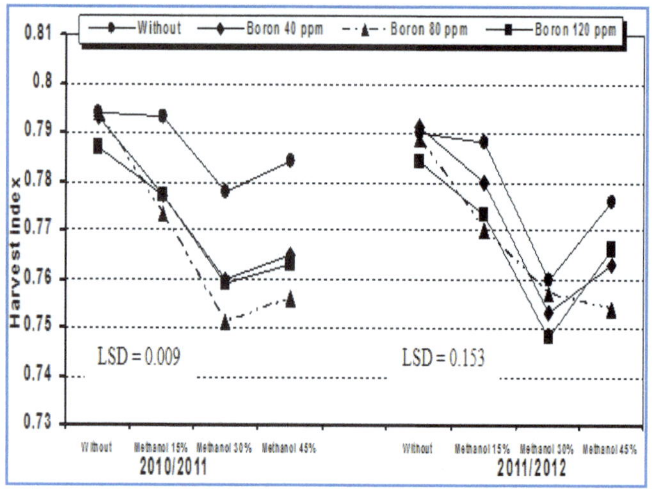

Fig. 11- Harvest index of sugar beet as affected by the interaction between methanol and boron treatments during 2010/2011 and 2011/2012 seasons.

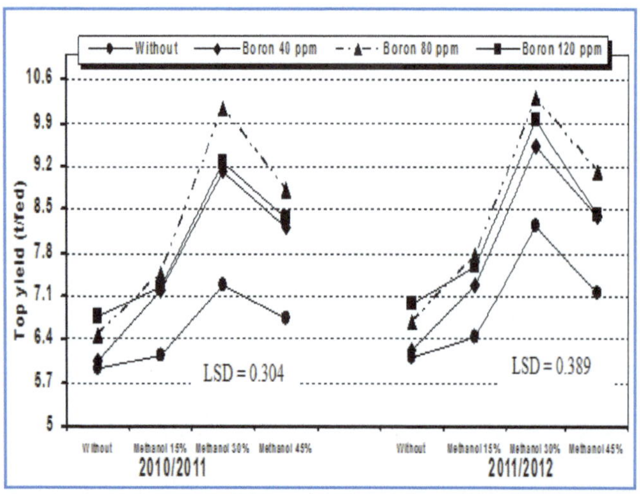

Fig. 9- Top yield (t/fed) of sugar beet as affected by the interaction between methanol and boron treatments during 2010/2011 and 2011/2012 seasons.

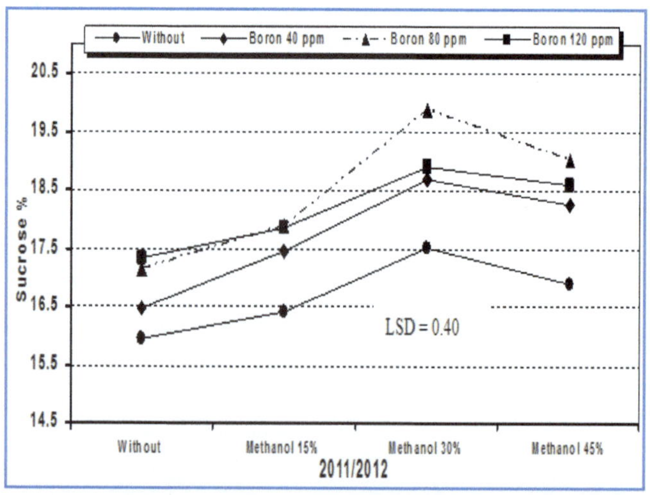

Fig. 12- Sucrose % of sugar beet as affected by the interaction between methanol and boron treatments during 2011/2012 season.

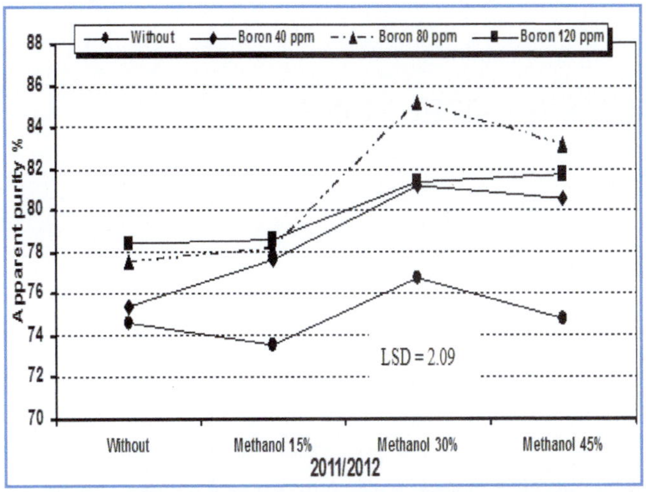

Fig. 13- Apparent purity % of sugar beet as affected by the interaction between methanol and boron treatments during 2011/2012 season.

It could be stated that highest averages of total chlorophyll, leaf area/plant, foliage and root fresh weight, foliage and root length, root diameter, sucrose (%), apparent purity percentages (%), top and root yield, sugar yield were obtained from foliar spraying of 30% methanol and 80 ppm boron. On the other hand, highest values of harvest index were recorded due to the interaction without methanol spraying.

It could be recommended that to maximize sugar beet yields and root quality achieved by foliar spraying 30% (v/v) methanol and 80 ppm boron.

References

[1] Abdel-Latif A., Schmieden U., Barakat S. and Wild A. (1996) *Plant Physiology and Biochemistry, Spec. Issue*, 133.

[2] Andres A.R., Lazaro J.J., Chueca A., Hermoso R. and Gorge J.L. (1990) *Physiologia Plantarum*, 78, 409-413.

[3] Armin M. and Asgharipour M.R. (2012) *American-Eurasian J. Agric. & Environ. Sci.*, 12(4), 444-448.

[4] Brown P.H. and Hu H. (1996) *Ann. Botany*, 77, 497-505.

[5] Carruthers A. and Oldfield J.E.T. (1960) *Int. Sugar J.*, 63, 72-74.

[6] Desclaux D., Huynh T.T. and Roumet P. (2000) *Crop Sci.*, 40, 716-722.

[7] Dordas C., Apostolides G.E. and Goundra O. (2007) *J. of Agric. Sci.*, 145, 377-384.

[8] Gerendas J. and Sattelmatcher B. (1990) *Plant Nutration Physiology and Application*, 33-37.

[9] Gezgin S., Hamurcu M. and Apaydin M. (2001) *Turkish J. of Agric. and Forestry*, 25, 89-95.

[10]Gobarah Mirvat E. and Mekki B.B. (2005) *J. of Applied Sci. Res.*, 1(5), 373-379.

[11]Gomez K.N. and Gomez A.A. (1984) *Statistical Procedures for Agricultural Research, 2nd Ed., John Wiley & Sons, Inc.*, New York, 95-109.

[12]Gout E., Aubert S., Bligny R., Rébeillé F., Nonomura A.R., Benson A.A. and Douce R. (2000) *Plant Physiology*, 123, 287-296.

[13]Hellal F.A., Taalab A.S. and Safaa A.M. (2009) *Ozean J. of Applied Sci.*, 2(1), 1-10.

[14]Kristek A., Stojic B. and Kristek S. (2006) *Agriculture-Scientific and Professional Review*, 12(1), 22-26.

[15]Lee H.S., Madhaiyan M., Kim C.W. Choi S.J., Chung K.Y. and Sa T.M. (2006) *Biology and Fertility of Soils*, 42, 402-408.

[16]Loomis W.D. and Durst R.W. (1992) *BioFactors*, 3, 229-239.

[17]Makhdum M.I., Malik M.N.A., Din S.U., Ahmad F. and Chaudhry F.I. (2002) *Journal of Researches in Sci.*, 13, 37-43.

[18]Nadali I., Paknejad F., Moradi F., Vazan S., Tookalo M., Jami Al-Ahmadi M. and Pazoki A. (2010) *Australian J. of Crop Sci.*, 4 (6), 398-401.

[19]Nonomura A.M. and Benson A. (1992) *Proceedings of National Academy of Science, USA*, 89, 9794-9798.

[20]Sadeghi-Shoae M., Paknejad F., Kashani A., Vazan S. and Nooralvandi T. (2012) *Resilience of Agricultural Systems Against Crises*, Tropentag, Gottingen- Kassel/Witzenhausen.

[21]Snedecor G.W. and Cochran W.G. (1980) *Statistical Methods, 7th ed.*, USA.

[22]Tersahima I.and Evans J.R. (1988) *Plant Cell Physiol.*, 29, 143-155.

[23]Zbieć I. and Karczmarczyk S. (1997) *Romanian Agricultural Res.*, 7-8, 45-49.

[24]Zbieć I., Karczmarczyk S. and Podsiadlo C. (2003) *Electronic-Journal of Polish Agricultural Universities, Agronomy*, 6, 1-7.

[25]Zbieć I.I., Karczmarczyk S. and Koszański Z. (1999) *Folia Universitatis Agriculturae Stetinensis, Agricultura*, 73, 217-220.

VALIDATION OF TOMATO POWDERY MILDEW FORECASTING MODEL USING METEOROLOGICAL DATA IN EGYPT

BAKEER A.R.T.[1]*, ABDEL-LATEF M.A.E.[2], AFIFI M.A.[2] AND BARAKAT M.E.[1]

[1]Botany Department, Faculty of Agriculture, Fayoum University, 63414- Fayoum, Egypt.
[2]Plant Pathology Institute, Agricultural Research Center, Egypt.
*Corresponding Author: Email- bakeer91@yahoo.com

Abstract- This study was initiated to validate a tomato powdery mildew, caused by *Leveillula taurica* (L´ev.) Arn., forecast model in Egypt. Environmentally acceptable control may be achieved using a weather based forecasting system due to the reduced use of agrochemicals and their residues in both environment and food. A computerized disease forecast model for tomato powdery mildew (Egy. Tom-PM) as named by the authors was validated during the 2006 and 2007 growing seasons. The micrometeorological parameters monitored hourly in tomato canopy were ambient air temperature, relative humidity, leaf wetness duration, wind speed and global radiation using a wireless real-time automatic Agro-weather station model (Adcon Telemetry model A733, add wave). Results indicated that Egy.Tom-PM forecast model accurately defined disease risk appearance and reduced number of sprays in both years compared with the routine application (10D) and (7D) sprays in 2006 and 2007 respectively. Basic roles of system analysis for model validation and evaluation are discussed.

Keywords- *Leveillua taurica*, *Lycopersicon esculentum*, Egy.Tom-PM forecast model

Introduction

Tomato (*Lycopersicon esculentum* Mill.) is very important vegetable crop in Egypt for both local consumption and exportation purposes. Powdery mildew, caused by *Leveillula taurica* (L´ev.) Arn., is the most important disease for tomato in both greenhouses and open fields. The anamorphic stage of *Leveillula taurica* found in tomato crop is known as *Oidium taurica* (L´ev.) Salmon [1]. Besides tomato, powdery mildew infects other vegetable crops such as pepper, eggplant, and cucumber [2]. The most common symptoms of tomato powdery mildew are yellow and necrotic spots on the upper leaf surface and powdery white spots on the lower side of the leaves [3].

Environmental conditions have a positive influence on tomato powdery mildew and greatly affect the disease development operation. Disease monitoring, early detection and prevention of tomato powdery mildew is critical. By the time tomato powdery mildew is detected in open fields many more leaves are already infected but do not show any disease symptoms or signs. In addition, The commonly observed symptoms of powdery mildew disease caused by *L.taurica* as light green to yellowish spots or blotches on the upper tomato leaf surface. While on the lower surface, fine whitish grey to purplish growth of the causal agent arises from the lower stomata. Later, infected areas of leaves dry up and turn into dark brown. Finally, the leaves may defoliate. Generally, the symptoms are considerably more visible on older leaves than on younger ones [4-7]. On the other hand, this disease does not infect the stem or fruit but causes general declines in growth, chief foliage damage, reduce the quality of fruits and finally, it sharply decreases tomato yield [8-11]. Moreover, increasing resistance to fungicides by pathogens has greatly increased pre-and post harvest losses [12,13]. Tomato growers need to follow an intensive disease prevention plan because it is very important that powdery mildew never gets out of hand. Increasing concern for environment and food safety against fungicides application motivates farmers to optimize powdery mildew control by reducing chemical applications.

Monitoring of weather conditions within the plant canopy is possible by automated weather stations, real-time weather data can be transmitted from automated weather stations at remote sites to personal computers via telephone modems or by radio telemetry. Many attempts have been made to forecast tomato powdery mildew, or to provide systems that support the spray decision and replace the routine prophylactic spraying. All forecasts are based on determining the probability of weather that will favor infection and use of daily disease infection potential (DDIP) to produce risk warning and spray recommendations [14,15]. An electronic warning system for grape downy mildew, based on models for the infection of leaves of *Vitis lambrusca*, production of sporangia by *Plasmopara viticola* in lesions, and sporangial survival, was tested over 7 years in Ohio. Grapevines were sprayed with metalaxyl plus mancozeb (Ridomil MZ58) when the warning system indicated that environmental conditions were favorable for sporulation and subsequent infection. Effective control of downy mildew, therefore, can be achieved with the use of the warning system with fewer sprays than with a standard schedule [16].

The computerized model (Egy.Tom-PM) simulates the development of powdery mildew on the basis of short-term observation of microclimate factors. The input variables to the model are: temperature, Relative humidity, leaf wetness, wind speed and global radiation for the disease infection potential (DIP).

The objective of this research was to validate Egy.Tom-PM, the first Egyptian computerized forecasting model for better disease control with minimum fungicide use. It works by forecasting the most appropriate spray intervals to manage effective control of tomato powdery mildew, compared with routine fungicide application.

Materials and Methods

Disease Forecasting Model to Avoid Epidemics of Tomato's Powdery Mildew caused by *Leveillula taurica*

Weather Monitoring

Automated portable and solar powder in-crop weather stations with high technology named (Adcon Telemetry model A733add wave, [Fig-1] was used to monitor the microclimatological conditions. Every 15 minutes, digital data for temperature, relative humidity, wind speed, leaf wetness and global radiation within the canopy of tomato plants were transmitted via PC using radio waves up to 20 km in distance with no geographic obstacles. The meteorological data are transmitted automatically by radio from the weather station via data collection center (receiver) to receive and save the collected data for up to 50 days and advantage software which allows simple interpretation of the readings using a graph and detailed list of chosen items [Fig-2].

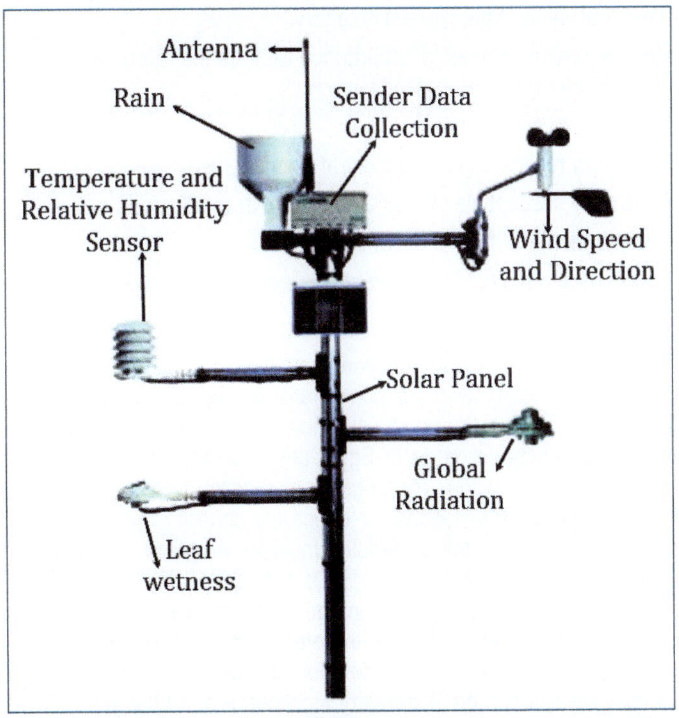

Fig. 1- The last generation of remote measuring agriculture weather stations for real time data around the world named Adcon Telemetry model A733.

Tomato Powdery Mildew Forecast Model

A computerized simulation software model named by the authors Egy.Tom-PM [Fig-3] was designed and tested several times under laboratory conditions before switching to a validation phase under field conditions at El-Khawagat village, Yousef El-Seddik County, Fayoum governorate, Egypt during the 2006 and 2007 Nile tomato growing seasons. These experiments were used to figure out how well such a model forecast disease infection potential (DIP), which revised and refined based on these findings. Frequently, disease control (times of sprays) of grapes treated according to the model recommendations were compared to the disease managed by traditional routine spray schedules.

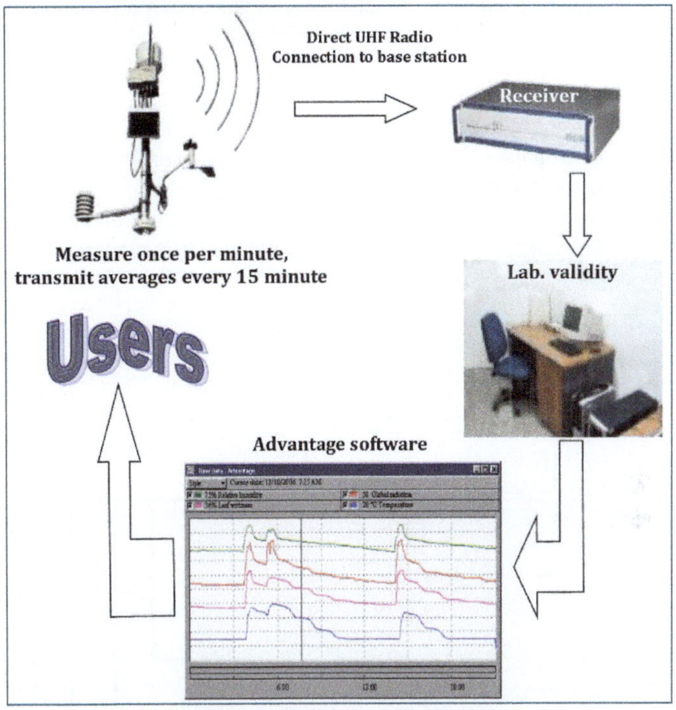

Fig. 2- Different stage of Data collection from weather station, throughout PC and advantage software to users.

Fig. 3- Interface of Egy.Tom-PM program, the first record of disease forecast and early warning computerized model all over the Middle East area.

The Computerized Model System Analysis

Egy.Tom-PM model, presented as a flowchart [Fig-4] is a creation and designation of a computerized model, based on short-term observations (hourly weather data) analyzing the correlation between the input variables of microclimatological factors such as: temperature, relative humidity, wind speed, leaf wetness and global radiation and its effect on the disease daily infection potential

(DDIP) of tomato powdery causal agent *Leveillula taurica*, to be calculated. Then, the model outputs a daily announcement as a warning message (spray or don't spray) to guide the fungicides application for perfect disease control in the appropriate time. The model evaluation follows the basic rules of system analysis to identify events (A and B). The model starts after the emergence phase is activated, and it tries to detect the event A, which is defined as: Event A is triggered when the model detects at the least X accumulated dynamic summation hours of RH >75 and temperature between Y and Z according to the data tabulated in [Table-1]. Then the model looks for event B which is continues hours of RH > 85 with temperature of the same value between Y and Z. The B event must happen within the 24h of the A event, but not later than 24h after an A event. Moreover, using the model system analysis brings an extension to the model that applies the same rules not only to identify the first critical phase of the season, but also to issue warnings through the whole growing season (DIP).

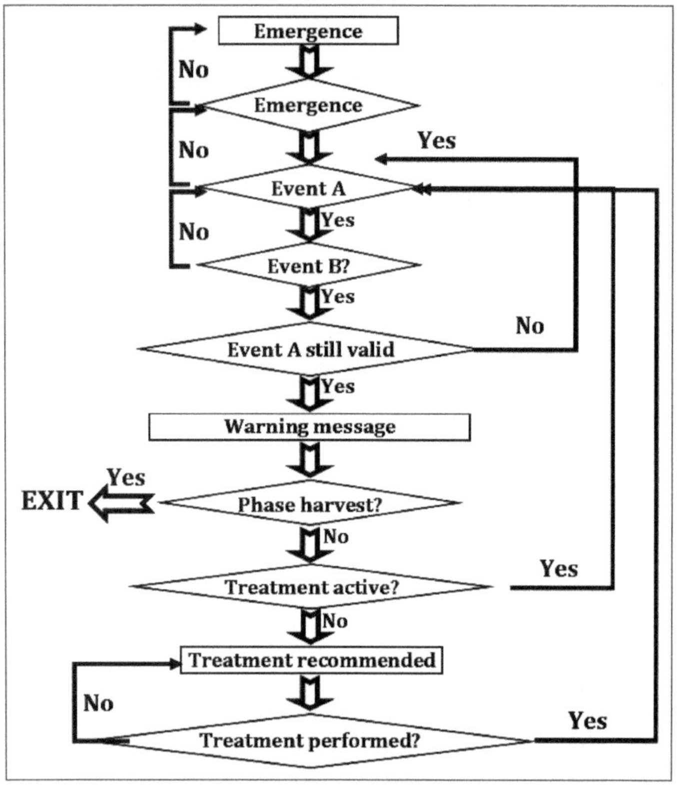

Fig. 4- Flowchart of system analysis for tomato powdery mildew computerized disease forecast model.

Table 1- Detection of X accumulated dynamic summation hours of R.H. and Temperature (event A)

Y = Temp from	Z = Temp. to	X = 6 hrs. of RH	X = 8 hrs. of RH	X = 10 hrs. of RH
10.0°C	15.0°C	Event A detect		
15.0°C	20.0°C		Event A detect	
20.0°C	25.0°C			Event A detect

Model Validation

Field experiment was carried out at El-Khawagat village, Yousef El-Seddik County, Fayoum governorate, Egypt through both 2006 and 2007 tomato Nili growing seasons. The highly susceptible Hybrid 765 tomato seedlings were transplanted in the field in a randomized block design with three replicates per treatment. The following three treatments were tested: (i) a full-schedule fungicide program, in which plants were sprayed every 10 days; (ii) a full-schedule fungicide program, in which plants were sprayed every 7 days; (iii) spraying when nominal use of forecasting system indicated that a spray application was needed and at least 10 days had elapsed since the previous fungicide application. In addition, sulphur 250 gm/100 L water and protectant fungicide (sumi-8) 35 cm3/100 L water were used alternatively with some specific guidance (sprays should cover tomato plants, particularly on the under surface of the foliage and the lower plant canopy and should not be used during very warm, sunny weather to prevent phytotoxicity) were used for disease control as a recommended effective chemicals. Weekly tomato field inspection was initiated to ensure clear detection of the early signs of powdery mildew. Microclimate weather data were automatically collected 24h a day by the advanced telemetry weather station (Adcon A 733), established within the tomato field canopy.

Disease Assessment

The percentage of disease severity was calculated using the equation suggested by Townsend & Heuberger [17] as follows:

$$P = \frac{\sum [\text{a (rating no.)} \times \text{b (no. leaves in rating category)}]}{\text{N (Total no. leaves)} \times \text{K (highest rating value)}} \times 100$$

where :

P = percentage of disease severity.

a = number of leaves in each category.

b = numerical value of each category.

N = total number of leaves in sample.

K = numerical value of the highest category.

A scale of scores was used in accordance with the leaf area affected, proposed by Ullasa, et al [18] of which:

1. Resistant (no symptoms).
2. Moderately resistant (10% of the leaf area affected).
3. Moderately susceptible (11-20% of the leaf area affected).
4. Susceptible (21-50% of the leaf area affected).
5. Highly susceptible (51% or more of the leaf area affected).

Results and Discussion

The results of entire study indicated that Egy.Tom-PM program (Powdery mildew early warning computerized model) for tomato powdery mildew, which designed using the roles of system analysis, was successfully validated in Egypt under tomato field conditions allowing the best timing for fungicide application to reduce the number of fungicide applications for effectively control of powdery mildew compared with full schedule-fungicide program (routine application) with both (7D) and/or (10D) schedule. At the same time, the message of early warning model can forecast the probability of powdery mildew outbreak therefore avoid the exorbitant crop loss in such case with a consideration of powdery mildew epidemics had been recorded in deferent locations around the world. The results are expressed in an early warning message (spray and do not spray) in agreement with weekly field inspection. Data in [Table-2] show that at 2006 season, the warning message of Egy.Tom-PM program was to spray during September, October, November and December where they had one warning message for every one of them individually. In September, disease severity recorded 2.7% when temperature recorded 25.3, 25.8 and 26.6°C and R.H. 76.0, 80.0 and 83.0% in 23, 24 and 25 September, respectively.

Table 2- Implementation of Egy.Tom-PM program conjugate with daily average of microclimate factors during 2006 at Yousef El-Seddik county, Fayoum governorate

Date	Temp (°C)	R.H (%)	Leaf wet. (%)	Wind speed (Km/h)	Global Radiation (watt/m2)	Disease Severity
September-2006						
01-Sep	32.3	47	24	3	51	
02-Sep	31.6	45	23	2	49	
03-Sep	30.4	43	21	4	46	
04-Sep	31.7	46	24	3	50	
05-Sep	32.5	50	27	1	53	
06-Sep	33.2	54	29	2	56	
07-Sep	31.6	49	26	4	54	0%
08-Sep	30.2	47	25	3	51	
09-Sep	30.8	48	24	4	53	
10-Sep	29.5	45	22	5	50	
11-Sep	28.7	46	23	3	47	
12-Sep	29.3	50	25	2	49	
13-Sep	28.1	48	24	3	48	
14-Sep	27.6	45	22	4	46	0%
15-Sep	26.8	47	21	3	44	
16-Sep	27.5	50	23	3	45	
17-Sep	28.9	52	25	4	48	
18-Sep	27.3	56	27	3	52	
19-Sep	25.4	60	24	2	54	
20-Sep	26.6	64	22	1	56	
21-Sep	27.9	68	23	2	55	1.8%
22-Sep	26.5	72	25	2	54	
23-Sep*	25.3	76	24	1	56	
24-Sep*	25.8	80	22	1	55	
25-Sep*	26.6	83	25	1	58	
26-Sep	27.7	78	24	2	56	
27-Sep	28.8	74	22	1	54	
28-Sep	29.4	70	21	2	51	2.7%
29-Sep	28.2	67	23	1	48	
30-Sep	28.6	65	25	1	50	
October-2006						
01-Oct	29.3	66	35	2	52	
02-Oct	28.2	63	33	3	54	
03-Oct	29.5	61	31	4	56	
04-Oct	28.6	62	32	2	53	
05-Oct	27.8	58	30	4	51	
06-Oct	28.4	60	31	3	52	
07-Oct	29.5	64	33	1	54	0%
08-Oct	30.1	67	34	2	56	
09-Oct	28.7	65	32	3	53	
10-Oct	27.3	68	34	1	50	
11-Oct	26.1	66	33	2	48	
12-Oct	27.4	70	35	1	49	
13-Oct	26.5	68	34	2	47	
14-Oct	27.1	71	33	3	48	0.9%
15-Oct	26.6	74	35	2	46	
16-Oct*	25.2	77	37	1	44	
17-Oct*	25.5	81	38	1	43	
18-Oct*	25.8	83	37	2	42	
19-Oct	26.4	84	36	3	44	
20-Oct	27.7	82	34	4	46	
21-Oct	28.5	84	33	3	48	2.5%
22-Oct	29.3	81	31	4	49	
23-Oct	28.1	78	30	3	47	
24-Oct	27.6	73	27	4	45	
25-Oct	26.2	70	26	3	43	
26-Oct	27.5	74	29	2	44	
27-Oct	28.8	72	27	4	46	1.3%
28-Oct	27.4	70	26	3	45	
29-Oct	26.9	68	25	4	43	
30-Oct	27.2	72	27	3	44	
31-Oct	28.5	76	29	2	46	

Table 2-Continue...

Date	Temp (°C)	R.H (%)	Leaf wet. (%)	Wind speed (Km/h)	Global Radiation (watt/m2)	Disease Severity
November-2006						
01-Nov	28.8	80	31	1	47	
02-Nov	28.3	78	29	2	46	
03-Nov	27.1	76	28	1	44	
04-Nov	26.7	74	26	2	43	
05-Nov	27.4	77	27	2	45	
06-Nov	28.6	80	29	4	47	
07-Nov	27.9	82	29	3	46	1.1%
08-Nov	27.2	84	30	2	45	
09-Nov	26.4	81	28	2	43	
10-Nov*	25.1	83	29	1	41	
11-Nov*	25.7	86	31	2	41	
12-Nov*	25.4	88	32	1	42	
13-Nov	24.3	84	30	3	40	
14-Nov	23.6	81	28	2	38	2.9%
15-Nov	22.7	77	29	3	39	
16-Nov	23.5	74	26	4	37	
17-Nov	22.8	71	23	4	36	
18-Nov	23.9	73	24	3	34	
19-Nov	23.1	71	25	3	36	
20-Nov	24.3	72	26	4	37	
21-Nov	25.6	75	24	3	39	1.6%
22-Nov	24.7	72	21	4	38	
23-Nov	24.2	70	20	3	37	
24-Nov	25	68	18	4	39	
25-Nov	24.1	66	16	3	38	
26-Nov	24.7	64	15	4	40	
27-Nov	23.5	61	13	3	39	
28-Nov	24.8	63	14	3	41	0%
29-Nov	24.3	65	16	2	40	
30-Nov	23.6	62	15	3	38	
December-2006						
01-Dec	26.5	76	29	1	44	
02-Dec	25.1	74	27	2	42	
03-Dec	24.8	73	26	1	41	
04-Dec	24.3	70	24	2	40	
05-Dec	25.6	74	25	1	42	0.7%
06-Dec*	26.4	77	27	3	43	
07-Dec*	25.2	78	26	2	44	
08-Dec*	24.9	80	28	1	42	
09-Dec	24.5	77	26	3	41	
10-Dec	23.4	80	27	2	38	
11-Dec	23.1	83	29	2	36	
12-Dec	22.6	85	31	3	37	22%
13-Dec	22	82	29	2	36	
14-Dec	21.4	78	27	3	35	
15-Dec	20.5	74	26	3	36	
16-Dec	21.7	71	25	2	37	
17-Dec	20.3	68	23	1	33	
18-Dec	21.5	70	22	1	32	
19-Dec	20.6	67	24	2	34	1%
20-Dec	21.9	69	25	3	35	
21-Dec	22.4	72	22	1	37	
22-Dec	21.8	69	20	2	35	
23-Dec	22.1	66	18	1	36	
24-Dec	23.3	64	16	1	38	
25-Dec	22.5	62	15	2	36	
26-Dec	22.8	61	13	1	37	
27-Dec	21.4	58	11	1	38	0%
28-Dec	22.6	60	12	2	39	
29-Dec	22	62	14	3	37	
30-Dec	21.3	59	13	2	35	
31-Dec	20.5	56	15	3	33	

**Disease severity of tomato powdery mildew throughout daily alert forecast message program "SPRAY"*

Moreover, during October disease severity recorded 2.5% when temperature recorded 25.2, 25.5 and 25.8°C and R.H. 77.0, 81.0 and 83.0% in 16, 17 and 18 October, respectively. Moreover, disease severity recorded 2.9%, during November, when temperatures were 25.1, 25.7 and 25.4°C and R.H. were 83.0, 86.0 and 88.0% in 10, 11 and 12 November, respectively. During December, disease severity recorded 2.2% when temperature recorded 26.4, 25.2 and 24.9°C and R.H. 77.0, 78.0 and 80.0% in 6, 7 and 8 December, respectively.

On the other hand, data in [Table-3] show that at the 2007 season, the warning message of Egy.Tom-PM program was to spray during September once when temperature recorded 25.4, 25.9 and 25.1°C where R.H. were recorded 81.0, 84 and 87.0% at 14, 15 and 16 September, respectively, and disease severity was 3.5%. During October, warning messages were twice. The disease severity was recorded twice, the first 3.8% when temperatures recorded 25.6, 25.9 and 26.4°C and R.H. were 82, 85 and 88% at 19, 20 and 21 October, respectively. The Second disease severity was 2.4% when temperatures recorded 25.4, 25.8 and 26.2°C, and R.H was 90, 93 and 96%, throughout days 28, 29 and 30 days of October, respectively. During November disease severity, was 3.4% when temperatures were 25.4, 25.7 and 25.1°C and R.H. recorded 93, 94 and 92% during 2, 3 and 4 November, respectively. Moreover, during December, disease severity was only recorded once, being 3.2% when temperatures were 24.4, 25.1 and 24.8°C and R.H. 91, 89 and 92% during 1, 2 and 3 December, respectively.

As a matter of fact, in this study the application of tomato powdery mildew disease forecast and early warning model, using a designed computerized model adopted with role of system analysis has been done for the first time not only in Egypt but also all over Arabian and African countries. According to the literature, it is also the first description of powdery mildew all over the Middle East area. Results in this point showed that using advanced last generation of agrometeorological remote weather station to collect the daily microclimate elements (the disease forecast and early warning input data) was agreement with Hstgaard [19] who reported that a fully equipped weather station with wireless data transmission, the Hardi MetpoleReg., and its presentation software developed for use by farmers. The Hardi MetpoleReg, was equipped with sensors for collecting wind velocity, surface wetness, global radiation, air temperature, relative humidity and rainfall data at various heights above the ground. Data were transmitted by radio waves to a receiver at the farm and presented using the computer software. All data are transferable to ASCII-format files for use in computerized monitoring and forecasting systems. Powdery mildew of tomato caused by *L. taurica* in considered one of the most limiting diseases for tomato production and could be controlled effectively in Egypt for the first time with fewer fungicide applications than a slandered calendar – based schedule with system of disease forecast using computerized model Egy - Tom - PM. This model resulted in a noticeable reduction in fungicide application and consequently reduced the number of sprays. During the 2006 season there were four warning messages in September, October, November and December individually. On the other hand, in 2006 season there were one, two, one and one warning messages in September, October, November and December, respectively. The previous results may be due to the favorable weather conditions, especially under optimum temperature (20-25) and R.H. ranging between 75-90% in which tomato powdery mildew (*L. taurica*) needs to development and progress.

Table 3- Implementation of Egy.Tom-PM program conjugate with daily average of microclimate factors during 2007 at Yousef El-Seddik county, Fayoum governorate

Date	Temp (°C)	R.H (%)	Leaf wet. (%)	Wind speed (Km/h)	Global Radiation (watt/m2)	Disease Severity
September-2007						
01-Sep	30.6	65	36	3	44	
02-Sep	29.5	67	33	2	42	
03-Sep	28.7	70	31	1	40	
04-Sep	29.1	73	34	2	41	
05-Sep	30.4	77	37	1	43	
06-Sep	31.3	79	40	2	46	
07-Sep	29.8	81	37	1	44	0
08-Sep	28.5	78	35	3	42	
09-Sep	28.9	80	36	2	43	
10-Sep	27.6	77	34	3	41	
11-Sep	26.3	75	32	2	38	
12-Sep	26.7	77	33	3	39	
13-Sep	26.1	78	31	2	37	
14-Sep*	25.4	81	34	1	35	1.2
15-Sep*	25.9	84	36	2	36	
16-Sep*	25.1	87	38	1	38	
17-Sep	27	85	40	3	41	
18-Sep	28.3	87	38	2	43	
19-Sep	29.6	84	36	3	46	
20-Sep	28.8	79	34	4	44	
21-Sep	27.5	76	32	3	42	3.5
22-Sep	28.4	77	33	4	45	
23-Sep	29.1	79	35	4	47	
24-Sep	30.7	82	38	3	50	
25-Sep	30.2	80	36	2	48	
26-Sep	29	75	33	3	46	
27-Sep	28.4	70	30	4	44	
28-Sep	27	67	28	2	41	1.8
29-Sep	26.2	63	26	3	39	
30-Sep	27.5	66	28	2	37	
October-2007						
01-Oct	26.3	63	25	2	35	
02-Oct	27.5	65	26	3	37	
03-Oct	28.8	68	28	1	39	
04-Oct	27.4	66	27	2	38	
05-Oct	28.6	69	29	1	40	
06-Oct	29.1	71	30	3	43	
07-Oct	28.5	74	32	1	41	0
08-Oct	28	72	31	2	40	
09-Oct	27.3	69	28	3	39	
10-Oct	26.7	73	31	1	37	
11-Oct	25.5	70	29	2	35	
12-Oct	26.2	73	32	1	36	
13-Oct	25.4	69.1	28	3	34	
14-Oct	26	73	30	2	35	1.5
15-Oct	27.3	78	33	1	35	
16-Oct	26.1	75	31	2	33	
17-Oct	26.6	77	32	2	34	
18-Oct	25.2	80	34	1	34	
19-Oct*	25.6	82	36	1	35	
20-Oct*	25.9	85	38	1	36	
21-Oct*	26.4	88	37	2	37	
22-Oct	24.8	84	34	3	35	3.8
23-Oct	25.5	87	36	2	36	
24-Oct	26.1	89	37	2	37	
25-Oct	27.6	86	35	3	40	
26-Oct	26.3	83	33	3	42	
27-Oct	26.9	87	35	2	40	
28-Oct*	25.4	90	37	1	38	
29-Oct*	25.8	93	38	2	36	
30-Oct*	26.2	96	40	2	38	
31-Oct	27.5	94	38	3	40	2.4%

Table 3-Continue...

Date	Temp (°C)	R.H (%)	Leaf wet. (%)	Wind speed (Km/h)	Global Radiation (watt/m2)	Disease Severity
November-2007						
01-Nov	26.3	95	39	1	38	
02-Nov	25.4*	93	37	2	36	
03-Nov	25.7*	94	38	2	37	
04-Nov	25.1*	92	36	1	35	
05-Nov	24.5	88	35	3	37	
06-Nov	23.3	86	34	2	36	
07-Nov	22.6	82	31	3	34	3.4%
08-Nov	23.5	84	30	1	33	
09-Nov	22.2	80	29	3	34	
10-Nov	21	77	31	2	36	
11-Nov	22.4	75	30	3	35	
12-Nov	23.7	77	29	2	34	
13-Nov	24.8	76	27	3	32	
14-Nov	23.1	74	28	2	33	0%
15-Nov	23.5	72	26	3	31	
16-Nov	22.1	69	24	2	30	
17-Nov	21.7	66	22	3	28	
18-Nov	22.6	69	23	2	29	
19-Nov	23.2	72	24	1	30	
20-Nov	23.5	73	25	2	30	
21-Nov	22.7	70	26	3	32	0%
22-Nov	21.4	68	28	2	33	
23-Nov	20.8	66	29	1	32	
24-Nov	21.1	67	30	2	34	
25-Nov	22.2	69	32	3	33	
26-Nov	21.7	67	33	2	33	
27-Nov	20.4	64	34	2	31	
28-Nov	21.6	65	32	3	29	0%
29-Nov	22.2	67	30	2	28	
30-Nov	23	70	31	1	30	
December-2007						
01-Dec	24.4*	91	36	2	35	
02-Dec	25.1*	89	34	2	32	
03-Dec	24.8*	92	35	1	34	
04-Dec	23.5	88	33	2	33	
05-Dec	22	85	31	3	34	
06-Dec	21.4	82	30	3	32	
07-Dec	20.2	78	28	4	31	3.2%
08-Dec	20.8	81	27	2	30	
09-Dec	21.3	77	26	2	31	
10-Dec	19.7	73	28	3	33	
11-Dec	20.5	71	27	3	31	
12-Dec	21.1	74	26	1	30	
13-Dec	22.4	72	24	2	30	
14-Dec	21.6	70	25	3	31	0%
15-Dec	20.9	68	23	3	29	
16-Dec	20.2	65	21	4	27	
17-Dec	19.5	62	18	4	26	
18-Dec	20.1	66	20	3	27	
19-Dec	21	69	22	2	28	
20-Dec	21.6	71	21	1	27	
21-Dec	20.4	68	23	2	29	0%
22-Dec	19.7	65	25	3	30	
23-Dec	18.3	62	27	4	29	
24-Dec	18.9	64	28	3	31	
25-Dec	20.3	67	29	2	30	
26-Dec	19.2	64	30	3	31	
27-Dec	18.5	60	31	4	28	
28-Dec	19.6	61	30	3	26	0%
29-Dec	20.4	64	27	2	25	
30-Dec	21.1	67	29	1	27	
31-Dec	20	65	26	2	26	

**Disease severity of tomato powdery mildew throughout daily alert forecast message program "SPRAY"*

The results are in harmony with these obtained by Reuveni, et al [20], Fiume [21] and Guzman-Plazola, et al. Egy-Tom-PM development of certain disease into a computer-based model and then forecast the disease under different environmental conditions has been discussed in detail. The model was very accurate in simulating successful daily infection potential (DIP), because all the observed disease appearance in the tomato open field were correctly simulated by the model. In addition to the work presented here, Davis & Marois [14], Guzman-Plazola [15] and Seem [23] found that effective control of tomato powdery mildew could be achieved in California when protectant fungicides were applied according to a spray forecast model. Moreover, disease symptoms were never observed when the model did not signal a successful infection, even though the starting date for fungicide applications was accurately determined.

Conclusion

The results obtained have emphasised how for tomato protection applications, the use of disease forecasting program for tomato powdery mildew was applied after program validity test during two successive seasons 2006 and 2007 at El-Khawagat village, Yousef El-Seddik County, Fayoum governorate, Egypt. Powdery mildew forecast program reduced the number of application times of fungicides compared with control. Simulation models based on agrometeorological variables and the energy balance could therefore be an important tool for operational applications. Their integration with the development of disease simulation models is of great importance regarding the application time of fungicide sprays and reduce the number of total applications numbers [24] while maintaining a high quality of production with a low impact on the environment. As there was a good agreement between model outputs and actual disease severity, the model can be considered as a satisfactory simulator of the effect of environmental conditions on the progress of powdery mildew epidemics. These approaches have proved useful in describing control effectiveness and, in some cases, optimizing or changing control practices [25,26].

Conflicts of Interest : None declared.

References

[1] Whipps J.M., Budge S.P., Fenlon J.S. (1998) *Plant Pathol.*, 47 (1), 36-48.

[2] Palti J. (1988) *Botanical Review.*, 54, 423-535.

[3] Holliday P. (1980) *Fungus Diseases of Tropical Crops*, Cambridge University Press, New York, 607.

[4] Correll J.C., Gordon J.R., Elliott Y.J. (1988) *California Agric.*, 42 (2), 8-10.

[5] McGrath M.T., Shishkoff N., Bornt C., Moyer D.D. (2001) *Plant Dis.*, 85(10), 1122.

[6] Haroun S.H. (2002) *Studies on Powdery Mildew Disease on Tomatoes in Egypt*, Ph.D. Thesis, Fac. Agric. Moshtohor, Zagazig Univ., 166.

[7] Reis A., Lopes C.A., Paz-Lima M.L., Boiteux L.S. (2004) *Fitopatologia Brasileira, Fortaleza*, 29(3), 339.

[8] Natour R.M., El-Behadi A.H., Majeed M.G. (1971) *Plant Dis.*, 55 (2), 192.

[9] Jones W.B., Thomson S.V. (1987) *Plant. Dis.*, 71(3), 266-268.

[10] Schickedanz F. (1989) *Pflanzenschutzdienstes*, 41(12), 207-208.

[11]Farag F.M. (2003) *Studies on powdery mildew disease affecting solanaceous plant, M.Sc. Thesis, Fac. of Agric. Cairo Univ.*, 105.

[12]Braun U. (1987) *Beih Nova Hedwigia.*, 89, 1-700.

[13]Correll J.C., Gordon T.R., Elliott V.J. (1987) *Plant Dis.*, 71(3), 248-251.

[14]Davis R.M., Marois J.J. (1995) *Development of a spray forecast model for powdery mildew of tomato*, *California Tomato Research Institute Final Report*, 52-59.

[15]Guzman-Plazola R.A. (1997) *Development of spray forecast model for tomato powdery mildew* (*Leveillula taurica* (Lev) Arn.) *Ph.D. Thesis, University of California.*

[16]Madden L.V., Ellis M.A., Lalancette N., Hughes G., Wilson L.L. (2000) *Plant. Dis.*, 84, 549-554.

[17]Townsend G.R. and Heuberger J.W. (1943) *Plant Dis. Rep.*, 27, 340-343.

[18]Ullasa B.A., Rawal R.D., Sohi H.S., Singh D.P., Joshi M.C. (1981) *Plant Dis.*, 65, 600-601.

[19]Hstgaard M.B. (1995) 36*th Swedish Crop Protection Conference: Agriculture Pests Diseases/Weeds*, Uppsala, Sweden, 249-259.

[20]Reuveni R., Perl M., Rotem J. (1974) *Phytopathologische Zeitscrift*, 80(1), 79-84.

[21]Fiume F. (1997) 49*th International Symposium on Crop Prot., Gent, Belgium.*

[22]Guzman-Plazola R.A., Davis R.M., Marois J.J. (2003) *Crop Protection*, 22(10), 1157-1168.

[23]Seem R. (2001) *Plant Pathology*, Kyongiu, Republic of Korea.

[24]Magarey R.D., Russo J.M., Seem R.C., Gadoury D.M. (2005) *Agricultural and Forest Meteorology*, 128(1), 111-122.

[25]Rossi V. and Giosue S. (2003) *Bull. OEPP.*, 33(3), 389-396.

[26]Jeger M.J. (2004) *Ann. Rev. Phytopathol.*, 42, 61-82.

EFFECT OF EFFLUENT WATER ON POPULATION DYNAMICS OF COWPEA APHID, APHIS CRACCIVORA KOCH (HEMIPTERA: APHIDIDAE), ON COWPEA PLANTS

TIROESELE B.*, NKETSO T.H. AND TSHWENYANE S.O.

Department of Crop Science and Production, Botswana College of Agriculture, Private Bag 0027, Gaborone, Botswana.
*Corresponding Author: Email- btiroese@yahoo.com

Abstract- Cowpea, *Vigna unguiculata* (L.) is an important legume grown in Botswana and the main yield losses incurred are from cowpea aphid, Aphis craccivora (Koch). Agriculture in Botswana is hampered by limitation in water resources. The objective of this study was to determine the effect of effluent water on cowpea aphid reproduction and population growth. This was randomized completely block design pot experiment conducted under net shade environment. Four treatments of 100%, 60%, 30% of effluent dilution and the control being clean tap water were used. This study revealed that the presence of N and K in the effluent water did not affect the population growth of aphids on cowpea plants. There was no significant difference among the effluent water treatments and the clean tap water. Therefore, effluent water in this study did not show any effect on cowpea aphid reproduction and population growth.

Keywords- *Aphis crassivora,* cowpea plant, effluent water

Introduction

Today's agriculture is faced with a problem of decreasing availability for irrigation water since there is ever increasing competition for use of water in agriculture, manufacturing and industries. This competition has led to use of water of marginal quality for irrigation [1]. Sewage waste water is the component of domestic waste comprising bathroom and laundry effluents (grey water) and sewage waste (black water). Generally this wastewater is processed through waste water treatment plants to produce a definable quality of sewage wastewater that may be recycled for a range of purposes. The beneficial use of treated effluents for agriculture is the major reuse application worldwide. Currently, most countries still have no legal framework for wastewater reuse in agriculture. Effluent contains valuable resources, such as nutrients and organic matter, however, it also can contain concentrations of chemical contaminants, salts and pathogens that are potentially detrimental to soils or plant growth. Nutrients in effluent such as phosphorus, magnesium, nitrogen, sulfur, potassium and calcium are generally beneficial to plant growth [2].

Development of agriculture in Botswana is especially hampered by a number of challenges, the most significant of which is limitation in water resources. Botswana is an arid to semi-arid country and it receives little rainfall ranging from a high of 550 and low of 200 mm per year with an estimated annual average evaporation rate of 1400 mm. It is a water stressed country and with lack or little water harvesting strategies, the country has turned to the use of effluent water to meet demands of water for irrigation. Notwane river, in the capital city Gaborone, receives treated sewage water discharged from wastewater treatment plant in Gaborone. This water is used by farmers to water their vegetables and other crops because of the scarcity of water in the country and also because of its nutritive advantage. Some of the crops are attacked by insects. Whether the water nutritive quality may have any impact on insect abundance is not fully understood. Over the past decade, the once overwhelmingly forest area around the river has been gradually replaced by agricultural community mainly because of easy access and cheap water. With these developments especially along the Notwane River, the threat of pests appears ever greater than in the past. It is well documented that agricultural operations play a significant role in the production of pest around the world [3].

Cowpea, *Vigna unguiculata* (L.) Walp. (Fabaceae), is a widely cultivated and the third most important crop in Botswana after maize, *Zea mays* L., and grain sorghum, *Sorghum bicolor* (L.) Moench (both Poaceae). It is an annual legume crop that is eaten in the form of dry seeds, green pods, green seeds and tender green leaves, and also utilized for fodder under a wide range of conditions [4]. Cowpeas play a critical role in the lives of millions of people in Africa and other parts of the developing world. It is of nutritional importance because it is high in protein content, therefore it can be a good replacement for beef, chicken and fish. The cowpea grains and leaves have a high protein content of about 24.8%, fat content of 1.9%, fiber content of 6.3%, carbohydrate content of 6.3% and water content of 8-9% [5]. The crop is also a valuable and dependable commodity that produces income for many small holder farmers and traders in Sub-Saharan Africa [6].

One of the major biological constraints in cowpea production is their infestation and damage by various pests in the field and during storage [7]. Yield losses caused by field pests have been reported to range from 20 to 100% [8,9]. The cowpea aphid, *Aphis craccivora* Koch (Hemiptera: Aphididae), is one of the major insect pests of cowpea. This insect is considered an important field pest of cowpea in Africa, Asia and Latin America [10]. Its high infestation can result in stunted plants, abscission of pods and shriveled seed [11], with a yield loss of up to 50% in Botswana [9].

Effluent water has been reported to have nutrients such as phosphorus, nitrogen, potassium, iron, calcium, magnesium and zinc. It is, therefore, suspected to have high nutritional value to crop production because of its mineral composition [1,12]. Some of these nutrients present in the water have been reported to have positive effect and others have negative effect on aphid population growth. Effluent water is suspected to be capable of boosting insect reproduction due to its nutrient composition such as nitrogen [13,14]. Nitrogen has been reported to be capable of increasing the reproduction of aphids [15-17]. In contrary, the increased levels of potassium have been found to reduce the aphid populations [18-20]. There is no research work that has been done on the effect of effluent water on population growth of aphid. Therefore, this work is meant to establish the relationship between effluent water and aphid population growth. This information might be useful for decision making in integrated pest management.

Materials and Methods

Experimental Design and Layout

The experimental site was at Botswana College of Agriculture gardens under net shade conditions. Four similar net-shades were used for this experiment. Cowpea plants (ER7) were planted to evaluate the effect of effluent water on cowpea aphid population growth. The plants were sown in polythene plastic pot containing garden soil. Two seeds were sown per hole at the depth of 5 cm and were later thinned to one after emergence by choosing a healthy looking seedling. Seeds which failed to emerge were replanted three days after emergence. No insecticide or fertilizer was applied on the experimental plants. Nets shades were kept free from weeds by regular manual weeding throughout the 5 weeks of the study period.

The pots of the cowpea plants were arranged in a randomized completely block design. Four water treatments viz treatment 1 (100% effluent water + 0% tap water), treatment 2 (60% effluent water + 40% tap water), treatment 3 (40% effluent water + 60% tap water) and treatment 4 (0% effluent + 100% tap water) were used in the experiment and were replicated four times ending with a total number of thirty two experimental plants. One replicate consisted of eight potted plants for each treatment in each of the four net-shade houses. The pots were randomly placed in the net shade house. Effluent water which was used to irrigate the cowpea plants was collected from Notwane river. The water was sent to National Environmental Laboratory at the Department of Waste Management and Pollution Control for various quality parameters (analysis of nutrients and their concentrations levels).

Aphid Colony

The cowpea aphid colony used for this study was obtained from a net-shade culture reared from progeny of a single apterous female. The aphid was maintained on susceptible cowpea plants inside proof cages. To avoid crowding, aphids were continuously transferred to new plants until the start of the experiment.

Experimental Conditions

The experiment was started when aphid colony had enough nymphs to infest all the experimental cowpea plants. A camel's hair brush was used to individually place neonate nymphs on individual cowpea plants. These nymphs were approximately same age and weight. Each nymph was examined daily to record the time length from the onset of reproduction also known as pre-reproductive period. The number of progeny per adult was recorded daily for the duration of the experimental period.

Data Collection

The cowpea aphid counts per plant started to be taken from 3 days after inoculation with cowpea aphids up to the end of the experiment (five weeks). Sampling was done between 900 and 1200 h. The cowpea developmental stages were recorded on each sampling date. The cowpea vegetative V1: first node, V2: second node, V3: third node, and V(n) represent n number of nodes on the main stem with fully developed leaves) and reproductive growth stages (R1: beginning bloom, R2: full bloom, R3: beginning pod, R4: full pod, R5: beginning seed, and R6: full seed stage, respectively) were identified according to Fehr [21]. Data collection was done by in-situ observation (visual counts on the plant) of insects on the plant. The cowpea aphid numbers were determined by counting those present on five randomly selected plants per replicate for all the treatments. Counts were taken from the top five trifoliate leaves.

Data Analysis

All data were subjected to analysis of variance (ANOVA) using mixed model procedures (PROC MIXED IN SAS [22]). Multiple comparisons were made on least square means. Differences among treatment means were based on LSD and considered significant when $P \leq 0.05$.

Results and Discussion

Aphid Population Growth Pattern

The experiment was conducted for 32 days after inoculating cowpea plants with cowpea aphids. The [Fig-1] depicts the aphid population dynamics from day 4 of observation to the last day of observation for all the treatments.

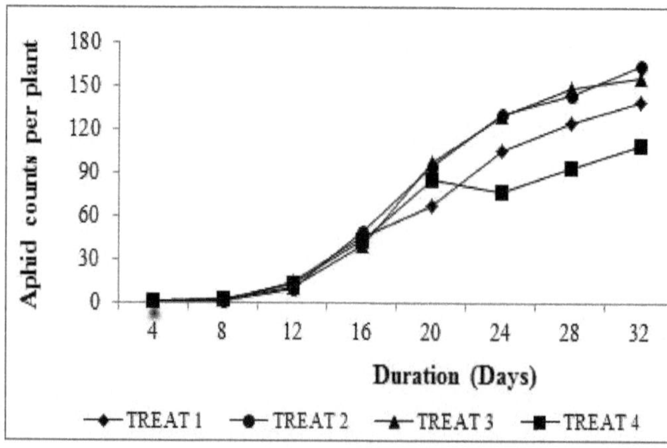

Fig. 1- Population Dynamics of Cowpea Aphid among the Treatments

There was no significant growth of aphid population from Day 4 to Day 12 for all the treatments. The aphid population started to increase after day 12 of observation for all treatments and still showed an increase trend at the last day of observation. Aphids have a tremendous capacity for reproduction and increase in population. Karley, et al [23] reported that aphids can double a population in just 3 days.

Comparison of Average Number of Aphids per Plant among the Treatments

The average population of cowpea aphids per plant from the first day of observation to last day of observation among all treatments (control inclusive) was found not to be significantly different ($F_{3,\ 1462}$ = 20.19; P < 0.8983) [Table-1]. The results of this study have shown that effluent water does not have impact on aphid population growth as this was evident when comparing with clean tap water as observed in [Table-1]. Effluent water has been reported to have relatively high nitrate contents (35.5 mg/l) [Table-4]. Therefore, this water was expected to increase the population growth of aphids in our experiment. This is because nitrogen has been reported to have a positive correlation with aphid growth [24]. Nitrogen is essential in insect development, aphids inclusive by increasing the abundance of free amino acids in the plant [25]. The aphid infestation per plant showed significant difference among the four treatments, control inclusive, at each day of observation ($F_{7,\ 86}$ = 11.79; P < 0.0001) [Table-2]. When looking at particular days chosen at random throughout the observation period, it was observed that treatments were significantly different in aphid infestation only at day 8 ($F_{3,8}$ = 0.75; P < 0.0001) [Table-2] where treatment 3 had significantly higher aphid infestation than other treatments. Treatment 1 and 4 were not significantly different in aphid infestation per plant after day 8. In addition, other treatments also did not show any significant difference in aphid infestation up to day 32 [Table-2]. However, from day 24 to day 32, there was least population growth on the control as compared to other treatment levels.

Table 1- Average Number of aphids per plant in all the treatments throughout the experimental period

Treatment	Mean ± SE
1	2.808 ± 1.331^a
2	3.754 ± 0.394^a
3	3.490 ± 0.397^a
4	3.454 ± 1.331^a

**Means within a column followed by the same small letter are not significantly different (P ≤ 0.05).*

**Treat 1 is 100% sewage water, treat 2 is 60% sewage water, treat 3 is 30% sewage water, and treat 4 is control (Clean tap water). Dilution was done with clean tap water.*

Table 2- The number of Aphids per plant after Infestation at a 4 day interval among the treatments

	Observation Day							
Treatment	4	8	12	16	20	24	28	32
1	1^a	1.25^{bc}	13.83^a	16^a	67.17^a	105.33^a	123.83^a	139.08^a
2	1^a	1.167^c	9.17^a	49.08^a	94.08^a	129.83^a	144.08^a	164^a
3	1^a	2.75^a	10^a	39.92^a	97.33^a	128.83^a	147.83^a	155.33^a
4	1^a	2^{bc}	12.83^a	43^a	85.17^a	76.83^a	93.75^a	108.42^a

** Means within a column followed by the same small letter are not significantly different (P ≤ 0.05).*

** Treat 1 is 100% sewage water, treat 2 is 60% sewage water, treat 3 is 30% sewage water, and treat 4 is control (Clean tap water). Dilution was done with clean tap water.*

Effect of Treatments Levels on Population Growth of Cowpea Aphid at Each Growth Stage

When comparing treatments at 4 days interval, it has been found that treatment 1 had lower aphid number per plant than other treatments, where significance difference was observed. The trend observed at [Table-3] is that population levels of aphid per plant increased as sewage water was diluted. Similarly, we expected vice versa to be the case for the reasons as above of effect of N on aphid growth. It should be noted that inoculation of cowpea aphid on the cowpea plants was done at V5 plant growth stage. The population of aphid remained low from V5 to V6 plant growth stage and a noticeable change started from V7 plant growth stage to the last day [Table-3].

Table 3- Effect of treatments levels to growth stage on population dynamics of cowpea aphid

Cowpea Growth Stage	Water Treatments Means ± SE			
	1	2	3	4
V5	1^a	1^a	1^a	1:00 AM
V6	0.672±12.55^a	1.00±12.55^a	1.8685±12.37^a	1.382±12.50^a
V7	1.167±2.55^a	2.853±12.25^a	0.6266±12.71^a	1.065±12.55^a
V8	9.0145±11.76^a	9.339±11.93^a	9.3611±11.77^a	11.964±11.79^a
R1	32.427±28.33^a	17.393±38.85^a	21.907±28.34^a	57.333±32.19^a
R2	66.05±11.27^a	83.300±11.38^b	79.22±11.29^{ab}	65.2634±11.29^a
R3	85.271±22.58^a	81.149±22.42^a	78.95±25.79^{ab}	89.8784±21.29^a
R4	105.83±15.48^a	128.73±13.79^a	131.92±14.70^a	82.1806±14.87^a
R6	140.08±21.23^a	158.58±38.85^a	108.62±25.79^a	45.961±28.33^b
R7	110.88±21.23^a	167.83±18.84^b	164.85±18.32^b	94.528±22.39^a

** Means within a column followed by the same small letter are not significantly different (P ≤ 0.05).*

** Treatment 1 is 100% effluent water, treatment 2 is 75% effluent water, treatment 3 is 50% effluent water, treatment 4 is 25% effluent water and treatment 5 is the control which is 100% tap water.*

Table 4- Composition of effluent water from Glen Valley sewage ponds

Parameter	Symbol	Concentration (mg/l)
Nitrate	NO_3^-	35.5 ± 0.43
Potassium	K	23.46
Magnesium	Mg	14.34
Chloride	Cl^-	671 ± 2.09
Calcium	Ca	33.07
Total Dissolved Solids	TDS	442.86

Source: National Environmental Laboratory Building (Department of Waste Management and Pollution Control), Gaborone, Botswana.

The average population of cowpea aphids per plant from first day of observation to last day of observation for the treatments at each growth stage was found to be significantly different ($F_{3,1437}$=36.52; P < 0.0001). During the plant growth stages from V1 to R4, the average aphid counts per plant among all treatments were observed to be not significantly different [Table-3]. Major differences in aphid numbers per plant were showed at plant reproductive stages R6 and R7. At plant growth stage R3 and R4 aphid population increased and this collided with cowpea plant having succulent pods and vining nodes [Table-3]. Fernandes, et al [26] found that higher aphid population feeding was distributed more on the pods and nodes than on the leaves mainly because of high nitrogen concentration.

Generally, this study has revealed that eventhough the nitrate content in the sewage water was at significant amounts, it did not influence or affect the population growth of aphids. In addition, it was also found from the sewage water analysis that the water had noticeable amounts of potassium (23.46 mg/l) [Table-4]. Potassium has been reported to negatively affect the population growth of insect. It has been reported to reduce population of aphids by reducing concentrations of free amino acids (Gash 2012). Therefore, it can be said that the presence of N does not always imply positive aphid growth especially when other elements are available at certain concentration levels such as potassium.

Conclusion

This study has shown that effluent water at all concentrations does not show any distinct and significant effect on cowpea aphid reproduction and population growth compared with clean tap water. Therefore, its use by farmers will not affect the population levels of aphids on cowpea crops.

Acknowledgement

The authors are thankful to Botswana College of Agriculture for allowing them to use their field to carry out the experiment.

Conflict of Interest : None declared.

Reference

[1] Emongor V.E., Ramolemana G.M., Machacha S., Khonga V.E., Marumo K. (2005) *Journal of Applied Science*, 5, 451-454.

[2] Department of Environment and Conservation (2004) *Environmental Guidelines: Use of Effluent by Irrigation*, Sydney, Australia.

[3] Mulla M.S., Mian L.S. and Gratz N.G. (1987) *Journal of Agricultural Entomology*, 4, 97-131.

[4] Machacha M., Obopile M., Tshegofatso A.B.N., Tiroesele B., Gwafila C. and Ramokapane M. (2012) *International Journal of Tropical Insect Science*, 32, 189-193.

[5] Dominic J., Udoh B.A. and Asequo N.U. (2005) *Journal of Stored Products Research*, 37, 216-217.

[6] Langyintuo A.S., Lowenberg-DeBoer J., Faye M., Lambert D., Ibro G., Moussa B., Kergna A., Kushwaha S., Musa S. and Ntoukam G. (2003) *Field Crops Research*, 82, 215-231.

[7] Jackai L.E.N. and Daoust R.A. (1986) *Annual Review of Entomology*, 31, 95-119.

[8] Youdeowei A. (1989) *The Search for Sustainable Solutions to Crop Pest Problems in Africa*, Cotonou (Benin), 31-50.

[9] Obopile M. (2006) *African Plant Protection*, 12, 111-115.

[10]Singh S.R. and Jackai L.E.N. (1985) *Cowpea Research, Production and Utilization*, 217-231.

[11]Ofuya T.I. (1997) *Integrated Pest Management Review*, 2, 199-207.

[12]John S.S. and McConchie D.M. (1994) *Australian Journal of Agricultural Research*, 45, 160-171.

[13]Noma T., Gratton C., Garcia M.C., Brewer M.J., Mueller E.E., Wyckhuys K.G., Heimpel G.E. and O'nea M.E. (2010) *Environmental Entomology*, 39, 31-41.

[14]Dixon A.F.G. (1998) *Aphid Ecology: An Optimization Approach*, 2nd ed., Chapman and Hall, London.

[15]Awmack C.S. and Leather S.R. (2002) *Annual Review of Entomology*, 47, 817-844.

[16]Nevo E. and Coll M. (2001) *Journal of Economic Entomology*, 94, 27-32.

[17]Honek A. (1991) *Journal of Plant Disease Protection*, 98, 655-660.

[18]Marwat G.N., Ahmad I. and Khatoon S. (1985) *Sarhad Journal of Agriculture*, 1, 223-229.

[19]Havlickova H. and Smetankova M. (1998) *Rostlinna Vyroba*, 44, 379-383.

[20]Myers S.W., Gratton C., Wolkowski R.P., Hogg D.B. and Wedberg J.L. (2005) *Journal of Economic Entomology*, 98, 113-120.

[21]Fehr W.R., Caviness C.E., Burmood D.T. and Pennington J.S. (1971) *Crop Science*, 11, 929-931.

[22]SAS Institute (2003) *SAS Users Guide: Statistics*, Cary, North Carolina.

[23]Karley A.J., Parker W.E., Pitchford J.W. and Douglas A.E. (2004) *Ecological Entomology*, 29, 383-388.

[24]de Valpine P. and Rosenheim J.A. (2008) *Ecology*, 89, 532-541.

[25]Throop H.J., Hollandw E.A., Partonz W.J. and Keoughz C.A. (2004) *Global Change Biology*, 10, 1092-1105.

[26]Fernandes G.W., Coelho M.S. and Luttge U. (2010) *Brazilian Journal of Biology*, 70, 723-728.

[27]Gash A.J.F. (2012) *Journal of Agronomy*, 2, 1-13.

GENOTYPIC DIFFERENCES BETWEEN TWO MUNGBEAN VARIETIES IN RESPONSE TO SALT STRESS AT SEEDLING STAGE

KANDIL A.A.[1], ARAFA A.A.[2], SHARIEF A.E.[1*] AND RAMADAN A.N.[3]

[1]Department of Agronomy, Faculty of Agriculture, Mansoura University, Egypt.
[2]Department of Agricultural Botany, Faculty of Agriculture, Mansoura University, Egypt.
[3]Ministry of Agriculture, Egypt.
*Corresponding Author: Email- shariefali42@gmail.com

Abstract- In order to study performance of mungbean to salinity stress, a laboratory experiment was carried out to investigate salinity tolerance of two mungbean varieties (Kawmy-1 and IV 2010) to eight salinity levels i.e. 0, 2, 4, 6, 8, 10, 12 and 14 dS/m of NaCl concentrations. Mungbean (*Vigna radiata* (L.) Wilczek) varieties were compared for germination efficiency and seedling characters. Results indicated that mungbean IV 2010 variety recorded the highest percentages of final germination, germination index %, seedling vigor index, energy of germination, mean germination time (day), coefficient velocity, abnormal seed percentage, root and shoot length, root fresh and dry weight and seedling height reduction. Whilst, mungbean Kawmy-1 variety recorded the highest weight of fresh and dry shoot and relative dry weight. Highest germination parameters and seedling characters of the two studied varieties under control treatment. The obtained results suggested that the two varieties registered a decrease in the percentage of germination and seedlings growth at higher NaCl concentrations. Results clearly indicated that mungbean Kawme-1 variety appeared to be more tolerant to salt stress than IV 2010 variety recording higher germination parameters and seedling characters. Increasing salinity concentrations significantly reduced germination percentage, seedling vigor index, coefficient of velocity, mean germination time, shoot and root length, shoot and root fresh and dry weight. It could be concluded that germination efficiency i.e. final germination percentage, germination index, energy of germination, mean germination time, abnormal seed percentage, root and shoot length, seedling total fresh and dry weight, dry weight reduction and shoot length reduction were gradually decreased significantly when salinity increased.

Keywords- Mungbean varieties, Salinity stress, Germination parameters, Seedling characters, Dry weight reduction and Shoot length reduction

Introduction

Salinity became a serious problem for agriculture all over the world. Salinity is one of the major stress factors which limit crop production in most of the arid and semi-arid regions of the world [4]. The extreme increase in population in Egypt needs to increase the total yield of legume crops in order to overcome the lack in protein through summer mungbean cultivation in the newly reclaimed lands especially under saline conditions of such soil. Mungbean (*Vigna radiata* (L.) Wilczek) is a summer legume crop with short duration (70-90 days) with high nutritive value. In Egyptian, there was shortage in legume summer crop production. Mungbean can be easily grown in the newly reclaimed sandy soil and irrigate by saline water, it can tolerant drought and soil salinity [18]. Therefore, this study was carried out to investigate the varietal seed germination as an important parameter under salinity stresses. Mungbean seed germination is an important parameter which is affected by various factors for example varieties and salinity stress. Crop productivity can be enhanced with optimization germination factors which increase germination rate. The most critical stage in seedling establishment is seed germination that determine fruitful crop production. It is reported by several researchers where there were large differences between introduced varieties in Egypt [5,7,18]. Naher and Alam (2010) concluded that highest germination % was observed in the control treatment of all studied varieties. Germination and seedling growth were reduced under salinity with varying responses for mungbean cultivars [20,26,28]. Moreover, salinity may affect mungbean seed germination by producing an outside osmotic potential that avoids water uptake or due to toxic effects of Na^+ & Cl^- ions during seed germination [19,14]. Increasing salinity levels during mungbean seed germination significantly reduced germination characters [2,6,24,26,28] and seedling characters [2,16,17,20,28]. Keeping in view the above mentioned facts, the present work was conducted to assess performance of some mungbean varieties under different salinity levels and their effect on germination parameters and seedling characters.

Materials and Methods

This research was conducted in Agronomy Department laboratory, Faculty of Agriculture, Mansoura University, Egypt during March and April 2011 to study the performance of some mungbean (*Vigna radiata* (L.) Wilczek) varieties to germination under salinity stress and their interactions.

Treatments and Experimental Design

The experiment was arranged in two factors in factorial experiment in randomized complete block design (RCBD), the first factor included two mungbean (Kawmy-1 and IV 2010) varieties. Seeds of Mungbean (*Vigna radiata* (L.) Wilczek) varieties i.e. Kawmy-1 and IV 2010 were obtained from National Agricultural Research Centre, Egypt, whereas the second variety was produced from imported lines. The second one included eight different concentrations (NaCl) as presented in Table 1. Seeds of cultivars were surface sterilized by immersion for 5 minutes in sodium hypochlorite solution, then repeatedly washed with deionized water. Twenty five seeds of uniform size in each treatment for each variety were allowed to germinate on a filter paper in 9 cm diameter petri dishes. Each filter paper was moistened with a water solution at eight different NaCl concentrations. Distilled water was used as the control solution (0 dS/m). The NaCl concentrations were 0, 2, 4, 6, 8, 10, 12 and 14 dS/m of NaCl which electrical conductivity was measured with a conductivimeter (dS/m = deciSiemen per meter). Thus, the whole experiment comprised 64 Petri dishes arranged in factorial experiment randomized complete block design.

Table 1- Salt experimental treatments (concentrations and weight of NaCl)

Treatments of salt concentration	Weight of NaCl	E C dsm-1
0 (distilled water as the control)	–	0
2 dS m^{-1} NaCl	1.14 gm/L (distilled water)	2
4 dS m^{-1} NaCl	2.280 gm/L (distilled water)	4
6 dS m^{-1} NaCl	3.440 gm/L (distilled water)	6
8 dS m^{-1} NaCl	4.590 gm/L (distilled water)	8
10 dS m^{-1} NaCl	6.000 gm/L (distilled water)	10
12 dS m^{-1} NaCl	6.900 gm/L (distilled water)	12
14 dS m^{-1} NaCl	7.980 gm/L (distilled water)	14

After pre-treatments of mungbean, seeds were germinated for 14 days, incubating in a growth chamber adjusted to 20 ± 1 °C in a dark. For each treatment, 25 seeds were counted at random and placed on a moist germination paper in a Petri dish. There were four replicates for each treatment. Dishes were inspected daily and sterilized water added as required. The number of normal seedlings, abnormal seedlings, dead seeds and hard seed in each replicate was recorded. Every 24 hours the number of germinated seeds was counted. Seeds were categorized as germinated (radical 2 mm), hard (no imbibition or swelling) or nonviable (abnormal, dead or infected seeds) as described by ISTA (1993).

Hard seeds can be illustrious from dead seeds by pushing down on each no germinated seed with the flat part of a pencil eraser. If the seed does not crush with gentle pressure, it is considered hard. All seeds which had taken up no water in six days were considered as hard seed. Seeds which at the end of test period are neither hard nor have produced any part of a seedling were considered dead.

Normal seedlings show the potential for continued development into acceptable plants when grown in good quality soil and under promising conditions of moisture, temperature and light, normal and abnormal seedling can documented according to ISTA rules.

Abnormal seedlings do not show the possibility to grow into a standard plant when developed in good quality soil and under positive conditions of moisture, temperature and light.

Studied Characters

Germination Parameters

Seedlings were evaluated for:

Final germination percentage: It was calculated according to the germination count taken after 14 days and expressed as percentage according to equation No. (1) which described by Ellis and Roberts (1981) and Ruan, et al (2002):

$$FGP = \frac{\text{Number of Germinated Seeds after 14 Days}}{\text{Number of Germinated Seeds}} X\ 100 \quad (1)$$

Germination Index (GI): It was calculated according to equation No.2 [13]:

$$GI = \frac{\text{Germination Percentage in each Treatment}}{\text{Germination Percentage in the Control}} X\ 100 \quad (2)$$

Seedling vigor index (VI): It was calculated according to Abdul-Baki and Anderson (1973) in the equation No 3:

$$(VI) = (\text{Average shoot length+Average root length}) \times \text{Germination Percentage} \quad (3)$$

Energy of germination was recorded at the fourth day. Energy of germination was the percentage of germinating seeds 4 days after sowing relative to the number of seeds tested [23].

Mean germination time (MGT): It was determined according to equation No. 4 [8]:

$$MGT = \sum dn / \sum n \quad (4)$$

Where (n) is the number of seeds which were germinated on day (d), and (d) is the number of days counted from the beginning of germination.

Coefficient of velocity (CV), a unit less parameter determined by a mathematical manipulation that incorporates the number of seeds germinated and the velocity of germination was calculated equation No. (5): $C\,V = 100[\sum Ni / \sum Ni\ Ti]$ (5)

Where N is the number of seeds germinated on day i and T is the number of days from sowing [25]. In general, a higher CV value reflects increased of seed germinated at shorter germination time.

Percentage of abnormal seedling = It was calculated according equation No. (6)

$$\text{Abnormal seedling} = \frac{\text{Number of abnormal seedling}}{\text{Total number of seeds}} \quad (6)$$

Seedling Characters

1. Shoot length: The length of ten seedlings from the seed to the tip of the leaf blade were recorded and expressed in centimeters as the shoot length.
2. Root length: The root length of ten seedlings from the seed to the tip of the root was recorded and expressed in centimeters (cm) as the root length
3. Shoot fresh weight: The weight of ten seedlings was measured and expressed in milligram as the shoot fresh weight.
4. Root fresh weight: The weight of ten seedling roots was measured and expressed in milligram (mg) as the root fresh weight.
5. Root dry weight: The weight of ten seedling roots was recorded and expressed in milligram after oven drying at 70 ° C for 72 h.
6. Shoot dry weight: The weight of ten seedlings was recorded and expressed in milligram after oven drying at 70 °c for 72 h.
7. Seedling height reduction: The seedling height reduction (SHR) was calculated using the equation No. 7 [12]:

$$\text{SHR (\%)} = \frac{\text{Plant Height at Control - Plant Height at Saline Condition}}{\text{Plant Height at Saline Condition}} \times 100 \quad (7)$$

8. Relative dry weight the relative dry weight (RDW) was calculated by using the equation No. 8 [12]

$$\text{RDW (\%)} = \frac{\text{Total Dry Weight under Saline Condition}}{\text{Total Dry Weight under Control Condition}} \times 100 \quad (8)$$

Statistical Analysis

All obtained data were statistically analyzed according to the technique of analysis of variance (ANOVA) for factorial experiment in randomized complete block design (RCBD) as published by Gomez and Gomez (1984) with using means of "MSTAT-C" computer software package. Least Significant Difference (LSD) method was used to test the differences between treatment means at 5 and 1% level of probability as described by Snedecor and Cochran (1980).

Result and Discussion

Performance of Varieties

Mungbean IV 2010 variety recorded highest values of final germination percentage, germination index, and seedling vigor index, energy of germination, coefficient velocity, root and shoots length, shoot dry weight, In contrast, mungbean Kawmy-1 variety recorded highest values of mean germination time, abnormal seed percentage, shoot fresh weight, shoot dry weight and relative dry weight. However, root dry weight, root fresh weight and seedling height reduction insignificantly affected by mungbean varieties (Tables 2 and 3). Mungbean IV 2010 variety exceeded Kawmy-1 variety by 10.07%, 7.32%, 23.1%, 10.07%, 1.36%, 16.96%, 27.25%, 16.28%, 9.64%, 16.28% and 10.05%, resp. in final germination percentage, germination index, seedling vigor index, energy of germination, mean germination time, coefficient velocity, abnormal seed percentage, shoot length, in root length, shoot fresh weight and seedling height reduction, respectively.

Table 2- Averages of germination percentage (%), germination index (%), seedling vigor index, energy of germination, mean germination time, coefficient velocity and abnormal seed percentage as affected by cultivars and salinity stress.

Treatments	Final Germination percentage (%)	Germination Index (%)	Seedling Vigor Index	Energy of germination (%)	Mean germination time (day)	Coefficient velocity (%)	Abnormal Seed Percentage(%)
A-Cultivars Performance							
KAWMY-1	72.5	75.5	977	68.5	3.07	33	38.31
IV 2010	80.62	81.47	1271	79.62	2.62	39.74	27.87
F- test	**	**	**	**	**	**	**
B- Salinity levels:							
Control	97.5	100	2178	96	2.57	44.12	8.5
2 dS m^{-1} NaCl	90.5	92.77	1816	87.5	2.46	41.32	13.5
4 dS m^{-1} NaCl	83.5	85.57	1524	84	2.43	40.72	17.5
6 dS m^{-1} NaCl	79	81.01	1290	79	2.58	37.06	22
8 dS m^{-1} NaCl	75.5	77.35	883	75	2.86	35.77	36.75
10 dS m^{-1} NaCl	69	70.73	672	64.5	2.83	34.7	42
12 dS m^{-1} NaCl	61.5	63.05	405	61	3.15	31.48	54.5
14 dS m^{-1} NaCl	56	57.42	223	45.5	3.87	25.55	70
F- Test	**	**	**	**	**	**	**
L.S.D 5%	3.47	3.35	192.75	4.07	0.21	2.86	10.78
L.S.D 1%	4.63	4.48	257.45	5.44	0.28	3.82	14.4
C. Interaction Effects							
F- test	N. S.	N. S.	N.S	**	**	**	N.S

*N. S.= Not significant, *= significant at 5%, **= significant at 1%*

Table 3- Averages of shoot length (cm), root length (cm), shoot and root fresh weight (gm), shoot and root dry weight (gm), seedling height reduction (%) and relative dry weight (%) as affected by mungbean cultivars and salinity stress.

Treatments	Shoot length (cm)	Root length (cm)	Shoot fresh weight (gm)	Root fresh weight (gm)	Shoot dry weight (gm)	Root dry weight (gm)	Seedling height reduction (%)	Relative dry weight (%)
A-Cultivars Performance:								
KAWMY-1	6.07	6.37	2.64	0.596	0.441	0.061	107.12	77.91
IV 2010	7.65	7.05	2.21	0.643	0.325	0.087	96.35	69.61
F- test	**	*	**	N.S	**	N.S	N.S	**
B- Salinity levels								
Control	11	11.32	2.91	0.99	0.517	0.189	0	0
2 dS m^{-1} NaCl	9.73	10.23	2.84	0.95	0.47	0.095	14.6	90.66
4 dS m^{-1} NaCl	8.48	9.76	2.61	0.86	0.432	0.09	33.21	84.55
6 dS m^{-1} NaCl	7.64	8.63	2.47	0.63	0.412	0.082	46.41	79.94
8 dS m^{-1} NaCl	6.13	5.55	2.38	0.55	0.376	0.065	95.5	71.28
10 dS m^{-1} NaCl	5.24	4.38	2.24	0.38	0.337	0.04	124.76	65.51
12 dS m^{-1} NaCl	3.99	2.53	2.08	0.31	0.277	0.024	181.01	53.37
14 dS m^{-1} NaCl	2.68	1.29	1.89	0.24	0.239	0.01	319.1	44.77
F- Test	**	**	**	**	**	**	**	**
L.S.D 5%	0.98	1.38	0.3	0.27	0.059	0.089	91.87	10.1
L.S.D 1%	1.31	1.85	0.4	0.37	0.079	0.119	122.72	13.5
Interaction Effects: C-								
F- test	N.S	N.S	N.S	N.S	N.S	N.S	**	N.S

*N. S.= Not significant, *= significant at 5%, **= significant at 1%*

Results indicated that mungbean kawmy-1 variety exceeded IV 2010 by 16.28%, 26.30%, 10.65% and 10.65% in relative dry weight, shoot fresh weight, shoot dry weight and relative dry weight, respectively. Several investigators such as Ashour, et al (1992); El Kramany (2002); Mohamed and El Kramany (2005). Naher and Alam (2010) reported that there were great differences between introduced varieties and concluded that highest germination % was observed in the control treatment of all studied varieties. Win, et al (2011) suggested that genetically diverse arrangements resistant to salt stresses within mungbean varieties due to substantial practical value for studying the mechanism of salt tolerance and for the delivery of genetic resources for salinity in breeding program. So, the differences between mungbean cultivars in final germination index might be due to the genetically factors and heredity variation among the two cultivars under study. These results are in good accordance with those found by Salah Uddin, et al (2009); Singh, et al (2006).

Salinity Stress Effects

The results clearly indicate that increasing salinity levels significantly increased abnormal seed percentage and seedling height reduction (Tables 2 and 3). The results showed that increasing salinity levels to 2, 4, 6, 8,10,12 and 14 dS/m NaCl significantly reduced the final Seedling vigor index by 16.6, 30.0, 40.7, 59.4, 69.1, 81.4 and 89.7%, resp. as compared with the control treatment. Increasing salinity levels to 2, 4, 6, 8,10,12 and 14 dS/m NaCl significantly reduced energy of germination by 18.22, 12.50, 17.70, 21.80, 32.81, 36.45 and 52.60%, resp. compared with the control treatment. Moreover, increasing salinity levels to 2, 4, 6, 8,10,12 and 14 dS/m NaCl significantly reduced main germination time compared with the control treatment. Increasing salinity levels to 2, 4, 6, 8, 10, 12 and 14 dS/m NaCl significantly reduced coefficient velocity by 1.66, 5.47, 7.18, 15.20, 14.80, 24.53 and 38.60%, resp. compared to the control treatment. In addition, increasing salinity levels to 2, 4, 6, 8,10,12 and 14 dS/m NaCl significantly reduced final abnormal seed percentage by 13.50, 17.50, 22.00, 36.75, 42.00, 54.50 and 70.0%, resp. as compared with the control treatment. Saline condition reduces the ability of seed to absorb water, causing rapid reduction in germination rate and induces many metabolic changes. Germination reduction under salinity stress may be due to this fact that dormancy increases in crop seeds under salinity stress [15]. On the other hand, mungbean seed germination may be affected by salinity through either creating external osmotic potential or toxic effect of Na^+ and Cl^- ions [14,19]. Babbar and Dhingra (2007) reported that seed germination was decreased with increasing level of salinity, reduction being maximum in Km-4 variety and least in Km-2 variety. Length of plumule and radicle also got reduced with increasing salinity and decrease was maximum in Asha and minimum in Km-2 variety for both parameters. These results are in good accordance with those reported by Ahmed (2005); Ruan, et al (2002); Salah Uddin, et al (2009) and Singh, et al (2006).

Increasing salinity levels to 2, 4, 6, 8, 10, 12 and 14 dS/m NaCl significantly reduced root length by 9.62, 13.78, 23.76, 50.97, 61.30, 77.65 and 88.60%, resp. as compared to the control treatment. Increasing salinity levels to 2, 4, 6, 8, 10, 12 and 14 dS/m NaCl significantly reduced shoot length by 11.5, 22.90, 30.54, 44.27, 52.36, 63.7 and 75.6%, resp. as compared with the control treatment. Increasing salinity levels to 2, 4, 6, 8, 10, 12 and 14 dS/m NaCl significantly reduced shoot fresh weight by 2.40, 10.30, 15.12, 18.20, 23.00, 28.52 and 35.0%, resp. as compared with the control treatment. Furthermore, increasing salinity levels to 2, 4, 6, 8, 10, 12 and 14 dS/m NaCl significantly reduced root fresh weight by 4.0, 13.13, 36.36, 44.44, 61.61, 68.68 and 75.75%, resp. as compared with the control treatment. Increasing salinity levels to 2, 4, 6, 8, 10, 12 and 14 dS/m NaCl significantly reduced shoot dry weight by 9.0, 16.44, 20.30, 27.27, 34.81, 46.42 and 53.77%, resp. as compared with the control treatment. Moreover, increasing salinity levels to 2, 4, 6, 8, 10, 12 and 14 dS/m NaCl significantly reduced root dry weight by 49.7, 52.3, 56.6, 65.6, 78.8, 87.3 and 94.7%, resp. as compared with the control treatment. However, increasing salinity levels to 2, 4, 6, 8,10,12 and 14 dsm^{-1} significantly increased seedling height reduction. Furthermore, increasing salinity levels to 2, 4, 6, 8,10,12 and 14 dS/m NaCl significantly reduced relative dry weight by 9.34, 18.45, 20.06, 28.72, 34.49, 46.63 and 58.23%, resp. as compared to the control treatment. It seems that reduction in seedling fresh weight may be due to decreasing water uptake by seedling under salinity. Salt decreases the osmatic water potential, creating a water stress in seedlings. Reduction in mungbean under salt stress resulted from combination of ions toxicity and altered water relation that caused large accumulation of sodium and magnesium ions and reduced calcium and potassium concentration, transpiration, stomatal conductance and hydraulic conductance decreased as salinity increased [22]. Naher and Alam (2010) reported that highest germination % was observed in the control treatment of all varieties. Higher NaCl concentrations and sowing BARI Mung 4 variety significantly reduced than other varieties. All the varieties showed similar performances at higher NaCl concentration considering yield contributing character. In addition, salinity may affect seed germination by generating an external osmotic potential that averts water uptake or due to toxic effects of Na^+ and Cl^- ions on seed germination [14,19]. These results are in agreement with those concluded by Ahmed (2005); Babbar and Dhingra (2007); Misra and Gupta (2006); Mohammed (2007) and Win, et al (2011).

Interaction Effects

With respect to the interaction between mungbean cultivars and salinity concentration, the results showed that there was a significant interaction effect between mungbean cultivars and salinity stress concentrations on energy of germination (%), mean germination time (day) coefficient velocity (%) and seedling height reduction (%). The results clearly showed that highest values of energy of germination, mean germination time and coefficient velocity were obtained from sowing IV 2010 variety and without salinity stress as shown in Fig.'s 1, 2 and 3. Seedling height reduction was significantly affected due to the interaction between mungbean cultivars and salinity concentrations. The results revealed that increasing salinity levels significantly reduced energy germination percentage and coefficient velocity percentage of the two varieties. IV 2010 variety recorded highest energy germination percentage and coefficient velocity (%) than Kawmy-1 under all salinity levels. However, Kawmy-1 recorded highest mean germination time (day) and seedling height reduction than IV 2010 variety under all salinity levels. However, increasing salinity levels significantly increased both mean germination time and seedling height reduction of both varieties after salinity level of 6dS/m NaCl. Minimum value of seedling height reduction was recorded with sowing IV 2010 variety and salinity level of 2 dS/m NaCl, which was 10.62. Meanwhile, a highest value of seedling height reduction was 205.02 reselected from sowing Kawmy-1 variety and salinity levels of 12dS/m NaCl (Fig. 4). This might be due to the varietal response of mungbean varieties to salinity. Hossain, et al (2005) reported that Binamoog-2 and mutant E5 M511 varieties showed better performance on almost all the parameters under the salinity treatments despite drastic reductions in those parameters. The mutant N5 J521 showed less tolerance to imposed salinity of 3.7 and 7.8 dS/m. Similar results were reported by Ahmed (2005); Naher and Alam (2010); Promila and Kumar (2000).

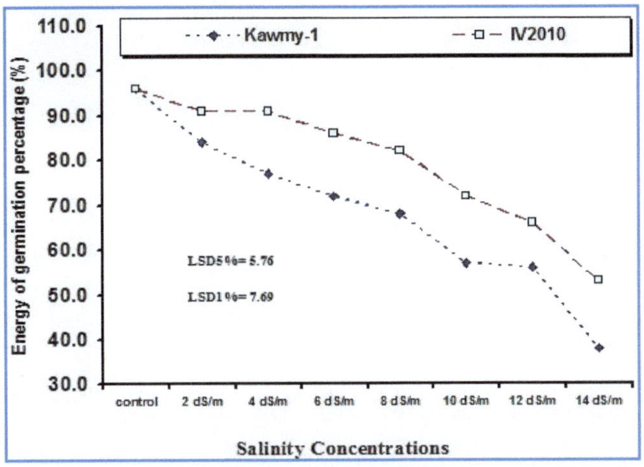

Fig. 1- Averages of energy of germination percentage (%) as effected by the interaction between cultivars and salinity stress

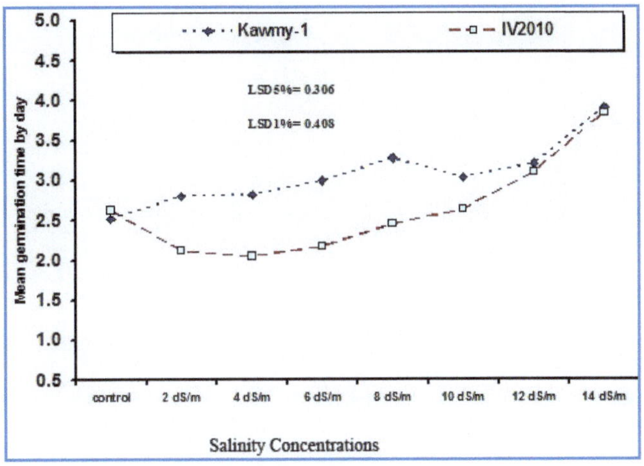

Fig. 2- Averages of mean germination time by day as effected by the interaction between cultivars and salinity stress

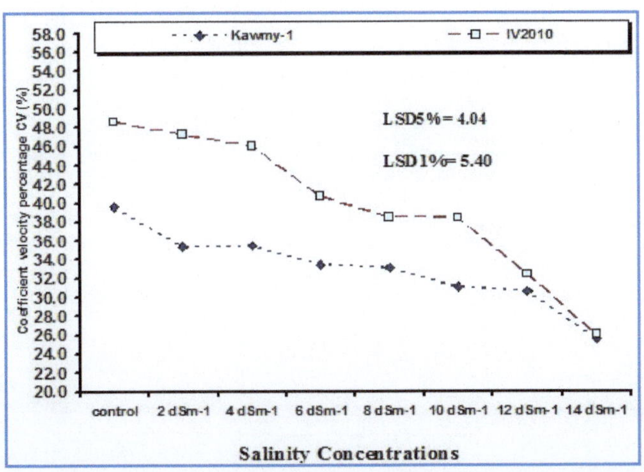

Fig. 3- Averages of coefficient velocity percentage (%) as effected by the interaction between cultivars and salinity stress

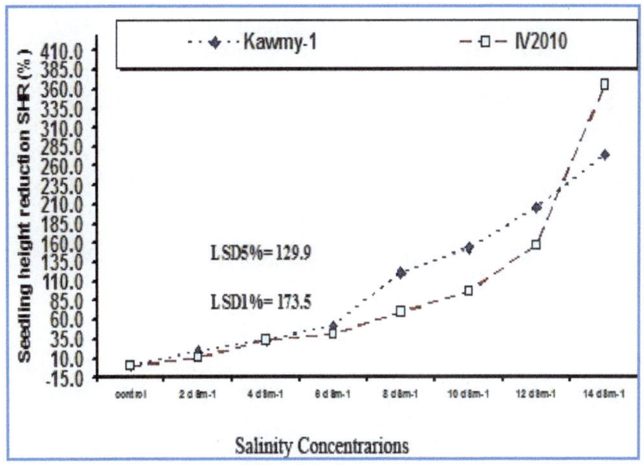

Fig. 4- Averages of seedling height reduction (%) as effected by the interaction between cultivars and salinity stress

Conclusion

So, from all the obtained results it could be concluded that maximum mungbean seed germination percentage and seedling parameters under salinity stress are produced by sowing kawmy-1 variety with increasing salinity concentrations levels up to 14 dS/m NaCl. This variety was more tolerant to salinity and recommended to be used in breeding program for enhancing mungbean cultivation in newly reclaimed soils.

References

[1] Abdul-Baki A.A. and Anderson J.D. (1973) *Crop Sci.*, 13, 630-633.

[2] Ahmad S., Wahid A., Rasul E. and Wahid A. (2005) *Bot. Bull. Acad. Sin.*, 46, 135-142.

[3] Ahmed S. (2005) *Int. J. Biol. Biotechnol.*, 2, 483-488.

[4] Anwer M., Hussain I., Alam S.S. and Baig F. (2001) *Pakistan Journal of Biological Sciences,* 4(2), 124-127.

[5] Ashour N.I., Behairy T.G., El Lateef E.M.A. and Selim M.M. (1992) *5th Conference on Agronomy*, Zagazig University, Egypt, 361-370.

[6] Babbar K.R.S. and Dhingra H.R. (2007) *Haryana Agric. Univ. J. Res.*, 37, 1-6.

[7] El Kramany M.F. (2002) *Egypt J. Applied Sci.*, 17, 218-227.

[8] Ellis R.H. and Roberts E.H. (1981) *Seed Sci. Technol.*, 9, 373-409.

[9] Gomez K.A. and Gomez A.A. (1984) *Statistical procedures for agricultural research*, 2nd edition, New York, 680.

[10]Hossain M.S., Islam M.T. and Prodhan A.K.M.A. (2005) *Bangladesh J. Train. Dev.*, 18, 123-127.

[11]ISTA (1993) *Seed Sci. Technol.*, 21, 1-285.

[12]Islam M.M. and Karim M.A. (2010) *The Agriculturists.*, 8(2), 57-65.

[13]Karim M.A., Utsunomiya N. and Shigenaga S. (1992) *Japan. J. Crop Sci.*, 61, 279-284.

[14]Khajeh-Hosseini M., Powell A.A. and Bingham I.J. (2003) *Seed Sci. Technol.*, 31, 715-725.

[15]Mahdavi B. and Sanavy S.A. (2007) *Pakistan Journal of Biological Sciences,* 10(2), 273-279.

[16]Misra N. and Gupta A.K. (2006) *American Journal of Plant Physiology,* 1(1), 1-12.

[17]Mohammed A.H.M.A. (2007) *Res. J. Agric. Biol. Sci.*, 3, 200-213.

[18]Mohamed M.H. and El Kramany (2005) *J. Applied Sci. Res.*, 1, 78-84.

[19]Murillo-Amador B., Lopez-Aguilar R., Kaya C., Larrinaga-Mayoral J. and Flores-Hernandez A. (2002) *J. Agronomy Crop Sci.*, 188, 235-247.

[20]Naher N. and Alam A.K.M.M. (2010) *Int. J. Sustain Crop Prod.*, 5, 8-11.

[21]Promila K. and Kumar S. (2000) *Biol. Plant.*, 43, 423-426.

[22]Raptan P.K., Hamid A., Khaliq Q.A., Solaiman A.R.M., Ahmed J.U. and Karim M.A. (2001) *Korean J. Crop Sci.*, 46, 387-394.

[23]Ruan S., Xue Q. and Tylkowska K. (2002) *Seed Sci. Technol.*, 30, 61-67.

[24]Salah Uddin A.K.M., Amin R., Ullah J. and Asaduzzman Md. (2009) *Am. Eur. J. Agron.*, 2, 180-184.

[25]Scott S.J., Jones R.A. and William W.A. (1984) *Crop Sci.*, 24, 1192-1199.

[26]Singh J., Mathur N., Bohra S., Bohra A. and Vyas A. (2006) *Am. Eur. J. Agric. Environ. Sci.*, 1, 48-50.

[27]Snedecor G.W. and Cochran W.G. (1980) *Statistical Methods,* 7th Ed., 507.

[28]Win K.T., Oo A.Z, Hirasawa T., Ookawa T. and Yutaka H. (2011) *Afr. J. Biotechnol.*, 10, 1615-1624.

BANANA AND PLANTAIN *(Musa* spp.) CULTIVAR PREFERENCE, LOCAL PROCESSING TECHNIQUES AND CONSUMPTION PATTERNS IN EASTERN DEMOCRATIC REPUBLIC OF CONGO

EKESA B.N.[1]*, KIMIYWE J.[2], DAVEY M.W.[3], DHUIQUE-MAYER C.[4], VAN DEN BERGH I.[5], KARAMURA D.[1] AND BLOMME G.[1]

[1]Bioversity International, Plot 106, Katalima Road, P.O. Box 24384, Kampala, Uganda.

[2]Kenyatta University, Department of Food, Nutrition and Dietetics, P.O. Box 43844, Nairobi, Kenya.

[3]Lab Fruit Breeding and Biotechnology, Department of Biosystems, Katholieke University of Leuven (KUL), de Croylaan 42, Heverlee, B-3000, Leuven, Belgium.

[4]Centre de Coopération Internationale en Recherche Agronomique pour le Développement (CIRAD), Dept. Persyst-UMR Qualisud TA B95/16 73 av. J.F Breton 34398 Montpellier Cedex 5, France.

[5]Bioversity International, Parc Scientifique Agropolis II 34397 Montpellier Cedex 5, France.

*Corresponding Author: Email- b.ekesa@cgiar.org

Abstract- This cross-sectional study carried out in North Kivu (NK) and South Kivu (SK) of the Eastern Democratic Republic of Congo aimed to identify the most preferred *Musa* cultivars, their processing/cooking methods and *Musa* consumption patterns among rural households. Our results showed that the preferred cooking banana varieties included yellow-pulped AAA-East African Highland bananas [EAHB] 'Nshikazi' (SK) and 'Vulambya' (NK), which were valued for their cooking qualities, large bunches and suitability for production of banana beer. The preferred plantains were orange-pulped and included 'Musheba' (SK) and 'Musilongo' (NK) and were preferred for their short maturation period, large bunches and higher market prices. Over 60% of the households examined had consumed EAHB within 24 hours, whereas <10% had consumed plantains. The most common cooking method was simple boiling of bananas/plantains and main accompaniments include beans and amaranth leaves. Over 69% of surveyed households that had consumed *Musa* products had consumed them boiled. The majority of the households (90%) obtained banana/plantains from their farms, and >55% of households from SK and NK consumed banana products between 2 and 4 times/week. This information will be used to help direct researchers on the *Musa* cultivars that could be further sampled for carotenoid analysis and those found to be rich could be used in the fast-tracking approach to reduce Vitamin A deficiency.

Keywords- Banana, plantain, preference, cooking, Democratic Republic of Congo (DRC)

Introduction

Plantain and bananas (*Musa* spp.) serve as important food crops in much of Africa. Together they provide over 25% of the carbohydrate needs and 10% of the daily calorie requirements for over 70 million people on the African continent [1]. Production of the plantains and bananas is concentrated in Eastern DRC and ranges between 75,000 and 80,000 tonnes per year, and they rank second in importance after cassava in the Democratic Republic of Congo (DRC) [2]. The majority of cultivated bananas and plantains are triploid varieties belonging to the Eumusa section of the genus *Musa*, family *Musaceae* [3]. These varieties evolved from intra- and inter-specific crosses involving two diploid ancestor species, *M. acuminata* Colla (genome AA) and *M. balbisiana* Colla (genome BB), which originated from Malaysia and India, respectively [4].

Bananas and plantains are one of the main dietary starchy sources [2], apart from this, studies have suggested that some cultivars have high pro-vitamin A carotenoids (pVACs) levels and are capable of providing up to half of the total human daily vitamin A requirement in a single fruit [5,6]. PVACs among cultivars of *Musa* fruits differ more than 200-fold, ranging from 0.23 - 59.56 µg/gdw, with very high levels found in plantains and lower levels in other cooking bananas and commercial dessert types [1]. PVAC levels are linked with orange fruit pulp color, with most dessert bananas having a light colored pulp (cream, ivory) and most plantain cultivars having more yellow-orange colored pulp [1].

Although there is no documentation on the prevalence of vitamin A deficiency (VAD) in Esstern DRC, based on WHO estimates, 61.1% of preschool children in the whole of DRC have VAD. According to the classification provided by the International vitamin A Consultative Group (known as the Annecy Accords), a prevalence of over 30% of the population with less than 20 µg/dl serum retinol defines a situation of severe VAD [7]. In addition, DRC has one of the highest under-five mortality rates (U5MR) in the world, with more than 200 deaths in every 1000 live births [2]. The most obvious result of VAD is progressive damage to the eyes, eventually leading to blindness. Sub-clinical VAD is a much more widespread problem, contributing to an increased risk of morbidity and mortality from common childhood infections [8]. Therefore, with *Musa* fruit consumption levels ranging from 137-to 174 kg/person/year in DRC [9], it is clear that *Musa* fruit could be an important source of carotenoids and other micronutrients to consumers.

Apart from the high banana and plantain production, DRC is endowed with a wealth of natural resources and has a substantial agricultural potential, capable of feeding the whole of Africa [10]. Unfortunately this potential has not been exploited to improve living standards for the general Congolese population [10]. Out of the 154 territories (districts) in the whole of DRC, more than 117 are moderately food insecure and 28 others are in a state of acute food insecurity and livelihood crisis. Most of the food-insecure households are concentrated in the eastern part of DRC and more specifically in North and South Kivu [10].

Despite the major role that bananas and plantains play in Eastern DRC, there has been limited research and hardly any documentation on the preferences for the various banana cultivars grown, their processing/cooking techniques and consumption patterns among small-holder farmer households. In addition, the nutrition value of the preferred *Musa* cultivars and in particular pVAC levels have yet to be determined to establish whether they could be used in approaches aimed at reducing VAD among households.

Methodology

A cross-sectional study was carried out in Eastern DRC using focus group discussions (FGDs) and household interviews (HHIs), during the month of September 2009. Two provinces, North Kivu (NK) and South Kivu (SK), were selected based on the high levels of food insecurity and their high dependency on bananas and plantains [11]. Multistage sampling was used to select one territory from each of the provinces and three localities from each of the two territories. Two territories/districts Beni and Bukavu were sampled from North Kivu and South Kivu, respectively. In Beni territory Mabuku, Kisungu and Rwakhwa localities were randomly sampled and in Bukavu territory, the localities randomly sampled were Kajeje, Murhesa and Miti. A listing of all households with preschool children in each locality was compiled and systematic random sampling was used to select the specific households to be interviewed. Sample size was calculated using Fisher's formula:

$$n = \frac{t^2 \times p(1-p)}{m^2}$$

where n = required sample size, t = confidence level at 95% (standard value of 1.96), p = estimated proportion of households with preschool children with regards to total households and m = margin of error at 5% (standard value of 0.05) [12]. This gave a sample size of 163 and 208 households with preschool children from Beni territory and Bukavu territory, respectively. Focus group discussions were carried out at the locality level, to obtain information on the most popular and preferred *Musa* cultivars within the community and the general processing (e.g., cooking) methods. Information obtained through FGDs was further verified using household interviews by which structured questionnaires were administered to mothers (caregivers) of children below 5 years old. FAO studies confirm that rural women make a tremendous contribution to food and agricultural production. They also play a crucial role in determining and guaranteeing food security and well-being for the entire household [13], it was therefore necessary that questions requiring details on household consumption patterns were addressed to women (caregivers); women respondents were purposefully selected for the household interviews.

Apart from verification, household interviews were also used to collect information on household food consumption patterns with regards to bananas and plantains. A ranking system based on traits identified as important to the community was used to establish the preference level among the different *Musa* cultivars. After identification of the two most preferred cultivars from each site (NK and SK), agronomists working on banana research identified mature plants of each of the popular cultivars from farmer fields and marked them. A mature bunch was harvested and fruit samples obtained using standardized sampling techniques where two fruits were obtained from the middle hand of the bunch at ripening stage 1 (hard and green) [14]. Pre-screening of fruit pulp color for carotenoid content was carried out using a standardized color strip developed by HarvestPlus reflecting the total carotenoid content of starchy staples. Descriptive statistics were computed using the statistical package for social sciences (SPSS) package and data were summarized in frequency tables, charts and graphs.

Results

Demographic Data

Of the 371 respondents from SK (N=208) and NK (N=163), 92% were women, 83% were monogamously married. Forty seven percent of the respondents from both NK and SK were aged between 25-34 years [Fig-1] and the youngest respondent was 15 years old.

The majority of the households (61% N=371) consisted of 5 to 8 members with an average household size of 6.4 members in NK and 7.1 in SK. Forty five percent of the respondents had not received any form of formal education, 32% had received incomplete primary education, 11% had completed primary education, while only 0.5% had gone on to a vocational college and none of the respondent had any college or university diploma/degree. Most of the households (76%) interviewed depended on agricultural production as the main source of income, 14% were involved in casual labor, while 5.3% and 4.3% were involved in small-scale business and formal employment, respectively. The mean monthly household income was $US20-50 per month, but over 80% of the households had a monthly average income of less than $US30.

Most Popular and Preferred *Musa* Cultivars

Through 14 FGDs in SK(8) and NK (6), a list of all *Musa* cultivars grown within the community was compiled. A total of 16 and 19 *Musa* cultivars were listed in SK and NK, respectively. The participants then arranged the named cultivars in order of priority. [Table-1] gives the physiological descriptions, synonyms and botanic characteristics used in the identification of the four most popular *Musa* cultivars in SK, Bukavu territory and NK, Beni territory identified by participants of the FGDs. In SK, Bukavu territory, the four most popular cultivars consisted of three East African Highland cooking banana (AAA-EAHB) cultivars and one plantain (AAB) cultivar while in NK, Beni territory the four included two East African Highland cooking banana (AAA-EAHB) cultivars and two plantain (AAB) cultivars. The cultivars were then scored on a scale of 1 to 5 by FGD participants for cooking quality, market price, bunch size (yield) and plant disease resistance in order to identify the two most popular and preferred from each of the two sites. In SK, the EAHB 'Nshikazi' also called 'Magizi' had the best overall score, with 4.65 for cooking quality, 3.93 for market price, 4.02 for bunch size and 2.39 for disease resistance [Table-2]. The second highest score was obtained by the plantain 'Musheba', with 3.98 for cooking, 4.60 for market price, 3.43 for bunch size (yield) and 2.08 for disease resistance. In NK, the EAHB 'Vulambya' received the highest score, followed by the plantain 'Musilongo', the traits that made 'Vulambya' and 'Musheba' popular were similar to those observed in 'Nshikazi' and 'Musheba' respectively from SK. The principal weakness of the popular cultivars was their low disease resistance [Table-2].

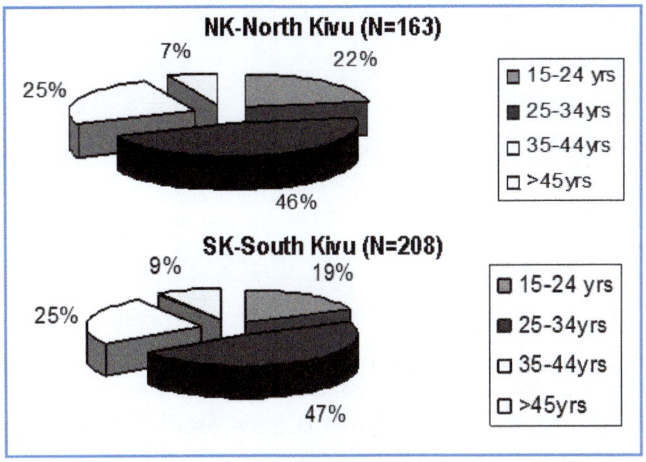

Fig. 1- Age structure of survey respondents in Eastern Democratic Republic of Congo

Table 1- Description and Botanic characteristics used in identification of four most popular cultivars from Beni territory (North Kivu) and Bukavu territory (South Kivu)

Local name	Origin	Synonyms	Genomic group	Sub-group	Botanic characteristic used to identify the cultivar				
					Pseudostem	leaves	Bunch	Fruit	Male bud
Vulambya	Beni (DRC)	Malambya[2], Bulambya[2], Nyalambya[2], Nyalambi[2] (Rukonjo (Uganda)[3]	AAA-EA[2,]	Lujugira-Mutika[2]	≥ 3metres light green with black blotches[2]	Dirty green[2]	Hangs vertically, Compact, cylindrical, maximum 11 hands[2]	Straight, Medium (15cm-20cm long), pointed tips, maximum 18 fruits in 3rd hand, naked tips[2]	Ovoid shaped, at harvest about 23cm width and 36cm length, white male flower[2]
Musilongo	Beni (DRC)	Kasilongo[2], Munzabo[2] Kibeda	AAB[2, 4]	Plantain (French horn)[2]	≥ 3 metres Red-purple[2]	Medium green[2],	Hangs vertically, Lax, assymetrical, maximum number of hands 9[2]	Sharp curved long fruit, pointed tip about 14 fruits in 3rd hand, persistent floral remains on fruit tips[2]	Lanceolate shaped, yellow male flower[2]
Nguma	Beni (DRC)	Mbaguma[2] Apakumo[2] Asongbe[2]	AAB[24,]	Plantain (French horn)[2]	≥ 3 metres Light Green almost yellowish[2]	Medium green[2]	Hangs vertically, lax, slightly truncated cone shape	Fruits almost straight and a bit inflated. Slightly pointed tips with persistent old style[2]	Ovoid, large, red purple and persistent, yellow male flower[2]
Kiware	Beni (DRC)	Ndabaware[2], Maware[2] Ndyabawali[2] Ngagara (Ug)[3] Nandigobe (Ug)[3]	AAA-EA[2,]	Lujugira-Mutika[2]	≥ 3 metres light green with black blotches[2]	Dirty green[2]	Hanging vertically, cylindrical, lax. Rachis with semi persistent floral parts, 10 hands	Fruits slender and angular slightly above 20cm, with persitent style on the fruit tip, 18 fruits in 3rd hand[2]	Intermediate Purple and pointed, yellow make flower[2]
Nshikazi	Bukavu (DRC)	'Magizi'= "bitter"[3] Ishika'(Rwanda),[3] Ensika',(Ug)[3] 'Emburansika' Omuburasika'(Ug[3]	AAA-EA[4]	Lujugira-Mutika	≥ 3 metres light green with black blotches[2]	Dirty green[2]	Oblique and compact cylindrical bunch[2]	Fruits medium and pulp astringent[2]	Ovoid shaped, at harvest about 23cm width and 36cm length, white male flower[2]
Musheba	Bukavu (DRC)	Busheba[3], misheba[3],	AAB[4]	Plantain (French)	≥ 3 metres Light green, almost yellowish[2]	Medium green[2]	Hangs vertically, lax, slightly truncated cone shape[2]	Fruits almost straight and a bit inflated. Slightly pointed tips with persistent old style[2]	Male bud large, red purple and persistent, yellow male flower[2]
Barhebesha	Bukavu (DRC)	Njakara, Incakara, Barabesha[3], Mutant of mudjuva	AAA-EA[4]	Lujugira-Mutika[2]	≥ 3 metres Green with black blotches[2]	Dirty green[2]	Hangs vertically, lax, Almost cylindrical[2],	Fruits > 25cm, almost curved towards the rachis, bottle necked tips[2]	Intermediate, Purple, pointed, white male flower[2]
Kamera	Bukavu (DRC)	Kamaramasenge, Kilore[2],	AAB[2,4]	Kamaramasenge[2]	Green yellow with bron rusty pigmentation	Medium green[2]	Asymmetrical-bunch axis nearly straight, compact, with bare rachis[2]	Short 15cm, almost perpendicular to rachis[2].	Purplish blue, imbricated and pointed2[1]

1=Information from Dowiya et al (2009), 2= Information from Université Catholique de Graben (UCG) Butembo collection, 3= Information from unpublished reports/personal communication, 4=Information from Simmonds et al (1995).

Table 2- Most popular and preferred Musa cultivars grown and consumed by households in North Kivu-NK (N=163) and South Kivu-SK (N=208), Eastern Democratic Republic of Congo.

Cultivar	Trait/Attribute (1=lowest score and 5= highest score) and standard deviation (S.D)								
Name (genome group)	Yield	S.D	Market price	S.D	Cooking quality	S.D	Disease resistance	S.D	Total Score
South Kivu Region									
Nshikazi (AAA)	4.02	0.68	3.93	0.67	4.65	0.49	2.39	0.57	15
Barhebesha (AAA)	3.46	0.77	2.44	0.58	3.83	0.69	2.26	0.59	11.99
Kamera (AAA)	2.2	0.69	2.24	0.73	1.4	0.58	1.39	0.55	7.25
Musheba (AAB)	3.43	0.81	4.6	0.53	3.98	0.75	2.08	0.91	4.08
North Kivu Region									
Vulambya (AAA)	4.02	0.68	4.03	0.71	4.66	0.49	2.41	0.57	15.12
Kiware (AAA)	3.35	0.74	2.71	0.69	4.52	0.71	2.21	0.65	12.79
Nguma (AAB)	2.86	1.02	3.9	0.75	2.32	0.79	2.57	0.69	11.65
Musilongo (AAB)	3.39	0.93	4.63	0.52	3.86	0.73	2.12	0.91	14

The results regarding the preferred cultivars obtained through FGDs were verified during the individual household interviews. Respondents confirmed that 'Nshikazi' and 'Musheba' in South Kivu and 'Vulambya' and 'Musilongo' in North Kivu were the most preferred *Musa* cultivars [Table-3]. In SK more than half of the households (69%) indicated 'Nshikazi' as the most preferred cultivar. The only popular and preferred plantain was 'Musheba'. In NK, 'Vulambya' was the most preferred cultivar overall (36%). The two popular plantain cultivars were; 'Nguma' and 'Musilongo' although more households preferred 'Musilongo' to 'Nguma' [Table-3]. [Table-3] also shows that the most preferred cultivars had pulp color ranging from ivory and yellow (EAHB cooking varieties) to orange (plantain varieties).

Table 3- Musa cultivars preferred by household members in North Kivu (N=163) and South Kivu (N=208), Eastern Democratic Republic of Congo

Cultivar local Name (genome group)	Pulp color At use	Harvest+ color Strip -carotene	Number of household indicating choice		
			1st Choice	2nd Choice	3rd Choice
South Kivu Region					
Nshikazi (AAA)	yellow	7/2 121U	143(68.8%)	9(4.3%)	1(0.5%)
Barhebesha (AAA)	yellow	½ 101U	6(2.9%)	36(17.3%)	6(2.9%)
Kamera (AAA)	Ivory	3/3 1205U	8(3.8%)	11(5.3%)	22(10.6%)
Musheba (AAB)	Orange	9/2 1355U	27(13.0%)	9(4.3%)	9(4.3%)
North Kivu Region					
Vulambya (AAA)	Yellow	7/2 121U	59(36.4%)	17(10.5%)	9(5.6%)
Kiware (AAA)	Yellow	-	20(12.3%)	29(17.9%)	6(3.7%)
Nguma (AAB)	Orange	9/2 1355U	5(3.1%)	2(1.2%)	9(5.6%)
Musilongo (AAB)	Orange	9/3 7507U	19(11.7%)	14(8.6%)	11(6.8%)

Popular Cooking and Processing Methods of the Preferred *Musa* Varieties Practiced by Households in Eastern DRC

It was established that the processing techniques applied to the *Musa* cultivars in North Kivu were similar to those in South Kivu [Table-4]. The two most popular EAHB cultivars ('Nshikazi' and 'Vulambya') were mostly utilized between stages 1 and 3 of ripening, i.e., before the fruit was ripe [15]. The fruits were generally peeled and boiled with only the addition of salt. Occasionally locally processed palm oil was added during the boiling process. The cooked banana was eaten either alone or with boiled beans or peas. In some instances the banana would be boiled together with beans and some amaranth leaves, 'lengalenga'. The plantain cultivars ('Musheba' and 'Musilongo') were also boiled but generally no salt was added because of the cultivars' natural sweetness and were eaten either alone or served with boiled green peas. 'Musheba' and 'Musilongo' were also used to make 'Nchimba' banana flour. To make 'Nchimba', the fruit was harvested at maturity, kept for about 1 week to reach stage 5 of ripening, all yellow with green tips [15], peeled, chopped into pieces, sun-dried on the ground and either pounded or milled into flour. This flour (Nchimba) was then used to make porridge, a weaning food. In some cases, cassava flour was added to the plantain flour at a ratio of 2:3. Occasionally, the plantain cultivars were deep fried in the local palm oil and eaten as 'banana fries'.

Table 4- Local cooking/processing methods applied to common Musa cultivars in Eastern Democratic Republic of Congo

Musa cultivar	Cooking method	Cooking ingredients	Common accompaniments
Nshika(AAA)	Boiling ('kitika')	Salt, water	Beans, peas, amaranths,
	Steaming	Salt, water,	Beans, meat[2], Fish[2]
Vulambya (AAA)	Boilin ('kisamunyu')	Salt and water	Beans, peas, amaranths,
Musheba (AAB)	Boiling ('ndizi')	Water	Peas, meat[2], chicken[2]
	Deep frying[3]	Palm oil	Meat[2], chicken[2]
	Roasting (snack)	-	-
	Porridge (flour)	Water, cassava flour[1]	-
Musilongo (AAB)	Boiling ('ndizi')	Water	Peas, meat[2], chicken[2]
	Deep frying[3]	Palm oil	Meat[2], chicken[2]
	Porridge (flour)	Water, cassava flour[1]	-

1 = Optional, 2 = Used occasionally, 3 = Prepared occasionally

Consumption Pattern of *Musa* Fruit

A 24-hour diet recall on all the food that had been consumed by the households indicated that 64 - 67% of all respondents had consumed EAHB cooking varieties within the last 24 hrs. (no significant difference between SK and NK) [Table-5]. The consumption of plantain cultivars was 4.8% and 10% in SK and NK, respectively. The majority of plantains, however, were not domestically consumed, but instead sold for cash due to their relatively high market price. Among cooking bananas, plantain bananas and desert bananas, cooking bananas (AAA-EAHB) were the most popular while desert bananas were the least popular [Fig-2] in both NK and SK regions.

Table 5- Consumption of cooking bananas and plantains in relation to other starchy staples available to communities in Eastern Democratic Republic of Congo

Starchy staples	South Kivu N= 208		North Kivu N=163	
	Frequency (%)	S.E.Mean	Frequency (%)	S.E.Mean
Millet	0.5	(±0.005)	0	(±0.000)
Sorghum	5.8	(±0.016)	0.6	(±0.006)
Maize	17.8	(±0.027)	14.2	(±0.028)
Wheat & products	0	(±0.000)	3.1	(±0.014)
Rice	1	(±0.007)	1.8	(±0.011)
Irish potatoes	0.5	(±0.005)	0.6	(±0.006)
Sweet potatoes	0.5	(±0.005)	0	(±0.000)
Cassava & products	86.6	(±0.026)	99.4	(±0.006)
Yam/Taro	0	(±0.000)	1.2	(±0.009)
Cooking banana	66.5	(±0.034)	64	(±0.038)
Plantain	4.8	(±0.015)	10	(±0.021)

Boiling was the most popular *Musa* fruit processing method in both regions [Table-6]. In SK, 75% of respondents consumed boiled bananas while in NK the figure was 69%. The varieties most used were the popular AAA-EAHB cultivars ('Nshikazi' in SK and 'Vulambya' in NK), which were boiled at ripening stages 1, 2 or 3 (when the mature fruits were still green). Although a significant amount of roasted banana was consumed in SK (47 %), none of the households in NK consumed the fruits this way. Other products were banana juice, beer and porridge in SK, and banana beer in NK. The beer banana cultivars were the AAA-EAHB 'Nshikazi' cultivar in SK and the AB 'Kisubi' cultivar in NK, which were kept up to ripening stage 6 (all yellow) before use [Table-6].

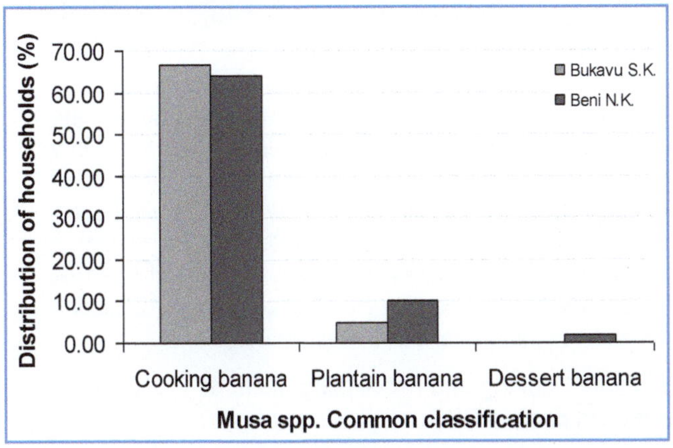

Fig. 2- Consumption rates of cooking, plantain and dessert bananas among households in Eastern Democratic Republic of Congo (SK-South Kivu: N=208, NK-North Kivu: N=163)

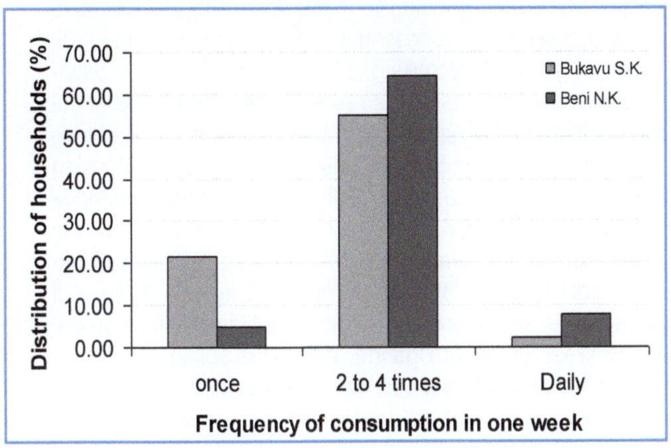

Fig. 3- Consumption rate of bananas and banana products among households in Eastern Democratic Republic of Congo (SK-South Kivu: N=208, NK-North Kivu: N=163)

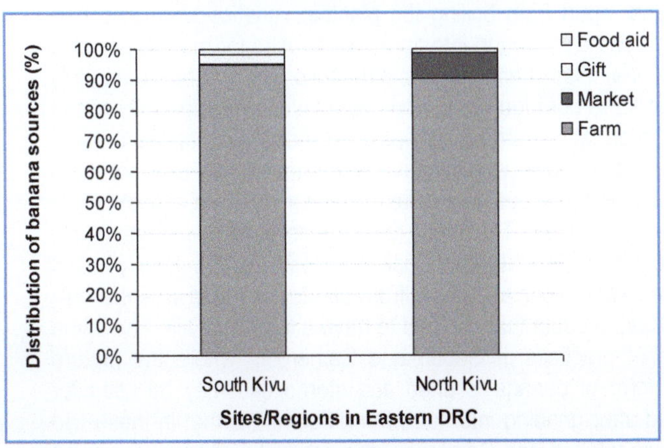

Fig. 4- Sources of bananas and plantains consumed by households in Eastern Democratic Republic of Congo (SK-South Kivu: N=208, NK-North Kivu: N=163)

Table 6- Consumption of local Musa products among households in South Kivu (N=208), North Kivu (N=163), Eastern Democratic Republic of Congo.

Musa product	Consumption freq. (%)	S.E. Mean	Cultivars commonly used	Genomic group	Ripening stage at use
Region/ Site South Kivu					
Steamed banana	1.2	(±0.00)	Nshikazi, Barhebesha	AAA	1-3(unripe)
Boiled banana	75	(±0.030)	Nshikazi, Barhebesha	AAA	1-3(unripe)
Roasted banana	47	(±0.035)	Nshikazi, Barhebesha	AAA	1-3(unripe)
Banana chips/crisps	0	(±0.000)	Musheba	AAB	5(ripe)
Banana beer/wine	6	(±0.000)	Nshikazi,	AAA	6(ripe)
Banana juice	13	(±0.000)	Nshikazi,	AAA	6(ripe)
Banana porridge	4	(±0.014)	Musheba	AAB	5(ripe)
Region/ Site North Kivu					
Steamed banana	0	(±0.000)	-	-	-
Boiled banana	69	(±0.036)	Vulambya,	AAA	1-3(unripe)
Roasted banana	0	(±0.000)	-	-	-
Banana chips/crisps	0	(±0.0000	Musilongo	AAB	5(ripe)
Banana beer/wine	6	(±0.019)	Kisubi,	AB	6(ripe)
Banana juice	1	(±0.006)	Banane		6(ripe)
Banana porridge	1	(±0.006)	Musilongo	AAB	5(ripe)

A large proportion of the household members (>55%) from both NK and SK had consumed *Musa* fruit (bananas and plantains) between 2 to 4 times in the 7 days before the interview and there were some households that had consumed *Musa* fruit every day of the last 7 days preceding the survey [Fig-3]. Although there were several geographic sources of food for the community, more than 90% of the households from NK and SK indicated having obtained the bananas and plantains from their farms [Fig-4].

Discussion

Demographic Data

FAO studies confirm that rural women make a tremendous contribution to food and agricultural production. They also play a crucial role in determining and guaranteeing food security and well-being for the entire household [13]. It was, therefore, necessary that questions requiring details on household consumption patterns were addressed to women (caregivers). This explains the high percentage of women respondents (91.7%). The majority of the women interviewed were between 25 - 35 years old, with 20% of them between the ages of 15 and 24 years, indicating that the women get married or become caregivers at a very early age. In DRC, the legal minimum age for marriage is 15 for women and 18 for men, and 74% of women between 15 and 19 years of age are already married [16]. One of the most common outcomes of marriage at an early age is the withdrawal of the girl from formal education, or no school enrollment at all. This could explain the low level of education observed in this study, which puts these women at a disadvantage since educated women are more likely to feel empowered to influence decisions affecting food choices, availability and accessibility [17]. In addition, the low level of education reduces the ability of these women to process information, acquire skills and model positive care giving practices to improve health, nutrition and education outcomes for their children [17]. The low literacy level could further be explained by the irregularity and inadequacy of teachers' salary payments and inadequate infrastructure (negatively affecting teaching morale) [2]. The reported absence of primary education fee waivers [2] and the observed low monthly average income (<US$30), could also be a contributor to high illiteracy. The girls may be forced to drop out of school due to a lack of money for tuition fees. The low income level observed in this study is also supported by a report which indicated that more than 80% of Congolese live on less than $1 per day [13].

Most Popular and Preferred *Musa* Cultivars

Findings indicated that an average of 16 and 19 *Musa* cultivars were being cultivated by the communities in SK and NK, respectively. This number was lower than that of a total of 29 and 32 *Musa* cultivars reported as being cultivated in SK and NK, respectively [18]. Both results show, however, that the diversity of *Musa* cultivars is slightly higher in NK as compared to SK, the higher number of cultivars reported in the earlier study could be due to the survey methodology used, where direct observation (transect walks) were carried out to establish the varied cultivars [18]. In addition more territories were included in the study as opposed to this study, where information was only collected through the FGDs and household interviews and only one territory/district from each of the two provinces (South Kivu and North Kivu) was included. In addition, not all villages were visited from the selected territories.

Although a beer banana cultivar 'Pisang Awak' (ABB), known locally as 'Kayinja' was not among the first four popular cultivars and did not even emerge in the preferred list, a study carried out in 2005 cites 'Kayinja' as the most popular cultivar occupying nearly ¾ of the land cropped with banana in Eastern DRC [19]. This could be because the participants of the FGDs were mostly local farmers and women of child-bearing age, and they seemed to prefer cultivars that could be used to make food. The most popular cultivars identified included three cooking cultivars and one plantain. According to the community members, 'Nshikazi'(AAA-EAHB) has high yield (big bunch) and good cooking quality. These findings confirm findings by [18], carried out in the same province (SK) although at different districts which also cited 'Nshikazi' as the most preferred *Musa* cultivar. According to [18], Nshikazi' was occupying 52% of the land under banana and apart from having the traits already mentioned (high yield, high market price and good cooking quality), 'Nshikazi' tastes better after cooking, and made good banana juice and banana beer. 'Nshikazi' is also grown in Rwanda where it is known as 'Ishika' and Uganda where it has different names depending on the regions i.e. 'Ensika', 'Emburansika' 'Omuburasika' or 'Nsika' [Table-1]. 'Musheba', a plantain cultivar whose name also means 'wedding' [Table-1], was the most popular plantain cultivar in SK. Apart from the reasons given by the community (high market price, good yield, and good cooking quality), other studies have reported that 'Musheba' is a popular cultivar because it matures faster and tastes very good [18]. In addition, 'Musheba' has social-cultural importance in times of marriage and the 'Mashi' tribe of SK had obtained it from Uganda as a form of dowry [18].

In North Kivu, the findings showed that the four most popular *Musa* cultivars were 'Vulambya', 'Kiware', 'Nguma' and 'Musilongo'. These four cultivars were also among the six most preferred *Musa* cultivars in NK identified in a previous study within the same region [18]. The reasons for Vulambya's popularity were similar to those of 'Nshikazi' in SK. Although 'Nguma' has been reported to be most popular cultivar in NK and 'Musilongo' didn't appear in the ten most popular cultivars [18], the findings of this study indicated that both 'Nguma' and 'Musilongo' were popular. In addition, the community members indicated that 'Musilongo' produces a bigger bunch and matures faster than 'Nguma'.

Among the above preferred plantains and cooking banana cultivars, the pulp color of the plantain cultivars was darker (orange) while the pulp color of the cooking banana varieties was lighter (yellow and ivory). These results are similar to those observed after pre-screening of cultivars from Uganda, Cambodia, Philippines and Hawaii, which revealed that more than 80% of the plantain cultivars (AAB) screened had an orange pulp while more than 80% of the cooking banana cultivars (AAA) had pulp color ranging from ivory to white [1]. Although the pulp color is not a definite indication of levels of carotenoids, it helps in filtering cultivars that could be further analyzed for enhanced carotenoids content. If a variety is found to contain considerable amounts of carotenoids, it could be used for fast-tracking approaches aimed at reducing VAD among *Musa* dependent rural households of Eastern DRC.

Popular *Musa* Cooking and Processing Methods Practiced by Households in Eastern DRC

As indicated earlier, in Eastern DRC bananas are a major part of the staple diet, complementing other sources of food. The methods of processing and cooking banana fruit range from simple boiling to fermentation, drying and grinding to make flour; The fruit can also be eaten as a dessert, or processed into juice, beer and wine [20]. The purpose of processing/cooking is to make the fruit more palatable and digestible. Processing also extends the storage life of the fruit, which is often highly perishable in its fresh condition. Findings of this study indicate that in both NK and SK, the bananas (cooking banana and plantains) were mostly boiled. Other cooking methods practiced included deep frying the plantains in local palm oil, but this was not so common since most of the harvested plantain bananas were sold to urban dwellers to generate income. Boiling of the banana fruit is also a common practice in Uganda, Cameroon, Burundi and West Indies. In Uganda, the fruit is boiled with beans and ghee then mixed with pepper, salt and onions to make a dish called 'akatogo'. In Cameroon, the green banana is boiled and served in a sauce of palm oil with fish, cooked meat, green beans and seasoning, whereas in the West Indies boiled green banana is served with salted fish or meat [21].

As noted, apart from boiling the plantain varieties were also harvested dried and ground into flour mostly used for making porridge for children less than 5 years old. According to the community members the porridge made from plantain flour is especially useful for children with diarrhea. It was also indicated that the banana flour could be mixed with cassava flour and not only used for making baby porridge but also boiled in water to make a thick paste locally called 'ugali.' The 'ugali' could be consumed with any relish, the most common one being, boiled cassava leaves, locally called 'sombe'. *Musa* flour especially from plantain banana was the only local *Musa* product that seemed to have a long shelf life. The same is true of practices in Gabon and Cameroon, where the mature green form of banana is dried and stored and may be used for cooking after grinding into flour, the difference is that in these two countries (Gabon and Cameroon) this product was mainly used as a famine reserve, not an everyday meal [21]. Steaming and roasting are very popular banana cooking methods in Uganda; the EAHB cultivars are mostly steamed in banana leaves and either pounded or eaten whole (a dish called 'matoke'), while the plantains are roasted and sold as street food. Despite this, findings of this study showed that these two cooking methods (steaming and roasting) are not popular in Eastern DRC.

Musa Fruit Consumption Patterns

Of all the staples considered in this study, cassava root was the most highly consumed. This finding is supported by the high cassava production of 74% observed by other researchers [2]. These researchers also indicated that DRC has the highest annual cassava consumption in the world, with an estimated 390 kg fresh root (equal to 1,100 kcal) per person per day[2]. Cassava is cultivated in around 50% of the arable land in Congo and provides to 70% of Congolese population 60% of food energy intake [2]. Just as reported [2], findings of this study showed that EAHBs were the second most consumed starchy staple recording a consumption rate of ≥70% among households in both SK and NK [Table-5] and [Fig-2]. Other researchers have also reported a EAHBs consumption rate of almost 80% in DRC [22]. Although the consumption of plantains has also been reported to be about 36% in SK and that plantains have traditionally been a starchy staple food of rural populations in the humid lowlands of DRC [22], in this study, the low level of plantain consumption observed (4.8% SK and 10% NK) could be explained by their current r high market value, thus farmers are increasingly selling plantains as a cash crop to urban consumers. With rapid urbanization and the growing prosperity of city dwellers, demand is outstripping supply, thereby leaving the rural poor households to depend on either cooking varieties or other starchy staples. The consumption of dessert bananas was even lower with consumption rates of 0% and 2% in SK and NK, respectively [Fig-2], because dessert bananas are mostly consumed as snacks and their production is also very low in Eastern DRC.

More than 90% of the households from both NK and SK obtained the bananas from their farms [Fig-2]. This is higher than values reported earlier [22], where only 64% of households in SK obtained their bananas from own production and 24% relied on purchase. This difference could be because in the reported study [22], the respondents included both rural and urban traders in *Musa* while in this study the respondents were exclusively small-holder farmers.

Conclusions and Recommendations

Bananas (cooking banana and plantains) are the second most popular starchy staple consumed by smallholder households in SK and NK of Eastern DRC. 'Nshikazi' (AAA-EAHB) and 'Musheba' (AAB-French plantain) are the most popular and preferred *Musa* cultivars in South Kivu while 'Vulambya' (AAA-EAHB) and 'Musilongo' (AAB-French plantain) are the most popular and preferred *Musa* cultivars in North Kivu. The AAA-EAHB cultivars ('Vulambya' and 'Nshikazi') are preferred because they produce big bunches, posses good cooking qualities and they make good banana beer. The plantain cultivars ('Musheba' and 'Musilongo') are preferred because they mature faster and have high market prices.

The fruit pulp color of the preferred *Musa* cultivars ranges from yellow to orange, an indication of potentially high carotenoids levels. A nutritional analysis should be carried out to verify whether they could indeed be a potential source of carotenoids. Of the three common *Musa* fruit classifications (cooking banana, plantain banana and desert banana), cooking bananas are the most consumed, having a consumption rate of more than 65% in both North Kivu and South Kivu.

Although the *Musa* fruit is normally processed/cooked using various methods such as boiling (with/without peel), steaming, roasting, drying and milling into flour, the most popular cooking method applied to the *Musa* cultivars consumed in Eastern DRC is boiling without peel. More than 50% of the households consume boiled bananas between two and four times per week, with 20% of the households consuming it on daily basis with 90% of the bananas obtained from the household's farm.

In the light of these results, it is recommended that thenutrition content of these preferred cultivars be examined to establish whether, the *Musa* fruits could be a good source of other micronutrients, especially provitamin A carotenoids. In addition, the *Musa* fruit is not consumed immediately after harvest, and in most cases, it is cooked into various products. Although maturation or ripening in *Musa* fruits after harvest is usually accompanied by enhanced carotenogenesis [23], some preliminary studies have shown that the impact of ripening is cultivar-specific, with pVACs contents remaining essentially unaltered, increasing or even slightly decreasing depending on the genotype [6].

Several reports have shown evidence that although results are still conflicting, retention or loss of carotenoids during storage/ripening and processing of *Musa* fruit is cultivar dependent [24, 23]. There is need to therefore carry out more research that will include a variety of *Musa* cultivars from Eastern and Central African to ascertain the content of carotenoids in these fruits, the stage in fruit development where carotenoid levels reach their peak and processing/cooking techniques that retain or avail more carotenoids. If the levels of carotenoids are found to be high in some of these local *Musa* cultivars, they could be used in the fast-tracking approach, to reduce Vitamin A deficiency among *Musa*-dependent populations of Eastern and Central Africa.

Acknowledgement

The authors acknowledge the Consortium for Improving Agriculture -based Livelihoods in Central Africa (CIALCA) for providing the necessary funding to carry out this study. Sincere gratitude goes to the team of persons who facilitated the data collection from the study areas. These include Mr. Charles Lwanga and Professor Ndung'o Vigheri of North Kivu, Beni territory, Mr. Jules Ntamwira and Mr. Charles Bisimwa of South Kivu, Bukavu territory, the enumerators and households" members from both NK and SK. Mr. Johnson Vincent science editor, Bioversity International in Montpellier- France is thanked for his assistance in editing this paper

References

[1] HarvestPlus (2007) *Annual Technical Report.*

[2] Luntala T., Mbile F., Jean Claude O. and M. Okulo (2000) *Country Presentation, DRC.*

[3] Tenkouano A., Vuylsteke D. and R. Swennen R. (2007) *Journal of Crop Science Improvement,* 20(1-2), 31-51.

[4] Stover R.H. and Simmonds N.W. (1991) Bananas, *Tropical*

Agriculture Series, 3rd ed., London, 468.

[5] Davey M.W., Keulemans J. and Swennen R. (2006) *Science Direct Journal of Chromatography,* A(1136), 176-184.

[6] Davey M.W., Van den Berge I., Markham R., Swennen R. and Johan K. (2009) *Journal of Food Chemistry,* 115, 806-813.

[7] Sub Committee on Nutrition (2010) *Sixth Report on the World Nutrition Situation, Geneva,* 2, 8-22.

[8] Mulokozi G.I. (2003*) Content and In vitro accessibility of Provitamin A carotenoids in some Tanzanian Vegetables and fruits. PHD Thesis,* Chalmers University of Technology, Goteborg, Sweden.

[9] Ministry of Planning (2005) *Piloting Unit of DSRP Process, Kinshasa-Gombe, DRC,* 129-155.

[10]World Food Program (2009) *Fighting Hunger World Wide. Democratic Republic of Congo.*

[11]World Food Program (2008) *Comprehensive Food Security and Vulnerability Analysis, Executive Brief: Democratic Republic of Congo.*

[12]Magnani R. (1997) *IMPACT Food Security and Nutrition Monitoring Project,* Arlington, Va.

[13]FAO (2003) *Food Security Responses to the Protracted Crisis Context of the Democratic Republic of Congo.*

[14]Dadzie B.K. and Orchard J.E. (1997) *Criteria and Methods, INIBAP Technical Guidelines,* 2, 63.

[15]Ngoh G.N., Tomekpe K., Fokou E. and Etoa F.X. (2009) *Global Science Books,* 3(1), 64-70.

[16]FIDH (2010) *Democratic Republic of the Congo: Breaking Impunity.*

[17]ACC/SCN (2004) *5th Report on the Worlds' Nutrition Situation,* Geneva, 5-14.

[18]Dowiya N.B., Rweyemamu C.L. and Maerere A.P. (2009) *Journal of Animal & Plant Sciences*, 4(2), 341-356.

[19]Ndungo V., Bakelana K., Eden-Green S. and Blomme G. (2005) *Infomusa,* 13, 43-44.

[20]Ndungo V., Fiaboe K.K.M. and Mwangi M. (2008) *Journal of Applied Bioscience,* 1(1), 1-7.

[21]FAO (1990) *FAO Corporate Document Repository.*

[22]Jagwe J., Ouma E., Van Asten P. and Abele S. (2008) *CIALCA Project Survey Report,* Bujumbura, 61.

[23]Ngoh G.N., Lusty C., Van den Bergh I., Akyeampong E., Davey M.W. and Tomekpe K. (2008) *Global Science Books,* 2(2).

[24]Lusty C., Akyeampong E., Davey M.W., Ngoh Newilah G. and Markham R. (2006) *InfoMusa*, 15(1-2).

SELECTIVE TILLAGE PRACTICES FOR IMPROVING *IN SITU* SOIL MOISTURE AND CROP YIELD IN COTTON BASED CROPPING SYSTEM

PAYAL S.D.[1]* AND SHINDE G.U.[2]

[1]Dept. of Soil & Water Conservation Engineering, College of Agricultural Engineering, V.N.M.A. University, Parbhani - 431 401, MS, India.
[2]Dept. of Farm Machinery and Power Engineering, College of Agricultural Engineering, V.N.M.A. University, Parbhani - 431 401, MS, India.
*Corresponding Author: Email- sandippayal35@gmail.com

Abstract- A field investigation entitled "Effect of tillage practices for soil moisture conservation and crop yield under cotton based cropping system in vertisol" was conducted during kharif season 2012-2013 at All India Co-ordinate Research Project (AICRP), for Dry Land Agriculture, Vasantrao Naik Marathwada Agricultural University, Parbhani, with Cotton and Soybean crop. The experiment was laid out in spilt plot design with four main treatments viz one ploughing followed by one harrowing across the slope up to 10 cm (T1) (control), up to 20cm depth (T2), up to 30cm depth (T3), and up to 40cm depth (T4) and sub main treatments with cropping systems viz sole cotton (C1), cotton + soybean (C2) (1:2) and sole soybean (C3). Treatment wise and replication wise sowing of cotton and soybean was done across the slope in the field to know the effect of tillage treatment on soil moisture conservation, replication wise soil sample were collected at 15cm, 30cm 45cm depth and the soil moisture was recorded fortnightly. Treatment and replication wise biometric observations viz. height of plant, number of bolls/pods and number of branches were also recorded fortnightly. Treatment and replication wise seed cotton and soybean yield was recorded. Among tillage practices one ploughing followed by one harrowing across the slope up to 30 cm depth (T3) recorded significantly higher soil moisture content (21.16 %), best vegetative growth, highest cotton equivalent yield. in sole cotton (18.73 q/ha). and intercropping (20.25 q/ha) and highest Gross Monetary Return (73047 Rs/ha), Net Monitory return (50201Rs./ha) and Benefit cost ratio (B:C) (3.19) over rest of the treatments.

Keywords- In-situ Soil Moisture, Tillage treatments

Introduction

Soil is one of the most important natural resource for crop production Soil, water and air play key role in plant growth and crop production. Soil in its natural state rarely provides the most favorable physical conditions for crop growth. Improvement of the soil physical condition by imparting tillage maintains desirable soil, air, water relations in seed bed. It changes the soil structure and moisture retention patterns over the growing seasons and affects crop growth. Soil tillage in general is one of the fundamental field operations in agriculture. To assure normal plant growth, the soil must be prepared in such conditions that roots can have enough air water and nutrient [1].

In rain fed agriculture, moisture conservation is very crucial for successful crop production. Tillage facilitates moisture in the soil and increases infiltration and water holding capacity of soil and also facilitates roots to penetrate easily in the soil. Limited availability of soil water in the profile at critical stage of crop growth period is major constraints of rain fed agriculture. Crops grown in rain fed conditions are prone to water stress, owing to rapid loss of soil water from profile resulting in low water availability for root growth. Modification of soil structure is necessary to facilitate the better germination of seed, root growth, more water intake and storage. Deep tillage increase the water holding capacity of the soil and allows the roots to penetrate to deeper soil layer for uptake of soil water and nutrient, especially for deep rooted crops like cotton. Rain water conservation practices adopted to reduce the loss of soil water through evaporation has recorded higher yield of rain fed crops [2]. Marathwada region of Maharashtra state have black cotton soils with medium deep. Cotton and soybean is grown in kharif under rain fed condition in the region. The cropping systems combination also important for better yield. Jat, et al [3] reported that maize + blackgram intercropping system was found effective in improving soil moisture in the root zone as compared to sole maize. Asewar & Jadhav [4] found that strip cropping of maize with legume (2:1) is effective for reduction of runoff and increases maize equivalent yield. Keeping in view the need of tillage practices with cropping system combination, the "study on effect of tillage practices for soil moisture conservation and crop yield under cotton based cropping system in vertisol" is undertaken with objectives: To study the effect of tillage practices on moisture conservation and crop yield under cotton based cropping system.

Materials and Methods

Soil and its Characteristics

The soil was medium black in texture having depth up to 60 cm. It has pH 8.02, EC 0.16 dsm^{-1}, bulk density 1.39 mgm^{-3}, field capacity 15%, permanent wilting point 7.1% and infiltration rate 13 mmh^{-1}. The soil had 0.62% organic carbon, 135.6 kgha^{-1} available N, 4 kgha^{-1} P and 829.4 kgha^{-1} exchangeable K. The experiment was laid out in split plot design.

Experiment Details

Field size	62 x 40.80 m
Main treatments	4
Sub-main treatments	3
Number of plots	12
Replications	3
Design	Split plot
Crop and Variety	Cotton - Banny BT
Soybean -	MAUS 61-2
Land slope	0.30%
Soil type	Medium black
Each Plot size	20 x 9.2= 184 m^2
Crop spacing	Cotton - 120 cm x 45 cm
Soybean -	30 cm x 5 cm
Fertilizer doses	Cotton:120:60:60 Soybean::30:60:30

Treatment Details

Main Treatments

T_1 - One ploughing followed by one harrowing across the slope up to10 cm depth (control).

T_2 - One ploughing followed by one harrowing across the slope up to 20 cm depth.

T_3 - One ploughing followed by one harrowing across the slope up to 30 cm depth.

T_4 - One ploughing followed by one harrowing across the slope up to 40 cm depth.

Sub Main Treatments

C_1 - Sole Cotton; C_2 - Cotton + Soybean (1:2); C_3 - Sole Soybean

Methodology

Soil Moisture Studies

In order to assess the changes in soil moisture status, soil moisture was measured at 15cm (D1), 30cm (D2) and 45cm (D3) depth in soil profile. The moisture content was measured gravimetrically by taking the soil samples from each plot with the help of screw auger. The initial weight was measured on the field at the time of sampling and final oven dry weights of soil samples were measured using sensitive weighing balance. For final weight, the soil samples were kept in oven at 105ºC for 24 hrs. Soil moisture content was calculated by oven dry method using following formula:

$$\text{Moisture content (\%)} = \frac{\text{Initial weight (gm) - Final weight (gm)}}{\text{Final weight (gm)}} \times 100$$

each plot were recorded. These plants were properly labeled and growth parameters were monitored on them.

Total Grain Yield

The total weight of both crops per plot was recorded after threshing and converted to total yield (qha^{-1}) from each treatment. The time required for ploughing with various depths and harrowing in different treatments was recorded and total time required per hectare was calculated. Accordingly the cost for ploughing and harrowing was recorded.

Results and Discussion

During the year 2011and2012 the total rainfall received was recorded as 456.7mm and 531.3mm respectively against the annual average rainfall to the station as 892mm.

Table 1- Mean soil moisture content as influenced by tillage practices under different cropping systems during kharif-2011 and 2012

Treatment	Percent soil moisture content								
Tillage Treatments	Soil sample collection depth(cm) 2011				2012				Pooled
	D_1	D_2	D_3	mean	D_1	D_2	D_3	mean	
T_1	23.48	23.32	22.06	22.95	23.36	23.36	22.9	23.53	23.24
T_2	23.35	23.04	22.87	23.02	24.21	23.5	23.77	23.83	23.42
T_3	24.75	24.16	24.44	24.45	25.06	24.54	24.51	25.07	24.76
T_4	25.03	24.67	24.78	24.83	25.85	24.62	24.79	25.03	24.93
Mean	24.15	23.8	23.74	23.9	25.12	23.99	24.24	24.45	24.08
SE		0.12				0.3			0.33
CD at 5%		0.35				0.88			1.14
Cropping System									
C_1	24.29	23.82	24.04	24.06	25.4	24.19	24.4	24.68	24.21
C_2	23.8	23.3	22.33	23.14	24.93	23.63	21.82	23.45	23.18
C_3	24.37	24.27	24.85	24.5	25.03	24.14	26.51	26.23	25.3
Mean	24.15	23.8	23.14	23.89	25.12	23.99	24.24	24.45	24.17
SE		0.1				0.26			0.33
CD at 5%		0.3				0.77			0.91
A x B									
SE		0.21				0.62			0.49
CD at 5%		NS				NS			NS

Moisture Content

Data presented in [Table-1] revealed that significantly higher soil moisture content (24.93%) was recorded with one ploughing followed by one harrowing across the slope up to 40 cm depth (T4) which was at par (24.76%) with ploughing across the slope up to 30 cm depth (T3) during two years of experiments and in pooled analysis.

Crop Yield

In case of cropping systems significantly higher soil moisture content was recorded with sole soybean (C3) over remaining cropping systems during two years of experiments and pooled analysis. This might be due to shallow root systems It has been also observed that due deep ploughing soil gets loosened and helps to improve the infiltration rate. It was recorded that there was improvement in T_4 as 7.27per-cent and in T_3 6.49 per-cent over the control treatment (T_1).

It was observed that the higher seed cotton equivalent yield in pooled analysis (1601 kg/ha)was recorded in the treatment one ploughing followed by one harrowing across the slope up to 30 cm depth ((T3). It was also found that treatment T4 was at par with treatment T3 [Table-2].

The effect of treatments and cropping systems with soil moisture is showed in [Fig-1]. where in treatment T-3 shows highest moisture recording at the root zone and in case of cropping system, intercropping system C2 uses highest soil moisture.

Table 2- Cotton seed yield (Banny BT) and Soybean (MAUS 61-2, Pratikar) as influenced by tillage practices and cropping system

Tillage Treatments	Seed Cotton equivalent (Kg/ha)		
One ploughing followed by one harrowing operation cycle across the slope	2011	2012	Pooled
T_1-up to 10cm depth (control)	1063	1028	1046
T_2 -up to 20cm depth	1282	1165	1224
T_3 -up to 30cm depth	1699	1503	1601
T_4 -up to 40cm depth	1608	1487	1547
SE+	25.01	15.83	25.63
CD at 5%	74.87	47.39	79.08
Cropping System			
C_1-Sole Cotton	1595	1232	1413
C_2-Cotton+Soyaben 1:2	1850	1711	1780
C_3-Sole Soybean	794	944	869
SE+	33.44	22.55	40.66
CD at 5%	101.31	67.51	125.46
Interaction A x B			
SE+	67.68	45.11	40.66
CD at 5%	202.62	135.03	125.46
Grand Mean	1413	1295	1354

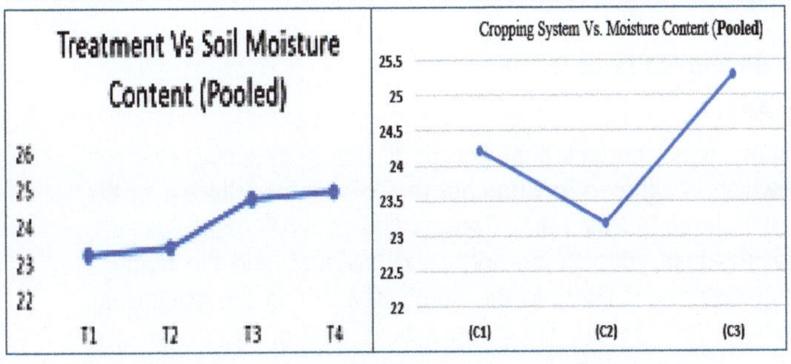

Fig. 1- Treatment and Cropping System Vs Soil moisture

In case of cropping systems the highest cotton yield was recorded in cotton + soybean (1:2) ratio (1780kg/ha) in pooled analysis.

The higher GMR, NMR and B:C ratio was recorded with one ploughing followed by one harrowing across the slope up to 30 cm depth (T3) over rest of the tillage treatments and cotton + Soybean (1:2) ratio in cropping system as mentioned in [Table-3].

Table 3- Mean Gross Monetary Returns, Net Monetary returns and Benefit Cost ratio as influenced different

Treatment	GMR(Rs/ha)		NMR(Rs/ha)		B:C ratio	
Tillage	2011	2012	2011	2012	2011	2012
T_1	37236	40092	16793	19760	1.76	1.91
T_2	44893	45470	24000	24576	2.07	2.15
T_3	59481	58626	38937	38082	2.83	2.74
T_4	56284	57993	34124	35835	2.48	2.57
SE+	875.4	617.49	875.49	659.3	0.04	0.05
CD at 5%	2620.7	1848.4	226.07	1973.7	0.12	0.16
Cropping System						
C_1	55840	48054	35232	24448	2.69	2.32
C_2	64770	66729	39664	41706	2.56	2.59
C_3	27810	36852	10494	19536	1.6	2.12
SE+	1184.5	879.64	1184.5	877.17	0.05	0.12
CD at 5%	3545.8	2633.2	3545.8	2625	0.15	0.36

Conclusion

Selective tillage practices such as one ploughing followed by one harrowing across the slope up to 30cm depth in vertisol. was found to be superior for *in situ* moisture conservation technique. The highest mean GMR, (Rs.59053) NMR (Rs.38509 and B: C ratio (2.78) of the seed cotton equivalent yield was recorded in the same treatment

Conflicts of Interest: None declared.

References

[1] Pandey A.K., Prakash V., Singh R.D. & Mani V.P. (2000) *Annals of Agricultural Research*, 21(4), 548-551.

[2] Shinde V.S., Deshmukh L.S., Shinde S.A. & Zade K.K. (2009) *Journal of Cotton Research and Development*, 23(1), 51-55.

[3] Jat M.L., Sammauria R., Balyan J.K. & Sharma R.K. (2010) *Indian Journal of Soil Conservation*, 38(2), 69-74.

[4] Asewar B.V. & Jadhav A.S. (2009) *Journal of Maharashtra Agricultural Universities*, 34(2), 229-230.

[5] Ahmadi H. & Mollazade K. (2009) *Agricultural Engineering International: The CIGR EJournal*, 11, 1-9.

[6] Bhanavase D.B., Deshpande A.N. & Tathe A.S. (2007) *Indian Journal of Soil Conservation*, 35(3), 230-233.

[7] Devakant & Tiwari P.N. (2000) *Bhartiya Krishi Anusandhan Patrika*, 15(1/2), 65-68.

[8] Khan F.U.H., Tahir A.R. & Yule I.J. (1999) *International Journal of Agriculture and Biology*, 1(3), 163-166.

[9] Kiran J.A. & Lingaraju B.S. (2010) *Karnataka Journal of Agricultural Sciences*, 18(2), 294-296.

[10]Selvraju R. & Ramaswamy R. (1997) *Indian Journal of Agronomy*, 42, 396-400.

LAND USE/LAND COVER CLASSIFICATION AND ACCURACY ASSESSMENT USING SATELLITE DATA - A CASE STUDY OF BHIND DISTRICT, MADHYA PRADESH

UPADHYAY R.[1], SINGH A.[2*], SHRIVASTAV P.[3] AND THAKUR S.[4]

[1]Department of Soil and Water Engineering, JNKVV, Jabalpur - 482 004, MP, India.
[2]Department of Soil Science & Agriculture, Chemistry, RVSKVV, Gwalior - 474 002, MP, India.
[3]Krishi Vigyan Kendra, Narsinghpur - 487 001, MP, India.
[4]Department of Soil and Water Engineering, JNKVV, Jabalpur - 482 004, MP, India.
*Corresponding Author: Email- akhileshsingh01@gmail.com

Abstract- Land is a finite natural resource and there is no scope to increase the area under cultivation. Moreover, this is becoming scarce resource due to immense agricultural and demographic pressure. Systematic study of any place requires information regarding the land use/ land cover of particular place to perform wide variety of tasks. Remote Sensing can provide important data for land use/land cover mapping. In the present study, satellite data IRS P6 LISS III for Bhind district, Madhya Pradesh was classified using supervised classification. Satellite data classification accuracy was also performed and resulted in overall accuracy as 95.75%.

Keywords- Land use/Land cover, Image Classification, Reference Data, Accuracy Assessment, Kappa Statistic

Introduction

Since long natural resources are being degraded due to population and poor management of land use. Natural resources [vegetation, water and soil] are responsive to human interaction and there together with terrain features determine the selection of proper land use pattern, which also in some way reflects the cultural, social and economic conditions. Utilization of land has lead to all areas of the Earth being modified [1]. The growth of population and the consequent demand for land are very high in Madhya Pradesh and per capita availability of land is very low. The indiscriminate use of available land causes the emergence of several environmental issues in many parts of Madhya Pradesh.

Land degradation is mainly due to population pressure, which leads to intense land use without proper management practices. Land development, sometimes-even over-development, leads degradation [2]. The land use/land cover system is highly dynamic which undergoes significant changes according to the changing socio-economic and natural environment. The change in any form of land use/land cover is highly related either with the external forces and the atmosphere built-up within the system [3]. So, the knowledge of spatial land cover information is essential for proper planning, management and monitoring of natural resources. Due to synoptic view, map like format and repetitive coverage, satellite remote sensing imagery is a viable source of gathering quality land use/land cover information at local, regional and global scales.

Remote Sensing can provide an important source of data for land use/ land cover mapping and environmental monitoring [4]. Image classification, which is the systematic grouping of remote sensing and other geographically referenced data by categorical or increasingly, fuzzy decision rules is considered the best known and most widely used information extraction technique in remote sensing [5].

The usefulness and success of land use and land cover mapping depends on the choice of appropriate classification scheme for feature extraction. To determine the quality of information derived from the classification process, accuracy assessment of the classification is implemented. Error matrix, which is primarily used in remote sensing for accuracy assessment, is typically based on an evaluation of the derived classification against some 'ground truth' or reference dataset. This study also accomplishes accuracy assessment which helps to identify the accuracy of land use/land cover data.

Materials and Methods

Study Area

The study was conducted at Bhind district of Madhya Pradesh state, which is situated in Chambal region in the northwest of the state. It is bounded by Agra, Etawah, Jalaun and Jhansi districts of Uttar Pradesh state to the north and the east, and the Madhya Pradesh districts of Datia to the south, Gwalior to the southwest, and Morena to the west. The geographical area of Bhind district is

4,459 km². It is situated between 250 54' 21" and 260 47' 49" N latitude and between 780 12' 47" and 790 08' 33" E longitude [Fig-1]. The temperature of study area varies between 80 and 460 and average annual rainfall is 668.3 mm.

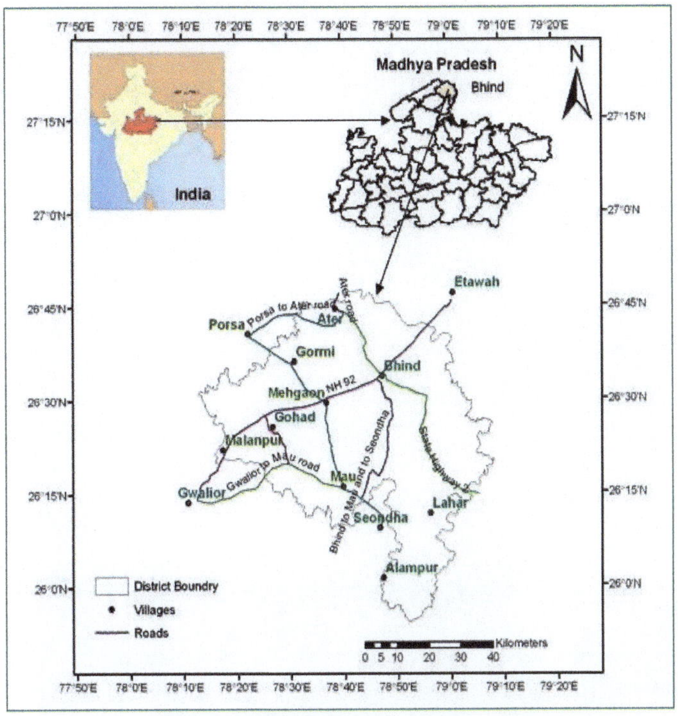

Fig. 1- Location Map of the Study area

Linear Imaging Self Scanning [LISS III] full scene geocoded satellite data for path 98 row 53 was acquired from National Remote Sensing Centre in Hyderabad dated 09th October 2008. Preparation of LULC map and their interpretation were achieved using ERDAS Imagine 9.1 and Arc GIS 9.3 software. The ancillary data Survey of India toposheet [54 J/2, 54 J/3, 54 J/6, 54 J/7, 54 J/9, 54 J/10, 54 J/11, 54 J/12, 54 J/13, 54 J/14, 54 J/15, 54 J/16, 54 K/13, 54 N/2, 54 N/3 and 54 N/4 [1:50000 scale] were used to perform the image processing and classification. These maps were also used in conducting a ground observation using GPS to verify the classification results from satellite imagery. The details of Satellite data used in the study are given in [Table-1].

Table 1- Details of Satellite Image used for the study

S. No.	Satellite	Sensor	Row/Path	Date of Passing
1.	IRS P6	LISS-III	98/53	09-Oct-2008

Preparation of LULC Map

The satellite imagery was interpreted using both digital and visual methods. The composite image was tested in order to choose the best band combination. The False Colour Composite [FCC] image of 1-2-3 [RGB] combination was used [Fig-2]. A classification scheme defines the land cover classes to be considered for remote sensing image classification. Sometimes a standard classification scheme such as Anderson's land use land cover classification system [6] is used, while at other times the number of land cover classes is chosen according to the requirements of the specific application. In this study, eight land cover classes were defined. The detail description of these classes along with their interpretative characteristics on the False Colour Composite [FCC] of LISS-III image is provided in [Table-2].

Table 2- Characteristics of land use/land cover classes

S. No	LULC Class	Description	Characteristics on LISS-III FCC
1	Forest	Trees cover covers, shrubs with partial grassland	Dark red/ dark brown to red
2	Agricultural/Other vegetation	Crop land and pasture, Orchards, groves, vineyard, nurseries, and ornamental horticulture area, other agriculture land	Red to Pink
3	Open/fallow/barren	Agricultural fields without crops, Exposed rocks without vegetation	Grey to green
4	Water body	Lakes, creeks, rivers, dams, forested wetland, non forested wetland	Blue to black
5	Waste land	Sparsely vegetated areas most often representative of bare earth or soil	White to whitish blue
6	Habitation	Commercial and residential areas, and with man made structure; road, railway lines, mixed urban built up areas, other urban or built up areas	Cyan to Light blue

Methodology of Supervised Classification

Supervised classification is the procedure most frequently used for quantitative analysis of remote sensing data; it rests upon using suitable algorithms to label the pixel in an image as representing particular ground cover types or classes [7].

Selecting training fields or samples is an important step in supervised classification. In this process, there will be selections for the pixels, which represent the different patterns based on the requirements. Then supervised classification is used, with parametric setting applied to maximum likelihood and it produces very good result.

In The Maximum Likelihood the program define the classification of pixels base on the probability that a pixel belongs to a particular class, assuming that probabilities are equal for all classes and that the input band have normal distribution. Image classification process is presented in [Fig-3].

Classification Accuracy Assessment

To determine the accuracy of classification, a sample of testing pixels is selected on the classified image and their class identity is compared with the reference data [ground truth]. The choice of a suitable sampling scheme and the determination of an appropriate sample size for testing data plays a key role in the assessment of classification accuracy [8].

The pixels of agreement and disagreement are generally compiled in the form of an error matrix. It is a c x c matrix [c is the number of classes], the elements of which indicate the number of pixels in the testing data. The columns of the matrix depict the number of pixels per class for the reference data, and the rows show the number of pixels per class for the classified image. From this error matrix, a number of accuracy measures such as overall accuracy, user's and producer's accuracy, may be determined [9]. The overall accuracy is used to indicate the accuracy of whole classification [i.e. number of correctly classified pixels divided by the total number of pixels in the error matrix], whereas the other two measures indicate the accuracy of individual classes.

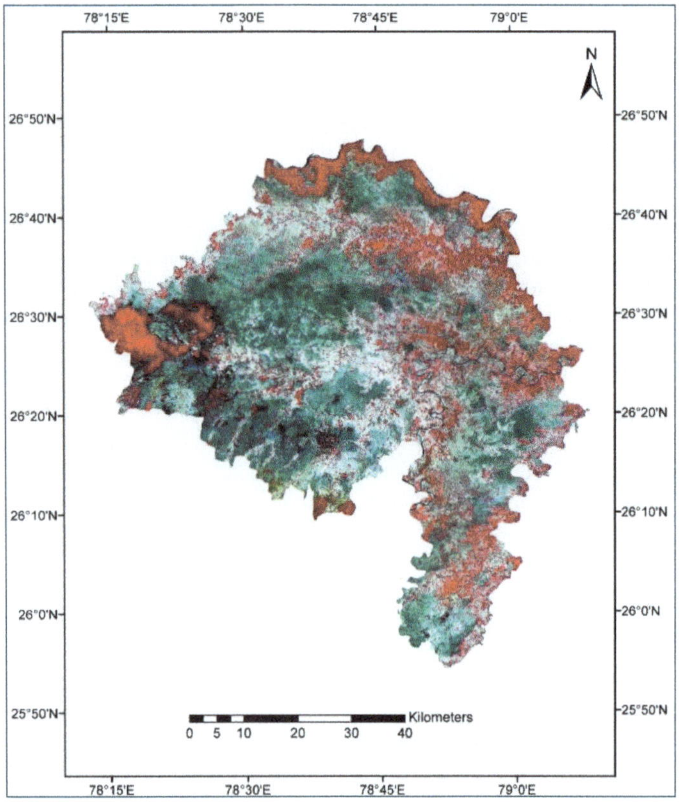

Fig. 2- False Colour Composite map of Bhind District

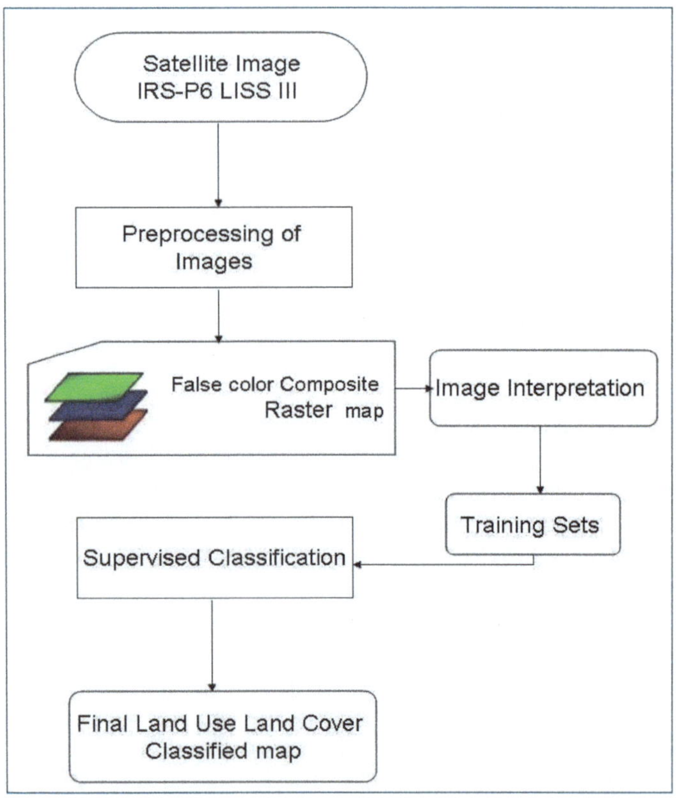

Fig. 3- Flow Chart of Land Use /Land Cover Map Preparation

User's accuracy is regarded as the probability that a pixel classified on the map actually represents that class on the ground or reference data, whereas producer's accuracy represents the probability that a pixel on reference data has been correctly classified.

Accuracy assessed through comparing classified land use/land cover with FCC using control point. Using stratified random method, where 400 points were specified as shown in [Fig-4].

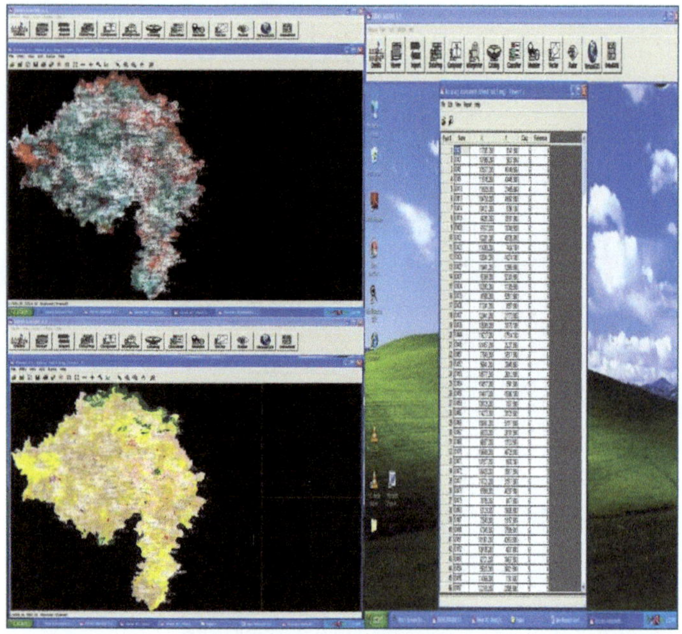

Fig. 4- Flow Chart of Land Use /Land Cover Map Preparation

It selects the random point from each class individually [the classes are weighted differently; hence the number of sample points differ from one class and another]. Then class value is assigned using "class value assignment option" and center value as the no majority option is used. Then, each point land cover type is identified by interpreting the underlying image. The report is generated which produces overall accuracy, user accuracy, producer accuracy and error matrix.

Results and Discussion

Land Use/Land Cover Classification

The result of classification is shown in the [Fig-5] which represents different land use/land cover classes i.e., forest, open/fallow/barren, agricultural/other vegetation, waste land, habitation, river/stream, canal and pond. Open/fallow/barren land occupies [57.15%] maximum area and pond occupies [0.16%] minimum area. Area statistics for each class obtained by the on screen visual and supervised classification of image is shown in [Table-3].

Table 3- Distribution of land use / land cover from classified image

S. No	Classes	Area [square km]	Area coverage [%]
1	River/Stream	32.28	0.73
2	Canal	13.86	0.31
3	Pond	7.21	0.16
4	Wasteland	338.61	7.61
5	Agriculture/Other Vegetation	1286.94	28.94
6	Open/Fallow/Barren	2541.78	57.15
7	Forest	183.87	4.13
8	Habitation	42.66	0.96
Total Area		4447.24	100

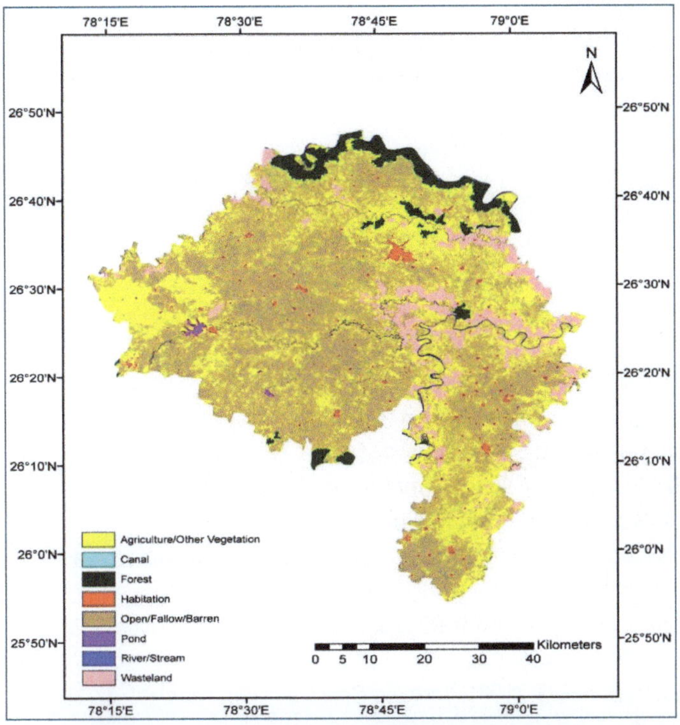

Fig. 5- Land Use /Land Cover Map of Bhind District

The dominant land use in the study area is open/fallow/barren land, which calls for the proper planning of the said land use, so that this land use can be utilized effectively. Also utilization of this area certainly improves the production of appropriate crop. Further, next to this class the agriculture is the majority area among the identified classes of land use/land cover.

Classification Accuracy Assessment Report

An error matrix is an appropriate beginning for many analytical statistical techniques, especially discrete multivariate techniques. Discrete multivariate techniques are appropriate because remotely sensed data are discrete rather than continuous. The data are also binomially or multinomially distributed, and therefore, common normal theory statistical techniques do not apply [10].

KAPPA is a discrete multivariate technique developed by [11] and has been utilized for land cover and land use accuracy assessment derived from remotely sensed data [12-14]. The result of performing a KAPPA analysis is the KHAT statistic [an estimate of KAPPA] which is another measure of accuracy or agreement. Values of KAPPA greater than 0.75 indicate strong agreement beyond chance, values between 0.40 and 0.79 indicate fair to good, and values below 0.40 indicate poor agreement [15]. Overall accuracy uses only the main diagonal elements of the error matrix, and, as such, it is a relatively simple and intuitive measure of agreement. On the other hand, because it does not take into account the proportion of agreement between data sets that is due to chance alone, it tends to overestimate classification accuracy [12,13,16].

KHAT accuracy has come into wide use because it attempts to control for chance agreement by incorporating the off-diagonal elements as a product of the row and column marginals of the error matrix [11]. Conceptually, k can be defined as:

$$K = \frac{\text{Observed accuracy - chance agreement}}{\text{1 - chance agreement}}$$

The error matrix showing producer's and user's, and overall classification accuracy, and including the Kappa coefficients is shown in [Table-4], [Table-5] and [Table-6] respectively.

Table 4- Classification accuracy error matrix for the land use /land cover map using reference data (ERROR MATRIX)

	Reference Data								
Classified Data	River/Stream	Canal	Pond	Wasteland	Agriculture/Other vegetation	Open/Fallow/Barren	Forest	Habitation	Row Total
River/Stream	3	0	0	0	0	0	0	0	3
Canal	0	0	0	0	0	0	0	0	0
Pond	0	0	3	0	0	0	0	0	3
Wasteland	0	0	0	28	0	0	0	0	28
Agriculture/Other vegetation	0	0	0	10	103	0	3	0	116
Open/Fallow/Barren	1	0	0	2	0	221	0	1	225
Forest	0	0	0	0	0	0	20	0	20
Habitation	0	0	0	0	0	0	0	5	5
Column Total	4	0	3	40	103	221	23	6	400

Table 5- Producers and User's with overall classification accuracy for the land use/ land cover map using reference data (ACCURACY TOTALS)

Class Name	Reference Totals	Classified Totals	Number Correct	Producers Accuracy [%]	Users Accuracy [%]
River/Stream	4	3	3	75	100
Canal	0	0	0	---	---
Pond	3	3	3	100	100
Wasteland	40	28	28	70	100
Agriculture/Other vegetation	103	116	103	100	88.79
Open/Fallow/Barren	221	225	221	100	98.22
Forest	23	20	20	86.96	100
Habitation	6	5	5	83.33	100
Totals	400	400	383		
Overall Classification Accuracy = [3+0+3+28+103+221+20+5]/400 = 95.75%					

Overall Classification *Accuracy* = [3+0+3+28+103+221+20+5]/400 = 95.75%

Table 6- kappa statistics for the land use/ land cover map (KAPPA [K^] STATISTICS)

Class Name	Kappa
River/Stream	1
Canal	0
Pond	1
Wasteland	1
Agriculture/Other Vegetation	0.8491
Open/Fallow/Barren	0.9603
Forest	1
Habitation	1

Overall Kaappa Statistics = 0.9297

In this study, the error matrix shows that the pixel classified for each training site are river/stream-3, pond-3, wasteland-28, agricultural/other vegetation-103, open /fallow/barren-221, forest-20 and habitation-5. The matrix of error shows that there is 1 cell which should be classified as river but classified as open /fallow/barren. There are 12 cells which should be classified as wasteland but classified as agricultural/other vegetation and open/fallow/barren. There are 3 cells which should be classified as forest but classified as agricultural/other vegetation. There is 1 cell which should be classified as habitation but classified as open/fallow/barren. The total accuracy in this classification accuracy is 95.75%. This means that the training sites selected are 95.75% spectral separable, and the training areas were classified very well.

Producer's accuracy refers to the how accurately the producer assigned the classes for the training sites. Producer's accuracy is computed by dividing the number of correctly classified pixels by the number of training sites pixels. The producer accuracy for river/stream is 75%, pond 100%, wasteland 70%, agricultural/other vegetation 100%, open/fallow/barren 100%, forest 86.96% and habitation 83.33%. User's accuracy refers to the accuracy that the pixel categorized in a certain class is truly representing that class on the ground. User's accuracy is calculated by dividing the number of correctly classified pixels by the total number of pixels that were classified in that class. For river/stream user accuracy is 100%, pond 100%, wasteland 100%, agricultural/other vegetation 88.79%, open/fallow/barren 98.22%, forest 100% and habitation 100%. The overall kappa statistics is 0.9297.

Conclusion

Land use land cover data is mostly derived from the satellite imaginary. This classified data used in wide variety of area such as planning, resource management, economic development, change detection etc. The update of this type of data is necessary. The present status of land use land cover in the Bhind district as evaluated by digital analysis of satellite data indicates that majority of area is open/fallow/barren land i.e. 57.15%. So, there is urgent need to use this open/fallow/barren land properly. Accuracy of assessment shows that overall accuracy is 95.75 percent which is good result and kappa statistics shows 0.9297 which shows good agreement between reference and classified image. This study clearly indicated that Remote Sensing and GIS is a novel tool to provide accurate spatial information on land use land cover of a region in a time and cost effective manner.

Conflicts of Interest: None declared.

References

[1] Harris J.A., Birch P. & Palmer J.P. (1996) *Land restoration and reclamation: principles and practice,* Addison Wesley Longman Ltd.

[2] Barrow C.J. (1991) *Land degradation: development and breakdown of terrestrial environments*, Cambridge University Press, 295.

[3] Bisht B.S. & Kothyari B.P. (2001) *Journal of the Indian Society of Remote Sensing,* 29(3), 137-141.

[4] Zhang Q., Wang J., Peng X., Gong P. & Shi P. (2002) *International Journal of Remote Sensing,* 23(15), 3057-3078.

[5] McDermid G.J., Franklin S.E. & LeDrew E.F. (2005) *Progress in Physical Geography,* 29(4), 449-474.

[6] Anderson J.M., Hardy E.E., Roach J.T. & Witmert R.E. (1976) *U.S. Geological Survey Professional Paper,* No. 964, Government Printing Office, Washington D.C.

[7] Richards J.A. (1993) *Remote sensing digital image analysis: an introduction,* Springer-Verlag, Berlin.

[8] Arora M.K. & Agarwal K. (2002) *Photogrammetry Journal of Finland,* 18(1), 33-43.

[9] Congalton R.G. (1991) *Remote Sensing of Environment,* 37(1), 35-46.

[10] Jensen J.R. (1996) *Introductory Digital Image Processing. A Remote Sensing Perspective*, Upper Saddle River, New Jersey: Prentice Hall.

[11] Cohen I. (1960) *A coefficient of agreement of nominal scales. Educational and Psychological Measurement,* 20(1), 37-46.

[12] Congalton R. & Mead R.A. (1983) *Photogrammetric Engineering & Remote Sensing,* 49(1), 69-74.

[13] Rosenfield G.H. & FitzpatrickLins K. (1986) *Photogrammetric Engineering and Remote Sensing,* 52(2), 223-227.

[14] Gong P. & Howarth P. (1990) *Photogrammetric Engineering and Remote Sensing,* 56(5), 597-603.

[15] SPSS Inc. (1998) *SYSTAT* - 8.0, Chicago, Illinois.

[16] Ma Z. & Redmond R.L. (1995) *Photogrammetric Engineering and Remote Sensing,* 61(4), 435-439.

STUDY OF SHELLED CORN SHRINKAGE IN A MICROWAVE-ASSISTED FLUIDIZED BED DRYER USING ARTIFICIAL NEURAL NETWORK

MOMENZADEH L. AND ZOMORODIAN A.

Agricultural Engineering Department, Shiraz University, Shiraz, Iran
*Corresponding Author: E-mail: zomorod@shirazu.ac.ir

Abstract- Grain drying is a vital unit operation in many processing plants. An undesirable change associated with this operation is shrinkage of dried product which results in decreased quality. Recently many attempts have been made to decrease the shrinkage of food stuff during drying. Microwave-assisted fluidized bed drying has particularly been proposed as a potentially effective method. In the present study, at each drying operating condition, the volume of shelled corn was calculated by measuring the three principal characteristic dimensions. The variation of the ratio of mean diameter of the kernel to its initial mean diameter was investigated for different operating conditions. It has been shown that employing microwave in fluidized bed drying reduces the shrinkage of particles considerably. Also, in this study, Artificial Neural Networks (ANN) analysis was employed to predict the extent of shelled corn shrinkage. In the construction of the network, three independent variables: microwave heat source, drying air temperature and moisture content were chosen as the input parameters and shrinkage of dried sample was set as the output parameter (dependent variable). The ANN model with 5 neurons was selected for studying the influence of transfer functions and training algorithms. It has been observed that back-propagation networks with logsig transfer function and trainlm algorithm were the most appropriate ANN configuration for predicting shrinkage. Results from the experiments and modeling showed good agreement. In order to test the ANN model the random errors were within an acceptable range of ±5% with a correlation coefficient (R^2) of 98%.
Keywords- Shelled corn; shrinking effect; Artificial Neural Network

Introduction

Drying of agricultural products is an important process in terms of storing and use in the food industry and post-harvest processing [7]. Drying is a process in which moisture migrates from interior of the drying object to the surface, resulting in some changes in physical properties of the dried sample [9].

There are several types of dryers such as solar, cabinet, rotary cylindrical, fluidized bed, microwave, infrared and etc. Among them, fluidized bed dryers assisted by microwave heating sources are very efficient due to efficient mixing and high heat and mass transfer rates between hot air and drying object [21].

An undesirable change associated with moisture diffusion is shrinkage. In general, shrinkage takes place as a result of volume reduction due to evaporation of the moisture contained in the drying object. The changes in dimensions of the solid could be monitored in most cases. Many researches have focused on changes in volume, area and shape. Shrinkage during drying is important not only from the viewpoint of product end-use but also for processing simulation [4].

Shrinkage of food stuff during drying processes has been of special interest during recent decades [1, 2, 4-6, 8, 10-13, 15, 17-19]. Several methods such as mathematical modeling, regression analysis, artificial neural networks and etc. are proposed for predicting shrinkage of agricultural products. Appropriate predicting models have to be selected based on several crucial parameters including the type of material, initial moisture content of sample and the drying procedure.

Artificial Neural Networks (ANN) is effective tools for modeling,

optimization and process control of several complex phenomena. It is believed that ANNs can be used to predict the shrinkage effect of drying samples in many cases [9].
In the present study, first shrinkage of corn cereal grain is measured in a fluidized bed dryer assisted by microwave. An ANN model is then developed and evaluated for the prediction of shrinkage behavior of shelled corn, based on measures of error deviation from experimental data

Materials and Methods

Specimen

Newly harvested shelled corn was used as test samples. The initial moisture content (dry basis) of shelled corn was approximately 26%.
Drying apparatus and procedures
An experimental apparatus consisting of a fluidized bed dryer assisted by a microwave energy source was designed and constructed. A cylindrical Pyrex column, 90 mm in diameter (100 mm outside diameter) and 280 mm high was used as the fluidized bed drying chamber. The chamber was placed in a domestic microwave oven (type: LG, MC- 2003TR (S)) with the frequency of 2450 MHz and approximate cavity volume of 0.075 m^3(outside dimensions of 574 mm × 376 mm × 505 mm). This oven was equipped with 5 power level settings of Low (180W), Medium low (360W), Medium (540W), Medium high (720W) and High (900W). Since the ratio of cylinder diameter to drying object diameter is much greater than 10, the wall effect was negligible. High pressure drying air was introduced at the bottom of the Pyrex column with constant air flow rate (650 lit/min) using a porous plate as distributor and an air compressor (type: TCS-SCLL, 7 bar pressure, 15 hp), to maintain the fluidization condition in the drying chamber. The air flow rates were measured by a rotameter with an accuracy of ±10 l/min. An electrical heating unit equipped with a thermostat (±1° C) was used to maintain different levels of drying air temperature of 30, 40, 50 and 60° C. A schematic diagram of employed apparatus is shown in Fig. 1.

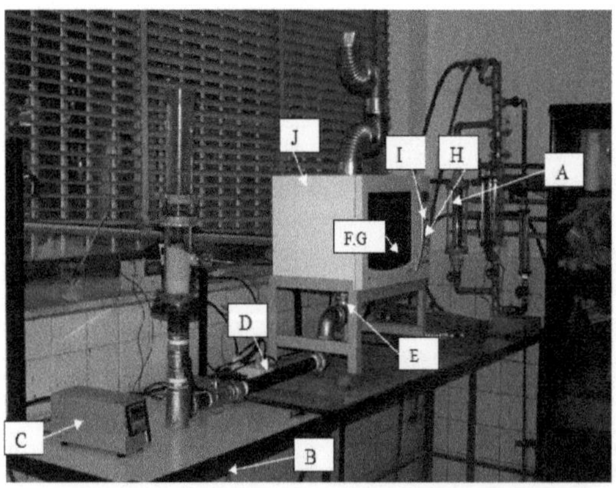

Fig. 1- Schematic diagram of the experimental apparatus

As shown in Fig. 1 three kernels of shelled corn were hung in the fluidized bed chamber as drying samples. Measurement of water loss from the samples was off-line. Sample weighting (in at most 10 seconds) was carried out using an electrical balance (balance type: MW-150t, Max weighing capacity of 150g, ±0.005g accuracy). Each sample weighting was done in triplicates and the average of the recorded weights was taken as an experimental data point.
Volume changes of shelled corn were calculated by measuring the three principal characteristic dimensions using a caliper (Mitutoyo, Japan, ±0.05mm). The geometric mean diameter was calculated using the following equation [14]:

$$D = \sqrt[3]{\frac{abc}{3}} \qquad (1)$$

Artificial Neural Networks (ANN)

The Artificial Neural Networks (ANN) are basically computational models, which simulate the function of biological neuron networks that are composed of neurons. ANN was developed using MATLAB v7.0. Each ANN has three layers of neurons: input, hidden and output (Tripathy and Kumar, 2008).

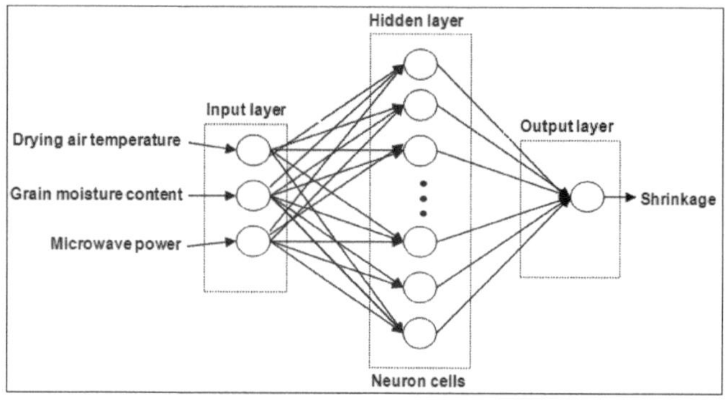

Fig. 2- Selected neural network structure

The ANN model was employed for investigating the shrinkage of shelled corn kernel. The ANN layout with three layers is shown in Fig. 2.
The input layer consists of important operating parameters namely: air drying temperature, grain moisture content and microwave power. Information from the input layer is then processed through one hidden layer with five neurons and finally the output layer shows the processing result as the value of shrinkage.
During training, weighting functions for the inputs were determined such that the predicted output data matched to the best experimental output from data set. Back propagation algorithms were used for minimizing the error of particular training pattern. A Log-Sigmoid activation function was applied and mathematical definition of the transfer function was trainlm (Levenberg–Marquardt back propagation) in the hidden layer.
Training was terminated when the Mean Absolute Error (MAE), Standard Error (SE) and Root Mean Square Error (RMSE) as given by equation 2-4 were less than ±0.05 and correlation coefficient (R^2), given by equations 5 was more than 95% [20].

$$MAE = \frac{1}{N}\sum_{i=1}^{N}|\overline{D}_{p,\exp,i} - \overline{D}_{p,cal,i}| \qquad (2)$$

$$SE = \frac{\sqrt{\sum_{i=1}^{N}(D_{p,\exp,i} - D_{p,cal,i})^2}}{N-1} \qquad (3)$$

$$RMSE = \left[\frac{1}{N}\sum_{i=1}^{N}(\overline{D}_{p,\exp,i} - \overline{D}_{p,cal,i})^2\right]^{1/2} \qquad (4)$$

$$R^2 = \sqrt{\frac{\sum_{i=1}^{N}(D_{p,\exp,i} - \overline{D}_{p,\exp,i})^2 - \sum_{i=1}^{N}(D_{p,\exp,i} - D_{p,cal,i})^2}{\sum_{i=1}^{N}(D_{p,\exp,i} - \overline{D}_{p,\exp,i})^2}} \qquad (5)$$

In this study, the available experimental data was partitioned into two parts, 120 data points for training and 30 data points for validation of the model.

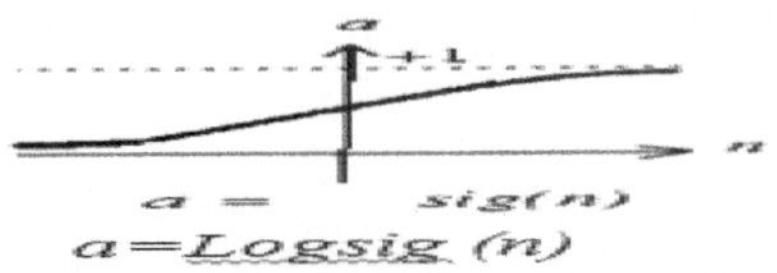

Fig. 3- Log-Sigmoid Transfer Function

The Log-sigmoid (logsig) activation function is shown in Fig. 3. This function takes the input (which may have any value between plus and minus infinity) and determines the value of output inside the range 0 to +1, as given equation 6 [3].

logsig(n) = 1 / (1 + exp(-n))

a = logsig(n)

In order to study the effects of different parameters on network performance, the model was run with changing an important parameter while keeping the others constant.

Results and Discussion

Shrinkage Effect

In order to show the effects of various parameters on the rate of shrinkage of shelled corn, several experiments were carried out under different operating conditions. In each experiment, the ratio of initial geometric mean diameter (D_0) to the geometric mean diameter of the kernel at a given condition (D) was determined and plotted verses grain moisture content, as shown in Figs. 4 and 5.

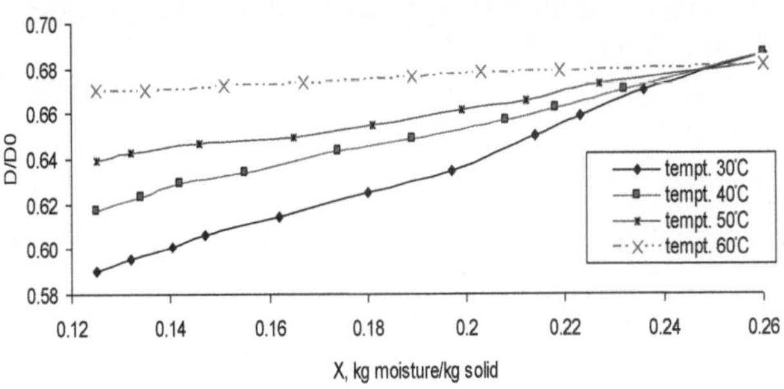

Fig. 4- Variation of D/D0 with drying sample moisture content for different air temperatures without microwave assisted.

It was concluded that drying air temperature and microwave power level were the main factors that showed significant effects on the grain shrinkage during the drying process. Analysis of the experimental data revealed that the variations of D/D_0 for the samples in a fluidized bed in both cases of drying with and without microwave heat source were well correlated as linear functions of the moisture content of drying samples.

As it was expected, increasing of the heating air temperature and microwave power resulted in decreasing of the shrinkage of the kernels. Indeed by increasing these parameters, the water vapor concentration on outer surface of the drying sample reached the equilibrium condition more rapidly. This shrinkage reduction can be interpreted in terms of a decrease in drying time so that the kernel could not find enough time to be shrunk. Similar results were reported by other researchers such as [6, 15 and 16].

The combined effect of microwave power and drying air temperature on shrinkage are showed in Fig. 5. The results indicated that by increasing the drying air temperature and using microwave energy power as an assisting heat source, the values of shrinkage increased. This increase can be due to the penetration of microwave energy into the sample and creation of a large vapor pressure difference between the core and the surface of grain.

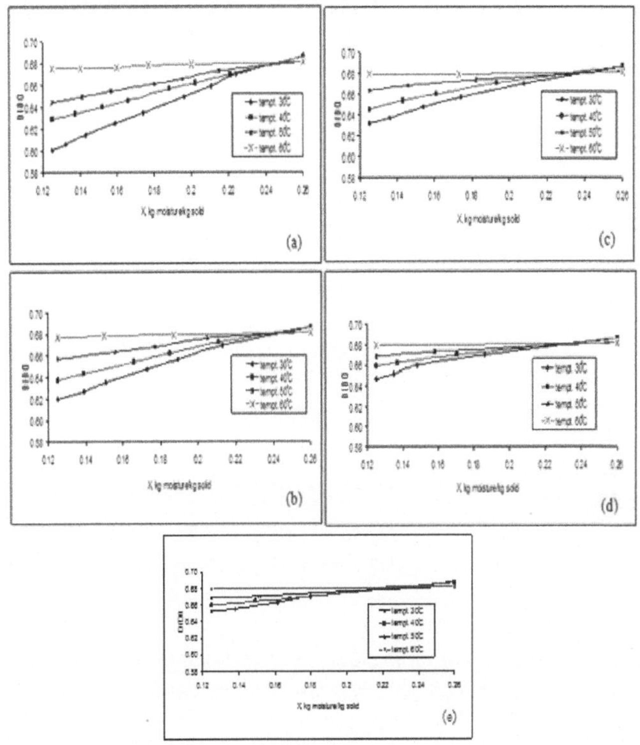

Fig. 5- Variation of D/D0 with drying sample moisture content for different air temperatures and different microwave power (a: 180, b: 360, c: 540, d: 720 and e: 900W).

Sensitive analysis in ANN prediction

In this Artificial Neural Network (ANN) model, three independent variables: level of microwave heat source (zero to 900 W), drying air temperature (30 to 60°C) and grain moisture content (26% to 12.5%) were chosen as the input parameters, and shrinkage of drying sample was regarded as the output parameter (dependent variable).

In order to determine the optimum structure of ANN, the rate of error convergence was checked by changing the number of hidden neurons (ANN with 1 to 10 neurons in hidden layer was tested) and adjusting the training algorithm. The data set was first normalized and then divided into two parts; one part was used for training the networks and the other for testing.

One of the most difficult tasks in ANN model development is to find the optimal network architecture. This network architecture can be selected among several network configurations containing the combination of various model parameters namely, the number of neurons in the hidden layers, different transfer functions and the training algorithms. A list of different transfer functions and training algorithms investigated during training network is summarized in Table 1 [20]. Table 2 represents the results of sensitivity analysis on shelled corn shrinkage experimental data. It can be seen that the ANN prediction results have a very strong dependence on input parameters. The present work showed a good agreement with the works of [9 and 18].

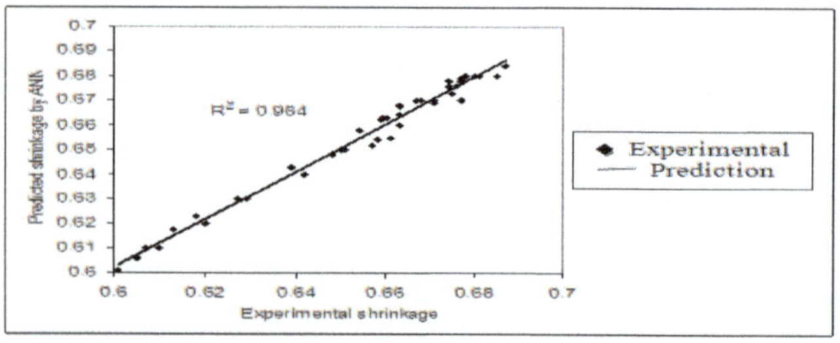

Fig. 6- Comparison of the experimental and predicting shrinkage values given by the proposed ANN model.

The accuracy of various proposed prediction models was checked by comparing of the predicted and the experimental drying sample shrinkage data. This comparison, showed a very good agreement, as shown in Fig. 6.

Table 1- List of transfer functions and back propagation training algorithms used in ANN training

Transfer function	Training algorithms
Logsig (Log sigmoid)	scg (Scaled conjugate gradient back propagation)
Tansig (Hyperbolic tangent sigmoid)	cgp (Polak–Ribiere conjugate gradient back propagation)
Poslin (Positive linear)	bfg (BFGS quasi-Newton back propagation)
Satlin (Saturating linear)	lm (Levenberg–Marquardt back propagation)
	rp (Resilient back propagation; Rprop)

Table 2- Measures of error in percent of shrinkage using the ANN model considering Logsig transfer function and trainlm algorithm for 5 neurons.

Source of error	Measures of error
Mean Absolute Error (MAE)	0.432
Root Mean Square Error (RMSE)	0.589
Standard Error (SE)	0.11
Correlation coefficient (R^2)	0.984

Conclusions

- Shrinkage of shelled corn could be well correlated to the moisture content by linear equations. It is shown that in fluidized bed dryer assisted by microwave system, the shrinking was less than the corresponding shrinking values in fluidized bed drying system alone.
- Applicability of Artificial Neural Networks (ANN) in predicting the shrinkage of shelled corn was well accepted. It was observed that back-propagation networks with logsig transfer function and trainlm algorithm were the most appropriate ANN configuration for prediction capability of shrinkage.

Nomenclature

a, b, c	Maximum, intermediate, minimum diameter (mm)
D	Geometrical mean diameter at any time (mm)
D_0	Initial geometrical mean diameter (mm)
MAE	Mean absolute error
X	Moisture content
N	Number of observations
R^2	Correlation coefficient
RMSE	Root mean square error
SE	Standard error
$D_{p, exp, i}$	Average experimental drying time for the ith observation
$D_{p, cal, i}$	Calculated drying time for the ith observation

Acknowledgments

The authors wish to thank Dr. A. Tahavvor for his comments regarding the Artificial Neural Network technique and Eng. Ostvar for his editorial comments.

References

[1] Abbasi Souraki B., Mowla D. (2008) *Journal of Food Engineering.*, 88, 9-19.

[2] Achanta S., Okos M.R., Cushman J.H. and Kessler D.P. (1997) *Moisture transport in shrinkage gels during saturated drying. AIChE L.*, 43, 2112-2122.

[3] Farkas I., Remenyi P. and Biro A. (2000) *Computers and Electronics in Agriculture.*, 29, 99–113.

[4] Hatamipour M.S., Mowla D. (2002) *Journal of Food Engineering.*, 55, 247-252.

[5] Hatamipour M.S., Mowla D. (2003) *Drying Technology.*, 21(1), 83-101.

[6] Hatamipour M.S., Mowla D. (2003*) Journal of Food Engineering.*, 59, 221-227.

[7] Hawang S., Cheng Y., Chang C., Lur H., Lin T. (2008) *Journal of Cereal Science.*, 50, 36-42.

[8] Hernandez J.A., Pavon G., Garcia M.A. (2000) *Journal of Food Engineering.*, 45, 1-10.

[9] Kerdpiboon S., Kerr W. and Devahastin S. (2006) *Food Research International.*, 39(10), 1110-1118.

[10]Kilpatrick P.W., Lowe E., Van Arsdel W.B. (1955) *Advances in Food Research.*, 6, 313-372.

[11] Lang W, Sokhansanj S., Rohani S. (1994) *Drying Technology*, 12(7), 1687-1708.

[12] Lozano J.E., Rotstein E. and Urbicain M.J. (1983) *Journal of. Food Science*. 48, 1497-1502.

[13] McMinn W.A.M., Magee T.R.A. (1997) *Journal of Food Engineering*. 33, 37-48.

[14] Mohsenin N.N. (1996) *Physical properties of plants and animal materials. Gordon* 79-127.

[15] Ratti C. (1999) *Journal of Food Engineering*. 23, 91-105.

[16] Sanga E.C.M, Mujumdar A.S., Raghavan G.S.V. (2002) *Chemical Engineering and Processing*. 41, 487-499.

[17] Sablani S. and Rahman M. (2003) *Food Research International*, 36(6), 617-623.

[18] Satish S. and Pydi Setty Y. (2004) *International Communications in Heat and Mass Transfer.*, 32,539-547.

[19] Suzuki K., Kiyoshi K., Hasegawa T. and Hosaka H. (1976) *Journal of Food Science.*, 41, 1189-1193.

[20] Tripathy P.P., Kumar S. (2008) *International Journal of Thermal Science.*, 48, 1452-1459.

[21] Topuz A. (2009) *Advances in Engineering Software.*, 41(3), 464-470.

Permissions

All chapters in this book were first published in IJAS, by Bioinfo Publications; hereby published with permission under the Creative Commons Attribution License or equivalent. Every chapter published in this book has been scrutinized by our experts. Their significance has been extensively debated. The topics covered herein carry significant findings which will fuel the growth of the discipline. They may even be implemented as practical applications or may be referred to as a beginning point for another development.

The contributors of this book come from diverse backgrounds, making this book a truly international effort. This book will bring forth new frontiers with its revolutionizing research information and detailed analysis of the nascent developments around the world.

We would like to thank all the contributing authors for lending their expertise to make the book truly unique. They have played a crucial role in the development of this book. Without their invaluable contributions this book wouldn't have been possible. They have made vital efforts to compile up to date information on the varied aspects of this subject to make this book a valuable addition to the collection of many professionals and students.

This book was conceptualized with the vision of imparting up-to-date information and advanced data in this field. To ensure the same, a matchless editorial board was set up. Every individual on the board went through rigorous rounds of assessment to prove their worth. After which they invested a large part of their time researching and compiling the most relevant data for our readers.

The editorial board has been involved in producing this book since its inception. They have spent rigorous hours researching and exploring the diverse topics which have resulted in the successful publishing of this book. They have passed on their knowledge of decades through this book. To expedite this challenging task, the publisher supported the team at every step. A small team of assistant editors was also appointed to further simplify the editing procedure and attain best results for the readers.

Apart from the editorial board, the designing team has also invested a significant amount of their time in understanding the subject and creating the most relevant covers. They scrutinized every image to scout for the most suitable representation of the subject and create an appropriate cover for the book.

The publishing team has been an ardent support to the editorial, designing and production team. Their endless efforts to recruit the best for this project, has resulted in the accomplishment of this book. They are a veteran in the field of academics and their pool of knowledge is as vast as their experience in printing. Their expertise and guidance has proved useful at every step. Their uncompromising quality standards have made this book an exceptional effort. Their encouragement from time to time has been an inspiration for everyone.

The publisher and the editorial board hope that this book will prove to be a valuable piece of knowledge for researchers, students, practitioners and scholars across the globe.

List of Contributors

G. U. Todkari
Dept. of Geography, Shri Sant Damji Mahavidyala Mangalweda, Dist-Solapur, MS, India

S. K. Jain
Sorghum Research Station, Sardarkrushinagar Dantiwada Agricultural University, Deesa- 385535 (BK), Gujarat, India

P. R. Patel
Sorghum Research Station, Sardarkrushinagar Dantiwada Agricultural University, Deesa- 385535 (BK), Gujarat, India

Dhanalakshmi Sampathkumar
Research Scholar, Jawaharlal Nehru Technological University, Hyderabad-500 085, Andhra Pradesh, India Department of Chemistry, KLE Society's BVB College of Engineering & Technology, Hubli-580 031, Karnataka, India

Ramadevi Punyamurthy
Research Scholar, Jawaharlal Nehru Technological University, Hyderabad-500 085, Andhra Pradesh, India Department of Chemistry, KLE Society's BVB College of Engineering & Technology, Hubli-580 031, Karnataka, India

Basavaraju Bennehalli
Department of Chemistry, Alva's Institute of Engineering & Technology, Moodbidri-574 225, Karnataka, India

Srinivasa Chikkol Venkateshappa
Department of Mechanical Engineering, GM Institute of Technology, Davangere-577 006, Karnataka, India

B. M. More
Brahmdevdada Mane Institute of Technology, Belati, Solapur- 413002, MS, India

P. S Patwardhan
Department of Economics, Dr. Babasaheb Ambedkar Marathwada University, Aurangabad- 431004, MS, India

S. S. Narwade
Department of Economics, Dr. Babasaheb Ambedkar Marathwada University, Aurangabad- 431004, MS, India

S. H. Ghatrehsamani
Agricultural Engineering Department, Shiraz University, Shiraz, Iran

A. Zomorodian
Agricultural Engineering Department, Shiraz University, Shiraz, Iran

D. M. Xavier
Departamento de Química Analítica e Inorgânica, Universidade Federal de Pelotas, Pelotas, Rio Grande do Sul, Brazil

A. S. Silva
Departamento de Química Analítica e Inorgânica, Universidade Federal de Pelotas, Pelotas, Rio Grande do Sul, Brazil

R. P. Santos
Laboratório de Engenharia de Materiais e da Computação de Sobral (LEMCS), Universidade Federal do Ceará, Sobral, Ceará, Brazil

M. F Mesko
Departamento de Química Analítica e Inorgânica, Universidade Federal de Pelotas, Pelotas, Rio Grande do Sul, Brazil

S. N. Costa
Departamento de Física, Universidade Federal do Ceará, Fortaleza, Ceará, Brazil

V. N. Freire
Departamento de Física, Universidade Federal do Ceará, Fortaleza, Ceará, Brazil

B. S. Cavada
Laboratório de Moléculas Biologicamente Ativas, Departamento de Bioquímica, Universidade Federal do Ceará, Fortaleza, Ceará, Brazil

J. L. Martins
Departamento de Química Analítica e Inorgânica, Universidade Federal de Pelotas, Pelotas, Rio Grande do Sul, Brazil

P. Assigbe
Institiut National des Recherches Agricoles du Benin, B.P. 226 Bohicon, Benin

B. Kone
Felix Houphouet Boigny University, Cocody, soil sciences department, Abidjan, Côte d'Ivoire

J. P. Bognonpke
Nagui Abrogba University, Department of plant biology and improvement, Abidjan, Côte d'Ivoire

A. Toure
Africa Rice Center (WARDA), Cotonou. 01 BP 2031 Cotonou, Benin

J. Huat
Africa Rice Center (WARDA), Cotonou. 01 BP 2031 Cotonou, Benin
CIRAD, UPR Hortsys, F-34398, Montpellier Cedex 05, France

A. Yao-Kouame
Felix Houphouet Boigny University, Cocody, soil sciences department, Abidjan, Côte d'Ivoire

A. A. Kandil
Department of Agronomy, Faculty of Agriculture, Mansoura University, Egypt

A. E. Sharief
Department of Agronomy, Faculty of Agriculture, Mansoura University, Egypt

W. A. E. Abido
Department of Agronomy, Faculty of Agriculture, Mansoura University, Egypt

M. M. Ibrahim
Ministry of Agriculture, Egypt

S. J. Awate
Dept. of Geography, Walchand College, Solapur, MS, India

G. U. Todkari
Geography Dept. Shri Sant Damaji College, Mangalwedha, Solapur, MS, India

S. A. A. M. Al-Saadi
Department of Biology, College of Science, University of Basra, Basrah, Iraq

S. S. Al-Taie
Department of Biology, College of Science, University of Misan, Amarah, Iraq

G. Bora
Division of Biotechnology, Defence Research Laboratory (DRL), Tezpur- 784 001, Assam, India
Department of Biotechnology, Gauhati University, Guwahati- 781 014, Assam, India

H. K. Gogoi
Division of Biotechnology, Defence Research Laboratory (DRL), Tezpur- 784 001, Assam, India

P. J. Handique
Department of Biotechnology, Gauhati University, Guwahati- 781 014, Assam, India

M. N. E. E. Elgimabi
Department of Biology, Taif University, Saudi Arabia

S. S. Musale
Sinhgad College of Engineering, Pune- 411 004, MS, India

P. M. Patil
RMD Sinhgad Technical Institutes Campus, Pune- 411 058, MS, India

G. U. Todkari
Dept. of Geography, Shri Sant Damji Mahavidyala Mangalweda, Dist-Solapur, MS, India

M. M. Maharlouie
Shiraz University, Shiraz, Iran

M. Loghavi
Shiraz University, Shiraz, Iran

S. Kamgar
Shiraz University, Shiraz, Iran

B. Yongkhamcha
School of Biology, Institute of Science, Suranaree University of Technology, Nakhon Ratchasima 30000, Thailand

K. Indrapichate
School of Biology, Institute of Science, Suranaree University of Technology, Nakhon Ratchasima 30000, Thailand

S. H. Ghatrehsamani
Department of Agricultural Engineering, Collage of Agriculture, Shiraz University, Shiraz, Iran

M. Dadashzadeh
Department of Agricultural Engineering, Collage of Agriculture, Shiraz University, Shiraz, Iran

A. Zomorodian
Department of Agricultural Engineering, Collage of Agriculture, Shiraz University, Shiraz, Iran

Ismini Nakouti
School of Pharmacy and Biomolecular Sciences, Liverpool John Moores University, Byrom Street, Liverpool, L3 3AF, UK

Glyn Hobbs
School of Pharmacy and Biomolecular Sciences, Liverpool John Moores University, Byrom Street, Liverpool, L3 3AF, UK

A. A. Kandil
Department of Agronomy, Faculty of Agriculture, Mansoura University, Egypt

A. E. Sharief
Department of Agronomy, Faculty of Agriculture, Mansoura University, Egypt

M. S. Sheteiwy
Department of Agronomy, Faculty of Agriculture, Mansoura University, Egypt

S. K. Ray
Department of Soil Science and Agricultural Chemistry, Uttar Banga Krishi Viswavidyalaya, Pundibari, CoochBehar- 736165, WB, India

D. Mukhopadhyay
Department of Soil Science and Agricultural Chemistry, Uttar Banga Krishi Viswavidyalaya, Pundibari, CoochBehar- 736165, WB, India

A. A. Kandil
Department of Agronomy, Faculty of Agriculture, Mansoura University, Egypt

A. E. Sharief
Department of Agronomy, Faculty of Agriculture, Mansoura University, Egypt

E. S. E Nassar
Central Administration for Seed certification (CASC), Ministry of Agriculture, Egypt

S. S. Bora
Department of Agricultural Biotechnology, Assam Agricultural University, Jorhat-785 013, Assam, India

J. P. LAHAN
Department of Agricultural Biotechnology, Assam Agricultural University, Jorhat-785 013, Assam, India

Madhumita Barooah
Department of Agricultural Biotechnology, Assam Agricultural University, Jorhat-785 013, Assam, India

Rupjyoti Sarmah
ICAR Research Complex for NEH Region, Arunachal Pradesh Centre, Basar-791101, Arunachal Pradesh, India

R. L. Ram
Soil and Land Use Survey of India, Mrida Sarvekshan Bhavan, Rajendra Nagar, Hyderabad- 500030, AP, India

P. Jha
Soil and Land Use Survey of India, Mrida Sarvekshan Bhavan, Rajendra Nagar, Hyderabad- 500030, AP, India

P. K. Sharma
Department of Soil Science and Agricultural Chemistry, Institute of Agriculture Sciences, BHU, Varanasi- 221005, UP, India

N. Ahmed
Division of Soil Science and Agricultural Chemistry, IARI, New Delhi- 110012, India

R. KUMAR
Soil & Land Use Survey of India, IARI Buildings, New Delhi- 110012, India

Sharanappa
Soil and Land Use Survey of India, Mrida Sarvekshan Bhavan, Rajendra Nagar, Hyderabad- 500030, AP, India

D. A. L. Vasudeo Rao
Soil and Land Use Survey of India, Mrida Sarvekshan Bhavan, Rajendra Nagar, Hyderabad- 500030, AP, India

V. Singh
Soil and Land Use Survey of India, Mrida Sarvekshan Bhavan, Rajendra Nagar, Hyderabad- 500030, AP, India

A. A. Kandil
Department of Agronomy, Faculty of Agriculture, Mansoura University, Egypt

A. E. Sharief
Department of Agronomy, Faculty of Agriculture, Mansoura University, Egypt

M. S. Sheteiwy
Department of Agronomy, Faculty of Agriculture, Mansoura University, Egypt

Reena
Dryland Research Sub-Station, SKUAST-J, Dhiansar, Bari Brahmana- 183311, Jammu, J&K, India

Ram Singh
Department of Entomology, CCS Haryana Agricultural University, Hisar- 125 004, India

B. K.Sinha
Division of Biochemistry and Plant Physiology, FOA, SKUAST-J, Chhatha- 180009, India

V. Pal
Department of Agricultural Economics, B.A. College of Agriculture, Anand Agricultural University, Anand - 388110, Gujarat, India

K. S. Jadav
Department of Agricultural Economics, B.A. College of Agriculture, Anand Agricultural University, Anand - 388110, Gujarat, India

G. Devi
Department of Agricultural Economics, B.A. College of Agriculture, Anand Agricultural University, Anand - 388110, Gujarat, India

Y. C. Zala
Department of Agricultural Economics, B.A. College of Agriculture, Anand Agricultural University, Anand - 388110, Gujarat, India

S. G. Pawar
Department of Agricultural Process Engineering, Dr. Punjabrao Deshmukh Krishi Vidyapeeth, Akola - 444 104, MS, India

G. U. Shinde
Department of Farm Machinery and Power Engineering, Vasantrao Naik Marathwada Krishi Vidyapeeth, Parbhani - 431 402, MS, India

S. U. Khodke
Department of Agriculture Process Engineering, Vasantrao Naik Marathwada Krishi Vidyapeeth, Parbhani - 431 402, MS, India

E. M. Draghici
University of Agronomical Sciences and Veterinary Medicine, Faculty of Horticulture, Bucharest, Romania

M. Pele
University of Agronomical Sciences and Veterinary Medicine, Faculty of Horticulture, Bucharest, Romania

S. A. Mir
Krishi Vigyan Kendra (Farm Science Centre), Kulgam, SKUAST Kashmir, J&K, India

M. Qasim
Krishi Vigyan Kendra (Farm Science Centre), Poonch, SKUAST Jammu, J&K, India

Y. Arfat
Krishi Vigyan Kendra (Farm Science Centre), Narkura Srinagar, SKUAST Kashmir, J&K, India

T. Mubarak
Krishi Vigyan Kendra (Farm Science Centre), Kulgam, SKUAST Kashmir, J&K, India

Z. A. Bhat
Mountain Research Station Khudwani, SKUAST Kashmir, J&K, India

J. A. Bhat
Krishi Vigyan Kendra (Farm Science Centre), Ganderbal, SKUAST Kashmir, J&K, India

S. A. Bangroo
Krishi Vigyan Kendra (Farm Science Centre), Kulgam, SKUAST Kashmir, J&K, India

T. A. SOFI
Krishi Vigyan Kendra (Farm Science Centre), Kulgam, SKUAST Kashmir, J&K, India

C. K. Hindumathy
Department of Biotechnology, Vinayaka Missions University, Salem 636 308 TN, India

J. S. Awati
ETC department, RIT, Sakharale, Sangli, MS, India

V. S. Patil
SCOE, Panhala, Kolhapur, MS, India

S. B. AWATI
DKTE, Ichalkarangi, Kolhapur, MS, India

Apu Das
Central Institute of Fisheries Education (CIFE), Mumbai-400061, Maharashtra, India

N. R. Kumar
Central Institute of Fisheries Education (CIFE), Mumbai-400061, Maharashtra, India

Biswajit Debnath
Central Institute of Fisheries Education (CIFE), Mumbai-400061, Maharashtra, India

S. C. Mandal
College of Fisheries, CAU, Lembucherra, Agartala, 799210, Tripura, India

U. Thapa
Department of Vegetable Crops Faculty of Horticulture, Bidhan Chandra Krishi Viswavidyalaya, Mohanpur, Nadia -741252, WB, India

R. Rai
Department of Vegetable Crops Faculty of Horticulture, Bidhan Chandra Krishi Viswavidyalaya, Mohanpur, Nadia -741252, WB, India

M. K. El-Nemr
Department of Agricultural Engineering, Faculty of Agriculture, Kafrelsheikh University, Egypt

K. F. Egbadzor
West Africa Centre for Crop Improvement, University of Ghana, Legon, Accra, Ghana
CSIR - Plant Genetic Resources Research Institute, Bunso, E.R. Ghana

M. Dadoza
Ministry of Food and Agriculture, Ketu North District, V.R. Ghana

E. Y. Danquah
West Africa Centre for Crop Improvement, University of Ghana, Legon, Accra, Ghana

M. Yeboah
West Africa Centre for Crop Improvement, University of Ghana, Legon, Accra, Ghana

S. K. Offei
The Biotech Centre, University of Ghana, Legon, Accra, Ghana

K. Ofori
Department of Crop Science, University of Ghana, Legon, Accra, Ghana

B. Demuyakor
Department of Biotechnology, Faculty of Agriculture, University for Development Studies, Box TL 1882, Tamale, Ghana

I. Galyuon
Department of Molecular Biology and Biotechnology, University of Cape Coast, Cape Coast, Ghana

S. Kyereh
Department of Agronomy, Faculty of Agriculture, University for Development Studies, Box TL 1882, Tamale, Ghana

M. Ahmed
Department of Agronomy, Faculty of Agriculture, University for Development Studies, Box TL 1882, Tamale, Ghana

S. E. Seadh
Agronomy Department, Faculty of Agriculture, Mansoura University, Egypt

A. R. T. Bakeer
Botany Department, Faculty of Agriculture, Fayoum University, 63414- Fayoum, Egypt

M. A. E. Abdel-Latef
Plant Pathology Institute, Agricultural Research Center, Egypt

M. A. Afifi
Plant Pathology Institute, Agricultural Research Center, Egypt

M. E. Barakat
Botany Department, Faculty of Agriculture, Fayoum University, 63414- Fayoum, Egypt

B. Tiroesele
Department of Crop Science and Production, Botswana College of Agriculture, Private Bag 0027, Gaborone, Botswana

T. H. Nketso
Department of Crop Science and Production, Botswana College of Agriculture, Private Bag 0027, Gaborone, Botswana

S. O. Tshwenyane
Department of Crop Science and Production, Botswana College of Agriculture, Private Bag 0027, Gaborone, Botswana

A. A. Kandil
Department of Agronomy, Faculty of Agriculture, Mansoura University, Egypt

A. A. A Araf
Department of Agricultural Botany, Faculty of Agriculture, Mansoura University, Egypt

A. E. Sharief
Department of Agronomy, Faculty of Agriculture, Mansoura University, Egypt

A. N. Ramadan
Ministry of Agriculture, Egypt

B. N. Ekesa
Bioversity International, Plot 106, Katalima Road, P.O. Box 24384, Kampala, Uganda

J. Kimiywe
Kenyatta University, Department of Food, Nutrition and Dietetics, P.O. Box 43844, Nairobi, Kenya

M. W. Davey
Lab Fruit Breeding and Biotechnology, Department of Biosystems, Katholieke University of Leuven (KUL), de Croylaan 42, Heverlee, B-3000, Leuven, Belgium

C. Dhuique-Mayer
Centre de Coopération Internationale en Recherche Agronomique pour le Développement (CIRAD), Dept. Persyst-UMR Qualisud TA B95/16 73 av. J.F Breton 34398 Montpellier Cedex 5, France

I. Van Den Bergh
Bioversity International, Parc Scientifique Agropolis II 34397 Montpellier Cedex 5, France

D. Karamura
Bioversity International, Plot 106, Katalima Road, P.O. Box 24384, Kampala, Uganda

G. Blomme
Bioversity International, Plot 106, Katalima Road, P.O. Box 24384, Kampala, Uganda

S. D. Payal
Dept. of Soil & Water Conservation Engineering, College of Agricultural Engineering, V.N.M.A. University, Parbhani - 431 401, MS, India

G. U. Shinde
Dept. of Farm Machinery and Power Engineering, College of Agricultural Engineering, V.N.M.A. University, Parbhani - 431 401, MS, India

R. Upadhyay
Department of Soil and Water Engineering, JNKVV, Jabalpur - 482 004, MP, India

A. Singh
Department of Soil Science & Agriculture, Chemistry, RVSKVV, Gwalior - 474 002, MP, India

P. Shrivastav
Krishi Vigyan Kendra, Narsinghpur - 487 001, MP, India

S. Thakur
Department of Soil and Water Engineering, JNKVV, Jabalpur - 482 004, MP, India

L. Momenzadeh
Agricultural Engineering Department, Shiraz University, Shiraz, Iran

A. Zomorodian
Agricultural Engineering Department, Shiraz University, Shiraz, Iran